Catecholamines and the Heart

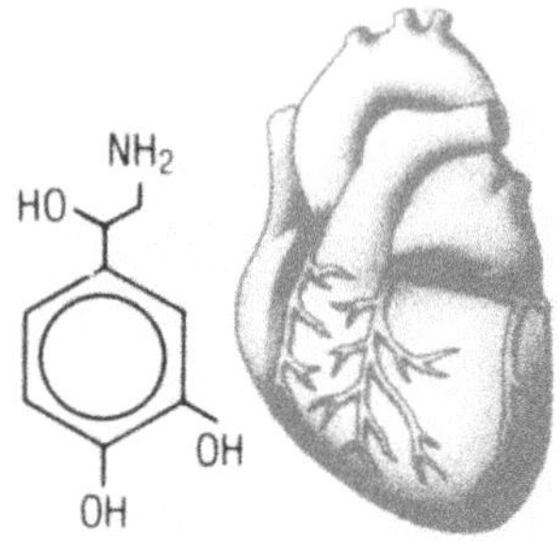

International Symposium
Munich, May 28–30, 1981

Catecholamines and the Heart

Recent Advances in
Experimental and Clinical Research

Editors
W. Delius, E. Gerlach, H. Grobecker, W. Kübler

With 149 Figures and 49 Tables

Springer-Verlag
Berlin Heidelberg New York 1981

Prof. Dr. W. Delius

Krankenhaus Kempfenhausen der Stadt München, Medizinische Abteilung

Prof. Dr. E. Gerlach

Physiologisches Institut der Universität München

Prof. Dr. H. Grobecker

Lehrstuhl für Pharmakologie der Universität Regensburg

Prof. Dr. W. Kübler

Abteilung Innere Medizin III (Kardiologie)
Medizinische Klinik der Universität Heidelberg

ISBN-13:978-3-642-68323-7 e-ISBN-13:978-3-642-68321-3
DOI: 10.1007/978-3-642-68321-3

Library of Congress Cataloging in Publication Data.
Main entry under title: Catecholamines and the heart. Includes bibliographies and index. 1. Heart—Diseases—Congresses. 2. Catecholamines—Physiological effect—Congresses. 3. Heart—Innervation—Congresses. 4. Adrenergic mechanisms—Congresses. I. Delius, W. (Wolfram), 1938– . [DNLM: 1. Catecholamines—Physiology—Congresses 2. Catecholamines—Therapeutic use—Congresses. 3. Heart—Physiopathology—Congresses. 4. Heart diseases—Drug therapy—Congresses. WK 725 C3575 1981] RC681.A2C36 616.1'2061 81-18308 AACR2

Softcover reprint of the hardcover 1st edition 1981

2119/3020/543210

In memoriam:

Professor Dr. med.

PETER HOLTZ

1902–1970

Preface

An International Symposium "Catecholamines and the Heart" was held in Munich in May 1981, which was organized in cooperation with the Council on Cardiac Metabolism of the International Society and Federation of Cardiology and with the Microcirculation Working Group of the European Society of Cardiology.

It was the aim of the organizers to evaluate in this Symposium the recent advances made in the field of sympathetic control of heart function. Particular attention was given to neural sympathetic pathways, to mechanisms and consequences of adrenoceptor stimulation and blockade as well as to clinically relevant implications mainly regarding the significance of catecholamines and of adrenoceptors in various heart diseases. The broad scope was covered in 34 papers presented by physiologists, pharmacologists and cardiologists during a two day meeting which was attended by an international audience.

This book includes all papers as well as the discussions following each presentation. To accomplish rapid publication the manuscripts are published as submitted by the authors. The discussions were edited making use of the written questions and answers of the discussants.

The Symposium was generously sponsored by Astra Chemicals GmbH, Wedel/Holstein, Germany, which thus made possible a very fruitful exchange of information, ideas and concepts between specialists in various disciplines. The organizers gratefully acknowledge this support once again and would like to thank Dr. T. Kreuzer, the representative of Astra Chemicals, for his personal engagement.

For their valuable assistance in the completion of this book the editors are particularly grateful to Dr. B. F. Becker and to Mrs. K. Schaipp, Munich. The editors are also indebted to Springer-Verlag for the prompt publication of the Symposium.

W. Delius
E. Gerlach
H. Grobecker
W. Kübler

List of Chairmen, Speakers, and Authors

A. Ablasser
Medizinische Klinik der Universität, Abteilung Innere Medizin III (Kardiologie)
Bergheimer Str. 58, 6900 Heidelberg (FRG)

I. Amende
Medizinische Hochschule Hannover, Abteilung für Kardiologie,
Karl-Wiechert-Allee 9, 3000 Hannover (FRG)

U. Baandrup
Department of Medicine I, Sahlgren's Hospital, S-413 45 Göteborg (Sweden)

E. Bassenge
Lehrstuhl für Angewandte Physiologie der Universität
Hermann-Herder-Str. 7, 7800 Freiburg/Br. (FRG)

G. Baumann
I. Medizinische Klinik der Technischen Universität
Ismaninger Str. 22, 8000 München 80 (FRG)

M. E. Bertrand
Service de Cardiologie et Hemodynamique, Hôpital Cardiologique
F-59.037 Lille (France)

H. Blömer
I. Medizinische Klinik der Technischen Universität
Ismaninger Str. 22, 8000 München 80 (FRG)

E. Carlsson
AB Hässle, S 431 83 Mölndal (Sweden)

C. Campbell
Department of Physiology and Pharmacology, University of Strathclyde
Royal College, George Street, Glasgow GI IXW (UK)

F. M. A. Correa
Laboratory of Clinical Science, National Institute of Mental Health
Bethesda, MD 20205 (USA)

J. Czachurski
I. Physiologisches Institut der Universität
Im Neuenheimer Feld 326, 6900 Heidelberg (FRG)

W. Delius
Krankenhaus Kempfenhausen der Stadt München, Medizinische Abteilung
8131 Berg 1 (FRG)

K. Dembowsky
I. Physiologisches Institut der Universität
Im Neuenheimer Feld 326, 6900 Heidelberg (FRG)

R. Dietz
Medizinische Klinik der Universität, Abteilung Innere Medizin III (Kardiologie)
Bergheimer Str. 58, 6900 Heidelberg (FRG)

J. G. Dobson, Jr.
Department of Physiology, University of Massachusetts Medical School
Worchester, MA 01605 (USA)

P. G. Dominiak
Lehrstuhl für Pharmakologie, Universität Regensburg
Universitätsstr. 31, 8400 Regensburg (FRG)

S. Dowinsky
Zentrum der Inneren Medizin, Abt. für Kardiologie, Klinikum der Universität
Theodor-Stern-Kai 7, 6000 Frankfurt/Main 70 (FRG)

L.-G. Ekelund
Department of Clinical Physiology, Karolinska Hospital
Box 60500, S-104 01 Stockholm (Sweden)

H. J. Engel
Medizinische Hochschule Hannover, Abteilung für Kardiologie,
Karl-Wiechert-Allee 9, 3000 Hannover (FRG)

D. Erhardt
Institut für Experimentelle Chirurgie der Technischen Universität
Ismaninger Str. 22, 8000 München 80 (FRG)

O. Fagbemi
Department of Physiology and Pharmacology, University of Strathclyde
Royal College, George Street, Glasgow GI IXW (UK)

E. O. Feigl
Department of Physiology and Biophysics, University of Washington
Medical School, Seattle WA 98195 (USA)

B. Felix
I. Medizinische Klinik der Technischen Universität
Ismaninger Str. 22, 8000 München 80 (FRG)

J. Fernandez-Pardal
Laboratory of Clinical Science, National Institute of Mental Health
Bethesda, MD 20205 (USA)

J. B. Furness
Laboratory of Clinical Science, National Institute of Mental Health
Bethesda, MD 20205 (USA)

E. Gerlach
Physiologisches Institut der Universität, Pettenkofer Str. 12, 8000 München 2 (FRG)

H. Grobecker
Lehrstuhl für Pharmakologie, Universität Regensburg
Universitätsstr. 31, 8400 Regensburg (FRG)

P. Guicheney
Laboratory of Clinical Science, National Institute of Mental Health
Bethesda, MD 20205 (USA)

E. Hackenthal
Medizinische Klinik der Universität, Abteilung Innere Medizin III (Kardiologie)
Bergheimer Str. 58 6900 Heidelberg (FRG)

P. Harris
Cardiothoracic Institute, University of London
2 Beaumont Street, London W1N 2DX (UK)

A. Hedberg
AB Hässle, S-431 83 Mölndal (Sweden)

A. Hjalmarson
Department of Medicine I, Sahlgren's Hospital, S-413 45 Göteborg (Sweden)

F. Hofmann
Pharmakologisches Institut der Universität
Im Neuenheimer Feld 326, 6900 Heidelberg (FRG)

J. Holtz
Lehrstuhl für Angewandte Physiologie der Universität
Hermann-Herder-Straße 7, 7800 Freiburg/Br. (FRG)

R. Hopf
Zentrum der Inneren Medizin, Abt. für Kardiologie, Klinikum der Universität
Theodor-Stern-Kai 7, 6000 Frankfurt/Main 70 (FRG)

H. Ibel
Physiologisches Institut der Universität, Pettenkoferstr. 12, 8000 München 2 (FRG)

K. H. Jakobs
Pharmakologisches Institut der Universität
Im Neuenheimer Feld 326, 6900 Heidelberg (FRG)

W. Kaltenbach
Zentrum der Inneren Medizin, Abt. für Kardiologie, Klinikum der Universität
Theodor-Stern-Kai 7, 6000 Frankfurt/Main 70 (FRG)

Y. N. Ke
Medizinische Klinik der Universität, Abteilung Innere Medizin III (Kardiologie)
Bergheimer Str. 58, 6900 Heidelberg (FRG)

G. Klein
I. Medizinische Klinik und Poliklinik der Technischen Universität,
Ismaninger Str. 22, 8000 München 80 (FRG)

E. Klinge
Division of Pharmacology, Department of Pharmacy, University of Helsinki
Kirkkokatu 20, SF-00170 Helsinki 17 (Finland)

G. Kober
Zentrum der Inneren Medizin, Abt. für Kardiologie, Klinikum der Universität
Theodor-Stern-Kai 7, 6000 Frankfurt/Main 70 (FRG)

W. Kübler
Medizinische Klinik der Universität, Abteilung Innere Medizin III (Kardiologie)
Bergheimer Str. 58, 6900 Heidelberg (FRG)

K. Kyösola
Department of Anatomy, University of Helsinki
Siltavuorenpenger 20, SF-00170 Helsinki 17 (Finland)

J. M. Lablanche
Service de Cardiologie et Hemodynamique, Hôpital Cardiologique
F-59.037 Lille (France)

S. Langlois
Division of Clinical Pharmacology, Montreal General Hospital, Montreal, Quebec (Canada)

L. Leclerc
Cardiothoracic Institute, University of London, 2 Beaumont Street, London W1N 2DX (UK)

R. Lefkowitz
Department of Medicine, Duke University Medical Center, Durham, North Carolina 27710 (USA)

H. Leinberger
Medizinische Klinik der Universität, Abteilung Innere Medizin III (Kardiologie)
Bergheimer Str. 58, 6900 Heidelberg (FRG)

P. Lichtlen
Medizinische Hochschule Hannover, Abteilung für Kardiologie
Karl-Wiechert-Allee 9, 3000 Hannover (FRG)

U. Loher
I. Medizinische Klinik der Technischen Universität, Ismaninger Str. 22
8000 München 80 (FRG)

L. Ludwig
I. Medizinische Klinik der Technischen Universität, Ismaninger Str. 22
8000 München 80 (FRG)

R. McCarty
Laboratory of Clinical Science, National Institute of Mental Health
Bethesda, MD 20205 (USA)

W. Mäurer
Medizinische Klinik der Universität, Abteilung Innere Medizin III (Kardiologie)
Bergheimer Str. 58, 6900 Heidelberg (FRG)

H. Majewski
Pharmakologisches Institut der Universität
Hermann-Herder-Str. 5, 7800 Freiburg/Br. (FRG)

J. Manthey
Medizinische Klinik der Universität, Abteilung Innere Medizin III (Kardiologie)
Bergheimer Str. 58, 6900 Heidelberg (FRG)

B. H. Marks
Department of Pharmacology, Wayne State University, Medical School, Detroit
Michigan 48201 (USA)

H. Mattsson
AB Hässle, S-431 83 Mölndal (Sweden)

E. Merikallio
Department of Thoracic Surgery, University Central Hospital
SF-00290 Helsinki 29 (Finland)

R. Mewes
Pharmakologisches Institut der Universität
Im Neuenheimer Feld 326, 6900 Heidelberg (FRG)

L. H. Opie
Heart Research Unit, Department of Medicine, University of Cape Town Medical
School, Observatory 7925, Cape Town (South Africa)

J. Parratt
Department of Physiology and Pharmacology, University of Strathclyde
Royal College, George Street, Glasgow GI IXW (UK)

O. Penttilä
Department of Pharmacology, University of Helsinki
Siltavuorenpenger 10, SF-00170 Helsinki 17 (Finland)

J. Pispa
Department of Medical Chemistry, University of Helsinki
Siltavuorenpenger 10, SF-00170 Helsinki 17 (Finland)

K. H. Rahn
Department of Medicine, University of Limburg
St. Annadal Hospital, NL-6200 MD Maastricht (The Netherlands)

R. Rangno
Division of Clinical Pharmacology, Montreal General Hospital
Montreal, Quebec (Canada)

H. Reuter
Pharmakologisches Institut der Universität
Friedbühlstr. 49, CH-3010 Bern (Switzerland)

G. Riess
I. Medizinische Klinik der Technischen Universität
Ismaninger Str. 22, 8000 München 80 (FRG)

G. A. Rooke
Department of Physiology and Biophysics, University of Washington
Medical School, Seattle WA-98195 (USA)

B. Rouot
Laboratory of Clinical Science, National Institute of Mental Health
Bethesda, MD-20205 (USA)

J. M. Saavedra
Laboratory of Clinical Science, National Institute of Mental Health
Bethesda, MD-20205 (USA)

A. Schömig
Medizinische Klinik der Universität, Abteilung Innere Medizin III (Kardiologie)
Bergheimer Str. 58, 6900 Heidelberg (FRG)

J. Schrader
Physiologisches Institut der Universität, Pettenkoferstr. 12, 8000 München 2 (FRG)

G. Schultz
Pharmakologisches Institut der Universität
Im Neuenheimer Feld 326, 6900 Heidelberg (FRG)

W. Schulz
Klinikum der Universität, Zentrum Innere Medizin, Abteilung Kardiologie
Theodor-Stern-Kai 7, 6000 Frankfurt/Main 70 (FRG)

F. Schwarz
Medizinische Klinik der Universität, Abteilung Innere Medizin III (Kardiologie)
Bergheimer Str. 58, 6900 Heidelberg (FRG)

H. Seller
I. Physiologisches Institut der Universität
Im Neuenheimer Feld 326, 6900 Heidelberg (FRG)

P. Siltanen
Cardiovascular Laboratory, First Medical Clinic, University Central Hospital
SF-00290 Helsinki 29 (Finland)

K. Starke
Pharmakologisches Institut der Universität
Hermann-Herder-Str. 5, 7800 Freiburg/Br. (FRG)

G. L. Stiles
Department of Medicine, Duke University Medical Center
Durham, North Carolina 27710 (USA)

E. Strasser
Medizinische Klinik der Universität, Abteilung Innere Medizin III (Kardiologie)
Bergheimer Str. 58, 6900 Heidelberg (FRG)

K. Swedberg
Department of Medicine I, Sahlgren's Hospital, S-413 45 Göteborg (Sweden)

J. A. Thomas
Department of Pharmacology, Wayne State University, Medical School
Detroit, Michigan 48201 (USA)

F. P. Thieuleux
Service de Cardiologie et Hemodynamique, Hôpital Cardiologique
F-59.037 Lille (France)

P. Y. Tilmant
Service de Cardiologie et Hemodynamique, Hôpital Cardiologique
F-59.037 Lille (France)

R. Tschada
Medizinische Klinik der Universität, Abteilung Innere Medizin III (Kardiologie)
Bergheimer Str. 58, 6900 Heidelberg (FRG)

D. V. Unverferth
Department of Pharmacology, Wayne State University, Medical School
Detroit, Michigan 48201 (USA)

F. Waagstein
Department of Medicine I, Sahlgren's Hospital, S-413 45 Göteborg (Sweden)

I. Wallentin
Department of Medicine I, Sahlgren's Hospital, S-413 45 Göteborg (Sweden)

B. G. Wallin
Department of Clinical Neurophysiology, University Hospital
S-750 14 Uppsala (Sweden)

L. Williams
Department of Medicine, Cardiac Unit, Massachusetts General Hospital
Boston, Mass. 02114 (USA)

A. Wirtzfeld
I. Medizinische Klinik und Poliklinik der Technischen Universität
München, Ismaninger Str. 22, 8000 München 80 (FRG)

H.-G. Zimmer
Physiologisches Institut der Universität, Pettenkofer Str. 12, 8000 München 2 (FRG)

Contents

I. Sympathetic Nerves and Adrenergic Receptors

II. Biochemical and Physiological Implications of Adrenergic Receptor Stimulation

III. Pathophysiological, Clinical and Therapeutic Aspects of Catecholamines and Adrenergic Receptors

I. Sympathetic Nerves and Adrenergic Receptors

Regulation of Sympathetic Outflow to the Heart [1]

H. Seller, J. Czachurski, K. Dembowsky

Summary

Recent experiments with anaesthetized cats revealed the following results as to the origin and regulation of the tonic background and reflex activity of efferent pre- and postganglionic sympathetic nerves to the heart:

1. The tonic background activity in postganglionic cardiac nerves is rhythmically modulated synchronously with the pulse pressure wave and with respiration. The pulse synchronous rhythm is induced by phasic inhibition from baroreceptor afferents and the respiratory modulation is caused by a central coupling between respiratory centres and sympatho-excitatory neurones in the brain stem which project to spinal preganglionic sympathetic neurones.

2. Stimulation of somatic spinal afferents (intercostal nerves) produce an early, spinally transmitted and a late, supraspinally transmitted reflex discharge in sympathetic efferents. During blockade of all descending and ascending fibre tracts in the spinal cord at the cervical level the late supraspinal reflex component is completely abolished and the early spinal reflex is markedly increased. The increase in the spinal reflex unmasks the existence of a supraspinally originating tonic inhibition in the intact cord. The spinal reflex inhibition can also be produced by the α_2-adrenoceptor agonistic agent clonidine and reduced by the α_2-antagonistic agent yohimbine.

3. Bilateral cold blockade of ventrolateral regions of the medulla resulted in a marked decrease of sympathetic background activity and blood pressure, an abolition of supraspinal, and an increase in spinal, somato-sympathetic reflexes. It is suggested that this area acts as the final centre of integration of all excitatory and inhibitory activity regulating the sympathetic outflow.

Introduction

It is the major task of this presentation to give a current view of the basic mechanisms of some excitatory and inhibitory phasic patterns of the sympathetic activity generated at the spinal or medullary level of the central nervous system. The work is based on experiments on anaesthetized animals but it has been shown that this basic pattern of activity occurs also in the conscious animal [14]. The work by *B. G. Wallin* during the past few years has demonstrated that many of the basic regulatory mechanisms of sympathetic activity are present equally in human subjects [17]. In the conscious animal or in human subjects there is of course a considerable influence from supramedullary levels of the central nervous system upon sympathetic activity, e.g. during various states of emotion or exercise. But the

1 This study was supported by the Deutsche Forschungsgemeinschaft within the SFB 90 "Cardiovasculäres System"

neurones in the brain stem and the spinal cord and the visceral and somatic reflex pathways at these levels constitute the basic organization upon which such supramedullary activity will be superimposed.

Methods

The experiments were performed on adult cats which were anaesthetized with alpha-chloralose (60–70 mg/kg). The femoral vein and femoral artery were cannulated for injection of drugs and recording of blood pressure respectively. The trachea was cannulated and the cats were paralyzed with hexacarbacholine and artificially ventilated. The postganglionic cardiac nerve or the white ramus at the third thoracic segment were prepared on the left side and cut peripherally. In some experiments the left phrenic nerve was dissected free from surrounding tissue and cut before its entry into the thoracic cavity. The desheathed nerves were placed on bipolar platinum electrodes for recording nerve activity. The activity was amplified and displayed on an oscilloscope or on a direct recording system. For recording of sympathetic reflexes the output of the amplifier was also connected to an averager, and 16 consecutive reflex responses were averaged every 1 or 2 min. The third and fourth intercostal nerves on the left side were severed peripherally and their central stumps placed on bipolar silver electrodes for electrical stimulation. Somato-sympathetic reflexes were evoked by using the following stimulation parameters: a single pulse or two pulses at an interval of 2–4 msec, 0.2–3.0 V for 0.5 msec. The stimulus was repeated every 3–4 sec. All exposed nerves were kept in a pool of warm paraffin oil.

A reversible blockade of all ascending and descending spinal pathways was performed by cooling the spinal cord between the second and third cervical segment. A laminectomy from C_2 to C_5 was performed and the ventral and dorsal roots of C_2 and C_3 were cut. A hollow metal ring, placed around the spinal cord, was perfused with cold ethanol to produce a cold block. In control experiments the dorsal funiculus cranial to the cold block was stimulated and the evoked activity was recorded at the level of T_3. At temperatures below 8.5 °C no evoked potentials could be recorded. This blockade was fully reversible when the spinal cord was rewarmed to body temperature by changing the perfusion from cold ethanol to warm water. A similar device with two elliptically shaped concavities was used for the cold blockade of ventral areas of the medulla oblongata. After surgical exposure of the ventral surface of the medulla this device was placed at the desired position by means of a micromanipulator.

Results

Tonic Background Activity

In the first recordings from cardiac sympathetic nerve fibres by *Bronk* and co-workers 1936 [3] it was demonstrated that tonic discharges in these nerves are modulated in a fast periodic rhythm in synchrony with the cardiac cycle and a superimposed slower rhythm synchronous with respiration (Fig. 1). From this work and many subsequent publications there is clear evidence that these grouped

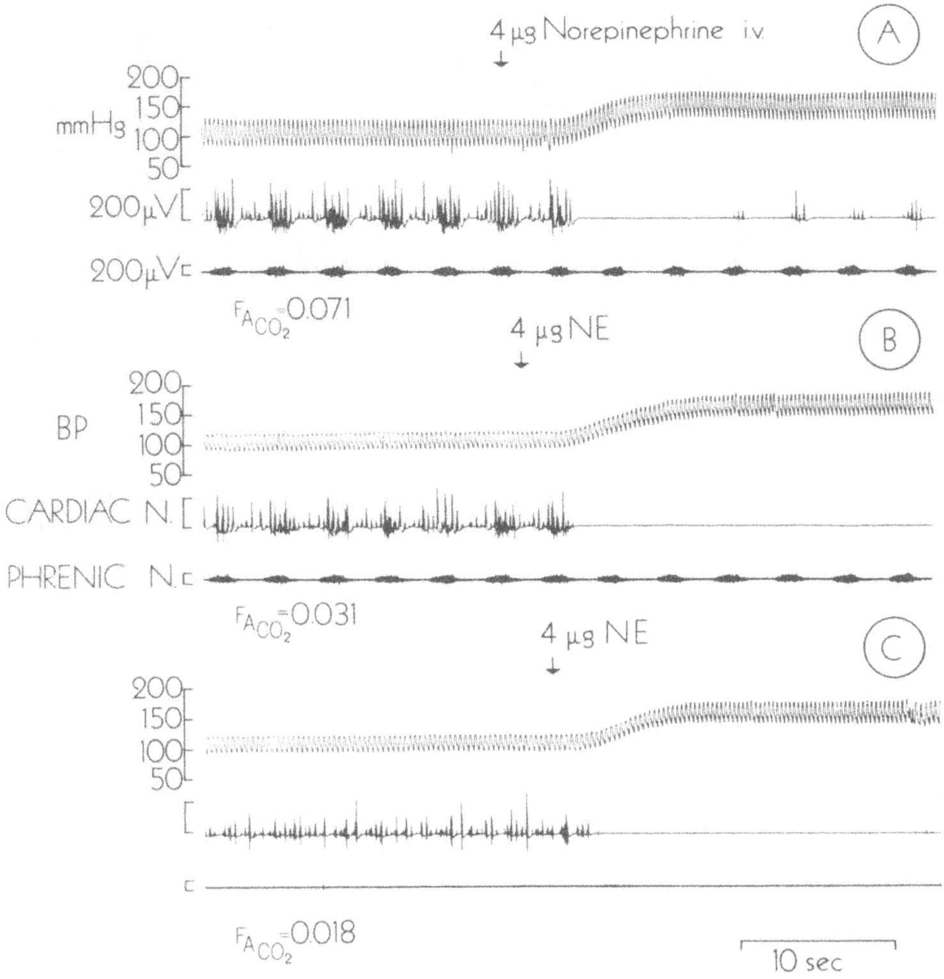

Fig. 1 A–C. Arterial blood pressure and activity in the left cardiac and phrenic nerve in an anaesthetized, paralyzed cat with artificial ventilation. Both vagal nerves are cut. In A, B and C the endtidal CO_2 is changed by various rates of artificial ventilation

discharges in synchrony with the pulse pressure wave are produced by the phasic inhibitory influence of baroreceptor afferents.

However, it was suggested by *Bronk* and co-workers [3] that the distension of lung stretch receptors during respiration was responsible for the respiratory modulation of sympathetic activity. This was shown not to be the case by experiments in which all afferents from the heart and lungs were excluded by transection of both vagal nerves [16]. In these experiments there remains a prominent increase of sympathetic activity in synchrony with the phrenic nerve activity (Fig. 1 A, B). This close correlation between sympathetic activity and phrenic nerve activity can be further demonstrated by variation of the central respiratory drive with different levels of endtidal CO_2 during artificial ventilation. High endtidal CO_2 levels during hypoventilation increase the augmentation of sympathetic activity in parallel with the phrenic nerve activity (Fig. 1 A), and the respiratory modulation of sympathetic activity is completely abolished in the

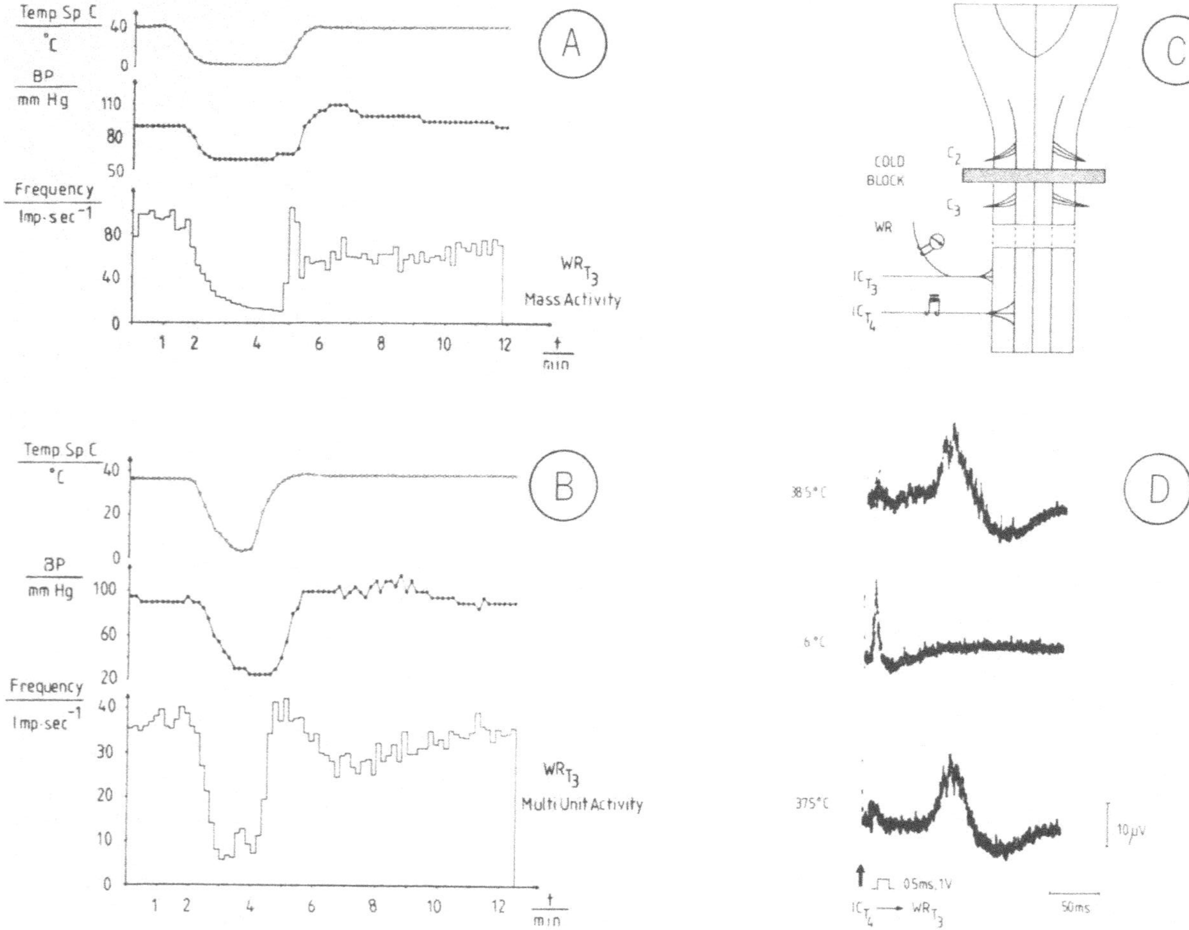

Fig. 2a–D. The variation of blood pressure, mass activity and activity of a multifibre strand of the T_3-white ramus, and the spinal and supraspinal somato-sympathetic reflex components during cold blockade of the spinal cord at C_2–C_3 (**C**) is shown in **A, B** and **D.** The temperature in **A, B** and **D** refers to the surface-temperature of the spinal cord at C_2–C_3. *IC*, intercostal nerve

absence of phrenic nerve discharges at low CO_2 levels during hyperventilation (Fig. 1 C). This finding implies a close central coupling between medullary inspiratory neurones and neurones in the brain stem which are responsible for activation of the sympathetic outflow to the heart.

With lesions of the spinal cord at the cervical level or with a reversible cold blockade of all descending and ascending fibre tracts (Fig. 2 C) it can be shown that nearly all activity in preganglionic sympathetic neurones depends on excitation which originates at supraspinal levels. After complete exclusion of all spinal descending pathways this activity is reduced to about 10% of its control value (Fig. 2 A, B). The origin of this autonomous activity at the spinal level is as yet unknown.

Somato-sympathetic Reflexes

In the early work by *Bronk* and co-workers [3] it was also demonstrated that stimulation of peripheral nerves increases the activity in cardiac sympathetic nerves. Many studies during the last few years have shown that afferent nerve stimulation produces a short latency-spinally transmitted and long latency-supraspinally transmitted reflex response (Fig. 2 D). During blockade of all ascending and descending fibre tracts in the spinal cord at the cervical level the late supraspinal reflex component is completely abolished (Fig. 2 D, middle trace). In addition, there is a marked increase in the amplitude of the early spinal reflex (Fig. 2 D), on average by about 200%. This increase of the spinal reflex during spinal cord blockade unmasks the existence of a supraspinally originating tonic inhibition of the transmission of spinal reflexes in the intact cord.

In a series of experiments much evidence was given that descending spinal noradrenergic or adrenergic fibre tracts are involved in this tonic inhibition of the spinal somato-sympathetic reflex [5]. The cold block induced increase in the amplitude of the spinal reflex is reduced by an intravenous injection of the α_2-adrenergic receptor agonistic agent clonidine (10–30 µg/kg), and this effect is reversed by the injection of the α_2-adrenergic receptor antagonistic agent yohimbine (0.3–1.0 mg/kg).

Medullary Origin of Tonic Sympatho-excitatory and Sympatho-inhibitory Systems

In 1871 *Owsjannikow* [12] showed that arterial blood pressure was unaffected by complete transection of the brain stem at the level of the inferior colliculi. Blood pressure, however, decreased following further transection at levels lower in the brain stem. These results were later confirmed by *Dittmar* [6] in 1873. By dorsal hemisections, *Dittmar* demonstrated that the area which was essential for the maintenance of normal blood pressure, i.e. the vasoconstrictor tone of the sympathetic nervous system, was restricted to the ventral half of the medulla.

In recent experiments we have studied the location of neurones in the brain stem which project to the preganglionic sympathetic neurones within the intermediolateral column at the thoracic level of the spinal cord [1]. The investigations were performed with the method of retrograde axonal transport of horseradish peroxidase which was injected into the intermediolateral column. These experi-

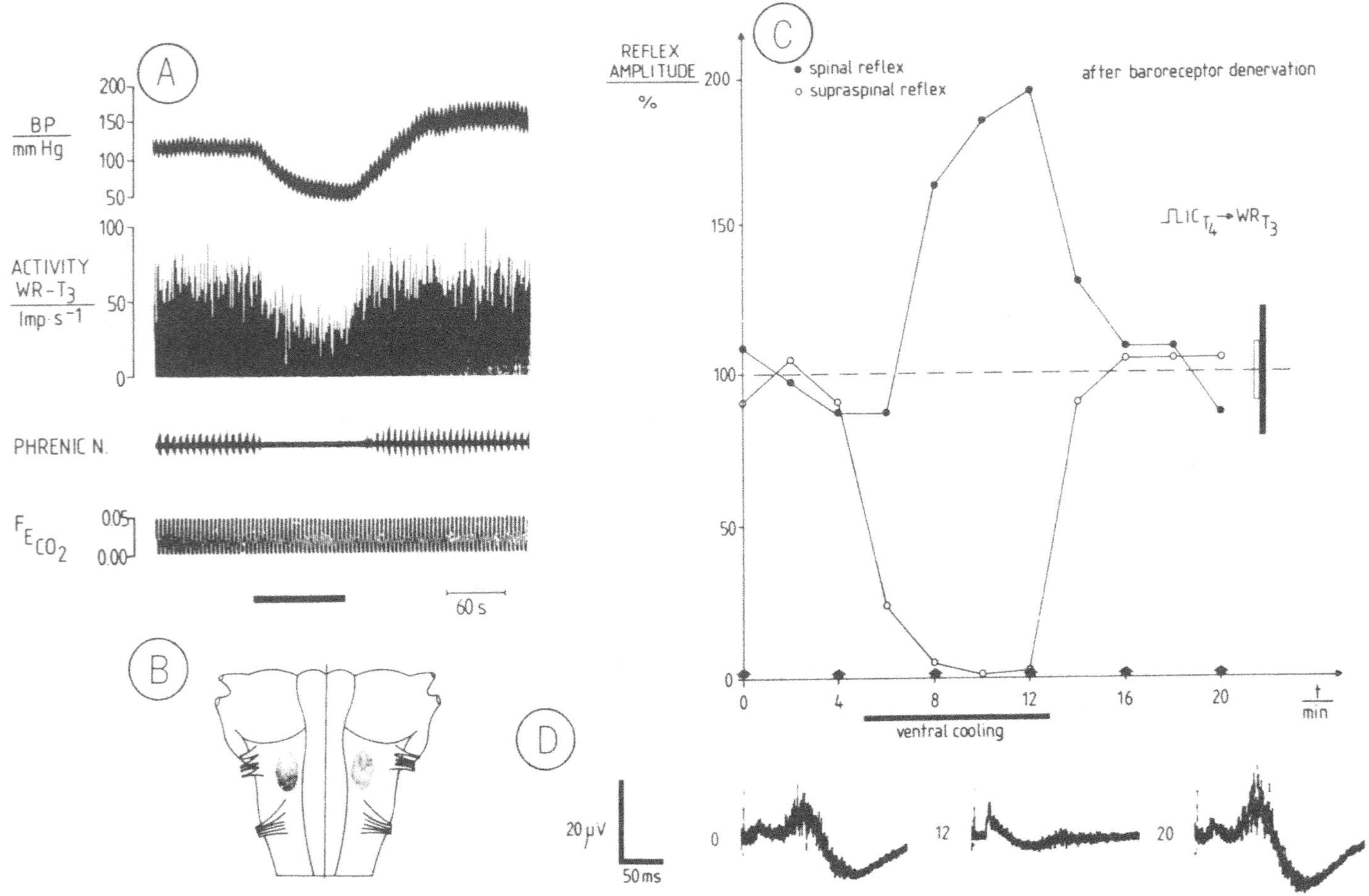

Fig. 3A–D. In **A, C** and **D** the variation of blood pressure, activity of T_3-white ramus, phrenic nerve activity and spinal and supraspinal reflex components of the somato-sympathetic reflex is shown during a cold blockade at the marked regions of the ventral surface of the medulla in **B**

ments revealed a population of spinally projecting neurones with their cell bodies in the ventrolateral reticular formation, beginning at the level of the obex and extending up to 8 mm cranial to the obex [1].

In a first series of experiments investigating the function of the neurones in this ventrolateral region of the medulla oblongata, we performed a bilateral cold block of this area (Fig. 3 B). During the blockade of neuronal activity in this area we observed a marked reduction of sympathetic activity and a fall in blood pressure (Fig. 3 A) which were comparable with those changes occurring after spinal cord blockade (Fig. 2 A, B). The same result was shown for the transmission of the supraspinal somato-sympathetic reflex: the supraspinal reflex component was almost completely abolished during the period in which the ventral medulla was cooled (Fig. 3 C, D). From these results we infer that the ventral area of the brain stem is the main final centre of integration for tonic and reflex excitatory activities which are then transmitted from this location to the preganglionic sympathetic neurones in the spinal cord.

In addition to the reduction of background activity and the extinction of the supraspinal reflex during cooling of the ventral area of the medulla, there was an increase in the spinal component of the somato-sympathetic reflex (Fig. 3 C, D). This increase in amplitude of the spinal reflex was never so extensive as the increase during spinal cord blockade, but it provides an indication that at least the major part of tonic inhibition of reflex transmission at the level of the spinal cord originates in this area of the brain stem. A further indication that this area is involved in tonic inhibition of spinal reflexes comes from the finding that catecholaminergic neurones are located in the same area of the ventral medulla [9]. This data is consistent with the results on the catecholaminergic nature of the inhibition of spinal reflexes mentioned above.

Discussion

The results of the medullary and spinal mechanisms regulating the sympathetic outflow to the heart are summarized in a simplified circuit diagram in Fig. 4. The main source of excitatory input to the preganglionic sympathetic neurones is in the

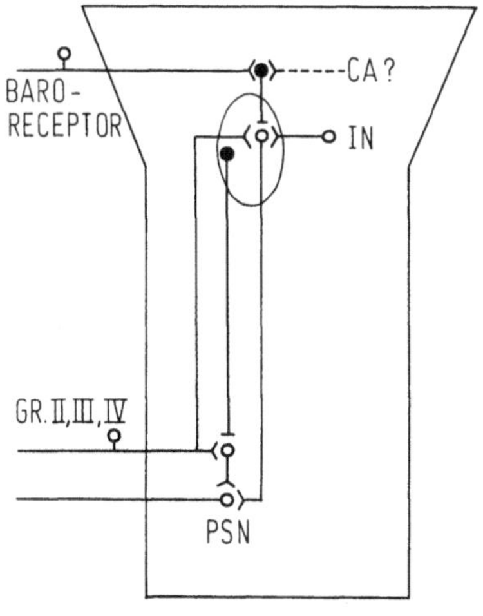

Fig. 4. A schematic diagram of excitatory (*open circles*) and inhibitory (*filled circles*) influences on preganglionic sympathetic neurones (*PSN*) at the intermediolateral column of the spinal cord. *IN*, medullary inspiratory neurones; CA, terminations of catecholaminergic fibres at the nucleus of the solitary tract. *GR. II, III, IV:* somatic afferents producing a spinal and supraspinal sympathetic reflex. The *encircled area* in the brainstem refers to the ventrolateral area (Fig. 3) which contains excitatory and inhibitory spinal descending neurones

lower brain stem and is closely coupled to the activity of inspiratory neurones (Fig. 4, IN). This implies that all chemoreflex respiratory stimuli, e.g. hypoxia or hypercapnia, must also be considered as strong excitatory influences on sympathetic outflow [4].

The baroreceptor inhibition of sympathetic activity and the baroreceptor reflex effects on the heart have been extensively studied [8]. With regard to baroreceptor reflex transmission at the first synaptic relay station at the nucleus of the solitary tract [15] quite new considerations are introduced by the finding of a dense termination of catecholaminergic fibres at this region [9] (Fig. 4, CA). This may conceivably represent the morphological basis for facilitation of the baroreceptor reflex by the centrally acting α_2-adrenoceptor agonistic agent clonidine [18].

By means of spinal cord blockade it has been shown that besides the medullary activity controlling the spinal sympathetic outflow there is some background and reflex activity autonomously present at the level of the spinal cord. The origin of this spinal autonomous activity remains so far obscure [13]. The work of *Malliani* during recent years has demonstrated that some myelinated and unmyelinated afferent fibres from the heart are spontaneously active under normal conditions [11]. This activity is capable of activating sympathetic efferents as a sympatho-sympathetic or cardio-cardiac spinal reflex. The activity in cardiac spinal afferents is increased during various states of ischaemia or myocardial infarction, and from this finding *Malliani* has postulated the existence of a positive feedback loop which might lead to deterioration in these pathological conditions. As yet it remains unknown whether the tonic inhibition of spinal somato-sympathetic reflexes by catecholaminergic fibres originating in the medulla applies equally to spinal sympatho-sympathetic reflexes. The question is of great interest because the inhibitory action of clonidine on these spinal reflexes suggests a possible means for interrupting the positive feedback loop.

Several investigators during the last few years have put forward the hypothesis that the ventrolateral region of the brain stem might serve as the final centre of integration of all excitatory and inhibitory activities transmitted to the spinal preganglionic sympathetic neurones [7]. This hypothesis is supported by many physiological and morphological findings, e.g. the location of inspiratory neurones in the close vicinity of this region [2] and the projection of fibres from the nucleus of the solitary tract [10] which might transmit the baroreceptor inhibition. But since an intensive neurophysiological exploration of this area is still lacking, the idea of an "all-important" cardiovascular centre in this region should not be emphasized too highly.

References

1. Amendt K, Czachurski J, Dembowsky K, Seller H (1979) Bulbospinal projections to the intermediolateral cell column: a neuroanatomical study. J Auton Nerv Syst 1:103–117
2. Batsel HL (1964) Localization of bulbar respiratory center by microelectrode sounding. Exp Neurol 9:410–426
3. Bronk DW, Ferguson LK, Margaria R, Solandt DY (1936) The activity of the cardiac sympathetic centers. Am J Physiol 117:237–249
4. Coleridge JCG, Coleridge HM (1979) Chemoreflex regulation of the heart. In: The cardiovascular system, vol. 1: the heart. Am Physiol Soc, Bethesda Maryland (Handbook of physiology, sec 2, pp)
5. Dembowsky K, Czachurski J, Amendt K, Seller H (1980) Tonic descending inhibition of the spinal somato-sympathetic reflex from the lower brain stem. J Auton Nerv Syst 2:157–182
6. Dittmar C (1873) Über die Lage des sogenannten Gefässcentrums in der Medulla oblongata. Ber Verh Sächs Ges Wiss, Leipzig. Math-Phys Kl 25:449–469
7. Feldberg W (1946) The ventral surface of the brain stem: a scarcely explored region of pharmacological sensitivity. Neuroscience 1:427–441
8. Kirchheim HR (1976) Systemic arterial baroreceptor reflexes. Physiol Rev 56:100–176
9. Lackner KJ (1980) Mapping of monoamine neurones and fibres in the cat lower brainstem and spinal cord. Anat Embryol (Berl) 161:169–195
10. Loewy AD, Burton H (1978) Nuclei of the solitary tract: efferent projections to the lower brain stem and spinal cord of the cat. J Comp Neurol 181:421–450
11. Malliani A, Lombardi F, Pagani M (1981) Functions of afferents in cardiovascular sympathetic nerves. J Auton Nerv Syst 3:231–236
12. Owsjannikow Ph (1871) Die tonischen und reflektorischen Centren der Gefässnerven. Ber Verh Sächs Ges Wiss Leipzig. Math-Phys Kl 23:135–147
13. Polosa C (1968) Spontaneous activity of sympathetic preganglionic neurons. Can J Physiol Pharmacol 46:887–896
14. Schad H, Seller H (1975) A method for recording autonomic nerve activity in unanesthetized, freely moving cats. Brain Res 100:425–430
15. Seller H, Illert M (1969) The localization of the first synapse in the carotid sinus baroreceptor reflex pathway and its alteration of the afferent input. Pflügers Arch 306:1–19
16. Seller H, Langhorst P, Richter D, Koepchen HP (1968) Über die Abhängigkeit der pressoreceptorischen Hemmung des Sympathicus von der Atemphase und ihre Auswirkung in der Vasomotorik. Pflügers Arch 302:300–314
17. Wallin BG, Delius W, Sundlöf G (1974) Human muscle nerve sympathetic activity in cardiac arrhythmias. Scand J Clin Lab Invest 34:293–300
18. van Zwieten PA (1975) Antihypertensive drugs with a central action. Prog Pharmacol 1:1–63

Discussion

Starke, Freiburg: Have you studied more selective drugs such as rauwolscine or prazosin?

Seller: As yet we have not performed experiments with prazosin and rauwolscine.

Kopin, Bethesda: Is there any evidence regarding the nature of the neurotransmitter involved in the influence of the respiratory center on sympatho-adrenal medullary secretory activity?
Have you tried morphine?

Seller: So far there is no indication concerning the nature of the neurotransmitters which are synthetized in respiratory neurones.
We have not tested morphine.

Starke, Freiburg: Why is the sympathetic tone of the efferent limb increased in congestive heart failure?

Seller: Malliani et al. (1981) have demonstrated that the activity in sympathetic afferent fibres is increased in these conditions and this in turn produces an activation in sympathetic efferent fibres (spinal cardio-cardiac reflex). Vagal afferents activated under such conditions show very rapid adaptation.

Relationship Between Sympathetic Outflow to Muscles, Heart Rate and Plasma Noradrenaline in Man

B. G. Wallin

Summary

Recordings of multi-unit sympathetic activity in human muscle nerves show that sympathetic impulses are grouped in pulse-synchronous bursts which occur preferentially during temporary reductions of diastolic blood pressure. There is, however, no relationship between arterial blood pressure levels and mean levels of sympathetic activity. Recordings from two extremities show that sympathetic outflow occurs in parallel in different nerves. In a given individual the level of activity is reproducible from day to day over several months but there are wide interindividual differences. The number of pulse synchronous bursts increases with age and there is a linear relationship between the level of activity and the plasma concentration of noradrenaline. Subjects with low levels of sympathetic activity tend to have higher resting heart rates than subjects with high sympathetic activity. Irrespective of the level of activity a statistical analysis showed that the occurrence of a sympathetic burst was time locked to a transient shortening of the RR-interval in the ECG. Manoeuvres influencing baroreceptor firing (postural changes, neck suction) caused changes of muscle sympathetic activity and RR-interval which were linearly related to initial levels and to each other. These findings suggest that there are interindividual quantitative differences in the balance between central nervous influence on the heart and the peripheral vascular beds. Adequate haemodynamic compensation to baroreceptor stimulation involves complex interaction between these neural outputs.

Introduction

Circulatory homeostasis depends on the integrated function of heart and peripheral vascular beds. Both the heart and the vascular beds are influenced by sympathetic nervous activity and traces of such activity can be detected in the blood as circulating noradrenaline. Sympathetic outflows to several vascular beds and vagal and sympathetic outflow to the heart are modulated by arterial baroreceptor mechanisms. Within this framework of normal physiology, common to all subjects, there are, however, signs of considerable interindividual differences in several cardiovascular parameters. For example, at rest under identical external conditions different age matched healthy subjects have different heart rates and different plasma levels of noradrenaline. Similarly, a manoeuvre such as a change of posture evokes different changes in heart rate and plasma noradrenaline in different subjects. This type of variability is rarely dealt with in human or animal studies and to a large extent it is unknown how and why such interindividual differences are brought about. We do not know if the differences should be regarded as uninteresting random variations or if they reflect interindividual differences in neural control of the circulation.

The aim of the present report is to present evidence related to this problem. I will describe the basic characteristics of sympathetic outflow to muscles in man, how that activity varies between subjects and how it relates to plasma noradrenaline and heart rate.

Methods

Nerve recordings were made in human subjects with tungsten microelectrodes with a tip diameter of a few microns. The recording electrode was inserted manually, through intact skin into an underlying nerve and the reference electrode was placed subcutaneously 1–2 cm away. Most recordings were made in the peroneal nerve at the fibular head but in a few cases in the median nerve at the elbow. The tip of the electrode was adjusted until a recording site with good signal-to-noise ratio for sympathetic impulses was found. Detailed descriptions of the technique as well as evidence for the sympathetic nature of the recorded impulses have been given previously [6, 8].

Results

General Characteristics of Sympathetic Outflow to Muscles

In nerve fascicles innervating skeletal muscles sympathetic impulses are grouped in pulse synchronous bursts which often occur in short sequences seperated by silent periods (Fig. 1), but sometimes in a more irregular staccato pattern [4]. Figure 1, left, illustrates that the bursts occur most frequently during temporary blood pressure reductions and disappear during peaks in the blood pressure curve [4, 7]. This behaviour is exactly what one would expect from vasoconstrictor impulses controlled by the arterial baroreflex (Fig. 1, right): a blood pressure fall is sensed by

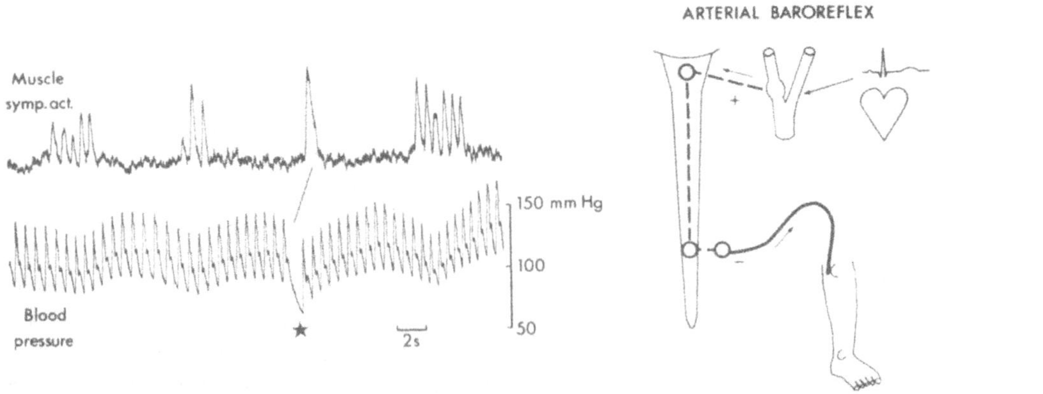

Fig. 1. *Left:* Relationship between spontaneous fluctuations of blood pressure and mean voltage display of muscle sympathetic activity recorded in the peroneal nerve at the fibular head (time constant of neurogram 0.1 s). *Right:* Baroreflex modulation accounts for pulse synchrony of sympathetic activity and inverse relationship to blood pressure fluctuations. * indicates diastolic blood pressure fall due to sudden AV-block. (Modified from *Wallin* et al. [10])

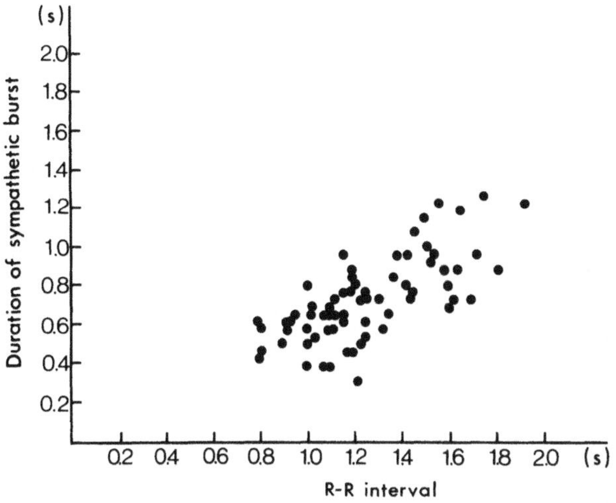

Fig. 2. Relationship between RR-interval and duration of corresponding sympathetic burst recorded in peroneal nerve at the fibular head in patient with atrial fibrillation. Measurements of burst duration taken from mean voltage neurogram (time constant 0.1 s)

the baroreceptors and counteracted by increased vasoconstriction. Figure 1 also illustrates that the activity corresponds to diastoles and occurs after a distinct latency. The blood pressure fall in the heart beat marked with an asterisk was due to a sudden AV-block and it was followed by a strong burst occurring after a reflex latency of approximately 1.3 s. Thus, the outflow of sympathetic impulses to muscle reflects in a striking way beat-to-beat variations of arterial diastolic blood pressure [9, 7].

Normally, durations of the sympathetic bursts do not vary much but in patients with atrial fibrillation who have abnormally wide variations of pulse interval there is a linear relationship between the duration of the pulse interval and the duration of corresponding sympathetic burst (Fig. 2). The finding agrees with the notion that outflow of sympathetic impulses initiated in conjunction with a diastole is interrupted by the inhibitory effect of the following systolic pressure wave.

In contrast to the intimate relationship between *dynamic variations* of blood pressure and sympathetic activity (present in all individuals) no systemic correlation was found if different subjects were compared with regard to their *mean levels* of diastolic blood pressure and sympathetic activity [7]. This suggests that muscle sympathetic activity is important for buffering acute changes of blood pressure but has little importance for setting the long term blood pressure level.

Relationship to Plasma Noradrenaline

Simultaneous recordings in two extremity nerves show that outflow of sympathetic impulses to muscles occurs in parallel in different parts of the body. If the level of sympathetic activity is quantitated by counting the number of pulse synchronous bursts, there are wide interindividual differences, the range being from less than 10 to more than 90 bursts/100 heart beats. In a given individual, however, the level is remarkably reproducible from day to day over many months [6], and there is a tendency for increasing values with increasing age [7]. The findings show that when recording sympathetic activity in a randomly chosen muscle nerve one obtains a measure of "muscle sympathetic tone" which is characteristic for the individual. Similar findings have been reported for plasma levels of noradrenaline, i.e. in a

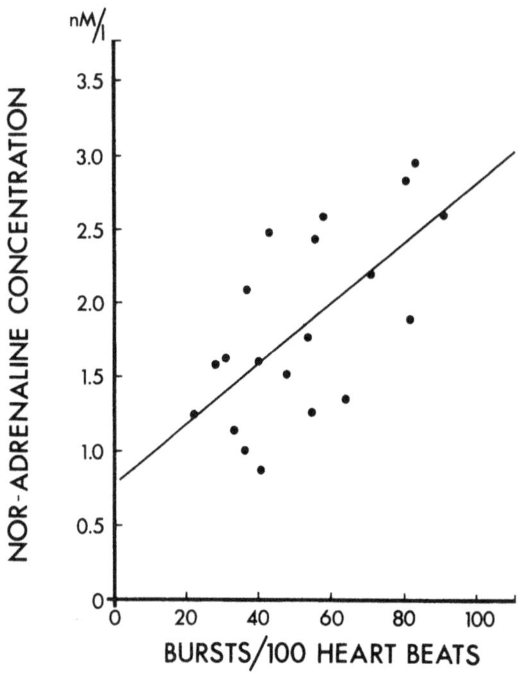

Fig. 3. Relationship between level of sympathetic activity in muscle nerves (expressed as number of sympathetic bursts/100 heart beats) and plasma concentrations of noradrenaline in forearm venous blood. Each point represents one healthy (normotensive) subject. $r = 0.65$, $P < 0.01$. (*Wallin* et al. [11])

given subject the noradrenaline level is fairly reproducible from day to day, there are wide interindividual differences and levels increase with age [5, 3]. Recently, a link between these two sets of findings was obtained when a positive correlation was found between the number of sympathetic bursts in muscle nerves and the plasma concentration of noradrenaline in forearm venous blood (Fig. 3). The relationship was demonstrated in spite of the fact that nerve recordings and blood samples were obtained 4–45 months apart [11]. In view of the fact that sympathetic outflow to different effector organs is controlled regionally by different mechanisms the relationship may seem surprising. A possible explanation may be that muscles, which comprise approximately 40% of total body weight contain so many adrenergic terminals that spill over from these will dominate plasma concentrations. Other factors may also contribute. Noradrenaline is metabolized in the liver and therefore the contribution to plasma from the splanchnic region may be less than could be expected on the basis of its size.

If plasma concentrations of noradrenaline at rest to a large extent are determined by sympathetic outflow in muscle nerves it becomes understandable why many authors find no relationship between plasma noradrenaline levels and blood pressure levels in normotensive subjects at rest. As mentioned above muscle sympathetic activity is modulated by dynamic arterial baroreceptor mechanisms and there is no relationship between mean levels of sympathetic activity and blood pressure. Consequently we are left with a situation where an important determinant of the plasma level of noradrenaline is impulse activity in sympathetic nerves supplying a vascular bed which is relatively unimportant for setting the long term blood pressure level.

Relationship to Heart Rate

Since arterial baroreceptors modulate muscle sympathetic activity and both sympathetic and parasympathetic outflow to the heart a systematic coupling between these neural outflows would seem logical. Cardiac autonomic activity cannot be measured directly in man and when searching for such evidence we have used variations in RR-intervals or heart rate as indicators of neural outflow to the heart. Several types of relationships have been found. Figure 4 illustrates that subjects with low levels of muscle sympathetic activity in the supine posture (expressed as bursts/100 heart beats) tend to have higher heart rates than subjects with high sympathetic activity. The finding is difficult to explain. Interindividual differences in arterial baroreceptor firing cannot be the cause since then the relationship in Fig. 4 would have had a positive slope.

A coupling between muscle sympathetic activity and heart rate can also be demonstrated by a statistical method similar to that used in "spike triggered averaging". The assumption is that the same reduction of baroreceptor input that leads to a sympathetic burst also changes cardiac autonomic activity so that RR-interval decreases. To show this in a recording at rest we started out by measuring the RR-interval in a heart beat associated with a sympathetic burst and then calculated how the RR-interval changed from that value during the succeeding 10 heart beats. The procedure was repeated for each beat associated with a burst that occurred during a period of approximately 10 min. With the aid of a computer the resulting mean change of RR-interval following a burst was determined. In agreement with the prediction there was a transient reduction of RR-interval in all 24 subjects analyzed in this way. Furthermore, the degree of shortening of the RR-

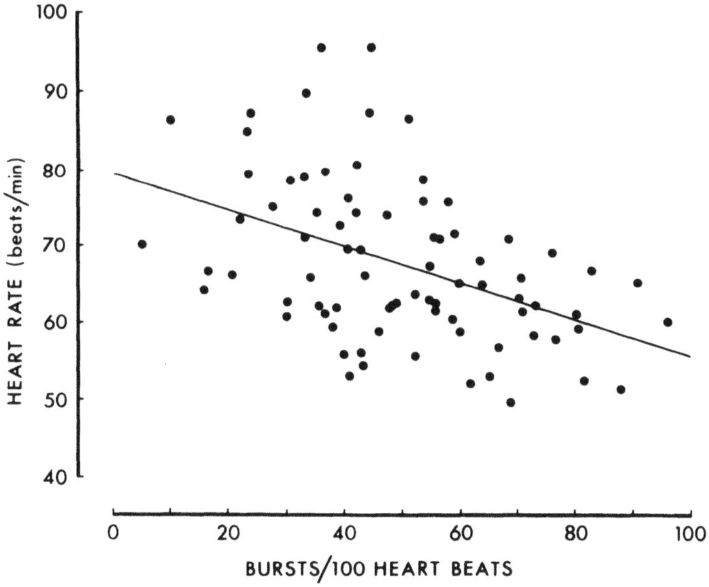

Fig. 4. Relationship between muscle sympathetic activity (expressed as number of sympathetic bursts/100 heart beats) and heart rate in healthy (normotensive) subjects resting in the recumbent posture. Each point represents mean values from one subject obtained during simultaneous recording of sympathetic activity and ECG during 10–30 min. $r=0.45$, $P<0.01$, $n=79$

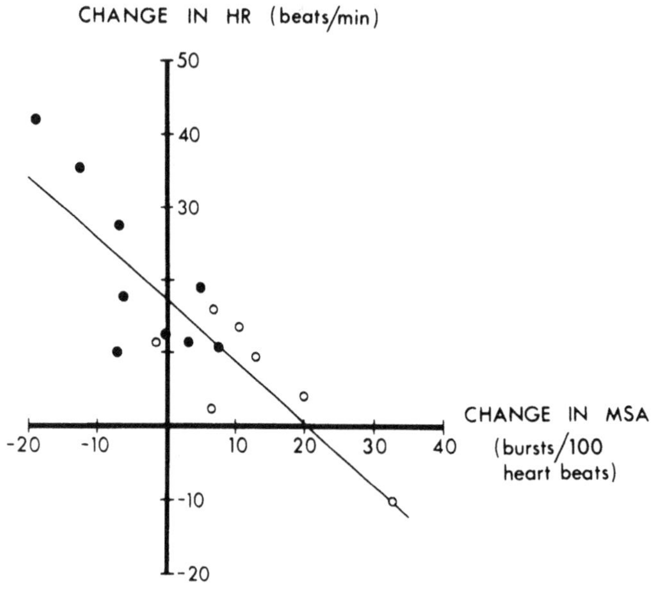

Fig. 5. Relationship between changes of number of sympathetic burst/100 heart beats and heart rate between lying and sitting (○) and between sitting and standing (●) postures in 8 healthy subjects. Data from *Burke* et al. [2]

interval was inversely related to the resting level of sympathetic activity (*Wallin*, unpublished observations).

A third finding is that in manoeuvres which involve baroreceptor stimulation *changes* of sympathetic activity and heart rate are related to basal levels and to each other [1, 2]. This is illustrated in Fig. 5 which comes from recordings of muscle sympathetic activity in different postures. The figure shows that subjects with large increases in the number of sympathetic bursts/100 heart beats have small increases of heart rate and those with small increases or decreases of sympathetic bursts/100 heart beats have large increases in heart rate. The results are reasonable from a general point of view but the underlying mechanism is unclear.

In conclusion, these examples demonstrate systematic relationships both between *resting levels* and *changes* of heart rate and sympathetic outflow to muscles. This indicates that there are interindividual quantitative differences in central nervous control of heart and peripheral vascular beds and that adequate haemodynamic compensation in response to baroreceptor stimulation is achieved by complex interaction of several autonomic reflexes.

Acknowledgement. Supported by Swedish Medical Research Council Grant no B81-14X-03546-10C.

References

1. Båth E, Lindblad L-E, Wallin BG (1981) Effects of dynamic and static neck suction on muscle nerve sympathetic activity, heart rate and blood pressure in man. J Physiol (Lond) 311:551–564
2. Burke D, Sundlöf G, Wallin BG (1977) Postural effects on muscle nerve sympathetic activity in man. J Physiol (Lond) 272:399–414
3. de Champlain J, Farley L, Cousineau D, van Ameringen M-R (1976) Circulating catecholamine levels in human and experimental hypertension. Circ Res 38:109–114

4. Delius W, Hagbarth K-E, Hongell A, Wallin BG (1972) General characteristics of sympathetic activity in human muscle nerves. Acta Physiol Scand 84:65–81
5. Lake CR, Ziegler MG, Kopin IJ (1976) Use of plasma norepinephrine for evaluation of sympathetic neuronal function in man. Life Sci 18:1315–1326
6. Sundlöf G, Wallin BG (1977) The variability of muscle nerve sympathetic activity in resting recumbent man. J Physiol (Lond) 272:383–397
7. Sundlöf G, Wallin BG (1978) Human muscle nerve sympathetic activity at rest. Relationship to blood pressure and age. J Physiol (Lond) 274:621–637
8. Vallbo ÅB, Hagbarth K-E, Torebjörk HE, Wallin BG (1979) Somatosensory, proprioceptive and sympathetic activity in human peripheral nerves. Physiol Rev 59:919–957
9. Wallin BG, Delius W, Sundlöf G (1974) Human muscle nerve sympathetic activity in cardiac arrhythmias. Scand J Clin Lab Invest 34:293–300
10. Wallin BG, Sundlöf G, Lindblad L-E (1980) Baroreflex mechanisms controlling sympathetic outflow to the muscles. In: Sleight P (ed) Arterial baroreceptors and hypertension. Oxford University Press, Oxford, pp 101–107
11. Wallin BG, Sundlöf G, Eriksson B-M, Dominiak P, Grobecker H, Lindblad L-E (1981) Plasma noradrenaline correlates to sympathetic muscle nerve activity in normotensive man. Acta Physiol Scand 111:69–73

Discussion

Dienstl, Innsbruck: Do you think that the interindividual differences you have observed in your experiments could have depended on environmental temperature? Could thermogenic shivering also influence the bursts you observed in muscle nerves?

Wallin: In a few experiments there were no significant changes of muscle sympathetic activity during moderate changes of ambient temperature. We have never recorded muscle sympathetic activity during shivering. Skin vasoconstrictor activity, on the other hand, increases when ambient temperature is lowered and decreases when ambient temperature is increased.

Bassenge, Freiburg: Do you think there is evidence for a differential response of the sympathetic outflow not only between organs but also within one particular organ, e.g. heart or kidney?

Wallin: I have no such evidence for heart or kidney. However, vasoconstrictor and sudomotor outflows to the skin are controlled differently.

Sleight, Oxford: Have you examined the individual sensitivity to infused noradrenaline in your subjects in order to explain the large interindividual differences in mean sympathetic activity you have found (cf. the work of *Philipp, Distler* and *Cordes, Lancet* 1979)?

Wallin: No, I have not.

Kübler, Heidelberg: In your paper plasma noradrenaline levels or heart rate were correlated with sympathetic outflow expressed as bursts/100 heart beats.
As heart rate partially depends on sympathetic activity, what is the rationale for using the dimension bursts/100 beats and what are the possible changes in your results if bursts/100 beats is replaced by the dimension bursts/time?

Wallin: Muscle sympathetic activity is almost always pulse synchronous, i.e. the maximal number of bursts is set by the number of heart beats. Therefore, bursts/100 heart beats will give a measure of how often central nervous mechanisms allow the outflow of sympathetic impulses in relation to available opportunities. This is a useful dimension when comparing interindividual differences in central autonomic control. If, on the other hand, one is interested in the effect of sympathetic activity on muscle vascular resistance the dimension bursts/min is more appropriate.
Replacing bursts/100 beats with bursts/min does not significantly alter the correlation between sympathetic activity and plasma noradrenaline (Fig. 3), but the results in Fig. 4 change and there is no correlation between bursts/min and heart rate.

Mattila, Helsinki: I wonder whether those bursts are influences by centrally active drugs, e.g. by diazepam which depresses cardiac sympathetic flow in animals and relaxes muscles by mechanisms so far poorly understood.

Wallin: With regard to diazepam I have no data concerning muscle sympathetic activity. However, skin sympathetic outflow is inhibited by i.v. injection of diazepam and also by induction of general anaesthesia by i.v. barbiturate.

Kopin, Bethesda: Several laboratories report that heart rate and norepinephrine levels in plasma are positively correlated although all agree that the heart contributes little to the circulating catecholamines. This observation appears to conflict with the inverse relationship of heart rate and muscle sympathetic fibre activity. Do you have any suggestions about this paradox?

Wallin: I am sorry, but I have no suggestion.

Szekeres, Szeged: Is the inverse relationship between heart rate and sympathetic outflow also valid in atropinized patients?

Wallin: We have not yet investigated such patients.

Koch, Berlin: I was interested in the correlation you showed between sympathetic outflow and plasma noradrenaline (NA). In our hands, the reproducibility of the resting (basal) NA level is rather poor; conversely, plasma NA during standardized stimulation as achieved, e.g. by exercise, is highly reproducible. I therefore wonder if you have data displaying the correlation between sympathetic outflow and plasma NA during states of increased adrenergic activity such as exercise.
I should also like to mention that in our experience there is a fairly good correlation between increases of heart rate (ΔHR) and of systolic blood pressure (ΔSBP), and increases of plasma NA (ΔNA) during different states of enhanced adrenergic activity, both in normal subjects and in subjects with cardiovascular disorders.

Wallin: We have started to look at the effect of isometric hand gripping, but so far we have no data.

Classification and Function of Adrenoceptors

E. Carlsson, A. Hedberg, H. Mattsson

Summary

The classification of adrenoceptors is based primarily on investigations of the effects of various pharmacological agents on functional parameters. In principle, potency orders of agonists and inhibitory effects of specific antagonists have provided the basis for the classification of α and β and the further subclassification into α_1/α_2 and β_1/β_2. During the last few years, direct receptor binding studies have been used for receptor characterization.

Both functional studies and receptor binding studies indicate that both α_1 and α_2 and β_1 and β_2 may coexist in the same target cell mediating the same response. The relative density of the two subtypes of α- and β-receptors may vary from organ to organ.

Receptor binding studies have made it possible to determine the total β-receptor density as well as the relative density of the β-receptor subtypes. Such studies have also provided information about receptor occupancy of β-adrenoceptor agonists at different functional effect levels.

Variation in the total receptor pool seems at least partly to be responsible for the variation in potency of the full agonist isoprenaline and the maximal response to the partial agonist prenalterol observed in the rabbit papillary muscle.

One week of treatment of cats with isoprenaline induced a reduction of the potency of isoprenaline and the maximal effect of prenalterol in the isolated papillary muscle. These changes were related to a decrease in the β-adrenoceptor density in the cardiac muscle.

Classification

The various effects of the endogenous catecholamines noradrenaline and adrenaline occur following interaction of these compounds with receptors on the external surface of the target cells. *Dale's* finding [7] that ergot alkaloids inhibited most of the excitatory, but not the inhibitory, effects of adrenaline indicated that at least adrenaline interacted with two different structures or receptors in the target organ. However, it was not until 1948 that Ahlquist published the criteria for distinction between the two catecholamine receptors and suggested the terminology α and β.

The concept of *Ahlquist* was more widely accepted only when the first compound, dichloroisoprenaline, able to inhibit β-receptormediated effects, was described [28]. Today, almost every paper dealing with adrenoceptors quotes this original work by *Ahlquist* [1].

In 1967, *Lands* and co-workers [16] showed that the beta-adrenoceptors were not homogeneous, but could be divided into β_1 and β_2. This classification was based on the same experimental principle as that used by *Ahlquist* for distinguishing α and β, namely potency rank order of a given series of agonists in different tissues. The potency order of the two endogenous catecholamines was noradrenaline $\geqq$ adrenaline for β_1-receptor mediated effects while adrenaline was much more potent than noradrenaline as regards β_2-receptor mediated effects.

More recently, the existence of at least two subtypes of α adrenoceptors has been recognized [17]. Originally, the α-adrenoceptor subtypes were classified on an anatomical basis as post- and presynaptic receptors. The terminology α_1 and α_2 was used as synonymes for the post- and presynaptic receptors, respectively. Subsequently, the existence of α-adrenoceptors that could be characterized as α_2 elsewhere than on nerve-terminals necessitated the use of the prefixes α_1 and α_2, regardless of location [4].

Distribution

The β-adrenoceptor subtypes β_1 and β_2 were, according to Lands et al. [16], characterized by absolute organ separation. However, experiments with selective and nonselective β-blockers both in vivo and in vitro gave indication of a coexistence of β_1 and β_2 in the same organ [5]. Figure 1 shows experiments in isolated cat hearts. Heart rate concentration-effect curves to noradrenaline and adrenaline were constructed before and after three doses of the β_1-selective blocker

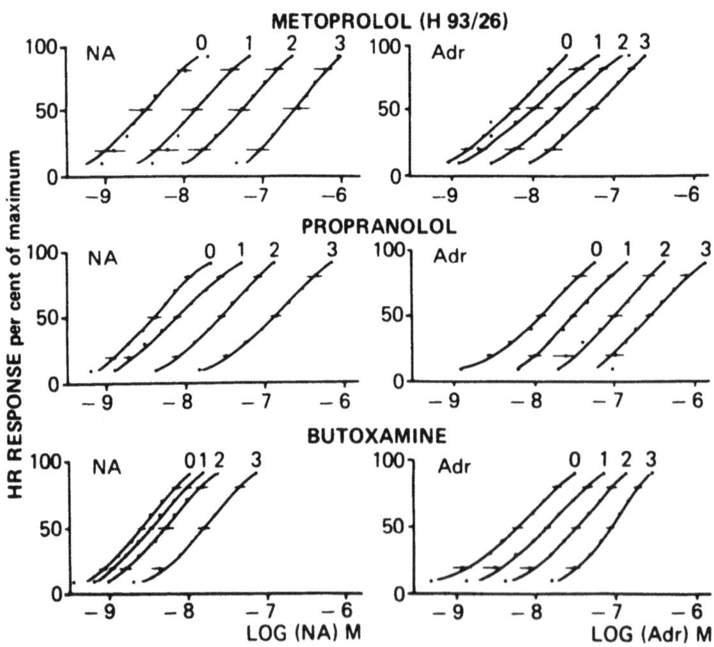

Fig. 1. Isolated perfused cat heart. Mean concentration-effect curves for chronotropic responses to noradrenaline and adrenaline before (0) and after (1, 2, 3) three increasing concentrations of metoprolol (0.01, 0.05 and 0.25 μg/ml), propranolol (0.001, 0.005 and 0.025 μg/ml) and butoxamine (0.1, 0.5 and 2.5 μg/ml). $n = 4–5$

Table 1. Mean apparent pA_2 values obtained from plots of log (dose ratio −1) versus log antagonist concentration

Agonist used	Apparent pA_2-values		
	Metoprolol	Propranolol	Butoxamine
Noradrenaline	8.0±0.12	8.4±0.05	5.8±0.18
Adrenaline	7.5±0.24	8.7±0.16	6.6±0.03

metoprolol, the nonselective blocker propranolol and the β_2-blocker butoxamine. Metoprolol shifted the concentration-effect curves to noradrenaline significantly more than the concentration-effect curves to adrenaline. Propranolol was an equipotential inhibitor of the responses to the two agonists, while butoxamine shifted the adrenaline concentration-effect curves more than the noradrenaline curves. The pA_2-values obtained from *Schild* analysis [2] of these experiments are shown in Table 1.

Qualitatively, the same results were obtained in isolated human right atrial tissue [31]. The differential blockade pattern obtained with the two selective β-antagonists is not consonant with a competitive interaction at one type of receptor. However, the results might be explained by assuming that there are both β_1- and β_2-receptors mediating the same response. The number of each type of receptor stimulated for a certain response depends on the relative affinities of the agonist to β_1- and β_2-receptors and the relative concentrations of the two subtypes of receptors. The heart rate response to noradrenaline seems to be a result of β_1-receptor stimulation depending both on the relatively high concentration of β_1-receptors and the selective affinity of noradrenaline for this receptor subtype. The heart rate response to adrenaline seems, at least partly, to be the result of β_2-receptor stimulation. There is now a large body of evidence supporting this concept [26, 27].

The first successful direct binding studies of β-adrenoceptors were reported in 1974 [3, 20, 21]. The binding technique has led to rapid progress in different fields of β-adrenoceptor research and it has been used for β-receptor characterization and studies of coexistence. The experimental idea is to analyse the displacement of a *nonselective* radiolabelled ligand with *selective* drugs having different affinities for β_1 and β_2. If in a tissue more than one population of receptors exists with different affinities for the selective drug, the relative proportion of each binding site and its affinity for the drug can be calculated.

Using this technique *Hedberg* et al. [12] showed that the cat right atrium contained about 80% of the β_1-adrenoceptor subtype and 20% of β_2. The left ventricle contained an apparently homogeneous population of β_1-receptors. Thus, in the cat heart, these results are in agreement with the results from classical pharmacological experiments and support the idea that the β_1- and β_2-receptors may occur on the same target cell. The data from the binding studies are also in accordance with findings that β_1-receptors predominately mediate the inotropic response in the left ventricle but that both β_1- and β_2-receptors mediate the chronotropic response in the cat heart. The consequence of this beeing a difference in the relative inotropic and chronotropic responses to selective β_1- and β_2-agonists [6].

Table 2 shows examples of other tissues with mixed populations of β_1- and β_2-adrenoceptors as assayed with binding studies [12, 27]. However, it should be

Table 2. Tissue distribution of β-adrenoceptor subtypes assayed with binding studies

Tissue	β_1 %	β_2 %
Guinea pig		
right atrium [a]	75	25
left ventricle [a]	100	0
Rabbit lung	80	20
Rat lung	20	80
Rat uterus		
oestrogen phase	20	80
progesteron phase	0	100

Data from *Hedberg* et al. 1980 [12][a] and *Nahorski* 1981 [27]

emphasized that these data must be verified in functional studies with a combination of selective agonists and antagonists. Biochemical studies give us information about heterogenicity of binding sites in a tissue but will not tell us whether or not the coexistence of receptor subtypes is due simply to heterogenicity of cell types within the tissue sample.

The coexistence of β_1- and β_2-adrenoceptors may have several pharmacological and clinical implications. From a pharmacological point of view it must be pointed out that studies of the potency order of agonists or studies of inhibitory patterns of selective antagonists with one agonist only are not sufficient for the characterization of the β-adrenoceptors of an organ. Selective agonists – β_1 and β_2 – must be combined with selective antagonists in order to be able to draw relevant conclusions.

Furthermore, the relative potencies of antagonists will differ depending on agonist used. From Table 1 it can be estimated that the potency of metoprolol and propranolol differs with a factor of 2.5 when noradrenaline is used as agonist while the difference is 15 when adrenaline is used. This difference seems to be of no clinical significance since the sympathetic control of cardiac function occurs via direct sympathetic innervation while circulating adrenaline is of little importance [9].

The question of whether or not there are more than two subtypes of β-adrenoceptors has often been raised. At present, there is at least one investigation to strongly support the idea that there are only two subtypes in mammalian tissue. *Minneman* et al. [25] have in binding studies determined the dissociation constants for a large series of selective and nonselective ligands in different mammalian tissues. The investigators found two distinct groups of K_D's for each selective compound, thus indicating only two receptor subtypes. Similar results have been discussed by *Nahorski* [27]. Although it is impossible to prove that there are only two subtypes, the available data strongly support this conclusion. The variation in apparent affinity constants for antagonists observed in different organs which has been taken as indicative of multiple β-adrenoceptor subtypes can be explained on the basis of coexistence of β_1 and β_2 and variability in the relative density of the two types in the different organs.

As mentioned above, the α-adrenoceptor subtypes originally were characterized as post- and pre-synaptic receptors. However, in 1979 *Lasch* and *Jakobs* [18]

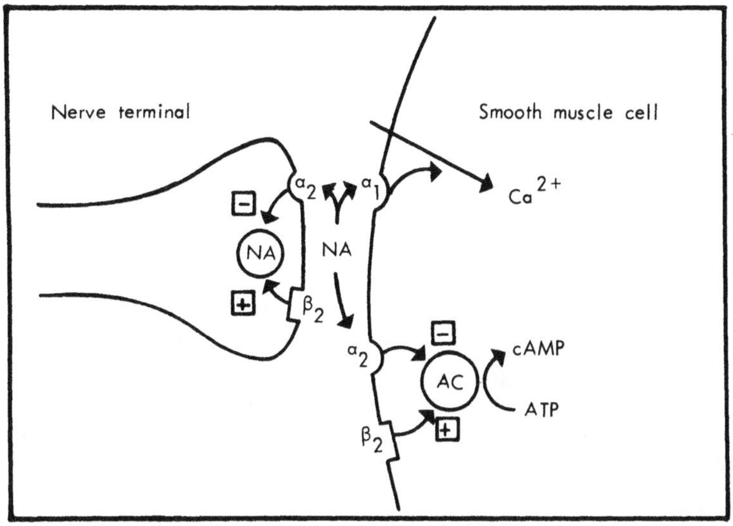

Fig. 2. Schematic representation of localisation and function of adrenoceptors at a sympathetic nerve-smooth muscle junction in blood vessels

reported results indicating that α-adrenoceptors on human blood platelets, mediating platelet aggregation, could be characterized as α_2. Furthermore, at the same time, *Drew* and *Whiting* [8] reported indications of both α_1- and α_2-adrenoceptors occurring postsynaptically in the smooth muscles of blood vessels, each subtype mediating vasoconstriction. Thus, similar to the β-adrenergic system where β_1- and β_2-receptors coexist, both α_1- and α_2-receptors may exist on the same target cell mediating the same response.

Figure 2 shows a schematic representation of present hypotheses concerning localisation and function of adrenoceptors at a sympathetic nerve-smooth muscle junction. It should be noted that both α_1- and α_2-adrenoceptors mediate vasoconstriction but that different mechanisms seem to be involved. α_1-Stimulation has been reported to be connected to increased turnover of the membrane phospholipid phosphatidylinositol and increased plasmamembrane permeability for extracellular calcium [24] while α_2 stimulation seems to inhibit adenylate cyclase, the same enzyme which is activated on β_2-adrenoceptor stimulation in the vasculature [29].

Functional Response, Adenylate Cyclase Activity and β-Adrenoceptor Occupancy

Prenalterol has been reported to be a partial β-adrenoceptor agonist with an intrinsic activity amounting to 80%–90% of that of isoprenaline in the anaesthetized cat heart [6]. In spite of the high intrinsic activity of prenalterol, the cAMP level in the myocardium at maximal stimulation were not significantly increased. This is in sharp contrast to the pronounced increase of cAMP levels observed after maximal stimulation with isoprenaline [14].

Figure 3 shows data from in vitro experiments in cardiac preparations from reserpine-pretreated cats [11]. The inotropic responses, measured as peak tension (T_{max}), were measured in electrically driven right ventricular papillary muscles.

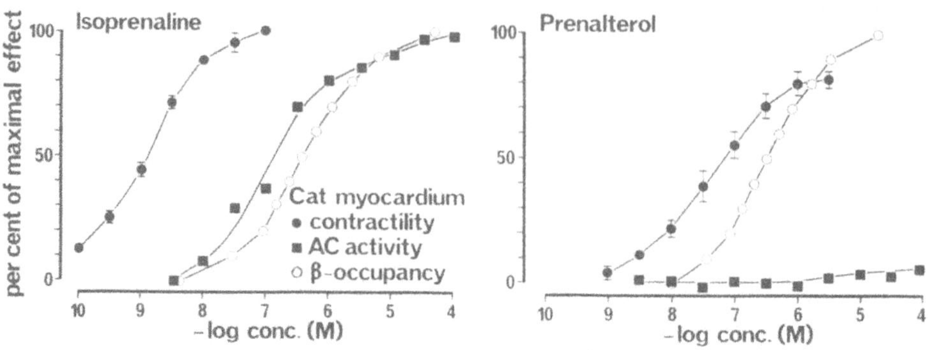

Fig. 3. Concentration-effect curves for the action of isoprenaline and prenalterol on isolated papillary muscle contractility, adenylate cyclase (*AC*) activity in ventricular muscle homogenate and β-adrenoceptor occupancy in ventricular crude membrane preparation. The hearts were taken from reserpine-pretreated cats. $n = 6$

Adenylate cyclase activity was measured in ventricular muscle homogenate and β-adrenoceptor occupancy was analysed in crude membrane preparations from ventricular muscles using radioligand binding technique. [^{125}I] Iodohydroxybenzylpindolol (IHYP) was used as the labelled ligand.

Regarding the functional response, isoprenaline was about 10 times more potent than prenalterol on a molar basis and the maximal increase of contractility was 20% higher after isoprenaline than after prenalterol. The adenylate cyclase activity showed a five-fold increase after isoprenaline but the dose-response curve was to the right of the inotropic dose-response curve. After prenalterol, there was only a marginal but statistically significant increase of the adenylate cyclase activity. The β-adrenoceptor occupancy curve for isoprenaline followed closely the adenylate cyclase dose-response curve while the β-adrenoceptor occupancy curve for prenalterol followed the functional dose response curve.

At first sight, these data may be interpreted to show that adenylate cyclase activation and cAMP are not involved in the inotropic stimulation with the β-adrenoceptor agonist prenalterol. Even if this possibility at the present state of knowledge cannot be excluded, there might be another explanation.

On maximal stimulation with prenalterol, there was a slight but statistically significant increase of the adenylate cyclase activity amounting to less than 10% of that measured after maximal stimulation with the full agonist isoprenaline. The isoprenaline data show that a maximal increase of contractility is reached at a concentration at which only about 10% of the total receptor pool is occupied and with only a fractional activation of the adenylate cyclase indicating a receptor reserve of about 90% for isoprenaline. At an effect level of 80%–90% of the maximum, i.e. corresponding to the maximal response to prenalterol, the increase of adenylate cyclase activity is slightly less than 10%. This level corresponds well with the adenylate cyclase activity on maximal stimulation with prenalterol. In accordance with the concept of spare receptors [30] there is no receptor-reserve for the partial agonist prenalterol.

Out of a pharmacological point of view prenalterol would be characterized as a partial agonist with an intrinsic activity of 80%–90% whereas, out of a biochemical point of view, prenalterol may be classified as a partial agonist with less than 10%

intrinsic activity or as a β-antagonist since it has been shown that prenalterol inhibits isoprenaline activated adenylate cyclase [11].

These data emphasize the importance of combining adenylate cyclase and binding studies with functional studies for relevant conclusions on mechanisms of drug action.

Variability in Responsiveness to β-Agonists

Studies with prenalterol in cat papillary muscle showed rather consistent values concerning intrinsic activity, i.e. the maximal response in relation to isoprenaline. In the isolated rabbit papillary muscle, however, a great variation was observed [23]. The maximal inotropic effect in relation to isoprenaline varied between 7% and 80% in different preparations. Figure 4 shows that the degree of intrinsic activity of prenalterol was positively correlated to the potency of isoprenaline. Thus, the lower the EC_{50} for isoprenaline the higher the intrinsic activity for prenalterol and vice versa. *Kenakin* and *Beek* [15] have shown a corresponding correlation when studying prenalterol and isoprenaline responses in different mammalian tissues.

How can this correlation be explained? Figure 5 shows a schematic representation of the observed phenomenon, i.e. the relation of a variation in the position of the dose response curve to a full agonist and the variation of the maximal response to a partial agonist.

The basic steps in agonist stimulation are shown at the top of the figure. The agonist is bound to the receptor which is activated and produces an initial stimulus (s). This stimulus is then amplified in a series of enzymatic reactions leading to a final stimulus (S) activating the effector system. The amplitude of the initial stimulus is determined by the concentration, affinity and efficacy (i.e. the ability of the drug to activate the receptor at binding) of the drug and the total receptor pool [10]. The amplitude of the final stimulus (S) (reaching the effector system) is

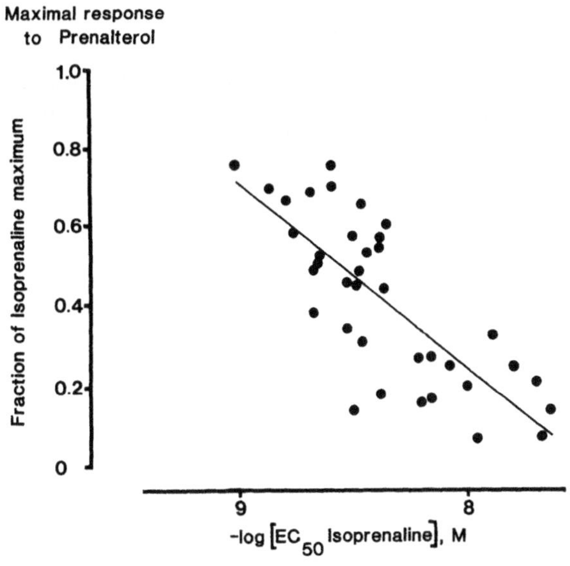

Fig. 4. Isolated rabbit papillary muscle. Correlation between the logarithm of isoprenaline EC_{50} and the corresponding intrinsic activity of prenalterol expressed as fraction of the maximal response to isoprenaline. $n = 37$

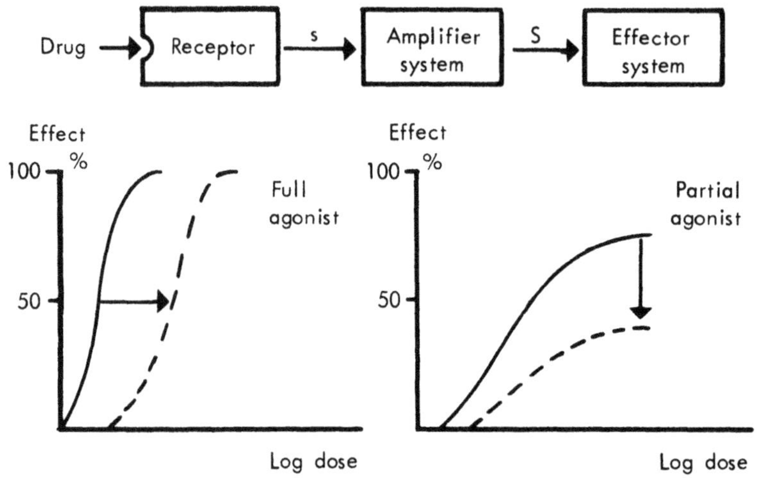

Fig. 5. Schematic representation of the relation of a variation in the position of the dose-response curve to a full agonist and the variation of the maximal response to a partial agonist. For further explanation, see text

determined by the amplitude of the initial stimulus and the efficiency of the amplifier system. Thus, the total receptor density and the efficiency of the amplifier system are two tissue-related factors which should be considered in relation to the observed variation in responsiveness to isoprenaline and prenalterol.

In a preliminary study β-adrenoceptor density has been determined in the rabbit ventricular myocardium by Scatchard analysis of IHYP binding data. Furthermore, in the same hearts the contractile responses to isoprenaline and prenalterol were studied in the papillary muscle. A good correlation between receptor density and intrinsic activity of prenalterol was found. It is therefore tempting to suggest that the variation in receptor density in the different rabbit papillary muscle preparations to a significant extent, even if not entirely, is responsible for the variation observed in the maximal responses to prenalterol and the sensitivity to isoprenaline.

Even if the variation in receptor density might be an important factor for explanation of the above reported results, it must be emphasized that similar variations in another organ or variations between organs concerning responsiveness to agonists may be related to variations in the efficiency of the amplifier system [15]. Combined functional, adenylate cyclase and receptor binding studies may shed light on this problem.

Similar pharmacological and biochemical studies will also give us new insight into the nature of the desensitization phenomena described in a variety of organs in connection with catecholamine exposure as well as changes in catecholamine responsiveness brought about by a variety of hormones such as thyroid hormones, progesterone and cortisone. Preliminary experiments in cats show that after one week of continuous administration of isoprenaline there is a more than ten-fold shift to the right of the concentration-effect curve to isoprenaline in the isolated papillary muscle as compared to the control. In addition, the maximal response to prenalterol was reduced from 90% to 25% of that of isoprenaline. Neither the basal nor the maximal responses to isoprenaline were changed. In the isoprenaline treated cats the β-receptor density was reduced by about 30% as compared to controls [13]. Even if the reduction of receptor density is significant, it is unlikely that this reduction in the total receptor pool is the only cause of the marked changes in agonist responsiveness.

Desensitization in connection with short term exposure to catecholamines seems to involve both changes in receptor density [19] and changes in the coupling mechanism between the receptor and the adenylate cyclase enzyme [22].

References

1. Ahlquist RP (1948) A study of the adrenotropic receptors. Am J Physiol 153:586–600
2. Arunlakshana O, Schild HO (1959) Some quantitative uses of drug antagonists. Br J Pharmacol 74:48–58
3. Aurbach GD, Fedak SA, Woodard CJ, Palmer JS, Hanser D, Troxler F (1974) The beta-adrenergic receptor: stereospecific interaction of an iodinated beta-blocking agent with a high affinity site. Science 186:1223–124
4. Berthelsen S, Pettinger A (1977) A functional basis for classification of α-adrenergic receptors. Life Sci 21:595–606
5. Carlsson E, Åblad B, Brändström A, Carlsson B (1972) Differentiated blockade of the chronotropic effects of various adrenergic stimuli in the cat heart. Life Sci 11:953–958
6. Carlsson E, Dahlöf CG, Hedberg A, Persson H, Tångstrand B (1977) Differentiation of cardiac chronotropic and inotropic effects of beta-adrenoceptor agonists. Naunyn-Schmiedebergs Arch Pharmacol 300:101–105
7. Dale HH (1906) On some physiological actions of ergot. J Physiol 34:163–206
8. Drew GM, Whiting SB (1979) Evidence for two distinct types of postsynaptic α-adrenoceptor in vascular smooth muscle in vivo. Br J Pharmacol 67:207–215
9. Folkow B, Heymans C, Neil E (1965) Integrated aspects of cardiovascular regulation. In: Hamilton WF, Dow P (eds) Handbook of physiology circulation, vol. 3: Am Physiol Soc, Washington, pp 1787–1823
10. Furchgott RF, Bursztyn P (1967) Comparison of dissociation constants and of relative efficacies of selected agonists acting on parasympathetic receptors. Am NY Acad Sci 144:882–899
11. Hedberg A, Mattsson H (in press) Beta-adrenoceptor interaction of full and partial agonists in the cat heart and soleus muscle. J Pharmacol Exp Ther
12. Hedberg A, Minneman KP, Molinoff PB (1980) Differential distribution of beta-1 and beta-2 adrenergic receptors in cat and guinea pig heart. J Pharmacol Exp Ther 212:503–508
13. Hedberg A, Mattsson H, Carlsson E (to be published) Changes in cat heart responsiveness to isoprenaline and prenalterol and β-adrenoceptor density after continuous administration of isoprenaline for 7 days.
14. Hedberg A, Carlsson E, Fellenius E, Lundgren B (in press) Cardiostimulatory effects of prenalterol, a beta-1 adrenoceptor partial agonist *in vivo* and *in vitro.* Correlation between physiological effects and adenylate cyclase activity. Naunyn-Schmiedebergs Arch Pharmacol
15. Kenakin TP, Beek D (1980) Is prenalterol (H 133/22) really a selective beta-1-adrenergic agonist? Tissue selectivity resulting from differences in stimulus-response relationships. J Pharmacol Exp Ther 213:406–413
16. Lands AM, Arnold A, McAnliff JP, Luduena FP, Brown TG (1967) Differentiation of receptor systems activated by sympathomimetic amines. Nature 214:597–598
17. Langer SZ (1974) Presynaptic regulation of catecholamine release. Biochem Pharmacol 23:1793–1800
18. Lasch P, Jakobs KH (1979) Agonistic and antagonistic effects of various α-adrenergic agonists in human platelets. Naunyn-Schmiedebergs Arch Pharmacol 306:119–125
19. Lefkowitz RJ (1979) Direct binding studies of adrenergic receptors: biochemical, physiologic and clinical implications. Ann Int Med 91:450–458
20. Lefkowitz RJ, Mukherjee C, Coverstone M, Caron MG (1974) Stereospecific ^{3}H(–) alprenolol binding sites, beta-adrenergic receptors and adenylcyclase. Biochem Biophys Res Commun 60:703–709

21. Levitzki A, Atlas D, Steer ML (1974) The binding characteristics and number of beta-adrenergic receptors on turkey erythrocyte. Proc Natl Acad Sci USA 71:2773–2776
22. March JD, Barry WH, Neer EJ, Alexander RW, Smith T (1980) Desensitization of chick embryo ventricle to physiological and biochemical effects of isoproterenol. Circ Res 47:493–501
23. Mattsson H, Hedberg A, Carlsson E (to be published) Correlation of intrinsic activity of prenalterol, sensitivity to isoprenaline and β-adrenoceptor density in the rabbit heart.
24. Michell RH, Kirk CJ (1981) Why is phosphatidylinositol degraded in response to stimulation of certain receptors. Trends Pharmacol Sci 2:86–89
25. Minneman KP, Hedberg A, Molinoff PB (1979) Comparison of beta-adrenergic receptor subtypes in mammalian tissues. J Pharmacol Exp Ther 211:502–508
26. Minneman KP, Pittman RN, Molinoff PB (1981) β-adrenergic receptor subtypes: Properties, distribution and regulation. Ann Rev Neurosci 4:419–461
27. Nahorski SR (1981) Identification and significance of beta-adrenoceptor subtypes. Trends Pharmacol Sci 2:95–98
28. Powell CE, Slater JH (1958) Blocking of inhibitory adrenergic receptors by dichloro analogue of isoproterenol. J Pharmacol Exp Ther 122:480–488
29. Rodbell M (1980) The role of hormone receptors and GTP-regulatory proteins in membrane transduction. Nature 284:17–22
30. Stephenson RP (1956) A modification of receptor theory. Br J Pharmacol 11:379–393
31. Åblad B, Carlsson E, Dahlöf C, Ek L, Hultberg E (1974) Cardiac effects of β-adrenergic receptor antagonists. Myocard Adv Cardiol 12:290–302

Discussion

Starke, Freiburg: Binding studies alone do not provide evidence for receptor location and/or function.

Carlsson: I agree. Heterogeneity in binding sites may just mirror heterogeneity of cell types in the tissue. For this reason I stressed that binding studies should be combined with functional studies making use of selective agonists and antagonists.

Role of Presynaptic and Postsynaptic Adrenoceptors in Cardiac Function

K. Starke, H. Majewski

Summary

At least three constituents of the heart are endowed with α- and β-adrenoceptors, namely the coronary vessels, the myocardium, and the postganglionic sympathetic axons. Only the two last-mentioned groups are discussed here. *Myocardial* α- and β-adrenoceptors, when activated, mediate a variety of mechanical and electrical changes, the best-known one being an increase in the force of contraction. The β-adrenoceptors predominate; it is not certain whether α-receptors play a physiological role. The myocardial α-adrenoceptors appear to be mainly α_1, the β-adrenoceptors mainly β_1. Activation of *presynaptic* α-adrenoceptors at the cardiac postganglionic sympathetic axons decreases, and activation of presynaptic β-receptors increases the release of noradrenaline evoked by nerve action potentials. The presynaptic receptors appear to be at least predominantly of the α_2- and β_2-type, respectively. The α-receptors are activated by previously released noradrenaline, mediate a negative feedback mechanism and, hence, are "autoreceptors". Presynaptic β-receptors, on the other hand, are little sensitive to noradrenaline and, normally, probably not links of a positive feedback mechanism. It seems likely that presynaptic α-adrenoceptors operate in vivo as well as in vitro and that, although they primarily affect transmitter release, this in turn leads to alterations in the postsynaptic response to nerve stimulation.

Introduction

Even recent textbooks reiterate the view that adrenoceptors in the heart are located only on the myocardial and the coronary smooth muscle cells. Moreover, while the coronary adrenoceptors are a mixed population of the α- and β-types, the myocardial receptors are still claimed to be exclusively β. Adrenoceptors in the coronary vascular bed are discussed in the accompanying paper by *Bassenge* and *Holtz* (1981). In the present article, the co-existence of myocardial α- *and* β-adrenoceptors will be shown briefly, and the occurrence and function of adrenoceptors at a third important constituent of the heart, namely the postganglionic sympathetic axons, discussed in more detail. Still other constituents such as, for instance, vagal fibers and fat cells, may, of course, possess adrenoceptors as well.

Myocardial α- and β-Adrenoceptors

α- and β-adrenoceptors mediate both mechanical and electrical changes in the myocardium (Table 1; for review, see [5, 41]). The β-adrenoceptors dominate. The physiological role of the α-adrenoceptors is uncertain. The best-studied ones, namely those that mediate an increase in the force of contraction, are surprisingly insensitive to noradrenaline in various heart preparations including human auric-

Table 1. Some responses mediated by myocardial α- and β-adrenoceptors

	α-Receptor	β-Receptor
Contractile force	Increased	Increased
Rate of beating	In general unchanged	Increased
Purkinje fiber action potential	Prolonged	Shortened
Atrial refractory period	Prolonged	Shortened
Atrial refractory period	Prolonged	Shortened
cAMP level	Unchanged	Increased

ular strips [60]. Moreover, in some types of preparation at least, α-adrenergic positive inotropic effects can be demonstrated only when the rate of beating is very low, for instance, 30 min^{-1} in guinea-pig left atria [59]. It seems doubtful, therefore, that the endogenous transmitter, at normal rates of beating, ever increases the force of contraction via activation of α-adrenoceptors. On the other hand, the α-adrenergic mechanism is sensitized in hypothyroidism and may then become operative in vivo [see 5, 41]. The receptors may also be targets of therapeutically used drugs, such as dopamine in cardiogenic shock [60].

Sites with properties of α-adrenoceptors have also been found in heart homogenates using radioligand binding techniques [42]. These findings are sometimes quoted as evidence for the hypothesis that α-adrenoceptors are located on heart muscle cells. However, although the findings are compatible with the hypothesis, they do not prove it, since, in the studies available, alternative locations have not been ruled out. Moreover, receptors sensu stricto can, until now, be identified only when a functional response – biochemical or physical – of the system is measured.

The contractility-enhancing α-adrenoceptors appear to be of the α_1-subtype, since they are blocked by low concentrations of prazosin [44]. The myocardial β-receptors are predominantly but not exclusively β_1 [7].

Occurrence of Presynaptic α- and β-Autoreceptors in the Heart

Guinea-pig atria and isolated rabbit hearts have been important in the detection of presynaptic α- and β-autoreceptors. α-Receptor blocking drugs such as phentolamine and phenoxybenzamine increased the overflow of noradrenaline elicited by sympathetic nerve stimulation; the increase could not be explained by decreased retention or degradation of noradrenaline within the tissue and was, hence, due to a facilitation of transmitter release from the sympathetic axons [28, 36, 50, 51]. Conversely, α-adrenergic agonists such as clonidine, phenylephrine, oxymetazoline, naphazoline and extracellular noradrenaline itself inhibited the nerve stimulation-evoked release of intraneuronal noradrenaline [36, 45, 46, 47, 61]. It was concluded that the terminal sympathetic axons possess α-adrenoceptors, the activation of which by either exogenous agonists or endogenous noradrenaline results, through an unknown series of events, in depression of transmitter release [28, 45]. The number of ^{3}H-dihydroergocryptine binding sites in membranes of rat heart ventricle is reduced after the sympathetic nerves have been destroyed by pretreatment with 6-hydroxydopamine; this observation supports the view that the release-inhibiting α-receptors as defined functionally are located on the axons themselves [56].

The release of noradrenaline is changed by drugs with affinity for β-adrenoceptors as well. In guinea-pig isolated atria, isoprenaline increased, and propranolol decreased, the nerve stimulation-evoked release [1]. These findings led to the suggestion that sympathetic terminal axons possess also β-adrenoceptors, the activation of which by either exogenous agonists or endogenous noradrenaline results, through an unknown series of events, in facilitation of transmitter release [1].

The rabbit heart has also been instrumental in the detection of differences between pre- and postsynaptic α-adrenoceptors and the subsequent general classification into α_1- and α_2-subtypes. The receptors mediating the inhibition of release were found to be sensitive to oxymetazoline and naphazoline, but rather insensitive to phenylephrine. Conversely, phenylephrine was a potent agonist at the receptors mediating an increase in the force of contraction, whereas oxymetazoline and naphazoline were inactive. Hence, a difference between the release-inhibiting (presynaptic) and the contractility-enhancing (postsynaptic) receptors was postulated [46]. Analogous differences were found in other tissues [26, 52, 53]. Using the terminology subsequently proposed [6, 27, 49], we can classify the contractility-enhancing receptors as α_1, as mentioned above; the presynaptic receptors belong to the α_2-subtype.

In most tissues examined, the presynaptic β-adrenoceptors appear to be β_2 (e.g., [63]).

Studies in which the release of noradrenaline was determined biochemically indicate that presynaptic α-autoreceptors occur in the hearts of dogs [64], cats [18], mice [19], guinea pigs (see above) and rabbits (see above), and presynaptic β-autoreceptors in the hearts of dogs [64], rats [34] and guinea pigs (see above).

Function of Cardiac Presynaptic α- and β-Autoreceptors

The interest raised by the detection of presynaptic autoreceptors stemmed mainly from the hypothesis that these receptors are links in local, release-controlling feedback circuits. When activated by previously released noradrenaline, they may facilitate (β-receptors; positive feedback) or depress (α-receptors; negative feedback) the release elicited by subsequent action potentials. For presynaptic α-autoreceptors, this hypothesis is widely accepted [4, 20, 24, 27, 38, 48, 54, 57, 62]. For the β-receptors, it is on less safe grounds. Being β_2, these receptors are little sensitive to noradrenaline, and in fact in some studies β-blocking drugs failed to reduce the release of noradrenaline even though facilitation by agonists revealed the existence of presynaptic β-receptors. This pattern has been observed, for instance, in guinea-pig atria [33; see, however, 1], rat atria [34] and in non-cardiac tissues such as human blood vessels [55] and the rat portal vein [63]. Moreover, in rat atria, exogenous noradrenaline, even in the presence of α-adrenoceptor blocking drugs, failed to facilitate the release of intraneuronal noradrenaline [34]. Thus, whereas the term "autoreceptor", i.e., a neurone's receptor for its own transmitter, may accurately describe the function of presynaptic α-adrenoceptors, it is perhaps less applicable for presynaptic β-adrenoceptors. Adrenaline is a more potent β_2-receptor agonist than noradrenaline. It has been proposed, therefore, that presynaptic β-adrenoceptors may be one of the physiological targets of adrenaline during enhanced adrenomedullary secretion [32, 51].

In this section we shall discuss the questions of whether the release of noradrenaline is modulated by cardiac presynaptic α- and β-receptors in vivo as well as in vitro, whether the changes in release are reflected in changes of the postsynaptic responses to stimulation, and whether the presynaptic cardiac receptors may play a therapeutic role.

Most of the evidence for cardiac presynaptic α- and β-autoreceptors summarized above was obtained in vitro. There is less biochemical in vivo evidence. In hearts of anaesthetized or spinal dogs, the α-adrenoceptor agonist clonidine as well as the β-receptor antagonist sotalol decreased, whereas the α-antagonists phenoxybenzamine, phentolamine and prazosin (the latter at relatively high doses which may perhaps have blocked α_2-receptors) and the β-agonist isoprenaline increased the release of noradrenaline [8, 9, 64]. The objection has been raised that the effects of isoprenaline and sotalol may have been, at least in part, secondary to their actions on the rate and force of contraction and the metabolic sequelae [12]. Further biochemical in vivo studies would be of great interest.

Presynaptic receptors primarily modulate neurotransmitter release. Secondary changes of postsynaptic responses to nerve action potentials would then be expected. In the heart, the effects of β-receptor agonists and antagonists at postsynaptic receptors will make any changes that are due to a β-adrenergic modulation of noradrenaline release difficult to detect, and, in fact, such changes have not been reported. Alterations in myocardial activity secondary to an α-adrenergic modulation of noradrenaline release can be demonstrated more easily because myocardial α-receptors are of minor importance, as mentioned above.

In vitro, α-adrenergic agonists (with little effect on β-receptors) such as clonidine, oxymetazoline, naphazoline and xylazine reduce the positive chronotropic and inotropic effects of sympathetic nerve stimulation [15, 22, 46, 58]. Figure 1 shows the parallel decrease, by naphazoline, of the presynaptic response (noradrenaline release) and the postsynaptic response (cardio-acceleration) in the rabbit heart. In most studies it has been found that antagonists, conversely, enhance the positive chronotropic and inotropic effect of nerve stimulation [11, 15, 23, 29, 36, 51]. α-Adrenergic agonists reduce the positive chronotropic effect in vivo as well [2, 8, 9, 10, 13, 16, 25, 31, 37, 40, 43, 64], and in most but not all in vivo studies α-adrenoceptor antagonists caused an increase [2, 9, 10, 13, 14, 30].

Although α-receptor antagonists uniformly facilitate the release of noradrenaline as determined biochemically, the postsynaptic response is enhanced less consistently. For instance, in guinea-pig left atria phentolamine failed to increase the positive inotropic effect of field stimulation of the sympathetic nerves in spite of the fact that it antagonized the inhibitory effect of xylazine [58]. In anaesthetized dogs, phenoxybenzamine increased the release of noradrenaline induced by nerve stimulation but not the myocardial responses [64]. From the small effect of phentolamine on the accelerator response to sympathetic nerve stimulation *Drew* (1980) concluded that "the physiological importance of the presynaptic α-inhibitory feedback system in regulating end-organ responses, rather than noradrenaline release, remains unclear". While this is true, it should be noted that in the experiments of

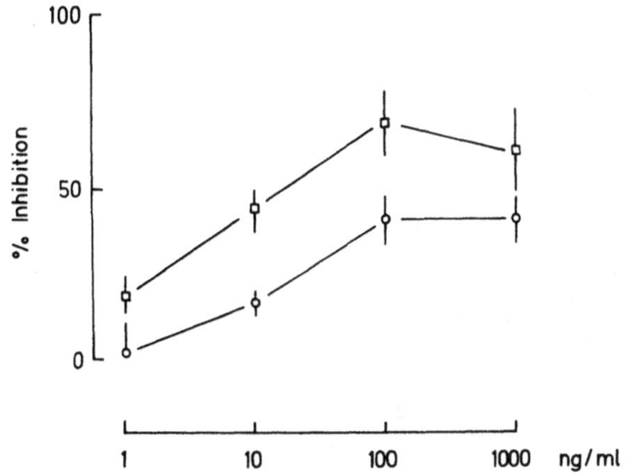

Fig. 1. Inhibition by naphazoline of the release of noradrenaline (□) and the increase in heart rate (○) caused by sympathetic nerve stimulation in the isolated perfused rabbit heart. The accelerans nerves were stimulated in each experiment twice for 1 min each at 5 Hz. Naphazoline was added 10 min before the second stimulation period. Changes obtained in the presence of naphazoline are corrected for changes that occurred in analogous experiments without naphazoline. Modified from *Starke* [46]

Drew [17] phentolamine decreased by 22% the accelerator response to exogenous noradrenaline; its facilitatory effect on the response to nerve stimulation was not corrected for this postsynaptic antagonist action and was, hence, underestimated. *Angus* and *Korner* [3] stimulated the sympathetic nerves of guinea-pig right atria with field pulses applied during the muscle refractory period. Trains of four pulses were given, one pulse following each of four consecutive atrial beats, yielding, with the atria beating at approximately 120 min^{-1}, a stimulation frequency of about 2 Hz. Phentolamine and yohimbine failed to enhance the ensuing positive chronotropic effect, suggesting "that presynaptic α-adrenoceptor modulation by synaptically released noradrenaline plays no part in cardiac sympathetic transmission, at any rate, at . . . physiological stimulus levels". *Rand* et al. [39] subsequently demonstrated, however, that 1.5 sec, i.e., the time needed to apply four pulses at 2 Hz, is too short an interval for the α-adrenergic inhibitory mechanism to become manifest. When four pulses were applied at 1 or 0.5 Hz, i.e., over 3 or 6 sec, or when stimulation at 2 Hz was extended to 7.5 sec, phentolamine did increase both the biochemically measured release of noradrenaline and the chronotropic effect. Since, in vivo, sympathetic nerves fire continuously, albeit slowly, stimulation with only four pulses over 1.5 sec as used by *Angus* and *Korner* [3] is probably rather non-physiological.

For the time being, the weight of evidence is in favour of not only the existence but also a physiological function of cardiac presynaptic α-autoreceptors. Yet, the increase in postsynaptic cardiac responses to nerve stimulation caused by α-receptor antagonists tends to be smaller than the increase in transmitter release. This may be due in part to the use of stimulation parameters that lead to near-maximal postsynaptic responses. Moreover, a small component in the postsynaptic effect may be mediated by myocardial α-receptors and, hence, blocked by the α-antagonists. Still, there seems to be a discrepancy, as pointed out previously [48]. Possibly, not all of the extra noradrenaline that is released after interruption of the feedback loop has access to the postsynaptic myocardial β-adrenoceptors.

Little is known about the activation or blockade of cardiac α- and β-autoreceptors in therapy. Inhibition of noradrenaline release may contribute to the antihypertensive effect of β-blocking drugs [32]. More specifically, the cardiac effects that β_2-selective agonists exert in vivo may be due not only to direct postsynaptic actions but also to facilitation of the release of noradrenaline.

Clonidine and related antihypertensive agents decrease the heart rate. Much of this bradycardia is probably central in origin and mediated by an increase in vagal and a withdrawal of sympathetic tone. However, inhibition of noradrenaline release via cardiac presynaptic α-autoreceptors may contribute. It has been shown, for instance, that minute doses of clonidine (e.g., 0.1 µg), injected into the sinus node artery of anaesthetized dogs, cause an immediate bradycardia [43] that is abolished by phentolamine [8]. Prazosin lowers the blood pressure with relatively little accompanying tachycardia. This has been attributed to the selectivity of prazosin for α_1-adrenoceptors and, hence, failure to interrupt the α-autoinhibition of noradrenaline release in the heart. In conscious rats, prazosin, for equal hypotensive effects, caused less increase in plasma noradrenaline levels than did phentolamine which blocks α_1- and α_2-receptors similarly, and the more pronounced effect of phentolamine on plasma noradrenaline was associated with a more marked tachycardia [21]. It seems possible, however, that additional effects help to minimize any tachycardia after prazosin such as a reduction of sympathetic outflow from the central nervous system [35].

References

1. Adler-Graschinsky E, Langer SZ (1975) Possible role of a β-adrenoceptor in the regulation of noradrenaline release by nerve stimulation through a positive feed-back mechanism. Br J Pharmacol 53:43–50
2. Algate DR, Waterfall JF (1978) Action of indoramin on pre- and postsynaptic α-adrenoceptors in pithed rats. J Pharm Pharmacol 30:651–652
3. Angus JA, Korner PI (1980) Evidence against presynaptic α-adrenoceptor modulation of cardiac sympathetic transmission. Nature 286:288–291
4. Bacq ZM (1976) Les contrôles de la libération des médiateurs aux terminaisons des nerf adrénergiques. J Physiol (Paris) 72:371–542
5. Benfey BG (1980) Cardiac α adrenoceptors. Can J Physiol Pharmacol 58:1145–1157
6. Berthelsen S, Pettinger WA (1977) A functional basis for classification of α-adrenergic receptors. Life Sci 21:595–606
7. Bryan LJ, Cole JJ, O'Donnell SR, Wanstall JC (1981) A study designed to explore the hypothesis that beta-1 adrenoceptors are "innervated" receptors and beta-2 adrenoceptors are "hormonal" receptors. J Pharmacol Exp Ther 216:395–400
8. Cavero I, Roach AG (1980) Effects of clonidine on canine cardiac neuroeffector structures controlling heart rae. Br J Pharmacol 70:269–276
9. Cavero I, Dennis T, Lefèvre-Borg F, Perrot P, Roach AG, Scatton B (1979) Effects of clonidine, prazosin and phentolamine on heart rate and coronary sinus catecholamine concentration during cardioaccelerator nerve stimulation in spinal dogs. Br J Pharmacol 67:283–292
10. Constantine JW, Weeks RA, McShane WK (1978) Prazosin and presynaptic α-receptors in the cardioaccelerator nerve of the dog. Eur J Pharmacol 50:51–60
11. Crews FT, Smith CB (1978) Presynaptic alpha-receptor subsensitivity after long-term antidepressant treatment. Science 202:322–324
12. Dixon WR, Mosimann WF, Weiner N (1979) The role of presynaptic feedback mechanisms in regulation of norepinephrine release by nerve stimulation. J Pharmacol Exp Ther 209:196–204
13. Docherty JR, McGrath JC (1979) An analysis of some factors influencing α-adrenoceptor feed-back at the sympathetic junction in the rat heart. Br J Pharmacol 66:55–63
14. Doxey JC (1977) Effect of clonidine on cardiac acceleration in pithed rats. J Pharm Pharmacol 29:173–174
15. Doxey JC, Roach AG (1980) Presynaptic α-adrenoceptors; in vitro methods and preparations utilised in the evaluation of agonists and antagonists. J Auton Pharmacol 1:73–99
16. Drew GM (1976) Effects of α-adrenoceptor agonists and antagonists on pre- and postsynaptically located α-adrenoceptors. Eur J Pharmacol 36:313–320

17. Drew GM (1980) Presynaptic modulation of heart rate responses to cardiac nerve stimulation in pithed rats. J Cardiovasc Pharmacol 2:843–856
18. Farah MB, Langer SZ (1974) Protection by phentolamine against the effects of phenoxybenzamine on transmitter release elicited by nerve stimulation in the perfused cat heart. Br J Pharmacol 52:549–577
19. Farnebo LO, Hamberger B (1974) Influence of α- and β-adrenoceptors on the release of noradrenaline from field stimulated atria and cerebral cortex slices. J Pharm Pharmacol 26:644–646
20. Gillespie JS (1980) Presynaptic receptors in the autonomic nervous system. In: Szekeres L (ed) Handbook of Experimental Pharmacology, vol. 54/I. Springer, Berlin Heidelberg New York, pp 353–425
21. Graham RM, Stephenson WH, Pettinger WA (1980) Pharmacological evidence for a functional role of the prejunctional alpha-adrenoreceptor in noradrenergic neurotransmission in the conscious rat. Naunyn-Schmiedebergs Arch Pharmacol 311:129–138
22. Ilhan M, Long JP, Cannon JG (1976) The ability of pimozide to prevent inhibition by dopamine analogs of cardioaccelerator nerves in cat hearts. Arch Int Pharmacodyn Ther 222:70–80
23. Kaumann AJ (1970) Adrenergic receptors in heart muscle: relations among factors influencing the sensitivity of the cat papillary muscle to catecholamines. J Pharmacol Exp Ther 173:383–398
24. Kirpekar SM (1975) Factors influencing transmission at adrenergic synapses. Prog Neurobiol 4:163–210
25. Kobinger W (1967) Über den Wirkungsmechanismus einer neuen antihypertensiven Substanz mit Imidazolinstruktur. Naunyn-Schmiedebergs Arch Pharmakol 258:48–58
26. Langer SZ (1973) The regulation of transmitter release elicited by nerve stimulation through a presynaptic feed-back mechanism. In: Usdin E, Snyder SH (eds) Frontiers in Catecholamine Research. Pergamon, New York, pp 543–549
27. Langer SZ (1974) Presynaptic regulation of catecholamine release. Biochem Pharmacol 23:1793–1800
28. Langer SZ, Adler E, Enero MA, Stefano FJE (1971) The role of the alpha receptor in regulating noradrenaline overflow by nerve stimulation. Proc XXVth Int Congr Physiol Sci, Munich, p 335
29. Langer SZ, Adler-Graschinsky E, Giorgi O (1977) Physiological significance of α-adrenoceptor-mediated negative feedback mechanism regulating noradrenaline release during nerve stimulation. Nature 265:648–650
30. Lokhandwala MF, Buckley JP (1976) Effect of presynaptic α-adrenoceptor blockade on responses to cardiac nerve stimulation in anesthetized dogs. Eur J Pharmacol 40:183–186
31. Lokhandwala MF, Coats JT, Buckley JP (1977) Effects of several catecholamines on sympathetic transmission to the myocardium: role of presynaptic α-adrenoceptors. Eur J Pharmacol 42:257–265
32. Majewski H, Rand MJ (1981) Adrenaline-mediated hypertension: a clue to the antihypertensive effect of β-adrenoceptor blocking drugs? Trends Pharmacol Sci 2:24–26
33. Majewski H, McCulloch MW, Rand MJ, Story DF (1980) Adrenaline activation of prejunctional β-adrenoceptors in guinea-pig atria. Br J Pharmacol 71:435–444
34. Majewski H, Rand MJ, Tung LH (1981) Activation of prejunctional β-adrenoceptors in rat atria by adrenaline applied exogenously or released as a co-transmitter. Br J Pharmacol 73:669–679
35. McCall RB, Humphrey SJ (in press) Evidence for a central depressor action of postsynaptic α_1-adrenergic receptor antagonists. J Auton Nerv System
36. McCulloch MW, Rand MJ, Story DF (1972) Inhibition of ^{3}H-noradrenaline release from sympathetic nerves of guinea-pig atria by a presynaptic α-adrenoceptor mechanism. Br J Pharmacol 46:523P–524P
37. Pacha W, Salzmann R, Scholtysik G (1975) Inhibitory effects of clonidine and BS 100–141 on responses to sympathetic nerve stimulation in cats and rabbits. Br J Pharmacol 53:513–516
38. Rand MJ, McCulloch MW, Story DF (1980) Catecholamine receptors on nerve terminals. In: Szekeres L (ed) Handbook of Experimental Pharmacology, vol 54/I. Springer, Berlin Heidelberg New York, pp 223–266
39. Rand MJ, McCulloch MW, Standford-Starr C, Story DF, Yang C (1981) The time course of the development and persistence of the autoinhibitory effect in noradrenergic trans-

mission. Abstract of the 8th International Congress of Pharmacology, Tokyo, Abstract No. S 13–5 p98
40. Robson RD, Antonaccio MJ (1974) Effect of clonidine on responses to cardiac nerve stimulation as a function of impulse frequency and stimulus duration in vagotomized dogs. Eur J Pharmacol 29: 182–186
41. Scholz H (1980) Effects of beta- and alpha-adrenoceptor activators and adrenergic transmitter releasing agents on the mechanical activity of the heart. In: Szekeres (ed) Handbook of Experimental Pharmacology, vol 54/I. Springer, Berlin Heidelberg New York, pp 651–733
42. Schümann HJ, Brodde OE (1979) Demonstration of α-adrenoceptors in the rabbit heart by [^{3}H]-dihydroergocryptine binding. Naunyn-Schmiedebergs Arch Pharmacol 308: 191–198
43. Scriabine A, Stavorski J, Wenger HC, Torchiana ML, Stone CA (1970) Cardiac slowing effects of clonidine (ST-155) in dogs. J Pharmacol Exp Ther 171:256–264
44. Skomedal T, Osnes JB, Øye I (1980) Competitive blockade of α-adrenergic receptors in rat heart by prazosin. Acta Pharmacol Toxicol 47:217–222
45. Starke K (1971) Influence of α-receptor stimulants on noradrenaline release. Naturwissenschaften 58:420
46. Starke K (1972a) Alpha sympathomimetic inhibition of adrenergic and cholinergic transmission in the rabbit heart. Naunyn-Schmiedebergs Arch Pharmacol 274: 18–45
47. Starke K (1972b) Influence of extracellular noradrenaline on the stimulation-evoked secretion of noradrenaline from sympathetic nerves: evidence for an α-receptor-mediated feed-back inhibition of noradrenaline release. Naunyn Schmiedebergs Arch Pharmacol 275: 11–23
48. Starke K (1977) Regulation of noradrenaline release by presynaptic receptor systems. Rev Physiol Biochem Pharmacol 77: 1–124
49. Starke K (1981) α-Adrenoceptor subclassification. Rev Physiol Biochem Pharmacol 88: 199–236
50. Starke K, Montel H, Schümann HJ (1971a) Influence of cocaine and phenoxybenzamine on noradrenaline uptake and release. Naunyn-Schmiedebergs Arch Pharmacol 270: 210–214
51. Starke K, Montel H, Wagner J (1971b) Effect of phentolamine on noradrenaline uptake and release. Naunyn-Schmiedebergs Arch Pharmacol 271: 181–192
52. Starke K, Montel H, Gayk W, Merker R (1974) Comparison of the effects of clonidine on pre- and postsynaptic adrenoceptors in the rabbit pulmonary artery. Naunyn-Schmiedebergs Arch Pharmacol 285: 133–150
53. Starke K, Endo T, Taube HD (1975) Relative pre- and postsynaptic potencies of α-adrenoceptor agonists in the rabbit pulmonary artery. Naunyn-Schmiedeberg's Arch Pharmacol 291:55–78
54. Stjärne L (1975) Basic mechanism and local feedback control of secretion of adrenergic and cholinergic neurotransmitters. In: Iversen LL, Iversen SD, Snyder SH (eds) Handbook of Psychopharmacology, vol 6. Plenum, New York, pp 179–233
55. Stjärne L, Brundin J (1975) Dual adrenoceptor-mediated control of noradrenaline secretion from human vasoconstrictor nerves: facilitation by β-receptors and inhibition by α-receptors. Acta Physiol Scand 94: 139–141
56. Story DD, Briley MS, Langer SZ (1979) The effects of chemical sympathectomy with 6-hydroxydopamine on α-adrenoceptor and muscarinic cholinoceptor binding in rat heart ventricle. Eur J Pharmacol 57:423–426
57. Vizi ES (1979) Presynaptic modulation of neurochemical transmission. Prog Neurobiol 12: 181–290
58. Vizi ES, Somogyi GT, Hadházy P, Knoll J (1973) Effect of duration and frequency of stimulation on the presynaptic inhibition by α-adrenoceptor stimulation of the adrenergic transmission. Naunyn-Schmiedebergs Arch Pharmacol 280:79–91
59. Wagner J, Knorr A, Schümann HJ (1977) On the presence of myocardial α-adrenoceptors in the atrium of the guinea pig. Arch Farmacol Toxicol 2: 83–92
60. Wagner J, Schümann HJ, Knorr A, Rohm N, Reidemeister JC (1980) Stimulation by adrenaline and dopamine but not by noradrenaline of myocardial α-adrenoceptors mediating positive inotropic effects in human atrial preparations. Naunyn-Schmiedebergs Arch Pharmacol 312:99–102

61. Werner U, Starke K, Schümann HJ (1970) Wirkungen von Clonidin (ST 155) and BAY a6781 auf das isolierte Kaninchenherz. Naunyn-Schmiedebergs Arch Pharmakol 266:474
62. Westfall TC (1977) Local regulation of adrenergic neurotransmission. Physiol Rev 57:659–728
63. Westfall TC, Peach MJ, Tittermary V (1970) Enhancement of the electrically induced release of norepinephrine from the rat portal vein: mediation by β_2-adrenoceptors. Eur J Pharmacol 58:67–74
64. Yamaguchi N, de Champlain J, Nadeau RA (1977) Regulation of norepinephrine release from cardiac sympathetic fibers in the dog by presynaptic α- and β-receptors. Circ Res 41:108–117

Discussion

Schümann, Essen: I would like to know whether the positive chronotropic effect observed after phentolamine is only due to its primary pre- and secondary post-synaptic effect, or whether the peripheral blood pressure decreasing effect plays a role, and, if so, what is its magnitude?

Starke: Under the conditions of the experiments that I mentioned [30], a reflex increase in heart rate seems unlikely because the autonomic cardiac nerves were largely cut, and the positive chronotropic effect of electrical stimulation of the sympathetic nerves was studied.

Kopin, Bethesda: Since circulating epinephrine is taken up by the heart, Michael Rand suggested that the epinephrine released from sympathetic nerves in hypertensives may enhance catecholamine release upon nerve stimulation and that β-blockers act as antihypertensives by preventing the enhanced release. Could you comment on this hypothesis?

Starke: Some of the assumptions in Rand's hypothesis have been verified experimentally. For instance, adrenaline is taken up into noradrenaline axons and can be released by action potentials. Moreover, a positive feedback mechanism facilitating transmitter release is then established which does not operate as long as the neurons store only noradrenaline. Furthermore, the group in Melbourne has shown that prolonged administration of adrenaline to rats induces hypertension which can be prevented by β-blockers. All this fits nicely together. On the other hand, there are assumptions that are still unproven, one being that adrenaline plasma levels are in fact increased early in the development of hypertension.

Reuter, Bern: I should like to make a few remarks concerning post-synaptic α-adrenoceptors. My co-worker Sanchez-Chopula has recently investigated 3 putative α-adrenoceptor agonists, phenylephrine, methoxamine and clonidine. The 3 different agonists had at least 3 different actions: phenylephrine had a strong β-adrenergic effect; methoxamine reduced force of contraction at high concentrations, but had no positive inotropic effect whatsoever; clonidine had only a histamine-like H_2-effect that could be inhibited by cimetidine. There may also be considerable species differences in so-called α-adrenoceptors.

Starke: As to the first point, a β-adrenergic component in the effect of phenylephrine is known. Therefore, it is necessary to additionally apply a β-receptor antagonist when one wants to study the α-effect of phenylephrine on the myocardium. With respect to the lack of an α-effect in a certain preparation and under certain conditions, one reason may be stimulation of the tissue at too high a frequency; the α-adrenergic positive inotropic effect, in contrast to the β-effect, is much higher at low heart rates. Finally, I may add that in my talk I gave a simplified picture, since, for instance, not only positive inotropic α-effects have been postulated (and electrophysiological effects such as a prolongation of Purkinje-fiber action potential) but also negative inotropic α-effects [44].

Ceremuzynski, Warsaw: In view of the mechanisms you have outlined would you kindly comment on the influence of β-blockers on coronary circulation.
You have mentioned the potency of clonidine to activate presynaptic α-receptors with subsequent reduction of the myocardial response to sympathetic stimulation. This might be a reason for clonidine to be used in patients with angina pectoris who frequently display high adrenergic drive. We tried clonidine clinically for this purpose and have had very encouraging results.

Starke: I am afraid that I cannot comment on the use of clonidine in patients with angina. As far as β-adrenoceptor antagonists in angina are concerned, their main effect is certainly the blockade of myocardial postsynaptic β-receptors.

Sympathetic Control of Coronary Circulation[1]

E. Bassenge, J. Holtz

Summary

Both α- and β-adrenergic receptors have been demonstrated in the coronary bed. Whereas a functional innervation of the α-receptors does exist, there is little evidence for the functional innervation of coronary β-receptors, which seem to be mainly activated by circulating catecholamines. In various segments of the coronary bed β_1- and/or β_2-receptors have been identified.

Alpha-receptor activity has been shown in large coronary conductance vessels (in which such activity may be of pathophysiological importance in the genesis of coronary spasm), and in coronary resistance vessels, which under physiological conditions regulate myocardial perfusion. The α-adrenergic constrictive control tends to keep myocardial perfusion at low levels and counteracts the metabolic coronary dilation. Surprisingly, this nervously induced constriction is still present and demonstrable under hypoxic coronary perfusion.

In dogs magnitude of the α-adrenergic constrictive control can be demonstrated convincingly by the hyperinflation reflex induced coronary dilation. The resulting excess perfusion (+100%) causing high coronary venous O_2-saturations is initiated by a sudden withdrawal of the α-constrictive activity. Whether this magnitude exists in the human coronary circulation has not been investigated to date. However, several reports point out the significance of α-adrenergic coronary innervation for cold pressor test-induced angina pectoris.

It should be possible to quantitatively describe the magnitude of α-receptor induced coronary constriction using α-antagonists. However, most of these blockers display a dual effect on both presynaptic α_2-receptors (thereby triggering excess release of NE competing at the vascular α-receptors) and postsynaptic α_1- (and α_2?) receptors. This dual effect was demonstrable by inducing hypotension with or *without* simultaneous increases in plasma-catecholamines, using a selective α_1-antagonist (prazosin) or a nonselective α-antagonist (phentolamine). Consequently, prazosin application led to a sudden blockade of the adrenergic constrictive tone, resulting in a simultaneous increase in myocardial perfusion (indicated by a rise in coronary venous O_2). Surprisingly, this excess perfusion decayed with time (though the α-antagonism was still demonstrable) and pointed to a readjustment by a complementary change in metabolic factors; thus the dominant control factor being the metabolic dilation, which in turn is modulated by the α-adrenergic coronary constriction.

Introduction

The physiological significance of sympathetic coronary innervation has been disputed for decades [3, 13]. It is difficult to demonstrate a substantial direct sympathetic regulation of coronary flow convincingly since local metabolic factors perfectly control myocardial perfusion, covering a large regulatory range. Also the

1 Supported by the Deutsche Forschungsgemeinschaft (Ba 408/10)

denervated heart regulates its own blood supply adequately. The difficulties to convincingly demonstrate direct sympathetic coronary effects originate from 3 problems, partly not yet settled:

1. Stimulation of sympathetic cardiac nerves results in substantial myocardial metabolic effects, so that the immediate response of local metabolic control overrides the possible direct coronary effects. In order to quantify these direct coronary effects, cardiac metabolic effects have to be minimized experimentally. In recent years this has often been done by pharmacological β-blockade, thereby obscuring possible direct β-adrenergic effects on the coronary vasculature.
2. The subtype of β-adrenoceptors existing in the coronary bed has been analyzed in various in vitro and in situ preparations. Both β_1- and β_2-receptors have been identified from experiments using selective agonists, therefore selective β_1-blockade cannot differentiate between β-adrenergic myocardial and coronary effects [see 3, 14, 23].
3. The direct sympathetic effects depend on the relative density of the α- versus β-receptor population in various segments of the coronary bed. Furthermore, they depend on whether these receptors are functionally innervated or not. Studies designed to demonstrate functional β-adrenergic innervation failed to show substantial effects [14, 25]. Yet, it was never proven conclusively that the coronary β-receptors cannot be activated neurogenically.

The first experiments designed to differentiate direct coronary effects of sympathetic nerve stimulation from myocardial metabolic effects were published in 1939

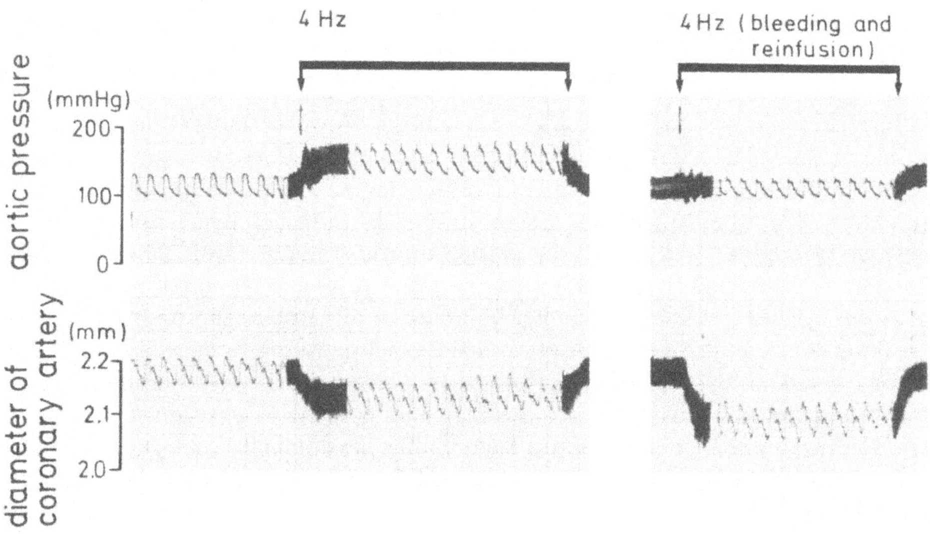

Fig. 1. *Constriction of coronary conductance artery in response to sympathetic stimulation.* Record from a dog in N_2O-pentobarbital anesthesia with heart rate kept constant (AV-block and ventricular pacing). The internal diameter of the circumflex branch of the left coronary artery is obtained by the intravascular resilient sensor of an absolute induction angiometer. Upon sympathetic stimulation, the coronary artery diameter declined in spite of a substantial rise in arterial pressure (*left hand panel*). When the changes in arterial pressure are minimized by appropriate bleeding and reinfusion, the decline in coronary artery diameter is more pronounced (*right hand panel*). The *arrows* indicate the duration of the stimulation of the decentralized left stellate ganglion. (*Bassenge* et al. [1])

by *Katz* and *Jochim* [17]. Using a fibrillating heart preparation, they found the predominant sympathetic coronary effect being constrictive. Later, *Berne* used a similar preparation to demonstrate the constrictive effect of norepinephrine [2] and of nerve stimulation [4] on myocardial oxygen balance by monitoring coronary venous P_{O_2}. When β-blockers became available, it was possible to demonstrate the constrictive effects of norepinephrine [8] and of sympathetic nerve stimulation [10] in the beating heart in situ. It was shown that this direct sympathetic coronary constriction could be activated by reflexes [24] and that it competed with metabolic coronary vasodilation [21]. Furthermore, this competition between sympathetic coronary constriction and metabolic vasodilation could still be demonstrated during myocardial underperfusion [11].

Recent studies have shown α-adrenergic coronary constriction in the large epicardial segments of the arteries, either by measuring pressure differences along the coronary bed [9, 18] or by directly recording coronary artery diameter during sympathetic nerve stimulation in the arrested heart [12], in the beating heart in situ (Fig. 1), or finally in the conscious dog [28]. Thus, functional innervation of coronary α-adrenoceptors in all segments of the coronary arterial bed has been convincingly demonstrated.

In accordance with the demonstration of a functional innervation of coronary vascular α-adrenoceptors, a tonic constrictive sympathetic control of coronary flow in conscious dogs has been repeatedly postulated [15, 29, 30]. However, α-adrenergic blockade by phentolamine did not result in the release of this postulated constrictive coronary control, when β-adrenergic effects were excluded by appropriate blockade [22]. In this study, we tested whether this discrepancy is explained by the presynaptic effects of phentolamine, interfering with the physiologic autoinhibition of norepinephrine release in vivo.

Methods

Mongrel dogs of either sex weighing 26±3 kg were used in this study. All dogs were adapted to the experimental surroundings and familiar to the experimenter and they were trained to rest quietly on the experimental table. Following the surgical implantation of instruments (see below) the dogs received antibiotics for 1 week (2 g/day ampicillin orally). Experiments were started 2 weeks after surgery, when the dogs' rectal temperature at rest was 39.0 °C or lower. During the experimental period the dogs were kept on standard diet containing 1–3 mEq/kg sodium and maintained their body weight.

Surgical Preparations

For analysis of vasomotion of coronary resistance vessels, the dogs were prepared in two steps. In the first operation an AV-block was produced via a right thoracotomy. Thus the heart rate of the dogs could be kept constant at any desired level by ventricular stimulation via implanted electrodes. Thereby the effect of a varying extravascular coronary resistance (due to changes in heart rate) on coronary flow could be avoided. The AV-block was brought about by injection of 0.1 ml formaldehyde into the bundle of His. An intracardiac electrode, introduced by punction of the right atrium, was used to identify the bundle of His and to inject the formalin. A complete AV-block developed within 10 s–3 min following the injection in all dogs and persisted throughout the experimental period. Electrodes were attached to the right ventricle. Postoperatively, the heart was paced at a rate of 100 bpm by a portable pacer. The second thoracotomy into the fifth left intercostal space was performed 3–5 weeks later. At the origin of the circumflex branch of the left coronary artery an electromagnetic flow transducer and a pneumatic occlusion cuff for mechanical zero reference were implanted. PVC-catheters with silastic tip were implanted into the pulmonary artery and the descending aorta. An additional catheter was implanted into the left common carotid artery for nicotine injection to stimulate the hyperinflation reflex.

The dogs used for the study of the hemodynamic, metabolic and hormonal effects of α-antagonists were also instrumented via a left thoracotomy in pentobarbital anesthesia (12 mg/kg) following premedication with atropine (0.05 mg/kg s.c.) and thalamonal (0.2 ml/kg i.m.). Two catheters were implanted into the pulmonary artery. A solid state blood pressure transducer (Konigsberg P4) was implanted into the aortic arch and a teflon catheter into the abdominal aorta via the left femoral artery.

Vasomotion of coronary resistance vessels was studied in the conscious, resting dogs. Circumflex coronary artery flow was measured by a sine-wave flowmeter and coronary diastolic vascular resistance was calculated beat to beat, while heart rate was kept constant. The continous, phasic analysis of coronary venous oxygen saturation was performed by reflection oximetry, using a flexible fiberoptic catheter in the coronary sinus [25]. The coronary sinus was catheterized in the conscious dogs from the external jugular vein, punctured under local anesthesia. The dogs were heparinized (500 U/kg) and the reflection oximetry was calibrated in situ by repeatedly withdrawing blood through the fiberoptic catheter for analysis of oxygen saturation and content.

For analysis of the hemodynamic effects of α-adrenergic antagonists, cardiac output was measured by the dye dilution technique, while aortic pressure and heart rate were obtained from the implanted solid state blood pressure transducer. Prior to each experiment, the pressure transducer was crosscalibrated against abdominal aortic pressure obtained from the femoral artery catheter. Thereafter, this catheter was used for withdrawal of arterial blood through a densitometer for estimation of cardiac output, while the dye was injected into the pulmonary artery catheter. The second pulmonary artery catheter was used for drug infusion and withdrawal of mixed venous blood for analysis of systemic arterio-venous oxygen difference. Total body oxygen consumption was calculated from cardiac output multiplied by arterio-venous oxygen difference. Alpha-antagonists (phentolamine und prazosin) were infused slowly (within 5 to 7 min) into the pulmonary artery in equihypotensive dosages.

The determination of plasma norepinephrine und epinephrine concentration was performed by a modification of the radioenzymatic method according to *DaPrada* and *Zürcher* [6]. Mixed venous blood samples from the heparinized dogs were collected via the pulmonary artery catheter into ice-chilled syringes and immediately centrifuged (10 min, 6000 rpm, 0 °C). Plasma was harvested and denaturated with an equal volume of 0.6 h perchloric acid containing EGTA (0.2%), $MgCl_2$ (0.2%) and 10 mM gluthadione. After centrifugation (10 min, 16 000 rpm, 0 °C) the supernatant was fractionated and stored at –30 °C. All measurements were done in duplicate.

Results

Reflex Vasomotion of Coronary Resistance Vessels

In ten dogs with constant heart rate (experimental AV-block) a reflex coronary vasodilation was elicited by stimulating a deep inspiration through injection of 1.5 µg/kg nicotine into the carotid artery. A coronary dilation was observed reaching a peak value within 15 s following the deep inspiration. In all trials mean diastolic coronary resistance fell below 50% of the control value (Fig. 2) and coronary venous O_2-saturation increased from 23%±4% saturation (mean±SD, $n=32$) to 56%±5% saturation ($P<0.001$) within 30 s following the injection of nicotine. In four of these dogs the femoral artery was punctured under local anesthesia and a fiberoptic catheter was introduced to monitor arterial oxygen saturation. In eight trials small fluctuations of arterial saturation above 94% were observed prior to, during and following the hyperinflation. In six dogs spontaneous deep inspirations were observed fortuitously, causing a comparable coronary dilation. Coronary venous O_2-saturation increased from 25%±5% to 52%±7% saturation ($n=9$, $P<0.01$) within 30 s following the spontaneous deep inspiration.

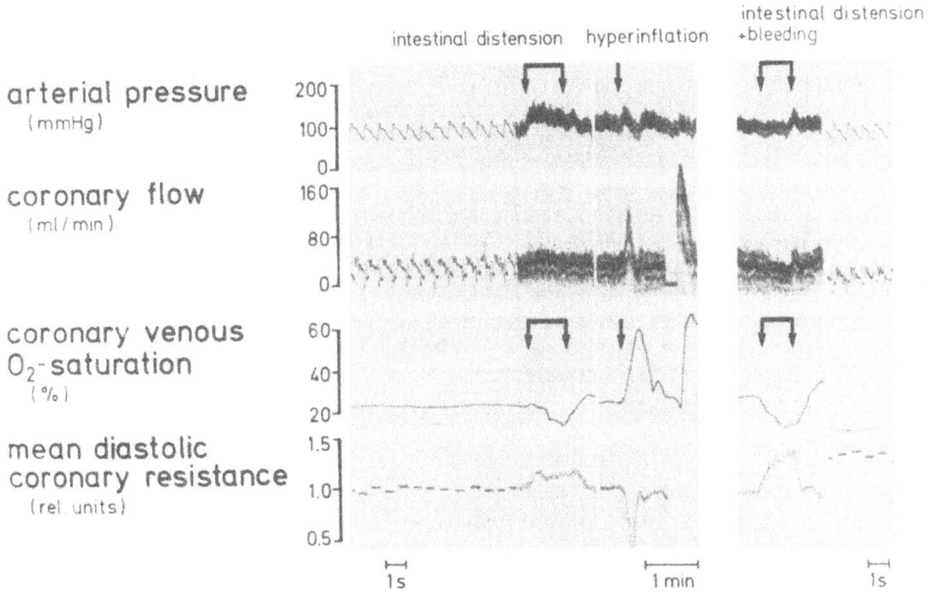

Fig. 2. *Vasomotion of coronary resistance vessels.* Record from a resting dog with heart rate kept constant (AV-block and ventricular pacing). Intestinal distension was produced by inflation of a 100 ml latex balloon chronically implanted in the jejunum. *Left-hand panel:* intestinal distension causes an increase in arterial pressure and relative coronary constriction (increase in diastolic coronary resistance, decline in coronary venous O_2-saturation, but no decline in coronary flow). *Middle panel:* Hyperinflation of the lung by a deep inspiration (elicited by injection of 1.5 µg/kg nicotine into one carotid artery to stimulate chemoreceptors) induces coronary dilation (increase in coronary flow and coronary venous O_2-saturation, decline in coronary resistance). Following decay of the coronary dilation, the coronary artery is occluded by an implanted cuff for zero flow control. *Right-hand panel:* During intestinal distension, the changes in arterial pressure are minimized by bleeding and reinfusion. Under this condition the intestinal distension causes a coronary constriction and a decline in coronary flow

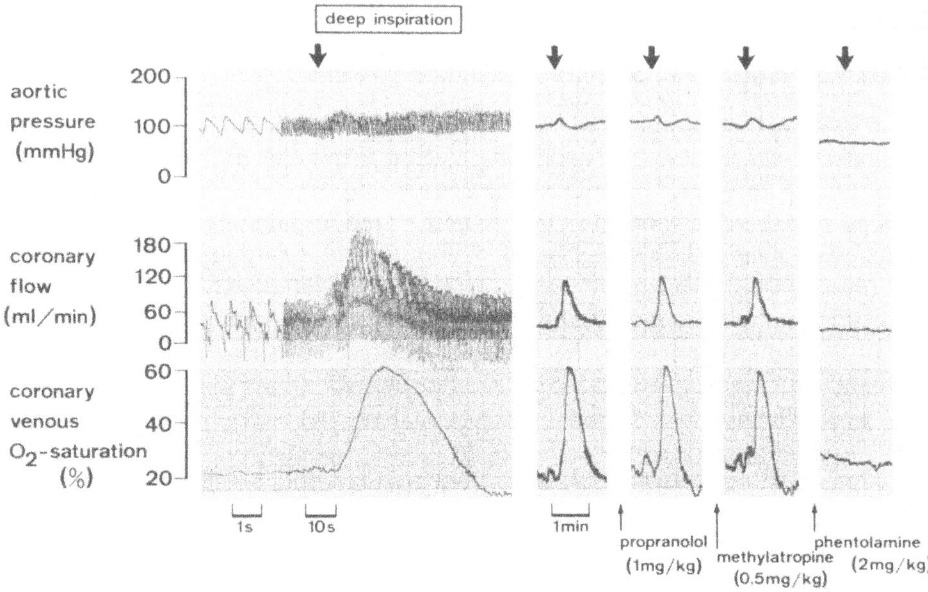

Fig. 3. *Pharmacologic characterization of hyperinflation-induced coronary dilation.* Record from a resting dog with heart rate kept constant (AV-block and ventricular pacing). *Arrows* in the top of the record indicate injection of 1.5 μg/kg nicotine into a carotid artery to induce a deep inspiration by stimulation of chemoreceptors. *First panel:* The deep inspiration causes a coronary dilation and a sharp increase in coronary venous O_2-saturation, indicating that the coronary dilation causes a hyperperfusion exceeding myocardial metabolic demands and thus cannot be caused metabolically. *Second panel:* The same reaction as in the first panel, but coronary flow and arterial pressure were averaged electronically. *Third panel:* Beta-adrenergic blockade by propranolol does not modify the coronary dilation. *Fourth panel:* Similarly, muscarinic blockade by methylatropine does not alter the coronary dilation. *Fifth panel:* Alpha-adrenergic blockade by phentolamine abolishes the hyperinflation-induced dilation. These data indicate that the coronary dilation induced by hyperinflation is due to reflex withdrawal of a constrictive sympathetic coronary control. However, phentolamine does *not* induce a coronary dilation which should result from pharmacologic suppression of this constrictive coronary control by an alpha-antagonist

In five of these dogs, a latex balloon, chronically implanted into the small intestine was inflated. This inflation caused an increase in mean arterial pressure by 26 ± 5 mmHg ($n = 12$, $P < 0.001$), no significant change in coronary flow, and a decline in coronary venous O_2-saturation from 22% ± 5% to 12% ± 2% saturation ($P < 0.01$). In three dogs, the left femoral artery was catheterized under local anesthesia to minimize the increase in arterial pressure during the intestinal distension by withdrawal of blood. Thus, in eight trials the increase in mean arterial pressure during inflation of the balloon could be kept below 5 mmHg, and coronary flow declined by 32% ± 8% ($P < 0.05$), mean diastolic coronary resistance increased by 36% ± 6% ($P < 0.01$) and coronary venous O_2-saturation declined from 24% ± 4% to 11% ± 2% saturation ($P < 0.01$). All dogs tolerated the inflation of the balloon without moving or crying.

In ten dogs, the hyperinflation reflex was studied following consecutive application of β-blockade by 1 mg/kg propranolol i.v., muscarinic blockade by 0.5 mg/kg methylatropine i.v., and α-blockade by 2 mg/kg phentolamine (Fig. 3). Propranolol and methylatropine (given 20 min later) did not significantly modify mean arterial pressure, coronary flow and coronary venous O_2-saturation. The coronary reflex dilation was not attenuated (Fig. 3). Phentolamine, given 20 min following methylatropine, caused a decline in mean arterial pressure by 18 ± 4 mmHg ($P < 0.01$), a fall in coronary flow by $22\% \pm 6\%$ ($P < 0.05$), but no change in coronary venous O_2-saturation. The coronary dilation following deep inspirations was abolished in seven dogs and attenuated in three dogs, further phentolamine application (up to 8 mg/kg) abolished the coronary dilation in these dogs also.

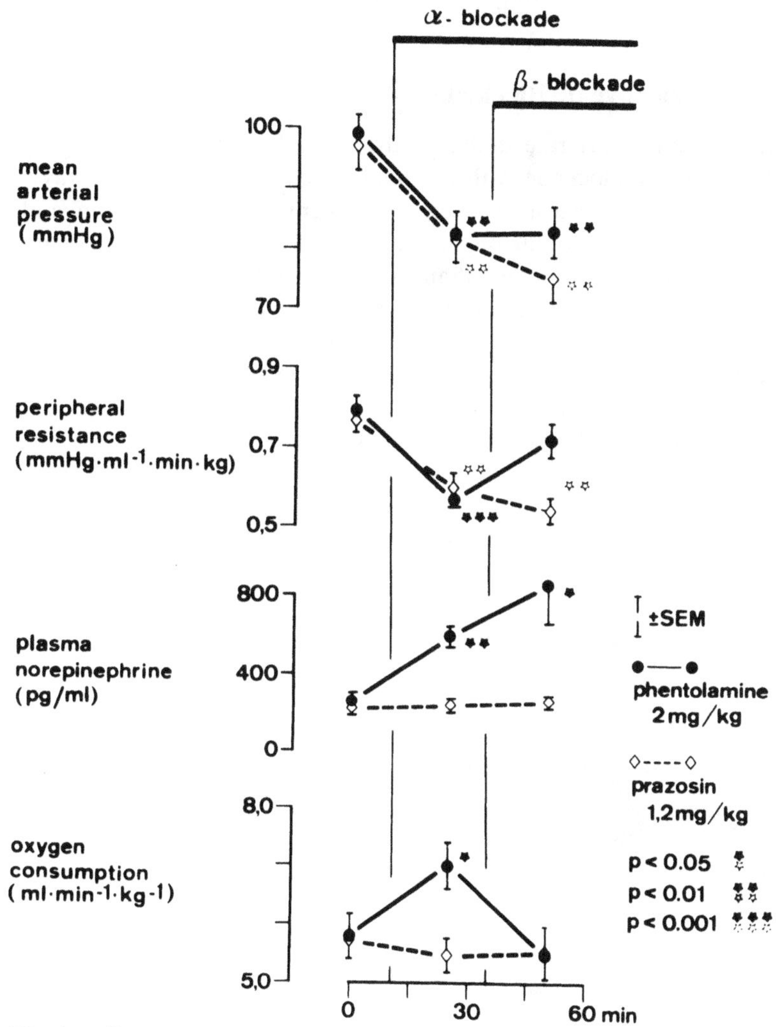

Fig. 4. *Effects of non-selective α-blockade (phentolamine) and selective α_1-blockade (prazosin) on hemodynamics, plasma-norepinephrine and total body oxygen consumption in conscious dogs.* Mean values ± SEM from eight dogs. The dosages of the alpha-antagonists were adjusted to induce comparable hypotension. Both antagonists cause systemic vasodilation (decline in peripheral resistance), but only phentolamine induces increments in plasma norepinephrine level and in oxygen consumption. Addition of β-adrenergic blockade (2 mg/kg nadolol i.v.) does not modify the effects of prazosin, but abolishes the phentolamine-induced vasodilation and elevation in oxygen consumption. The *asterisks* indicate significance of difference to the control value (analysis of variance)

Hemodynamic Effects of Non-selective α-Blockade and of Selective α_1-Blockade

In eight dogs the hemodynamic effects of 2 mg/kg phentolamine were compared to an equihypotensive dosage of the selective α_1-blocker prazosin (1.2 mg/kg i.v.). The results are summarized in Fig. 4. Though both agents caused comparable hypotension and vasodilation, only phentolamine induced significant elevations of plasma norepinephrine and total body oxygen consumption. The phentolamine-induced vasodilation and elevation in oxygen consumption were abolished by β-adrenergic blockade (2 mg/kg nadolol i.v.). Following phentolamine and nadolol, the total peripheral resistance was not significantly different from the control value.

Coronary Effects of Selective α_1-Blockade

In ten dogs with constant heart rate under β-adrenergic blockade with 1 mg/kg propranolol and muscarinic blockade with 0.5 mg/kg methylatropine, the intravenous injection of 1.2 mg/kg prazosin caused a transient coronary dilation and myocardial excess perfusion of variable degree. In two out of ten dogs the coronary venous O_2-saturation increased to more than 50% saturation, in three dogs the fluctuations of coronary venous O_2-saturation were less than 5% saturation following prazosin application. In all dogs, the myocardial excess perfusion was transient, coronary venous O_2-saturation returning to the control value within 6–22 min. Thereafter, the coronary dilation elicited by the hyperinflation reflex was tested and found to be attenuated by more than 50% (Fig. 5).

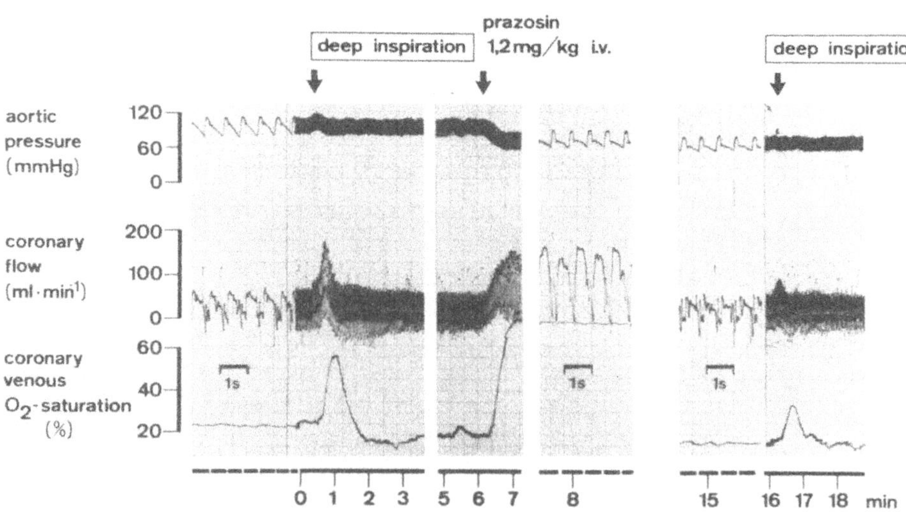

Fig. 5. *Effect of selective α_1-blockade (prazosin) on coronary flow and on hyperinflation-induced coronary dilation.* Record from a resting dog with heart rate kept constant (AV-block and ventricular pacing) under β-adrenergic and muscarinic blockade (1 mg/kg propranolol and 0.5 mg/kg methylatropine). *Left-hand panels:* Hyperinflation-induced coronary dilation and prazosin-induced coronary dilation. *Right-hand panel:* Seven minutes following prazosin application the coronary dilation (and the elevation in coronary venous oxygen saturation) has completely disappeared, but the hypotension persists. The hyperinflation-induced coronary dilation is attenuated but not abolished

Discussion

In conscious dogs with heart rates experimentally kept constant a surprisingly large range of *non*-metabolically induced vasomotion of coronary resistance vessels can be observed in the course of reflexes (Fig. 2). When comparing the peak values of the hyperinflation induced coronary dilation and of the coronary constriction during intestinal distension, the coronary vascular conductance varies roughly fourfold, resulting in extreme values of coronary venous O_2-saturation as low as 10% and as high as 60% saturation. However, it should be stressed that this range is derived from peak values during transient vasomotor reactions and not from steady state values.

The coronary constriction during intestinal distension does not necessarily indicate a specific intestinal-coronary reflex. In our opinion this coronary constriction is part of a general vasoconstriction during pain and might be provoked by other types of painful stimuli, too. The coronary dilation reflex elicited by hyperinflation of the lung was analyzed in detail by *Vatner* and *McRitchie* (29). The steep increase in coronary venous O_2-saturation indicates a myocardial perfusion substantially exceeding myocardial metabolic demands (Figs. 2, 3). The application of adrenergic blockers shows that this dilation is brought about by reflex withdrawal of an α-adrenergic constrictive tone (Fig. 3). The existence of a sympathetic constrictive coronary control in conscious dogs has also been concluded from analyses of other reflexes [30] and from studies using regional myocardial sympathectomy [15]. The values of myocardial excess perfusion following hyperinflation and following regional myocardial sympathectomy [15] are quantitatively in agreement.

However, the myocardial and metabolic effects of myocardial sympathectomy are not completely known, and disagreeing data on myocardial perfusion following myocardial sympathectomy have been published [see 3]. Furthermore, the α-blocker phentolamine, though abolishing the coronary dilation of the hyperinflation reflex, does not induce myocardial excess perfusion in dogs under β-blockade (Fig. 3), neither when given intravenously by a bolus injection or by a continuous infusion, nor when administered locally via intracoronary infusion (unpublished observation of our laboratory). A myocardial excess perfusion, however, should result from the pharmacologic blockade of a postulated α-adrenergic coronary control.

The physiologic role of α-blockade is not restricted to the postsynaptic effector cells. The autoinhibition of norepinephrine release by presynaptic inhibitory α_2-adrenoceptors, previously analyzed in vitro, is operative under in vivo conditions ([19, 26, 27] and the foregoing article by *Starke* in this proceedings). This feed-back inhibition lowers the synaptic (or junctional) transmitter concentration during sympathetic activation (Fig. 6) and this inhibition is suppressed by nonselective α-blockers like phentolamine [19, 23, 26, 27]. Due to its selectivity for α_1-adrenoceptors prazosin is assumed to have no or minimal presynaptic effects [6], while phentolamine has a dual action [23]: sympatholytic by postsynaptic blockade and sympathomimetic by presynaptic blockade. This difference in presynaptic activity might explain the great differences in plasma norepinephrine levels observed during identical levels of hypotension induced by these two agents (see Fig. 4, though additional reasons for this difference cannot be ruled out).

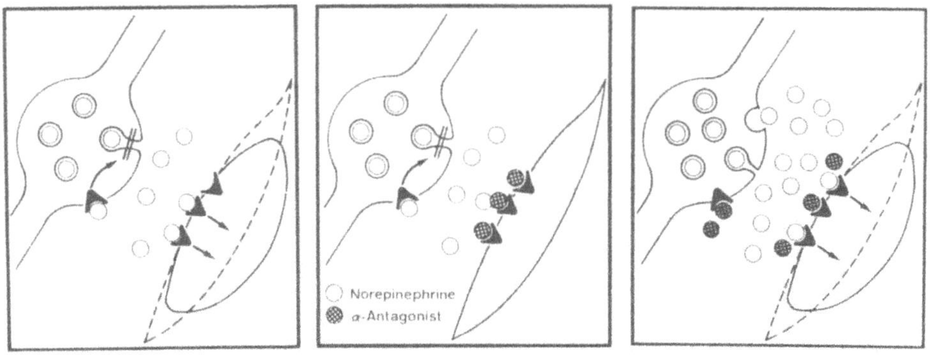

Fig. 6. *Effect of α-blockade on sympathetic neurovascular transmission. Left-hand panel:* The physiologic autoinhibition of norepinephrine release by stimulation of presynaptic α_2-adrenoceptors prevents excessive norepinephrine concentrations. *Middle-panel:* The experimental objective would be selective postsynaptic blockade for quantification of α-adrenergic vascular control. However, both $\alpha_1\alpha_2$-receptors have been demonstrated at vascular smooth muscle cells. Thus, selective and complete postsynaptic blockade may be impossible. *Right-hand panel:* Non-selective α-blockade has a complex biological effect: the postsynaptic blockade is counterbalanced by increased non-inhibited transmitter release due to the presynaptic blockade. Accidentally, these pre- and postsynaptic effects may compensate each other

Interestingly, the phentolamine-induced increase in catecholamine release causes β-adrenergic vasodilation and elevated oxygen consumption, which are offset by subsequent β-blockade (Fig. 4), while the prazosin-induced vasodilation is not attenuated. An inappropriate degree or duration of the postsynaptic vascular α-blockade cannot be the reason for this different effect of prazosin and phentolamine, as has been tested by agonist infusions [16]. It is reasonable to assume that the minimal, nonsignificant effect of phentolamine on peripheral resistance under β-adrenergic blockade (Fig. 4) results from a balance of pre- and postsynaptic effects of the drug as indicated schematically in Fig. 6.

This balance might explain the lack of coronary vasodilation by phentolamine in dogs under β-blockade (Fig. 3), but this balance may not be perfect during all experimental conditions. In fact, phentolamine has a coronary dilating effect in the exercising dog under β-blockade [22]. Selective postsynaptic α-blockade by prazosin causes coronary dilation and myocardial excess perfusion (Fig. 5) thus demonstrating the postulated constrictive coronary control. Yet, several aspects are not clear at present:

Firstly, the prazosin-induced coronary dilation was extremely variable between individual dogs and the hyperinflation reflex dilation was not completely abolished. This indicates a possible role of vascular α_2-adrenoceptors in the coronary bed. Secondly, the prazosin-induced coronary dilation was transient in all experiments and decayed rapidly though the hypotension still persisted and the effectiveness of the α_1-blockade was demonstrable by agonist infusion [16]. A hypothetic explanation might be a potential metabolic readjustment of the excess perfusion by washout of dilating metabolites, but this is difficult to substantiate experimentally. Finally, it seems not possible to exactly quantify the α-adrenergic coronary control in a certain physiological or clinical situation. Selective α_1-blockers cannot block postsynaptic coronary α_2-adrenoceptors; non-selective antagonists, though acting on all vascular α-adrenoceptors, have an "indirect sympathomimetic" activity in vivo due to their presynaptic action.

Thus, the physiological and, even more, the pathological role of coronary α-adrenoceptors is still a matter of debate. Several reports indicate that α-adrenergic coronary constriction might be the mechanism or the trigger of coronary spasms [see 20]. However, recent observations in patients with autotransplanted hearts and clinical protocols designed to reduce the frequency of spastic anginal attacks by α-blocker infusion failed to demonstrate that α-adrenergic constriction is a frequent or important trigger of coronary spasms [20].

References

1. Bassenge E, Holtz J, Müller C, Kinadeter H, Kolin A (1980) Experimental evaluation of coronary artery vasomotion: Possible significance for myocardial ischemia in coronary heart disease. Adv Clin Cardiol 1:300
2. Berne RM (1958) Effect of epinephrine and norepinephrine on coronary circulation. Circ Res 6:644
3. Berne RM, Rubio R (1979) Coronary circulation. In: Berne RM, Sperelakis N (eds) The cardiovascular system, vol I: the heart. Baltimore (Handbook of Physiology, Section 2, p873)
4. Berne RM, DeGeest H, Levy MN (1965) Influence of the cardiac nerves on coronary resistance. Am J Physiol 208:763
5. Constantine JW, Gunnel D, Weeks RA (1980) α_1- and α_2-vascular adrenoceptors in the dog. Eur J Pharmacol 66:281
6. Davey MJ (1980) Relevant features of the pharmacology of prazosin. J Cardiovasc Pharmacol 2:287
7. DaPrada M, Zürcher G (1976) Simultaneous radioenzymatic determination of plasma and tissue adrenaline, noradrenaline and dopamine within the femotomole range. Life Sci 19:1161
8. Doutheil U, tenBruggencate HG, Kramer K (1964) Coronarvasomotorik unter l-Noradrenalin und Isopropylnoradrenalin nach Blockierung der adrenergischen β-Receptoren durch Nethalide. Pfluegers Arch 281:181
9. Ertl G, Fuchs M (1980) Alpha-adrenergic vasoconstriction in arterial and arteriolar sections of the canine coronary circulation. Basic Res Cardiol 75:600
10. Feigl EO (1967) Sympathetic control of coronary circulation. Circ Res 20:262
11. Feigl EO, Buffington CW (1981) Adrenergic coronary vasoconstriction with coronary stenosis. Abstracts of 4th Sympos. on Coronary Heart Disease, Frankfurt 1981.
12. Gerova M, Barta E, Gero J (1979) Sympathetic control of major coronary artery diameter in the dog. Circ Res 44:459
13. Gregg DE, Fisher LC (1963) Blood supply to the heart. In: Hamilton WF, Dow P (eds) Circulation, vol II. Baltimore (Handbook of Physiology, sect. 2, p 1517)
14. Hamilton FN, Feigl EO (1976) Caronary vascular sympathetic beta-receptor innervation. Am J Physiol 230:6 1569
15. Holtz J, Mayer E, Bassenge E (1977) Demonstration of alpha-adrenergic coronary control in different layers of canine myocardium by regional myocardial sympathectomy. Pfluegers Arch 372:187
16. Holtz J, Saeed M, Sommer O, Bassenge E (in press) Alpha-adrenergic blockade (phentolamine) in conscious dogs: The decline in peripheral vascular resistance is mainly due to stimulation of beta-adrenergic receptors. In: Kenner T, Busse R, Hinghofer-Szalkay H (eds) Cardiovascular system dynamics: models and measurements. Plenum Publish Comp, New York London
17. Katz LN, Jochim K (1939) Observations on the innervation of the coronary vessels of the dog. Am J Physiol 126:395
18. Kelley KO, Feigl EO (1978) Segmental α-receptor-mediated vasoconstriction in the canine coronary circulation. Circ Res 43:908
19. Langer SZ (1977) Presynaptic receptors and their role in the regulation of transmitter release. Br J Pharmacol 60:481

20. Maseri A, Chierchia S, L'Abbate MD (1980) Pathogenetic mechanisms underlying the clinical events associated with atherosclerotic heart disease. Circulation 62(V):3
21. Mohrman DE, Feigl EO (1978) Competition between sympathetic vasoconstriction and metabolic vasodilation in the canine coronary circulation. Circ Res 42:79
22. Murray PA, Vatner SF (1979) α-adrenoceptor attenuation of the coronary vascular response to severe exercise in the conscious dog. Circ Res 45:654
23. Parratt JR (1980) Effects of adrenergic activators and inhibitors on the coronary circulation. In: Szekeres L (ed) Adrenergic activators and inhibitors. Springer, Berlin Heidelberg New York (Handbook of Experimental Pharmacology, vol 54/I, p 735)
24. Powell JR, Feigl EO (1979) Carotid sinus reflex coronary vasoconstriction during controlled myocardial oxygen metabolism in the dog. Circ Res 44:44
25. v Restorff W, Bassenge E (1976) Evaluation of a neurogenic rapid coronary dilatation during an excitatory response in conscious dogs. Pfluegers Arch 367:157
26. Starke K (1977) Regulation of noradrenaline release by presynaptic receptor systems. Rev Physiol Biochem Pharmacol 77:1
27. Starke K, Docherty JR (1980) Recent developments in α-adrenoceptor research. J Cardiovasc Pharmacol 2:269
28. Vatner SF (1980) Neural factors influencing perfusion. Am Heart Assoc, 53rd Sci Sessions, Postgraduate Seminar, 1980
29. Vatner SF, McRitchie RJ (1975) Interaction of the chemoreflex and the pulmonary inflation reflex in the regulation of coronary circulation in conscious dogs. Circ Res 37:664
30. Vatner SF, Franklin D, Van Citters RL, Braunwald E (1970) Effects of carotid sinus nerve stimulation on the coronary circulation of the conscious dog. Circ Res 27:11

Discussion

Parratt, Glasgow: Could you please summarize for us what is known regarding physiological (or pathophysiological) reflexes and α-adrenoceptor-mediated coronary vasoconstriction, apart from the effect of hyperventilation? Have you, for example, looked at hollow organ distension? Several years ago Glynne Moore and I (Basic Res. Cardiol.) found in anaesthetised cats that distension of the bladder or small intestine resulted in an increase in blood pressure and heart rate but a decrease in local myocardial blood flow. We felt the cause was an α-adrenoceptor-mediated effect, but without any experimental evidence for this hypothesis.

Bassenge: We have looked into a number of reflexes, among them also intestinal distension by a pneumatic device. This resulted in a coronary constriction, lowering coronary venous oxygen saturation to less than 10%. α-adrenergic blockade by phentolamine attenuated this constriction. However, this effect was rather small and we did not check it very carefully.

Feigl, Seattle: The most provocative result you have shown is the large transient increase in coronary blood flow following prazosin infusion. Could the result be due to a transient beta receptor stimulation by prazosin? Adrenergic antagonists often have partial agonist effects.

Bassenge: I perhaps did not mention that all those experiments were done under β-blockade. Thus there cannot be a substantial β-adrenergic contribution to the excess perfusion. As I already mentioned this excess perfusion varied greatly with the animal, what we cannot readily explain. To my knowledge additional β-stimulating effects of prazosin have not been reported as yet.

Starke, Freiburg: It now seems likely that prazosin, in contrast to what has been supposed for many years, lowers sympathetic outflow from the central nervous system, perhaps by blockade of central α_1-adrenoceptors. Is it therefore possible that part of the effect of prazosin you observed is not blockade of coronary smooth muscle α_1-adrenoceptors but a central action?

Bassenge: I agree that this possibility has to be considered and that our observations can – at least in part – be explained by this additional central action.

Kopin, Bethesda: The lack of increase of plasma norepinephrine after the marked hypotension induced by prazosin certainly is consistent with Dr. Starke's suggestion that prazosin has a central effect which blocks sympathetic outflow and thereby prevents baroreflex induced increase in plasma catecholamines.

Bassenge: This may have contributed to the missing increase in plasma catecholamines after prazosin administration in conscious dogs.

Berne, Charlottesville: 1. Was the drug prazosin given as a continuous infusion or only as a single bolus? If so, was the effect still transient and, if transient, why did the effect wear off so rapidly?
2. What is the physiological significance of a vasoconstrictor tone? Teleologically, it does not make much sense but since it exists there must be a reason as well as a mechanism to the phenomenon. What is your explanation?
Bassenge: 1. Prazosin was administered both as bolus and as continuous infusion. In both cases α-mediated effects were demonstrated to be completely blocked, yet the induced excess perfusion was only transient. The magnitude of the induced excess perfusion greatly varied from animal to animal for unexplained reasons.
2. α-Adrenergic constriction is only one of several constrictory mechanisms. Very likely it can be replaced by other constrictory mechanisms, as is the case with the substitution of various dilatory mechanisms. The physiological reason for constrictive mechanisms seems to be the control of an excessive metabolic dilation. Excessive dilation may interfere with the sudden tension development at the onset of isometric contraction because of the following considerations: Excessive dilation → substantial increase in intramyocardial vascular blood volume (from 10% to 30%) → squeezing out this blood volume from the myocardium (like a sponge) interferes with the fast tension development at the onset of isometric contraction (contraction under quick release condition).

Kuhn, Düsseldorf: You have shown that prazosin acts as a postsynaptic α_1-blocker. However, the dosage used in your experiments is about ten-fold that used in man. Did you also study prazosin in lower doses?

Bassenge: The effects of low dosages of prazosin are much less pronounced. In order to block α_1-mediated effects completely, we had to administer these high dosages. To my knowledge the rather low dosages used in patients are not apt to clarify quantitatively the role of α_1-mediated effects. Of course, you are trying to obtain a satisfactory drop in blood pressure, however, you have to keep in mind the patient's security.

Kuhn: Nevertheless, in man we find a pronounced vasodilating effect using, for example, 6 or 10 mg prazosin in a person weighing 80 kg.

Bassenge: It is very doubtful that this is the maximum vasodilation which can be obtained with an α_1-blocker.

Identification of Cardiac Alpha-Adrenergic Receptors by Direct Binding Studies: Clinical Implications

L. T. Williams

Summary

For over 15 years physiologists have debated the role of alpha adrenoceptors in myocardial function. The recent use of radioligand binding studies has demonstrated that $alpha_1$ adrenoceptors are present in homogenates of cardiac tissue from a variety of species and that the density of these receptor sites in membrane preparations is comparable to the density of beta-adrenergic binding sites. While it is possible that the alpha-adrenergic binding sites are from adipose or vascular tissue contaminating the membrane preparations, it seems likely that the binding sites are from myocardial cells. Whether these binding sites play a role in cardiac function under physiological conditions is still debatable. Under certain situations there may be a small positive inotropic influence of alpha-adrenergic stimulation. However, it is doubtful that this is of great importance in most clinical situations. Recently one group of investigators have reported that in a cat model of coronary reperfusion following occlusion, there is an increase in the number of alpha adrenergic binding sites associated with ventricular dysrhythmias which could be blocked by alpha antagonists. This observation and previous reports in the clinical literature suggest that some cardiac dysrhythmias may be exacerbated by cardiac $alpha_1$ stimulation and that alpha-antagonists may be effective antiarrhythmic agents in the setting of myocardial ischemia or infarction.

Introduction

The question of whether there are functional cardiac alpha-adrenergic receptors has been debated by physiologists for over a decade. Although this issue is still somewhat controversial, the majority of data suggest that cardiac alpha-adrenergic receptors exist. The more difficult question is whether these receptors are merely pharmacological curiosities which have survived evolution as vestiges of previously essential structures, or whether they play important roles in physiological or pathological situations. The purpose of this manuscript is to review the data from radioligand binding studies which have added further credence to the existence of cardiac alpha receptors, and to speculate on the clinical implications of our knowledge of the function of these receptors. Before discussing the more speculative areas I would like to first briefly summarize the pharmacology of adrenergic receptors; second, mention the direct binding techniques we and others have used to identify alpha receptors; and third, summarize some physiological studies which have suggested a functional role for these receptors.

Basic and Clinical Pharmacology of Alpha-Adrenergic Receptors

The catecholamines stimulate a diverse group of physiological responses. Ahlquist recognized a pattern in these responses when he noticed that they could be divided into two groups which he termed alpha and beta [1]. He assigned each response to the alpha or beta group according to the order of potency of a series of catecholamine analogues in stimulating the response. The alpha responses included contraction of smooth muscle in a variety of tissues and the beta responses included relaxation of smooth muscle as well as the cardiac inotropic and chronotropic responses to catecholamines. Based on his findings that these responses can be classified into two distinct groups, Ahlquist concluded that each group is mediated by a specific type of receptor.

Subsequently many pharmacologists have validated that alpha adrenergic responses are stimulated by catecholamines in the order of potency epinephrine $\geqq$ norepinephrine > phenylphrine $\gg$ isoproterenol, whereas beta-adrenergic responses are stimulated in the order of potency isoproterenol > epinephrine $\geqq$ norepinephrine $\gg$ phenylphrine. An important point is that none of these agonists is absolutely specific for alpha or beta receptors. Isoproterenol however is *relatively* specific for beta receptors especially at low concentrations, while phenylephrine is somewhat selective for alpha receptors. The antagonists are much more specific then the agonists and even permit subdivision of the alpha responses into $alpha_1$ and $alpha_2$ responses, prazosin being an $alpha_1$ antagonist and yohimbine being relatively specific as an $alpha_2$ antagonist.

In many tissues alpha and beta receptors mediate opposite responses; for example, in most smooth muscle the alpha response is contraction and the beta response is relaxation. Early pharmacological studies of cardiac tissue suggested that the heart has beta but not alpha adrenergic responses. A wealth of clinical and basic pharmacological data have confirmed that the inotropic, chronotropic, and metabolic effects of catecholamines on the heart are mediated predominantly by beta-adrenergic receptors. These concepts have dictated the clinical use of adrenergic agonists and antagonists. Thus, alpha-adrenergic agonists are used clinically as vasoconstrictors and most clinicians disregard any potential direct cardiac effects of alpha stimulation. The more "alpha specific" agents (phenylephrine) are often used as pressors in situations in which direct cardiac inotropic and chronotropic responses are undesirable (e.g. myocardial ischemia, hypertrophic cardiomyopathy with obstruction) and agents which are less specific (e.g. norepinephrine) are used when direct beta-adrenergic cardiac effects as well as vascular alpha stimulation are expected. $Alpha_1$ antagonists are used clinically to reduce systemic vascular resistance by $alpha_1$ receptor blockade. Thus prazosin is a useful antihypertensive agent because of its peripheral vascular effects. Prazosin has also been used to reduce peripheral vascular resistance in chronic congestive heart failure [2]. Again, in these cases the assumption has been made that there are few, if any, direct cardiac effects of alpha-adrenergic agents. Hence, classical pharmacological teaching as well as current clinical practice ascribe little importance to cardiac alpha-adrenergic receptors. This view may be superficial and may be revised on the basis of some of the data discussed below and elsewhere in this volume.

Radioligand Binding Studies

One approach to the question of whether there are alpha-adrenergic receptors in the heart is to label the receptors by direct binding studies. During the past 5 years we and many others have repeatedly demonstrated that radioactively labeled alpha-adrenergic agonists and antagonists can be used in a variety of tissues to label membrane receptor sites which have the characteristics expected of alpha-adrenergic receptors [15, 8]. For these studies the appropriate radioactively labeled agonist or antagonist is incubated with membranes prepared from the tissue containing the receptor sites. The bound radioactive compound is separated from the free compound by rapid filtration and the bound radioactivity is counted. In order to demonstrate that the binding of a radioactive compound is indeed to an alpha-adrenergic receptor, one must demonstrate that the specificity of the binding sites reflects the pharmacology of alpha-adrenergic receptors. For example, if the binding sites are alpha receptors, agonists should compete with the radioactive compound with an order of potency epinephrine $\geqq$ norepinephrine $>$ phenylephrine $\gg$ isoproterenol. The (–) or levo stereoisomers of the catecholamines should be more potent in competing for the sites than their corresponding (+) or dextro stereoisomers. The alpha-adrenergic antagonists such as prazosin or phentolamine should be potent competitors of binding whereas beta-adrenergic antagonists should be much weaker.

Assuming that the specificity of the binding sites is appropriate, these techniques can be used to assess the number and affinity of receptors in a given tissue. Straightforward scatchard analysis in which the ratio of bound to free ligand is plotted as a function of bound ligand permits the determination of the affinity of the ligand for receptor site and the number or density of the binding sites. Thus experiments can be performed in which changes in these parameters can be assessed following experimental manipulations.

Several investigators have been able to use these binding techniques to identify cardiac alpha-adrenergic receptors. *R. S. Williams* and *R. J. Lefkowitz* [16] used tritiated dihydroergocryptine to identify alpha-adrenergic receptors in membranes prepared from homogenates of rat heart. These sites appeared to have the specificity of $alpha_1$ adrenergic receptors. Few if any $alpha_2$ sites were present. Perhaps an even better approach to the identification of cardiac alpha receptors was taken by *Karliner* et al. [6] who used the specific, high affinity compound, tritated prazosin. This compound has been used in rats, guinea pigs, dogs and cats to identify specific $alpha_1$ receptor sites in membranes from homogenates of cardiac tissue.

These data have demonstrated that membrane preparations of cardiac tissue contain alpha-adrenergic receptor sites. It is possible that the receptors labeled by these techniques are not from myocardial cells but are from other cell types such as vascular smooth muscle cells or nerve cells which are present in the cardiac tissue when it is homogenized to make membrane fragments. This, however, seems unlikely since the number of alpha-adrenergic receptors is similar to the number of beta-adrenergic receptors in the same preparations [16], and it is generally accepted that the beta-adrenergic receptors which can be detected by binding studies are mostly from myocardial tissue. Studies currently underway in several laboratories on cultured cardiac myocytes may help to elucidate the issue of the cell type responsible for the binding of alpha ligands. Hence binding sites strongly suggest but do not conclusively prove that alpha-adrenergic receptors are present on myocardial cells.

Physiological Evidence of Cardiac Alpha-Adrenergic Receptors and Possible Clinical Implications

The presence of alpha receptors detected by binding studies is of little practical importance, unless there is a physiological or pathological function of the binding sites. The first evidence for a functional role for cardiac alpha-adrenergic receptors appeared in 1966 when *Wenzel* and *Su* [14] demonstrated that phentolamine abolished the positive inotropic effect of the synthetic alpha-agonist, phenylephrine, in a ventricular strip. *Grovier* [5] and later *Schümann* [11] showed that a portion of the positive inotropic effect of epinephrine was blocked by phentolamine. Specifically these investigators showed that the combination of beta-blocker (pindolol) and alpha-blocker (phentolamine) resulted in a blockade of epinephrine responses greater than that produced by beta-blocker alone. The conclusion was that the response to beta-receptor stimulation by epinephrine normally is so great that the alpha component becomes insignificant in comparison. If, however, the beta-adrenergic response is eliminated, alpha-receptor blockade can be demonstrated. These findings have been reproduced by a number of investigators using a variety of species. An excellent review of this work has recently been published [12].

It is more difficult to demonstrate an alpha-adrenergic component of contractility when norepinephrine, rather than epinephrine is used as the agonist [11]. Since norepinephrine is the endogenous transmitter and is the predominant catecholamine acting on the heart except in times of severe stress when adrenal medullary release of epinephrine becomes important, the role of the alpha-mediated inotropic response in most physiological situations must be very minor. Even when the level of epinephrine is high the overwhelming inotropic response in vivo is beta-mediated unless a very high concentration of beta-antagonist is present. It is difficult to conceive a clinical setting in which the alpha-mediated cardiac inotropic effect would be very important except perhaps in a hypotensive patient taking beta blockers who would require pressor agents. Even in this case the cardiac alpha effects would be relatively minor in importance compared with the peripheral effects.

It is also unlikely that the alpha-mediated inotropic response is clinically important in patients taking alpha antagonists. Although theoretically alpha antagonists might have a slight negative inotropic effect due to removal of alpha-mediated contractility, this effect has never been demonstrated clinically. In fact, the alpha antagonist, prazosin, is a useful drug in treating chronic congestive heart failure because of its afterload-reducing effect [2]. Hence the cardiac alpha-adrenergic effect is apparent under certain pharmacological conditions but is probably not a clinically important phenomenon.

A second group of physiological effects of cardiac alpha stimulation are the metabolic myocardial effects. *Keely* et al. [7] have demonstrated in rat heart that alpha stimulation by epinephrine of phenylephrine causes a small but significant decrease in protein kinase and phosphorylase activity. They also observed a doubling of the level of cardiac cyclic GMP during alpha stimulation, in observation which has been difficult to reproduce in other labs. Whether any of these metabolic effects have implications in physiological or pathological function in vivo is an unanswered question.

Finally the last and most important physiological effect of cardiac alpha-receptor stimulation is the apparent role of alpha-receptors in mediating ventricular dysrhythmias. Recently *Sheridan* et al. [13] occluded cat coronary arteries for 35 min and studied the ventricular dysrhythmias which occurred on reperfusion. They demonstrated that if the animals were pretreated with phentolamine, prazosin, or 6-hydroxydopamine the incidence of PVCs and ventricular fibrillation was dramatically reduced. Pretreatment with propranolol had little or no effect on these ventricular dysrhythmias. These investigators, in a subsequent study [3], measured the number of cardiac alpha-adrenergic receptors by using tritiated prazosin.

The results indicated that after 30 min of coronary artery occlusion there is a doubling of the binding sites for tritiated prazosin in membranes prepared from the ischemic region. These results suggest that coronary artery occlusion leads to an increase in the number alpha-adrenergic receptor sites. This increase in receptors may be involved in the mechanism of the dysrhythmias associated with reperfusion of occluded coronary arteries. The exact electrophysiologic mechanism of the antiarrhythmic effect of alpha antagonists in reperfusion arrhythmias is not known. However, it is tempting to speculate that prazosin or other alpha antagonists might be useful antiarrhythmic agents in the setting of acute myocardial infarction in man. In fact, one small clinical trial involving thirty-nine patients with acute myocardial infarction has been performed by *Gould* et al. [4]. In this study phentolamine conferred significant protection against PVCs in the setting of acute infarction. Clearly, larger scale studies should be performed in order to assess the efficacy of alpha blockers in preventing the dysrhythmic complications of acute infarction.

One situation in which these data may be particularly applicable is the new technique of thrombolytic therapy in the setting of acute myocardial infarction associated with thrombotic occlusion of a major coronary artery. For this technique streptokinase is selectively injected into a coronary artery which has acutely occluded [4, 9]. Several investigators have noticed that when lysis of the clot is successful and perfusion is suddenly restored to viable ischemic myocardium, ventricular dysrhythmias are likely to occur [4, 9]. Since this situation is analogous to the animal reperfusion experiments done by *Sheridan* et al. [13] one might speculate that the dysrhythmias are alpha mediated and might be prevented by alpha blockade. We and others are beginning experimental studies to address this issue. In addition, it may be appropriate to examine the efficacy of alpha blockers as antiarrhythmic agents in Prinzmetal's angina due to spasm of the coronary arteries, and in postinfarction patients (who might be at high risk for subsequent cardiac dysrhythmias).

In summary, a number of investigators have demonstrated using direct binding studies that cardiac tissue contains alpha-adrenergic receptors. Although we have not eliminated the possibility that the binding is due to vascular or nerve tissue contaminating the tissue preparations, it seems likely that the receptors are on myocardial cells. There is a positive inotropic effect of alpha stimulation which can be unmasked under certain pharmacological conditions. However, it is unlikely that this inotropic effect is ever clinically very important. Finally, animal experiments have suggested that reperfusion dysrhythmias are mediated by alpha-adrenergic receptors and can be blocked by alpha antagonists. These observations have suggested that in certain clinical situations alpha antagonists may be useful as anti-arrhythmic agents. Additional clinical and basic studies on the electrophysiological effects of $alpha_1$ antagonists should be undertaken in order to elucidate the role of alpha receptors in rhythm disorders and to investigate whether changes in receptor number or function can be implicated in the pathogenesis of ischemic-related dysrhythmias.

References

1. Ahlquist RP (1948) A study of the adrenotropic receptors. Am J. Physiol 135:586–600
2. Awan NA, Miller RR, Maxwell KE, Demaria AN, Neumann A, Mason DT (1977) Efficacy of ambulatory systemic vasodilator therapy with oral prazosin in chronic refractory heart failure. Circ 56:346
3. Corr PB, Shayman JA, Kramer JB, Kipnis RJ (1981) Increased alpha-adrenergic receptors in ischemic cat myocardium. J Clin Invest 67:1232–1236
4. Gould L, Raddy CVR, Weinstein T, Gomprecht RF (1975) Antiarrhythmia prophylaxis with phentolamine in acute myocardial infarction. J Clin Pharmacol 15:191
5. Govier WC (1968) Myocardial alpha adrenergic receptors and their role in the production of a positive inotropic effect by sympathomimatic agents. J Pharmacol Exp Ther 158:82–90
6. Karliner JS, Barnes P, Hamilton CA, Dollery CJ (1979) Alpha$_1$-adrenergic receptors in guinea pig myocardium: identification by binding of a new radioligand, [^{3}H]-prazosin. Biochem Biophys Res Commun 90:142–149
7. Keely SL, Corbin JD, Lincoln T (1977) Alpha adrenergic involvement in heart metabolism: effects on adenosive cyclic 3^1,5^1-monophosphate, adenosine cyclic 3^1,5^1-monophosphate-dependent protein kinase, guanosine cyclic 3^1,5^1-monophosphate, and glucose transport. Mol Pharmacol 13:965–975
8. Lefkowitz RJ, Hoffman BB (1980) Adrenergic receptors. Adv Lyc Nucl Res 12:37–47
9. Mathey DG, Kuck K, Tilsnor V, Krebber H, Bleifeld W (1981) Nonsurgical coronary artery recanalization after acute transmural myocardial infarction. Circulation 63: 489–497
10. Rentrop P, Blanke H, Farsch KK, Kaiser H, Kostering H, Leitz K (1981) Selective intracoronary thrombolysis in acute myocardial infarction and unstable angina pectoris. Circ 63:307–316
11. Schümann HJ, Endoh M, Wagner J (1974) Positive inotropic effect of phenylephrine in the isolated rabbit papillary muscle is mediated by both alpha- and beta-adrenoceptors. Naunyn Schmiedebergs Arch Pharmacol 284:133–148
12. Scholz H (1980) Effects of beta- and alpha-adrenoceptor activators. In: Szekeres L (ed) Adrenergic activators. Springer, Berlin Heidelberg New York (Handbook of experimental pharmacology, vol54, pp651–733)
13. Sheridan DJ, Penkoske PR, Sobel BE, Corr PB (1980) Alpha adrenergic contributions to dysrhythmia during myocardial ischemia and reperfusion in cats. J Clin Invest 65:161–171
14. Wenzel DG, Su JL (1966). Arch Int Pharmacodyn Ther 160:379
15. Williams LT, Lefkowitz RJ (1978) Receptor binding studies in adrenergic pharmacology. Raven, New York
16. Williams RS, Lefkowitz RJ (1978) Alpha-adrenergic receptors in rat myocardium. Circ Res 43:721–727

Discussion

Szekeres, Szeged: Owing to the short time elapsed after occlusion (15 min) in your experiments, I doubt whether the increase of total α-receptor concentration reflects a real production of these receptors. Could there be a methodological artifact?

Furthermore, are the α-receptor mediated dysrhythmias due to direct cardiac membrane changes produced by the α-agonists, or are they due to their hemodynamic actions?

Williams: It is difficult to rule out artifacts in any experiment involving receptor changes in membrane preparations. Since Sheridan et al. (1980) found no change in Na/K ATPase activity in the preparations showing an increase in α-receptor number, I think that an artifactual explanation of the data is unlikely.

As you will know, α-adrenergic agents cause direct membrane effects in cardiac tissue. These include an increase in action potential duration and refractory period. The relevance of these changes to the anti-arrhythmic effects of α-antagonists is unknown. There may be hemodynamic effects as well.

Palm, Frankfurt: If α-adrenoceptor density changes within 15–30 min under the experimental conditions described by Dr. Williams these results would show for the first time a rapid turnover or new synthesis of these receptors. I suspect that when adrenoceptor density is expressed as fmoles/mg protein most likely a change of protein concentration would lead to an "apparent change" of receptor concentration.

This would be true especially when homogenates rather than purified membranes are used in such experiments.

Williams: I agree that it is unlikely that the findings of Sheridan et al. showing an increased receptor density are due to increased synthesis of receptors. Other mechanisms such as release of membrane active enzymes uncovering "latent" receptor sites must be involved, but this is speculation. These investigators have used Na/K ATPase activity as their control to show that the amount of functional sarcolemma in their preparations from control and ischemic regions are the same.

Parratt, Glasgow: In the paper by Sheridan and Sobel, to which you referred, prazosin not only reduced reperfusion arrhythmias, but also prevented those ventricular arrhythmias that occurred when the occlusion was present. This protective effect, to which we will return tomorrow afternoon, thus occurred *before* there was evidence for an increase in the number of α-adrenoceptor binding sites.

I also have a more general point. This is a major symposium on catecholamines and receptors and I think we are getting sloppy in our terminology. Dr. Williams, paper, and the next two papers, talk of adrenergic receptors. The Detroit group, later this morning, talk (correctly in my view) of adrenoceptors. This is really a plea for correct terminology and for consistency. Also, one cannot antagonise any receptors, one can only block them. It is the effects of the agonist that are "antagonised".

Williams: I agree that Sheridan et al. reported ventricular dysrhythmia with occlusion. There was a protective effect of α-blockade (and β-blockade as well). These arrhythmias occurred within the first 32 min of occlusion. Whether the arrhythmias occurring during occlusion and reperfusion are caused by the same mechanism is unclear. All we know from the binding study is that between 15 and 32 min the density of binding sites has increased. This finding may be more relevant to reperfusion dysrhythmia than occlusion dysrhythmia.

I have no comment on the semantic discussion about the terms "adrenergic receptor", "adrenoceptor" and "antagonizing of receptors".

Kraupp, Vienna: The antiarrhythmic action of α-antagonists has long been known in the prevention of ventricular arrhythmias in the course of cyclopropane anesthesia. Do you think that these effects are mediated via myocardial α-receptors?

Williams: I am not aware of any data on this issue.

Scholz, Hannover: It is known that the positive inotropic response to α-adrenoceptor stimulating agents is relatively weak, if at all present, in guinea-pig papillary muscles as compared to ventricular tissues from cat, cow, rabbit or rat. Are there differences in the number of α-adrenoceptors in ventricular tissues from these species?

Williams: I am not aware of data demonstrating these marked species differences.

Identification and Characterization of Beta-Adrenergic Receptors in Myocardium Under Normal and Pathologic Conditions

G. L. Stiles[1], *R. J. Lefkowitz*[2]

Summary

Although beta-adrenergic receptors have been quantitated directly for only 6 years the development of our understanding of receptor function has grown exponentially in that time. We now know that receptors are dynamically regulated by a variety of pathophysiologic states, and that their distribution within a given tissue such as the heart is not homogeneous with respect to either subtype or density. Further, there are multiple components of the beta-adrenergic receptor-adenylate cyclase system that interact in a complex and sequential fashion. It appears that pathophysiologic conditions can alter one or more of these steps to produce changes in the transmembrane signal. Areas of future research are likely to include purification and characterization of each of the known components of the beta-adrenergic receptor system and how each of their interactions are altered in disease states. Whether other components presently unknown will be uncovered along the way remains to be seen.

Introduction

The stimulatory effects of catecholamines on the myocardium include an increase in heart rate, an increase in the force of contraction, an increase in excitability, enhanced oxygen consumption and induction of automaticity. The concept of a specific cellular receptor for catecholamines was first put forward by *Dale* in 1906 in what he termed the "receptive mechanism for adrenaline". In 1948, *Alhquist* [2] first suggested that there were two "distinct types of adrenotropic receptors". These were defined by their relative responsiveness to a series of sympathetic amines. The alpha receptor was characterized by a potency series of epinephrine > norepinephrine $\gg$ isoproterenol while for the beta receptors the potency series was isoproterenol > epinephrine > norepinephrine. In 1967 *Lands* et al. [16] provided evidence that two subtypes of beta receptors existed, which were termed $beta_1$ and $beta_2$. These receptors were defined by their relative affinities for epinephrine and norepinephrine; $beta_1$ receptors display approximately equal affinity for epinephrine and norepinephrine while $beta_2$ receptors have a considerably greater affinity for epinephrine than norepinephrine. Receptors were to remain mostly a physiologic concept until the mid-1970s.

---

1 *Gary L. Stiles* is supported by an NIH training grant HL 07101
2 *Robert J. Lefkowitz* is an investigator of the Howard Hughes Medical Institute

Characterization of Receptors by Ligand Binding

Early attempts to characterize beta-adrenergic receptors with a variety of tritiated catecholamine derivatives met with little success. It was not until 1974 that beta receptors were first directly labelled and in 1975 that canine myocardial membrane beta-adrenergic receptors were characterized using the antagonist [^{3}H]dihydroalprenolol ([^{3}H]DHA) [3]. These [^{3}H]DHA binding sites possessed the characteristics expected of interactions of the ligand with the physiological beta-adrenergic receptor.

[^{3}H]DHA had a high affinity for the receptors (nanomolar range) and the binding was saturable. The ligand demonstrated rapid kinetics both for association and dissociation with the receptor sites. The binding sites demonstrated a potency series of isoproternol > epinephrine $\geqq$ norepinephrine consistent with the beta-adrenergic receptor as defined previously by physiologic studies. Finally, the binding sites recognized catecholamines with the appropriate stereospecificity which had been a major problem in earlier studies. These same criteria have been fulfilled in myocardial tissue from a number of species.

More recent radioligand binding data have also suggested that different species have varying proportions of beta$_1$ and beta$_2$ receptors in their hearts [12]. Also, within the heart of several species there appears to be different regional distributions for both the subtype and density of beta-adrenergic receptors. The relative density of beta-adrenergic receptors in the various chambers of the canine heart [5] has been evaluated and the density shown to be greatest in the left ventricular free wall with decreasing quantities in the intraventricular septum, the right ventricle, the right atrium and the left atrium, respectively. This same distribution order is maintained if the relative blood flow to each chamber or the adenylate cyclase activity is assessed suggesting that the relative density differences may play an important physiologic role such as mediating contractility.

The most commonly used beta-adrenergic receptor ligands are (–)[^{3}H]DHA and (±)[^{3}H]hydroxybenzylpindolol. These do not distinguish between beta$_1$ and beta$_2$ receptors so that an indirect method must be applied to quantitative beta$_1$ and beta$_2$ receptor proportions. This has been accomplished by employing unlabelled drugs which have been shown to have selectively different affinities for beta$_1$ and beta$_2$ receptors and constructing competition curves with the radioligand [^{3}H]DHA. By applying various analytic techniques to these data one can determine the relative proportions of beta$_1$ and beta$_2$ receptors using a computer generated non-linear least squares curve fitting technique (see Fig. 1). *Hancock* et al. [12] found that rat ventricle contains exclusively beta$_1$ receptors whereas frog ventricle contained 80% beta$_2$ and 20% beta$_1$ receptors. In contrast, *Minneman* [20] using a different analytic technique found that the rat ventricle contained 83% beta$_1$ and 17% beta$_2$ receptors. There are several possible explanations for these differences. First, the analytic techniques were different. The *Minneman* group used a computer assisted technique to analyze data from a modified scatchard plot while *Hancock* et al. [12] employed computer modeling of untransformed binding data from competition curves with statistical analysis to define the most appropriate model. Second, it has been demonstrated [17] that in neonatal rat heart cultures fibroblast type cells are present which have beta$_2$ receptors while the muscle cells have beta$_1$ receptors. Thus, different membrane preparations may have different proportions of the two receptor subtypes on the basis of different recoveries of the two cell types.

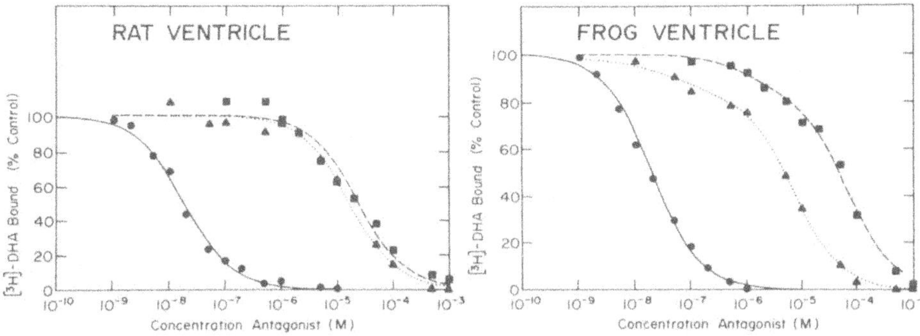

Fig. 1. Competition curves for specific DHA binding to frog and rat ventricle membranes by beta-adrenergic antagonists. The *ordinate* indicates the percent maximal specific DHA binding (0.0239 n*M*, frog; 0.0186 n*M*, rat). The *abcissa* is the molar concentration of various antagonists. The symbols are ▲, butoxamine; ●, propranolol; and ■, practolol. Propranolol is a nonsubtype selection antagonist while Butoxamine has slight $beta_2$ subtype and practolol $beta_1$ subtype selectivity. The ability of these three beta-blocking agents to compete for DHA binding was different in these two species. In the frog ventricle the potency order of propranolol:butoxamine:practolol was 2000:6.7:1. In rat myocardium butoxamine and practolol were approximately equipotent and ~ 1000fold weaker than propranolol. In the rat heart, the three antagonist competition curves differ only in their potency, i.e. all three curves are parallel, and steep. In the frog heart, however, the displacement of DHA by butoxamine and practolol is more complex (shallow) than that with propranolol. This indicates 2 components to the curves and by applying a nonlinear least squares curve fitting technique the relative proportion and affinity of these 2 affinity states can be assessed. (*Hancock* et al. [12])

The hearts of guinea pig and cat appear to contain both $beta_1$ and $beta_2$ receptors [13]. The distribution of these subtypes is not homogeneous. There appears to be a greater preponderance of $beta_1$ receptors in the ventricle (guinea pig 100%, cat 98%) while the atria of guinea pigs contain 23% $beta_2$ receptor and those of cats 22%.

The physiologic relevance of this differential distribution has been studied in these two species by using subtype selective agonists and antagonists. In the cat heart the chronotropic effects of catecholamines appear to be mediated by both $beta_1$ and $beta_2$ receptors, but the $beta_2$ receptors have a more significant role in the sinus node than in the ventricle [8]. This correlates well with the radioligand binding studies. The existence of different beta-adrenergic receptor subtypes preferentially mediating chronotropic and inotropic responses in the heart is of obvious clinical significance. A model such as this could be used to develop and test potentially useful drugs that could selectively alter heart rate or contractility. The human atria appears to function in a manner similar to the cat.

Receptor Regulation

With the advent of radioligand binding studies and the quantitation of beta-adrenergic receptors, it quickly became evident that receptors were not static entities but rather were under dynamic regulation. The number of receptors was found to be influenced by a wide variety of hormones and pathophysiologic states. Some of the conditions and/or treatments known to alter receptor number in the

Table 1. Alterations in beta-adrenergic receptor-adenylate cyclase system from myocardial tissue

Species	Condition	Receptor number	Adenylate cyclase activity	Ref.
Rat	Hyperthyroidism	Increase	Increase	23, 26, 28
Rat	Hypothyroidism	Decrease	Decrease	7, 23
Rat	Isoproterenol administration	Decrease	Decrease	25
Rat	Hypertension [a]	Decrease	Decrease	29
Rat	Hypertension [b]	Decrease	N. D.	30
Rat	Adrenalectomy	Increase, no change	Increase EC_{50}	1, 9
Rat	6-OH Dopamine	Increase	N. D.	27
Rat	Propranolol	Increase, no change	N. D.	10, 5
Rat	Guanethidine	Decrease	N. D.	11
Rat	Ethanol Treatment	Decrease	N. D.	6
Rat	Ethanol Withdrawal	Increase	N. D.	6
Rat	Hypertrophy [c]	Increase	N. D.	18
Rat	Diabetes [d]	Decrease	N. D.	22
Guinea Pig	CHF [e]	Increase	N. D.	14
Dog	Ischemia	Increase	↑ cAMP content	21
Chick Embryo	Isoproterenol	No change	Decreased isoproterenol [f] stimulation	19

The two entities under adrenalectomy and propranolol refer to different findings from separate groups. The EC_{50} refers to an increase in the concentration required to half maximally stimulate adenylate cyclase activity without effecting maximal activity.

[a] Hypertension induced by NaCl and deoxycorticosterone administration
[b] Spontaneous or Goldblatt kidney produced hypertension
[c] Hypertrophy induced by aortic banding
[d] Diabetes induced by Streptozoticin administration
[e] Congestive heart failure (CHF) produced by aortic banding
[f] Decreased stimulation in absence of added guanine nucleotide

N. D. not determined; ↑ cAMP content refers to total cellular concentration compared to controll cells heart are listed in Table 1. It is important to attempt to correlate the alterations in receptors to their possible physiologic relevance. It has long been known that exposure of many catecholamine responsive cells to catecholamines for a period of time leads to a state of decreased responsiveness. Conversely, depletion of catecholamines can lead to a hyper-responsive state. The decreased responsiveness has been termed desensitization or tachyphylaxis. In order to explore the mechanisms whereby catecholamines may produce desensitization it is important to briefly review the current concept of how catecholamines interact at the membrane level to activate adenylate cyclase.

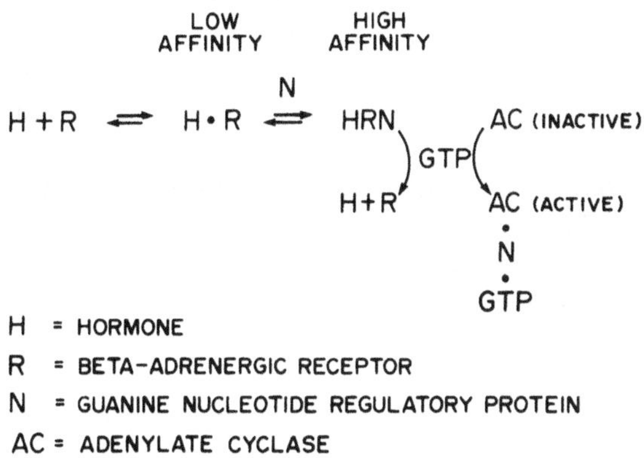

Fig. 2. Schematic representation for a proposed mechanism of adenylate cyclase activation by beta-adrenergic agonists and GTP

Catecholamines are thought to bind to the beta-adrenergic receptor to initially form a low affinity complex H · R, where H is a hormone and R is the beta-adrenergic receptor (see Fig. 2). When H is an agonist it induces a change in the receptor such that it can now interact with a membrane protein termed the guanine nucleotide regulatory protein (N) such that a high affinity *ternary complex* is formed, HRN. This nucleotide regulatory protein has been demonstrated in several cell types and has been shown to be required for the activation of adenylate cyclase. The agonist specific complex, HRN, promotes the association of guanine nucleotides such as GTP with the N protein. This association induces the dissociation of HRN and the N-GTP complex can then interact with the dormant catalytic unit (adenylate cyclase) thereby activating it and catalyzing the production of cAMP from ATP. cAMP can then interact with other intracellular components to promote a physiologic response. The intracellular cAMP concentration is regulated by the equilibrium between synthesis and degradation by phosphodiesterases. There are, therefore, many steps in the transmembrane signalling process where interactions could produce the decreased responsiveness observed in desensitization [15].

Tse et al. [25] investigated possible mechanisms for cardiac desensitization to catecholamines by administering isoproterenol to rats and studying various components of the myocardial beta-adrenergic receptor-adenylate cyclase system. They found that chronic administration of isoproterenol was associated with a reduction in both the sensitivity and maximal response of adenylate cyclase to activation by isoproterenol compared to control animals. They further noted that the number of beta-adrenergic receptors as assessed by [^{3}H]DHA binding was also decreased while the affinity of the receptors for the antagonist was unchanged. This suggests that in this model the *down regulation* of beta-adrenergic receptors may play a role in the desensitization process.

Fluoride ions are known to be an activator of adenylate cyclase in most eukaryotic systems. Although its direct mechanism of action is unknown, it is clear that the N protein must be present for F^- to be effective while a functional beta-adrenergic receptor is not required. In the same rat heart model noted above it was found that the F-stimulated adenylate cyclase activity was also decreased after chronic isoproterenol administration suggesting an additional lesion(s) distal to the beta-adrenergic receptor. The possibility of alterations in N protein interactions was suggested in this study. Thus, in competition binding studies with [^{3}H]DHA, the isoproterenol displacement curve in membranes derived from treated animals was shifted to the right and steepened. This alteration in agonist binding implies changes in receptor-cyclase coupling and will be discussed in detail below. These alterations in RN coupling are in addition to those in receptor number discussed above.

Regulation of a receptor by a hormone or drug that normally interacts with it has been termed homologous regulation. Another type of regulation is heterologous regulation where a hormone that does not directly interact with a receptor can regulate receptor number or properties. A well known model for this in the heart is the effect of hyper- and hypothyroidism on the number of beta-adrenergic receptors.

Altered sensitivity to catecholamine action in hyperthyroidism and hypothyroidism is well established [7, 26]. In the rat heart, hyperthyreoidism has been demonstrated to enhance phosphorylase activity and to increase the sensitivity of adenylate cyclase to catecholamines [26]. Conversely, myocardial membranes from hypothyroid animals have been shown to have decreased catecholamine stimulated adenylate cyclase [7]. We [23] have used this model to investigate the mechanisms for the altered sensitivity to catecholamines in the rat heart.

The demonstration in rat heart tissue of increased numbers of beta-adrenergic receptor in hyperthyroidism and decreased numbers of beta adrenergic receptors in hypothyroidism [23, 26, 28] has suggested that changes in receptor number may play a significant role in the observed altered sensitivity to catecholamines. This would certainly be similar to the findings noted above in the isoproterenol desensitization model. In addition, it is possible that changes in agonist binding to the receptor or the coupling of the receptor with adenylate cyclase might also contribute.

Agonist-Receptor Interactions

Recent advances in quantitative analysis of radioligand binding data now permit the direct assessment of agonist-receptor interactions and the indirect assessment of receptor cyclase coupling [15]. It has been demonstrated in several beta-adrenergic receptor systems that agonist competition curves are shallow and can be resolved into two receptor affinity states having respectively high (K_H) and low (K_L) affinity for agonists. This heterogeneity of binding appears to be related to the dynamic equilibrium between two hormone receptor complexes HR and HRN, where the HR complex is the low affinity component and HRN is the high affinity component. The HRN component is thought to be an agonist specific entity and antagonists are not capable of inducing its formation. This is in concert with the finding that competition curves of antagonists with radiolabelled antagonists are steep and reveal only a single dissociation constant suggesting antagonists bind with equal affinity to R and RN. Agonist competition curves in the presence of guanine nucleotides are also steep and are adequately described by a single dissociation constant indistinguishable from the K_L of the agonist curve in the absence of guanine nucleotides. This suggests that guanine nucleotides can convert high affinity state receptors into those at low affinity and that the affinity states of the receptor exist in a dynamic equilibrium.

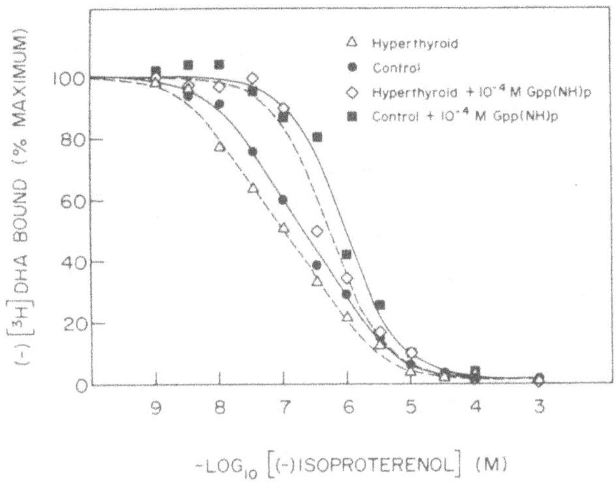

Fig. 3. Computerized curve fitting of binding data from competition of [^{3}H]DHA by (–) isoproterenol in myocardial membranes from control and hyperthyroid rats in the presence and absence of Gpp(NH)p, (0.1 m*M*). The membranes were prepared and radioligand binding was performed as described in reference 23. The data points represent the mean of 5 experiments (control) and 4 experiments (hyperthyroid) each performed in duplicate. Data were normalized for the purpose of graphic representation

Previous work [15] in the frog erythrocyte system has shown that the ratio of K_L/K_H for an agonist, reflecting the relative ability of that agonist to stabilize the high affinity form of the receptor correlates with the efficacy (intrinsic activity) for activation of adenylate cyclase. The ratio K_L/K_H may therefore be viewed as an indirect measure of the coupling of the receptor with other components of the adenylate cyclase system. Using these new techniques we investigated whether the unique properties of agonist binding to beta-adrenergic receptor in other systems could be demonstrated in mammalian heart and whether altered thyroid states modified these agonist specific binding properties [23].

We have found that agonist (isoproterenol) competition curves in normal myocardial membranes were shallow with a slope factor of 0.64 ± 0.06 (mean $\pm$ SEM) and had an EC_{50} (50% displacement) of 180 ± 40 n*M*. In the presence of guanine nucleotides [Gpp(NH)p (10^{-4} M)] the EC_{50} 910 ± 150 n*M* was significantly increased ($P < 0.001$) as was the slope factor (0.93 ± 0.12) which was now not significantly different from 1 (see Fig. 3). The isoproterenol curve in the absence of Gpp(NH)p was most appropriately described by a two state model with a K_H of 30 ± 9 n*M* and a K_L of 595 ± 50 n*M*. In the presence of Gpp(NH)p the isoproterenol curve was adequately described by a one state model with a single dissociation constant that was indistinguishable from the K_L of the isoproterenol curve in the absence of guanine nucleotide. The K_L/K_H ratio for isoproterenol in the control membranes was 20 ± 3 (Table 2).

Table 2. Computer-derived parameters from (–)isoproterenol displacement curves with (–)-[^{3}H]DHA

	Control ($n=5$)	Hyperthyroid ($n=4$)	P value *	Hypothyroid ($n=4$)	P value *
Slope factor without Gpp(NH)p	0.64± 0.06	0.54± 0.04	N.S.	0.73± 0.04	N.S.
Slope factor with Gpp(NH)p	0.93± 0.12	0.91± 0.07	N.S.	0.78± 0.06	N.S
EC_{50} (n*M*) without Gpp(NH)p	180 ± 40	80 ±20	0.016	150 ±20	N.S
EC_{50} (n*M*) with Gpp(NH)p	910 ±150	600 ±70	N.S.	640 ±70	0.04
K_H (n*M*)	30 ± 9	7 ± 2	<0.001	29 ± 9	N.S
K_L (n*M*)	595 ± 52	376 ±34	0.002	446 ±92	N.S
K_L/K_H	20 ± 3	54 ± 9	0.002	16 ± 8	N.S.

For each set of displacement curves (see Figure 1), the computer programs provided estimates of the slope factor, the EC_{50} (50% displacement) and the high (K_H) and low (K_L) affinity state dissociation constants.
n = number of experiments. * = compared to control. The results are expressed as mean ± SEE. The statistical significance was determined as detailed in reference 23.

In concert with previously reported observations we have found significant increases in the number of cardiac beta-adrenergic receptors in hyperthyroidism 43.5±4.2 fmol/mg, and a significant decrease in hypothyroidism 16.1±3.1 fmol/mg compared to control 25.5±2.1 fmol/mg. There was no change in antagonist ([^{3}H]DHA) affinity 1.2±0.2 n*M*.

In the hyperthyroid state, there were agonist-specific alterations in binding. The isoproterenol competition curve was shifted to the left with the EC_{50} changing from 180±40 n*M* to 80±20 n*M* ($P=0.016$) while there was no statistically significant change in the slope factor (Fig. 3). The K_H underwent a 4-fold decrease in the hyperthyroid state from 30±9 to 7±2 ($P<0.001$). The K_L (n*M*) was also slightly but significantly shifted from 545±52 to 376±34 ($P=0.002$). Of particular interest was the change in K_L/K_H ratio from 20±3 to 54±9 ($P=0.002$) suggesting an enhanced ability of the agonist to stabilize the high affinity state of the receptors.

In contrast to the findings in hyperthyroidism there were no significant changes in K_H or K_L/K_H in cardiac membranes from hypothyroid animals.

As outlined earlier there are multiple steps in the catecholamine-induced transmembrane signalling process. Alterations in any of these steps could account for changes in tissue sensitivity to catecholamines. In this model system we have found changes in rat myocardial beta-adrenergic receptor number in both hyper- and hypothyroidism which may well represent one mechanism for altered sensitivity.

Another factor not previously assessed quantitatively is that agonist binding and the subsequent activation of adenylate cyclase are altered. Agonist competition curves in myocardial membranes were found to be shallow as in other systems.

This agonist-specific heterogeneity of binding appears to be due to the ability of agonists but not antagonists to distinguish separate affinity states of the receptor (K_H and K_L). In simple erythrocyte systems, the high affinity state appears to represent a "coupled" agonist stabilized complex of receptor and N which is an intermediate required for hormone promoted activation of adenylate cyclase [15]. The binding of agonists but not antagonists to the beta-adrenergic receptor of rat reticulocytes, for example, has been directly shown to induce the coupling of R with N. When guanine nucleotides are present they bind to the N-site thus destabilizing the HRN complex which then dissociates into free hormone-receptor and GTP charged N protein. This leads simultaneously to the activation of adenylate cyclase and conversion of the high affinity state of the receptor into the low affinity state [15]. The ratio of the dissociation constants of an agonist for the high and low affinity states of the receptor "K_L/K_H" is an indirect of the coupling of the R with the other component of the adenylate cyclase system. Thus, an increase in this ratio would be expected to be association with a more stabilized high affinity state and more efficient hormone stimulation of adenylate cyclase.

Our findings in hyperthyroidism of a shift to the left in the agonist competition curve, with an increase in the affinity of isoproterenol for the beta-adrenergic receptor as manifested by a change in both K_H and K_L/K_H suggest an increased efficiency of coupling of HR with the other components of the adenylate cyclase system.

This finding suggests another potential mechanism whereby catecholamine stimulated production of cAMP is increased in hyperthyroidism. Thus both receptor number and coupling are likely involved in the supersensitivity to catecholamines. In hypothyroidism in the heart only receptor number appears involved. Further characterization of this system will require purification and characterization of its individual components. In another model system for studying hypothyroidism, the rat reticulocyte, multiple alterations in the transmembrane signalling process have recently been elucidated including a decrease in receptor number, a decreased affinity of the high affinity state of the receptor, decreased catecholamine and NaF stimulated adenylate cyclase as well as a decreased quantity of N protein [24]. Thus in order to fully understand the molecular mechanisms involved in such pathophysiologic states all components of this system must be evaluated.

References

1. Abrass TB, Scarpace PJ (1981) Glucocorticoid regulation of myocardial beta-adrenergic receptors. Endocrinology 108:977–980
2. Ahlquist RP (1948) Study of adrenotropic receptors. Am J Physiol 153:586–600
3. Alexander RW, Williams LT, Lefkowitz RJ (1975) Identification of cardiac beta-adrenergic receptors by (–)[^{3}H]alprenolol. Proc Natl Acad Sci USA 72:1564–1568
4. Baker SP, Potter LT (1980) Effect of propranolol on beta-adrenoceptors in rat hearts. Br J Pharmacol 68:8–10
5. Baker SP, Boyd HM, Potter LT (1980) Distribution and function of beta-adrenoreceptors in different chambers of the canine heart. Br J Pharmacol 68:57–63
6. Banerjee SP, Sharma VK, Khanna JM (1978) Alterations in beta-adrenergic receptor binding during ethanol withdrawal. Nature 276:407–408
7. Brodde OE, Schumann HJ, Wagner J (1980) Decreased responsiveness of adenylate cyclase system on left atria from hypothyroid rats. Mol Pharmacol 17:180–186

8. Carlsson E, Dahlof CG, Hedberg A, Persson H, Tangstrand B (1977) Differentiation of cardiac chronotropic and inotropic effects of beta-adrenergic agonists. Naunyn-Schmiedebergs Arch Pharmacol 300: 101–105
9. Davies AO, De Lean A, Lefkowitz RJ (1981) Myocardial beta-adrenergic receptors from adrenalectomized rats: impaired formation of high-affinity agonist-receptor complexes. Endocrinology 108:720–722
10. Glaubiger G, Lefkowitz RJ (1977) Elevated beta-adrenergic receptor number after chronic propranolol treatment. Biochim Biophys Res Commun 78:720–725
11. Glaubiger G, Tsai BS, Lefkowitz RJ, Weiss B, Johnston EM (1978) Chronic guanethidine treatment increases cardiac beta-adrenergic receptors. Nature 273:240–242
12. Hancock AA, De Lean AL, Lefkowitz RJ (1979) Quantitative resolution of beta-adrenergic receptor subtypes by selective ligand binding: application of a computerized model fitting technique. Mol Pharmacol 16: 1–9
13. Hedberg A, Minneman KP, Molinoff PB (1980) Differential distribution of $beta_1$ and $beta_2$ adrenergic receptors in cat and guinea pig heart. J Pharmacol Exp Ther 213: 503–508
14. Karliner JS, Barnes P, Brown M, Dollery CT (1980) Effects of chronic heart failure on myocardial $alpha_1$ and beta adrenoceptors assessed by radioligand binding. Clin Res 28:470 A
15. Kent RS, De Lean A, Lefkowitz RJ (1980) A quantitative analysis of beta-adrenergic receptor interactions: resolution of high and low affinity states of the receptor by computer modeling of ligand binding data. Mol Pharmacol 17: 14–23
16. Lands AM, Arnold A, McAuliff JP, Luderena FP, Brown JG (1967) Differentiation of receptor systems activated by sympathomimetic amines. Nature 214:577–598
17. Lau YH, Robinson RB, Rosen MR, Bilezikian JP (1980) Subclassification of beta-adrenergic receptors in cultured rat cardiac myoblasts and fibroblasts. Circ Res 47:41–48
18. Limas LT (1979) Increased number of beta-adrenergic receptors in the hypertrophied myocardium. Biochim Biophys Acta 588:174–178
19. Marsh JD, Barry WH, Neer EJ, Alexander RW, Smith TW (1980) Desensitization of chick embryo ventricle to the physiological and biochemical effects of isoproterenol. Circ Res 47:493–501
20. Minneman KP, Heystrand LR, Molinoff PB (1974) Simultaneous determination of $beta_1$ and $beta_2$ adrenergic receptors in tissues containing both receptor subtypes. Mol Pharmacol 16:34–46
21. Mukherjee A, Wong TM, Buja LM, Lefkowitz RJ, Willerson JT (1979) Beta-adrenergic and muscarinic cholinergic receptors in canine myocardium effects of ischemia. J Clin Invest 64: 1423–1428
22. Savarese JJ, Berkowitz BA (1979) Beta-adrenergic receptor decrease in diabetic rat hearts. Life Sci 25:2075–2078
23. Stiles GL, Lefkowitz RJ (1981) Thyroid hormone modulation of agonist-beta-adrenergic receptor interactions in the rat heart. Life Sci 28:2529–2536
24. Stiles GL, Stadel JM, De Lean A, Lefkowitz RJ (1981) Molecular mechanisms of beta-adrenergic hyporesponsiveness in hypothyroidism. Clin Res 29:508 A
25. Tse J, Powell JR, Baste CA, Priest RE, Kuo JF (1979) Isoproterenol-induced cardiac hypertrophy: modifications in characteristic of beta-adrenergic receptors adenylate cyclase and ventricular contraction. Endocrinology 105:246–255
26. Tse J, Wrenn RW, Kuo JF (1980) Thyroxine-induced changes in characteristics and activities of beta-adrenergic receptors and adenosine 3′,5′monophosphate and guanosine 3′,5′monophosphate systems in the heart may be related to reputed catecholamines supersensitivity in hyperthyroidism. Endocrinology 107:6–16
27. Yamada S, Yamamura HI, Roeske WR (1980) Alterations in cardiac autonomic receptors following 6 hydroxydopamine treatment in rats. Mol Pharmacol 18: 185–192
28. Williams LT, Lefkowitz RJ, Watanabe AM, Hathaway DR, Besch HR (1977) Thyroid hormone regulation of beta-adrenergic receptor number. J Biol Chem 252:2767–2769
29. Woodcock EA, Johnston CI (1980) Changes in tissue alpha- and beta-adrenergic receptors in renal hypertension in the rat. Hypertension 2: 156–161
30. Woodcock EA, Funder JW, Johnston CI (1979) Decreased cardiac beta-adrenergic receptors in deoxycorticosterone-salt and renal hypertensive rats. Circ Res 45:560–565

Discussion

Reuter, Bern: Have you done functional tests in the tissues you used for radioligand binding and adenylate cyclase assays, e.g. did you measure dose-response curves of isoproterenol for heart rate under the various thyroidal states?

Stiles: We have not performed in vivo experiments in our laboratory. However, *Guarnieri* et al. have found enhanced contractility in the rat ventricle in hyperthyroidism using a low concentration of isoproterenol and further have demonstrated that these changes were due to events proximal to cAMP production.

Palm, Frankfurt: Did you try ^{3}H-hydroxybenzylisoprenaline (^{3}H-HBI) as an agonist ligand in such experiments? In my opinion ^{3}H-HBI would be a better ligand than ^{3}H-dihydroalprenolol (^{3}H-DHA), since with the agonist ligands the high affinity state HRN* can be characterized directly, whereas with the antagonist ligand ^{3}H-DHA R and RN can be shown only indirectly in competition experiments using non-radioactive isoprenaline.

Stiles: We have attempted to use [^{3}H]-HBI as an agonist ligand, but there are several problems associated with its use in the rat ventricle. First, [^{3}H]-HBI is a relatively β_2-specific agent so that there is little binding. Secondly, there is fairly high unspecific binding in the heart ($\approx$40%–50%). This makes it doubly difficult to assess the small amount of specific binding. I agree that [^{3}H]-HBI binding in systems such as the frog erythrocyte is of great value in evaluating the formation at the high affinity state (HRN). We, therefore, employed the indirect methods of analysis in our system to assess the high affinity state.

* H, hormone (catecholamine); R, β-adrenergic receptor; N, guanine nucleotide regulatory protein

Catecholamines and β-Adrenergic Receptors in the Non-ischemic Myocardium After Coronary Occlusion

G. Baumann, G. Riess, S. B. Felix, W. D. Erhardt[1], *U. Loher, L. Ludwig, H. Blömer*

Summary

Left ventricular infarction (MI) was produced in guinea pigs and the contractile response to isoproterenol and tyramine tested in the isolated perfused heart (IPH). Adenylate cyclase activity (AC) and β-receptor-binding characteristics were determined in a sarcolemmal preparation (MP) of the right ventricle of the same hearts. Three days after MI the isoproterenol- and tyramine-induced positive inotropic effect was significantly attenuated to a similar extend as were stimulation rates of the adenylate cyclase activity (–90%). In the MP's β-receptor-binding studies revealed a decrease in receptor number (88%) and a 12 times lowered affinity (K_D) of the remaining receptors. In the same hearts which were refractory to stimulation with catecholamines no impairment of the histaminergic H_2-receptor-adenylate cyclase system was seen. Inotropic effects of histamine and two recently developed selective H_2-agonists, impromidine and dimaprit, were the same in the infarcted hearts compared to sham-operated controls. Similar results were obtained in the MP's revealing stimulation rates of the AC at the 3rd day after MI not different from controls. Furthermore, all above alterations could be effectively prevented if treatment had been made with reserpine or the β-blocking agents metoprolol (cardioselective) or Th 326 (non-cardioselective) during the whole postinfarction period. A reduction of infarct size could be established for metoprolol (–33%), for reserpine (–27%) and for Th 326 (–14%). It is concluded that the abnormalities observed in the non-ischemic surviving myocardium are the result of a specific damage of sarcolemmal β-receptors due to post-MI excessive levels of circulating catecholamines. This impairment can be effectively prevented by treatment with β-blocking agents. Furthermore, the possible risk of β-blocker-induced heart failure can be avoided if cardiac contractility is stimulated via the H_2-receptor-adenylate cyclase system which was found to exist in the human myocardium as well with very similar receptor kinetics and properties.

Introduction

After myocardial infarction the non-ischemic portion of the heart has been traditionally regarded as normal myocardium. Meanwhile there is striking evidence suggesting that remarkable changes in cardiac contractility and myocardial metabolism occur in diverse heart diseases causally involving considerable impairment of the β-adrenergic stimulation mechanism, e.g. a transient decline in contractility and depletion of cardiac catecholamine stores [14, 15]. The functional significance of these changes, however, remains largely unknown. Enhanced activity of the sympathetic nervous system leading to elevated plasma catecholamine levels and increased urinary excretion in the postinfarction period may reflect an important

1 Department of Experimental Surgery, Technical University of Munich, Germany

reserve mechanism for maintaining cardiac performance in the early phase [17, 24]. However, prolonged exposure to elevated concentrations of catecholamines cannot prevent development of spontaneous heart failure in many cases, but results in a subsequent loss of sensitivity to β-adrenergic stimulation. The underlying mechanism of these changes are poorly understood at present, but may involve sarcolemmal alterations, especially the membrane bound adenylate cyclase system which is thought to mediate the positive inotropic action of catecholamines [9, 19], histamine [1–5, 13] and other hormones [12, 23], and the resulting increases in tissue levels of 3′,5′-cyclic-AMP are assumed to be causally related to the response to these substances [23].

The functional effects of the hormones result from their primary interaction with a specific membrane associated receptor site on the outer cell surface, which then initiates activation of the adenylate cyclase at the inner site of the membrane [9,23]. The question, by which mechanism cyclic-AMP can trigger transmembrane calcium influx is presently not yet clear, but may involve phosphorylation of sarcolemmal proteins by a cyclic-AMP-dependent protein kinase [20].

A recent report from our laboratory indicates that a pronounced specific impairment of the β-adrenergic stimulation mechanism occurs involving the catecholamine-mediated stimulation of cardiac contractile force and adenylate cyclase activity in the whole surviving myocardium after coronary occlusion [4]. These effects have been shown to be most likely the result of a decreased number and affinity of sarcolemmal β-receptors. In contrast to this the response to histaminergic H_2-receptors was found to be essentially unaltered after myocardial infarction. Because of the well-known disadvantageous effects of histamine, e.g. bronchospasm, urticaria, anaphyllactic shock, which are most likely due to be mediated by specific H_1-receptors in the respective organ areas, the administration of histamine to enhance cardiac contractile force by use of the uninvolved sarcolemmal H_2-receptor adenylate cyclase system seems without any therapeutical value.

The development of specific histaminergic agonists, which are assumed to activate selectively H_2-receptors, led us to investigate their myocardial effects in the catecholamine refractory, surviving myocardium after coronary occlusion.

In order to elucidate whether the results obtained may also be relevant for the human myocardium, additional experiments were carried out in a sarcolemmal membrane preparation of human papillary muscles which were resected during operations for mitral valve replacement.

Materials and Methods

All substances used in this study have been purchased from sources listed elsewhere [4]. Impromidine and dimaprit were generous gifts by Dr. D. D. A. Owen, Welwyn Garden City Hertfordshire, Great Britain and Dr. B. Hein, Munich, Germany.

The operative techniques for ligation of the coronaries, anesthesia with carfentanyl-ethomidate according to *Neumann* et al. [16], and biventricular evaluation of cardiac contractile parameters in the isolated perfused heart preparation have been described in detail in a recent report [4]. Preparation of sarcolemmal membranes from the right ventricle of each heart was done according to *Drummond* and *Duncan* [8] with the modifications recently described [19]. A modified adenylate cyclase assay, separation of [^{32}P]-cyclic-AMP according to *Salomon* et al. [18] and β-receptor-binding characteristics with [^{3}H]-dihydroalprenolol were performed as reported [4]. Evaluation of infarct size followed a spectrophotometric method for determination of phosphocreatine kinase activity as described [4].

Statistical analysis of the data was performed using Student's unpaired t-test and differences were considered significant only if $P < 0.05$. When indicated, statistical analysis was performed among the different groups by the one-way analysis of variance (BMDP Biochemical, Computer Programms, University of California Los Angeles Press, 1979).

Results

Distinction Between Inotropic Effects Mediated by Myocardial β_1- and H_2-Receptors

The existence of two different and independent receptor sites can be basically proved in the isolated perfused biventricularly catheterized heart preparation. As can be seen from Fig. 1, bolus injections of isoproterenol and histamine induced a nearly identical transient increase of left and right ventricular contractile parameters at the given concentration. In the presence of the β_1-blocking agent metoprolol, however, the response of the heart to isoproterenol is completely abolished whereas the effect of histamine is not altered. In turn, if the H_2-receptor antagonist cimetidine was continuously infused, the positive inotropic and chronotropic effect of isoproterenol was not affected while that of histamine was found to be markedly reduced. After infusions of metoprolol and cimetidine were discontinued, the inotropic and chronotropic effects of both agonists were the same compared to those at the beginning of the experiment, thus indicating a reversible and competitive receptor-mediated mechanism.

To evaluate the order of potency of isoproterenol, histamine and the two H_2-agonists, impromide and dimaprit, the sensitivity of the heart to the inotropic effects of these agents were studied by bolus injections in increasing amounts and the positive inotropic response evaluated by simultaneously recording ventricular pressure and dP/dt_{max} in the left and right ventricle. In the left ventricle isoproterenol (K_A 3.3×10^{-11} moles) proved to be the most potent inotropic agent compared to histamine (K_A 2.4×10^{-9} moles) and the H_2-agonistic compounds impromidine (K_A 1.7×10^{-10} moles) and dimaprit (K_A 6.2×10^{-9} moles) the latter two curves differing from each other by nearly two orders of magnitude (Fig. 2). Though the position of the dose-response curves differs considerably with respect to the amount of the respective agonist, all curves reached a plateau phase which approached 6000 mmHg/sec in the left ventricle, except that of histamine, which gradually declined if quantities greater than 4×10^{-8} moles were applied. K_A-values for right ventricular dP/dt_{max} were 2.8×10^{-11} moles for isoproterenol, 9.1×10^{-10} moles for histamine, 1.4×10^{-10} moles for impromidine and 4.8×10^{-9} moles for dimaprit.

It should be noted, that simultaneous infusion of metoprolol (4×10^{-6} mol/liter) or cimetidine (2×10^{-5} mol/liter) selectively antagonized the inotropic effects of isoproterenol or histamine, impromidine and dimaprit, respectively, shifting the dose-response curves for all agonists to the right by more than two orders of magnitude (results not shown).

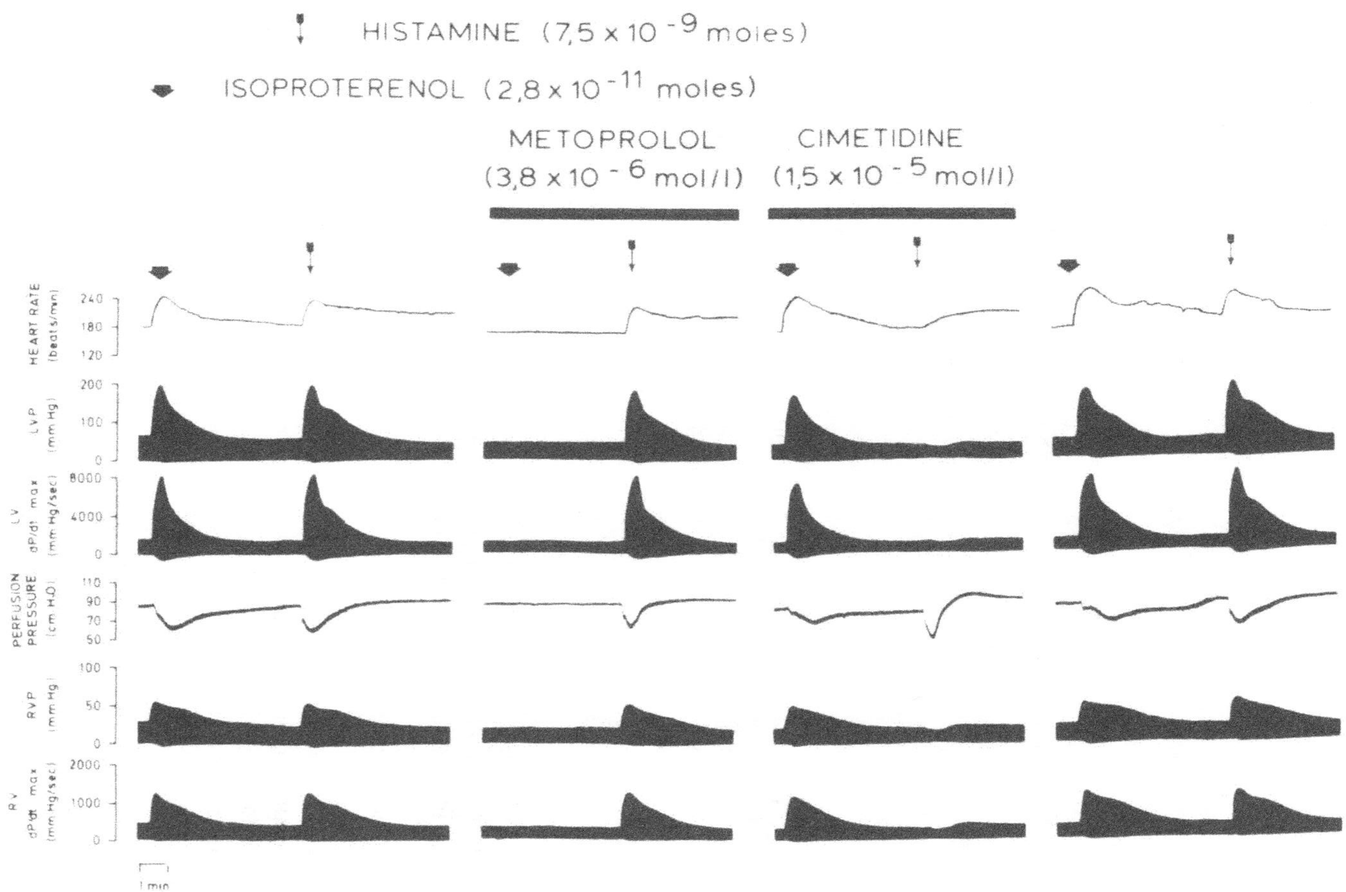

Fig. 1. Representative example for the positive inotropic effect of equipotent amounts of isoproterenol and histamine on left and right ventricular contractile parameters in the absence and presence of metoprolol or cimetidine. Compounds were applied either as bolus injection (15 µl, agonists) or infusion (< 100 µl/min, antagonist) directly into the perfusion stream. Volume constant perfusion: 12 ml/min

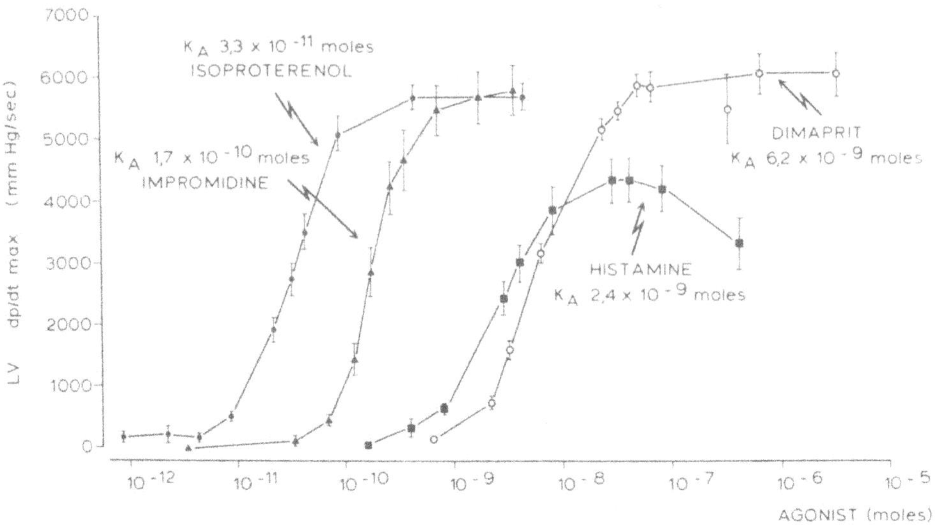

Fig. 2. Dose-response curves of the inotropic effects of isoproterenol ($n=8$), impromidine ($n=8$), dimaprit ($n=8$) and histamine ($n=15$) on left ventricular dP/dt_{max} in isolated perfused biventricular catheterized guinea pig hearts. All compounds were applied as bolus injections (15 µl). Volume constant perfusion: 12 ml/min. Corresponding curves were also obtained for right ventricular dP/dt_{max} (results not shown)

Effects of β- and H_2-Receptor Agonists on Adenylate Cyclase Activities and β-Receptor-Binding Characteristics in Sarcolemmal Membrane Preparations from Guinea Pig and Human Myocardium

In a particulate sarcolemmal membrane preparation of guinea pig ventricular myocardium agonist-induced stimulation rates of the membrane bound cyclase activity were examined and compared to results obtained in membrane preparations of human papillary muscles. As is evident from Fig. 3A, activation of adenylate cyclase by isoproterenol begins at 10^{-8} mol/liter and reaches its maximum at 10^{-5} mol/liter. Impromidine proved to be less potent followed by dimaprit. Similar results were obtained for the activation of human myocardial adenylate cyclase with nearly identical K_A-values for all agonists compared to the guinea pig. In these experiments effects of isoproterenol, impromidine and dimaprit could be selectively antagonized by metoprolol and cimetidine, respectively. In order to exclude nonspecific interference of the H_2-agonists with the myocardial β-receptor, kinetic binding studies were performed in a particulate membrane preparation with [^{3}H]-dihydroalprenolol as specific ligand. As can be seen from Fig. 3B stereospecific binding of [^{3}H]-dihydroalprenolol to sarcolemmal β-receptors is a rapid process and rapidly reversible, if (–) alprenolol is added to the incubation mixture under steady state conditions. No change in binding characteristics, however, could be detected after addition of the H_2-agonistic compounds, impromidine or dimaprit, neither in the guinea pig nor in the human sarcolemmal membrane preparation.

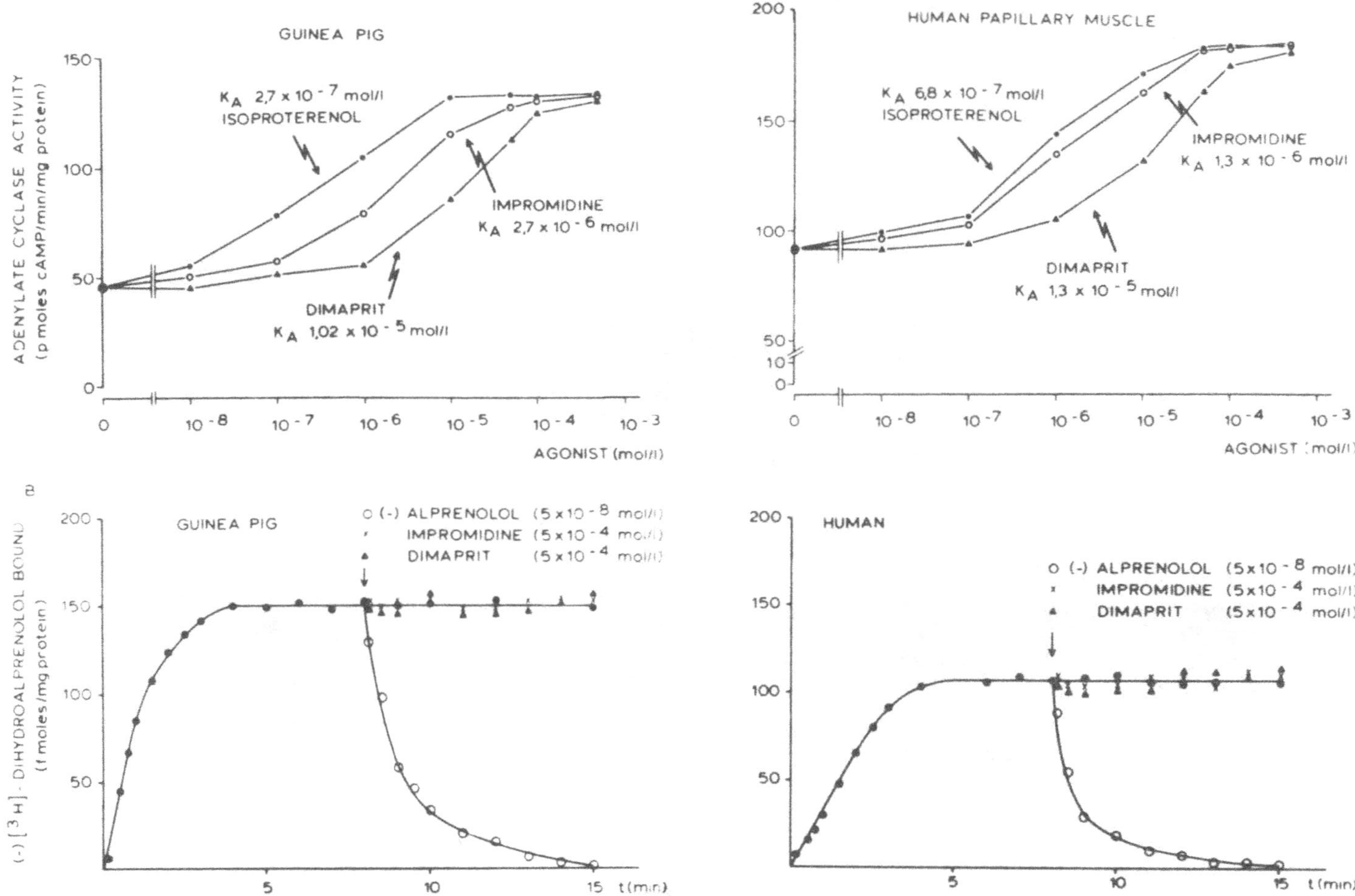

Fig. 3. A. Effects of isoproterenol, impromidine and dimaprit on cardiac adenylate cyclase activities of sarcolemmal membrane preparations from guinea pig ventricular myocardium (*left*) and human papillary muscle (*right*). Points represent mean values from triplicate determinations in pooled membrane preparations of 3 different hearts or 3 human papillary muscles, respectively. Membrane protein 80–100 µg, incubation time 6 min at 37 °C. **B.** Effect of impromidine, dimaprit and (–) alprenolol on stereospecific binding of (–)[^{3}H]-dihydroalprenolol (1,5 nmol/l, filled cycles) to cardiac β-receptors. (Same membrane preparations as above.) Membrane protein 500 µg, 37 °C. Values represent means of triplicate experiments

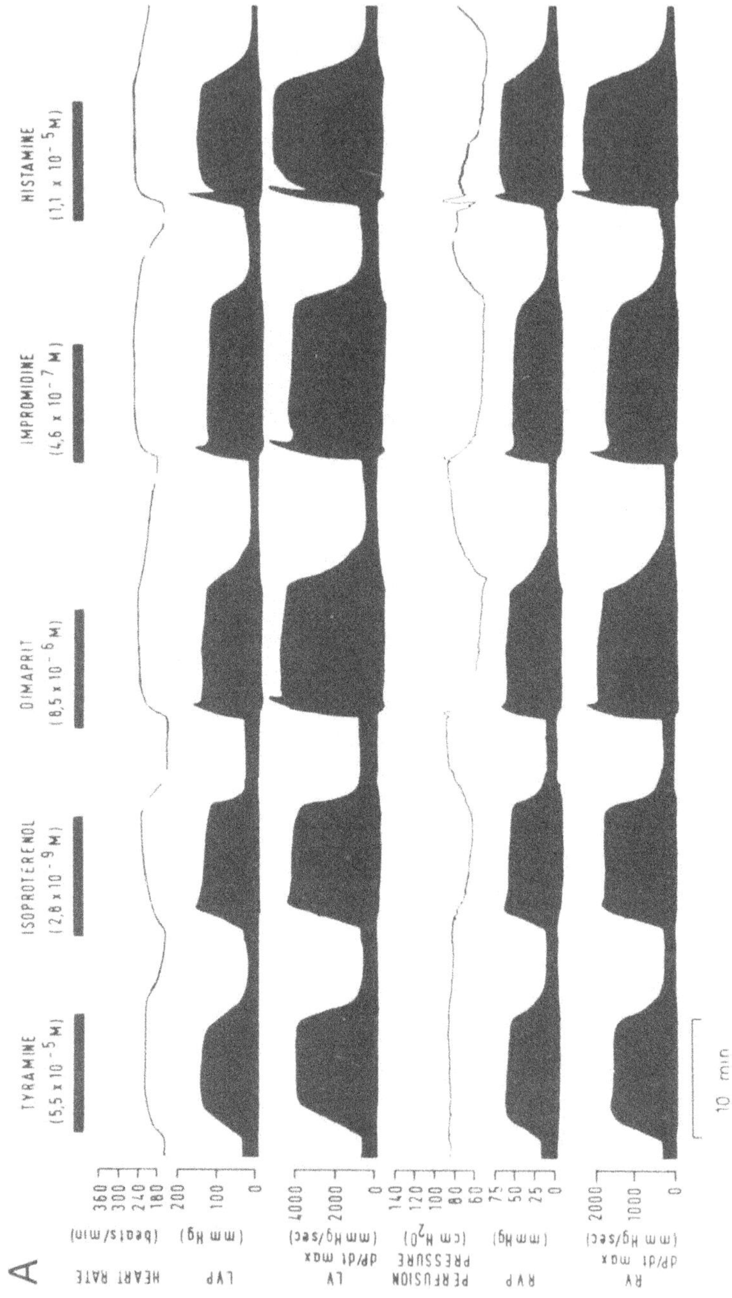
A
HEART RATE (beats/min)
360
300
240
180
LVP (mm Hg)
200
100
0
LV dP/dt max (mm Hg/sec)
4000
2000
0
PERFUSION PRESSURE (cm H_2O)
140
120
100
80
60
RVP (mm Hg)
75
50
25
0
RV dP/dt max (mm Hg/sec)
2000
1000
0
TYRAMINE (5,5 x 10^{-5} M)
ISOPROTERENOL (2,8 x 10^{-9} M)
DIMAPRIT (8,5 x 10^{-6} M)
IMPROMIDINE (4,6 x 10^{-7} M)
HISTAMINE (1,1 x 10^{-5} M)
10 min

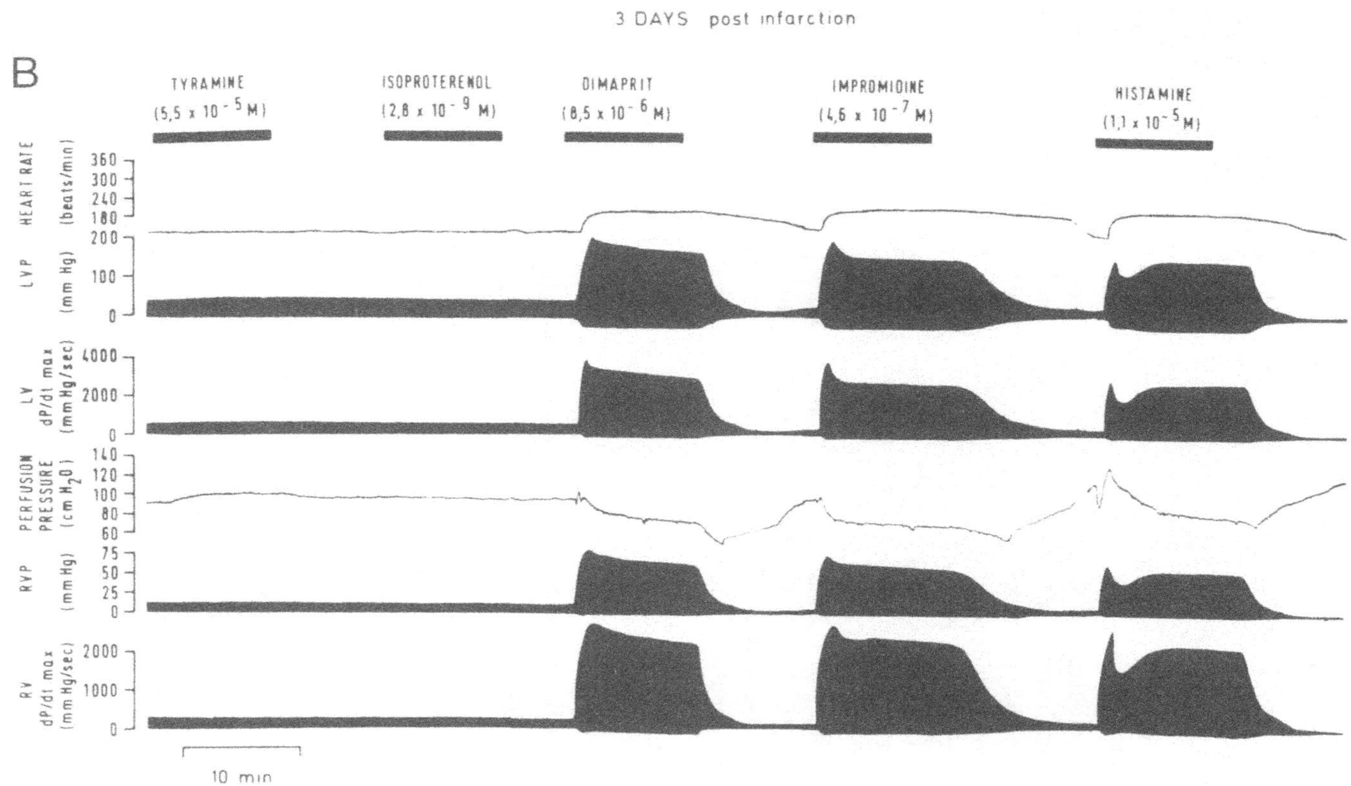

Fig. 4. Original recordings of the inotropic effect of tyramine, isoproterenol, dimaprit, impromidine and histamine in **A** a sham-operated control heart and **B** a heart 3 days after myocardial infarction. The infarcted area is confined to the free wall of the left ventricle without affecting the interventricular septum or right ventricle. Size of infarction was 42% of total left ventricular myocardium in this heart. Volume constant perfusion: 12 ml/min

Evaluation of Left Ventricular Infarct Size

Size of infarction was determined by measuring total creatine kinase activity in left ventricular (LV) homogenates of all 131 hearts analysed in this study. Forty-six did not have biochemical evidence of myocardial infarction and were termed sham-operated controls, since creatine kinase activity was the same compared to a group ($n = 19$) which had not undergone the operative procedure. The 66 infarcted hearts demonstrated considerable loss in left ventricular creatine kinase activity indicating a size of infarction ranging from 15% to 45% of total left ventricular myocardial mass. The infarcted area was strictly localized in the free wall of the left ventricle and did not affect the interventricular septum (Fig. 4), since it could be shown in a different series of 15 experiments that the septum did not have histologic or electron-microscopic evidence of necrosis. The following creatine kinase activities were measured (expressed as IU/mg supernatant protein, mean values ± 1 S.E.M.) in the 5 groups: (a) sham-operated group was 15.81 ± 0.61 ($n = 46$); (b) 3-day post-infarction, no treatment group was 9.72 ± 0.28 ($n = 32$); (c) 3-day postinfarction, reserpine-treated group was 11.42 ± 0.27 ($n = 10$); (d) 3-day postinfarction, meto-prolol-treated group was 11.75 ± 0.18 ($n = 18$); and (e) 3-day postinfarction, Th 326-treated group was 10.56 ± 0.22 ($n = 6$). Consequently, these values correspond to an inhibition of infarct-extension by 28% in the case of treatment with reserpine, 33% in the case of treatment with the cardio-selective β_1-blocking agent metoprolol and 14% in the case of treatment with the non-cardioselective β-blocking agent Th 326, all compared to the untreated group 3 days after coronary occlusion.

Cardiac Contractility of the Surviving Myocardium After Coronary Occlusion

The sensitivity of the hearts to β- and H_2-receptor stimulation after myocardial infarction was studied in two experimental arrangements. In one series of experiments equipotent doses of each compound were given as infusions. An actual recording of a representative experiment is shown in Fig. 4 for a sham-operated control (A) and an infarcted heart (B) 3 days after coronary occlusion. In the control heart isoproterenol (2.8×10^{-9} mol/liter) and tyramine (5.5×10^{-5} mol/liter), an indirectly acting sympathomimetic drug, as well as histamine (1.1×10^{-5} mol/liter) and the H_2-agonists, impromidine (4.6×10^{-7} mol/liter) and dimaprit (8.5×10^{-6} mol/liter) induced comparable enhancement of ventricular pressure and dP/dt_{max} in both, the right and the left ventricle. Furthermore it seems of interest that coronary perfusion pressure is decreased by impromidine and dimaprit to a similar extent compared to isoproterenol, whereas histamine has a biphasic effect with a pronounced coronary constriction, a fact which may be due to its H_1-agonistic action and also explain the decline of the dose-response curve for cardiac contractility occuring within higher concentration ranges (Fig. 2). In contrast, 3 days after coronary occlusion cardiac contractile force proved to be refractory to direct and indirect sympathetic stimulation with isoproterenol and tyramine (Fig. 4B). The inotropic response to impromidine, dimaprit and histamine, however, was essentially unaltered in the same heart. Left ventricular contractile parameters do, of course, not reach control values according to the infarcted area strictly localized in the free wall of the left ventricle.

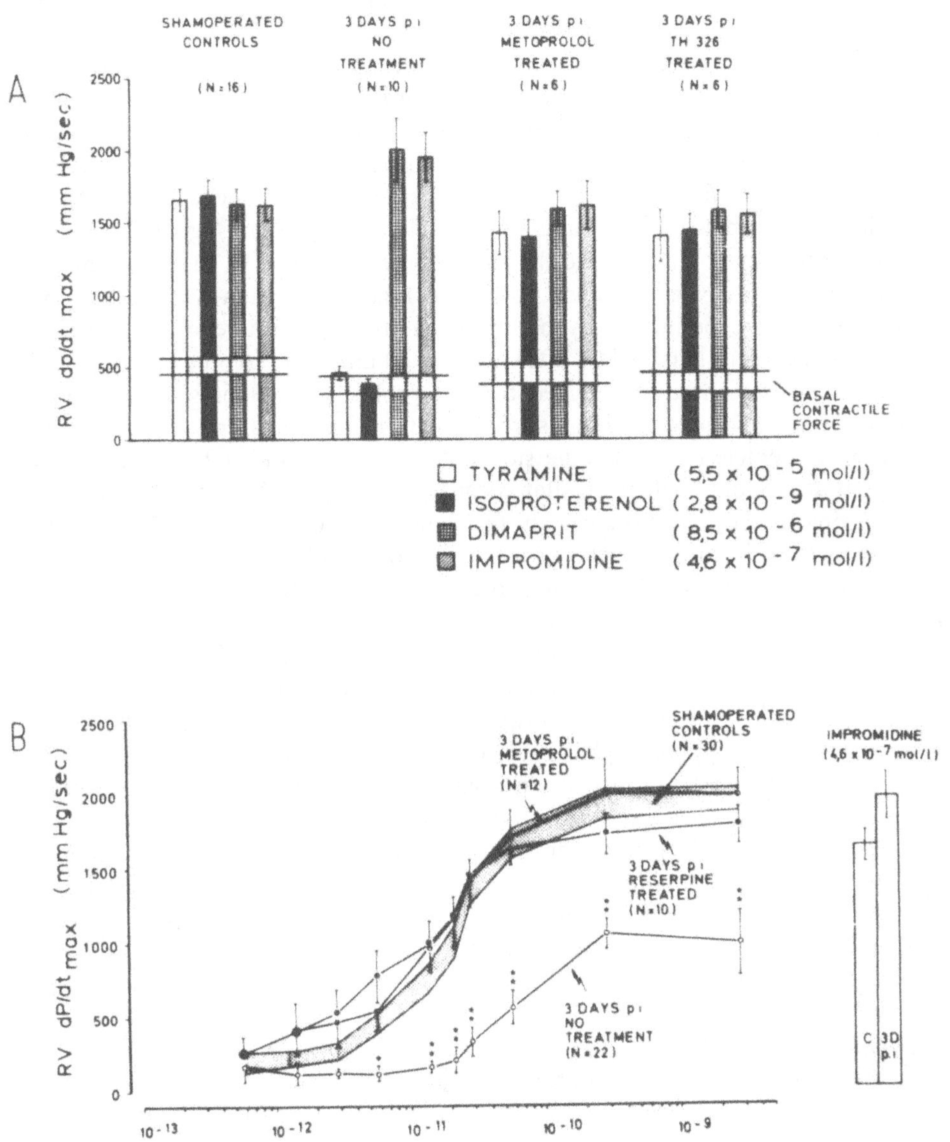

Fig. 5. A. Evaluation of the inotropic action of tyramine, isoproterenol, dimaprit and impromidine (applied as infusions) on right ventricular dP/dt_{max}. Volume constant perfusion: 12 ml/min. Mean values ± 1 S.E.M. Values for tyramine and isoproterenol in the untreated group 3 days postinfarction differed by $P<0.0001$ from their respective control values. No significant differences were seen in the other groups. *p.i.*, post-infarction. **B.** Dose-response curves for the isoproterenol-induced increase of right ventricular dP/dt_{max} in the sham-operated control hearts and in infarcted hearts analyzed 3 days after coronary ligation with and without pretreatment with metoprolol or reserpine. Isoproterenol was applied as bolus injection in increasing amounts. *Columns* illustrate the contractile response to impromidine (applied as infusion) in hearts of the sham-operated control group (C) and of the untreated group 3 days postinfarction (3 days p.i.). Average basal right ventricular dP/dt_{max} was 490 mm Hg in all groups and has been substracted from all values. Constant pressure perfusion: 60 cm H_2O. Mean values ± 1 S.E.M. Statistical difference from control values: * $P<0.01$ and ** $P<0.001$. *p.i.*, post-infarction

Results of these experiments for the non-ischemic right ventricular myocardium 3 days postinfarction are summarized in Fig. 5 A. The average loss of β-adrenergic stimulation by isoproterenol and tyramine in the uninvolved right ventricle was more than 90% whereas stimulation rates by impromidine and dimaprit were not different compared to the sham-operated control group. Treatment with β-blocking agents during the postinfarction period are obviously effective in preventing the observed decline in β-adrenergic response, irrespective if treatment was made with a cardioselective (metoprolol) or a non-cardioselective β-blocking agent (Th 326).

In order to further characterize the role of catecholamines in the postinfarction period, the sensitivity of the heart to β-adrenergic stimulation was studied in a second series of experiments by bolus injections of isoproterenol in increasing amounts in hearts isolated 3 days after myocardial infarction. In one group, cardiac β-receptors were constantly blocked by metoprolol. In another group, coronary occlusion was performed in animals treated with reserpine 24 and 4 h prior to the operation to deplete catecholamine stores. A third group did not receive medical treatment at any time before or after coronary occlusion. As is shown in Fig. 5 B, the inotropic effect to isoproterenol in the sham-operated control hearts was dose-dependent (shaded area), beginning at 10^{-12} moles and reaching a maximum at 10^{-9} moles isoproterenol. Three days after coronary ligation, however, the dose-response curve for isoproterenol was markedly shifted to the right and significantly flattened, yielding maximal stimulation rises which were only 40% compared to the sham-operated controls. The inotropic effect of impromidine (right columns in Fig. 5 B) on right ventricular myocardium of the same hearts was not impaired at any time. Interestingly, both pharmacological interventions were equally effective in preventing the observed loss of sensitivity to β-adrenergic stimulation: position and steepness of the dose-response curves 3 days postinfarction did not differ from sham-operated controls, if treatment was made by either reserpine or metoprolol. Similar results were obtained for the left ventricle, except that the dose-response curves for the reserpine and metoprolol treated hearts remained flattened corresponding to the loss of contractile myocardium.

Adenylate Cyclase Activities and β-Receptor-Binding Properties in Sarcolemmal Membrane Preparations of the Surviving, Uninvolved Ventricular Myocardium

Adenylate Cyclase Activities. The influence of isoproterenol, dimaprit, impromidine and NaF on cardiac adenylate cyclase activity was examined in the particulate membrane preparation of the right ventricle of each heart (Fig. 6 A). Basal adenylate cyclase activity did not differ significantly in all 4 groups (horizontal bars). Isoproterenol elicited a 3-fold stimulation of the enzyme activity in the sham-operated hearts. Three days after coronary ligation, however, isoproterenol only weakly increased basal adenylate cyclase activity, yielding an average loss of stimulation by about 90%. If the animals had been treated with metoprolol or Th 326, the isoproterenol mediated increase of adenylate cyclase activity was nearly identical compared to controls. It should be noted that in all groups stimulation properties of impromidine, dimaprit and NaF were practically the same. As shown in Table 1, the degree of β-adrenergic stimulation correlated well with the size of infarction. Hearts with relatively moderate necrosis (15–25%) exhibited a 40% loss

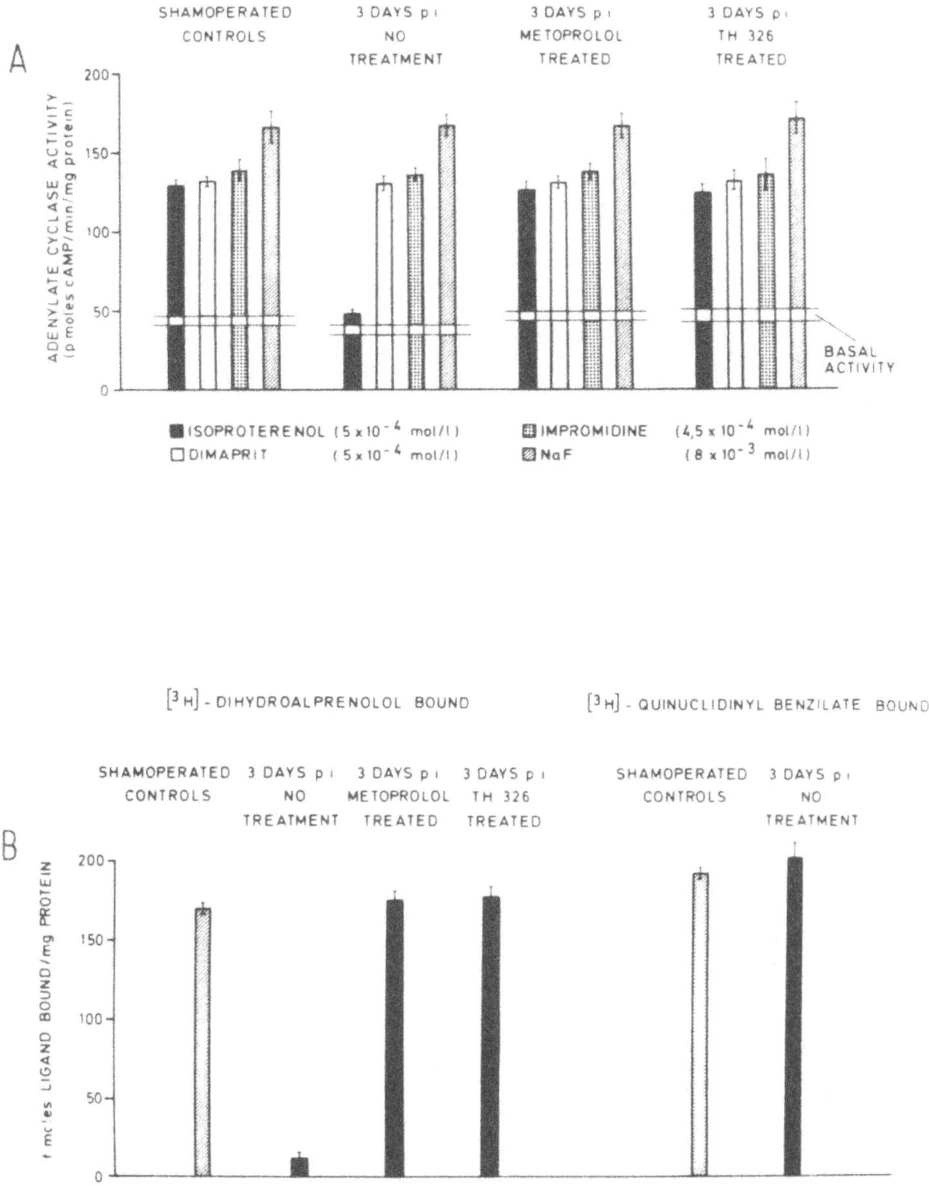

Fig. 6. A. Cardiac adenylate cyclase activity of right ventricular membrane preparations in the absence of agonist (= basal activity, *horizontal bars*) and presence of isoproterenol, dimaprit, impromidine and NaF in the respective groups. Values resemble mean ± 1 S.E.M. of duplicate determinations for each heart. The isoproterenol-induced stimulation in the untreated group, 3 days postinfarction differed by $P < 0.0001$ from respective control value. All other differences were not statistically significant. *p.i.,* post-infarction. **B.** Total (–)[^{3}H]-dihydroalprenolol binding capacity of right ventricular sarcolemmal membrane preparations of the respective groups. Quantitative uptake studies were performed for each heart preparation in duplicate experiments and the means of B_{max}-values calculated according to Scatchard. K_D-values are specified in Table 1. Mean values ± 1 S.E.M. *p.i.,* post-infarction. A one-way analysis of variance was performed among the different groups, revealing an F-value of 141.2179 with a tail area probability of $P < 0.00005$. *Columns* on the *right* illustrate binding data of [^{3}H]-quinuclidinyl benzilate to cholinergic muscarinic receptors in membrane preparations of the same hearts

Table 1. Influence of the size of left ventricular infarction on isoproterenol-induced stimulation rates of adenylate cyclase activity, (–)[^{3}H]-DHA binding capacity and K_D-values in a particulate membrane preparation of the right ventricles of 32 hearts 3 days postinfarction. Analysis of data from Fig. 6

Size of infarction [% of left ventricular myocardium]	*n*	Isoproterenol-stimulated adenylate cyclase activity (basal activity substracted) [pmoles cAMP/min/mg protein]	Loss of stimulation [% of control]	Total(–)[^{3}H]-DHA binding-capacity (B_{max})[a] [f moles bound/mg protein]	Loss of binding [% of control]	K_D-value[a] [moles/liter]
0	46	98.7±4.3	0	170.1± 4.4	0	7.6×10^{-9}
15–25	9	62.1±4.9*	−37	67.3± 6.8**	−60	3.2×10^{-8}
25–35	9	41.3±5.1**	−58	45.7±10.3**	−73	6.1×10^{-8}
>35	14	4.7±2.2**	−95	12.4± 5.8**	−93	9.8×10^{-8}

Significantly different from control: * $P<0.01$; ** $P<0.001$

[a] B_{max}, total(–)[^{3}H]-DHA bound and K_D, dissociation constants derived from Scatchard analysis

Mean values± 1 S.E.M.

of stimulation, whereas the ability of isoproterenol to enhance adenylate cyclase activity was nearly abolished, if the infarct size exceeded 35% of total left ventricular myocardium.

β-Receptor-Binding Properties. β-Receptor-binding studies with radioactively labeled (–)[^{3}H]-dihydroalprenolol were performed in the same membrane preparations, in which determinations of adenylate cyclase activities were made (separately for each heart). As shown in Fig. 6B, binding of (–)[^{3}H]-dihydroalprenolol to cardiac membranes of the sham-operated controls was 170 fmoles/mg protein. Three days after coronary ligation, a highly significant decrease of (–)[^{3}H]-dihydroalprenolol binding occured. Treatment of the animals with either metoprolol or Th 326 proved to be equally effective in preventing the loss of binding capacity. Analysis of these data with respect to size of left ventricular infarction provided good correlation (Table 1) with maximal loss (approximately 95%) of (–)[^{3}H]-dihydroalprenolol binding sites. When binding data were plotted according to Scatchard, a straight line was obtained in all groups, indicating a single class of binding sites (results not shown). From the slope of the curves (direct measure for the affinity of the remaining receptors) the dissociation constant (K_D-value) can be easily calculated. Data in Table 1 also suggest a close correlation between left ventricular infarct size and change of affinity (K_D-value) of the remaining β-receptors in the surviving myocardium. Whereas the K_D-value was 7.6×10^{-9} mol/liter) in the sham-operated control group 9.8×10^{-8} mol/liter was calculated in those hearts, in which size of left ventricular infarction was greater 35%. This decreased affinity indicates, that agonist concentrations 15 times higher were required to saturate 50% of the remaining β-receptors 3 days after myocardial infarction.

In contrast to the observed changes in β-receptor number and affinity cholinergic muscarinic receptors were not affected (Fig. 6 B), as revealed by binding studies with their specific ligand [^{3}H]-quinuclidinyl-benzilate. Receptor number (and affinity) did not differ from the sham-operated controls 3 days postinfarction.

It should be noted, that all above alterations concerning the postinfarction decreased inotropic effects of isoproterenol, the diminished β-adrenergic response of myocardial adenylate cyclase as well as the loss of affinity and number of sarcolemmal β-receptors seem to be of a reversible nature, since – without any pharmacologic intervention – all parameters already showed a clear tendency to normalization 7 days after the infarction occurred, though for complete recovery 3–4 weeks were necessary.

Discussion

Impaired cardiac contractility after acute myocardial infarction has been described in numerous reports in both clinical and experimental studies. Depression of cardiac output, elevated left ventricular enddiastolic pressure and hypotension are common observations, and many of these studies have been made primarily attempting to quantify the infarcted area or to reduce the extension of necrosis by various pharmacologic interventions. Little information, however, is available about the functional state of the surviving uninvolved portion of the heart which has been conventionally considered as normal myocardium, unless coexisting stenosis of additional vessels caused ischemia of the non-infarcted ventricular segments. Earlier experimental investigations demonstrated reversible decline in cardiac norepinephrine stores at sites remote from the infarcted area [14, 15]. The functional significance of these changes is far from being understood at present. However, in view of the central role which catecholamines play in contractility, energy supply and metabolism of the myocardial cell, support is given to the hypothesis that activation of the sympathetic nervous system may reflect an important reserve mechanism to compensate for the loss of functioning tissue in the early phase after acute myocardial infarction. But elevated plasma catecholamine levels occurring in the postinfarction period cannot prevent heart failure in many cases, and infusions of exogeneous β-sympathomimetic drugs often fail to exert their usual positive inotropic effect as well. The discrepancy between degree of myocardial injury and impaired global cardiac performance indicates that several mechanisms may be involved in the pathogenesis of events occurring in the postinfarction period.

The present study demonstrates a reversible decline of β-adrenergic response in the surviving non-infarcted right ventricular myocardium following experimental coronary occlusion. These findings include a diminished positive inotropic response to isoproterenol in the intact heart preparation and, following the molecular mechanism of action of catecholamines, further include the decreased ability of isoproterenol to stimulate adenylate cyclase in the particulate sarcolemmal membrane preparation of the non-infarcted ventricle of the same hearts. In these right ventricular preparations β-receptor-binding studies revealed a considerable loss of binding sites and a markedly lowered affinity of the remaining receptors for their ligand. All above parameters correlate with the size of infarction and thereby lead to (reversible) damage of the β-receptor system.

In this context it is of interest that pharmacologic interventions, such as administration of β-blocking agents or reserpinization are effective in preventing the observed impaired function of the β-adrenergic system. These findings emphasize the specifity of the process and strongly suggest that this mechanism is triggered directly by catecholamines and only involve cardiac β-receptors, resulting in a decrease of their number and affinity.

This assumption is further supported by a number of other observations: (a) The effect of histamine, impromidine and dimaprit, which exert their myocardial effects via specific H_2-receptors, proved to be unchanged. This implies that the sarcolemmal H_2-receptor adenylate cyclase system is not involved in the process. (b) The functional state of the adenylate cyclase itself is also not responsible for the observed disturbance since basal activity of this enzyme and stimulation rates by NaF, which is thought to activate the enzyme at the regulatory protein of the catalytic unit, were practically the same in all groups. (c) Studies of [^{3}H]-QNB binding to cardiac membranes 3 days after coronary occlusion revealed no changes concerning both, number and affinity of muscarinic cholinergic receptors in the same hearts in which [^{3}H]-DHA binding sites and K_D-values were found to be markedly reduced.

It is noteworthy that the observed effects are not due to catecholamine-mediated increase in phosphodiesterase activity. In the absence and presence of isoproterenol, histamine and NaF the activity of this enzyme was found essentially unaltered regardless of pretreatment of the animals with β-blocking agents or reserpine. This has been carefully checked in a previous study [4]. Thus it can be ruled out that enhanced degradation of cAMP by phosphodiesterase can account for the depressed β-adrenergic inotropic response of the non-infarcted myocardium.

The observed loss in inotropic response to catecholamines could, at least in part, be the result of depletion of high energy phosphates in the surviving myocardium. Analysis of ATP and creatine phosphate tissue concentration in such hearts ($n = 10$) demonstrated declines of about 20% and 35% respectively, compared to sham-operated controls (results not shown). This moderate depletion, however, can not account for the loss of the isoproterenol-induced inotropic effect. This is supported by some of our findings which clearly demonstrate, that the positive inotropic effect mediated by impromidine, dimaprit and histamine was not impaired in the same hearts in which β-adrenergic stimulation was nearly abolished.

Several biogenic amines exert their positive inotropic effect on the myocardium by causing a release of endogeneous catecholamines from the terminal sympathetic nerve endings, e.g. tyramine. But this mechanism can by no means account for the myocardial effects of histamine and the two H_2-receptor agonistic compounds since metoprolol effectively antagonizes the inotropic action of tyramine, but did not alter the response of the heart to either H_2-agonist or histamine. Cimetidine in turn, did not influence the response of the heart to tyramine, but completely antagonized those of impromidine, dimaprit and histamine.

Clinical Implications

Certain pathophysiological implications of our findings should be considered. Results of the present study strongly suggest, that in the non-ischemic myocardium a causality exists between the observed decrease in inotropic action of isoproterenol, the diminished β-adrenergic response of myocardial adenylate cyclase and loss of affinity and number of sarcolemmal β-receptors. These changes in the surviving myocardium most likely resemble catecholamine-induced specific detriments of the sarcolemma brought about by postinfarction increased circulating and from the necrotic area released catecholamines. Possibly these findings reflect a pathophysiological mechanism according to which the surviving myocardium progressively loses its pump output ability to adapt to stress situations and to increased requirements of the body in the critical phase of the first 72 h after coronary occlusion. In this regard, the work of *Thomas* and *Marks* seems to be of special interest. These authors demonstrated, that β-adrenergic receptors are "desensitized" in patients with chronic artereosclerotic heart disease, congestive cardiomyopathy and hypertensive coronary heart disease [22]. They showed that the "desensitization" contributes to the observed alterations in myocardial contractility, and that plasma norepinephrine concentrations were directly related to the degree of LV dysfunction in patients with congestive heart failure.

There are obvious differences between human myocardial infarction and infarction produced by acute coronary ligation in animals concerning nature of pathogenesis and degree of collateral circulation to the ischemic region. However, it should be allowed to speculate on the basis of our findings that therapeutical administration of β-sympathomimetic drugs, e.g. to cure cardiogenic shock, may be disadvantageous since plasma levels of catecholamines in heart failure have been shown to be elevated anyway, irrespective of the genesis of the underlying disease [6, 17, 21, 22, 24]. Consequently, additional treatment with dopamine, dobutamine or any other β-stimulator probably accelerates the catecholamine-induced impairment of the physiological β-receptor response, though transient improvement of hemodynamic parameters have been reported [10]. Seen from this point of view the positive inotropic effect mediated by the cardiac H_2-receptor adenylate cyclase system may be of considerable therapeutical value, especially under those pathophysiological conditions where the myocardium is refractory to catecholamine stimulation. Furthermore, treatment in the postinfarction period with β-blocking agents can obviously prevent the impairment of all above parameters and has been considered beneficial also in clinical trials [7, 21, 25], but is usually avoided because of the fear to worsen already depressed cardiac contractility. The possible risk of β-blocker induced heart failure may than be successively overcome, if cardiac contractile force is stimulated by simultaneous infusions of H_2-agonists.

Acknowledgement. The authors thank Miss Beate Pichotka and Miss Christina Kaiser for their expert technical assistance. Furthermore, we are indebted to Dr. Raymonde Busch, Technical University of Munich, for her help in statistical analysis of the presented data. The studies were supported by the Deutsche Forschungsgemeinschaft (Ba 666/1 and Ba 666/2-2).

References

1. Baumann G, Schrader J, Gerlach E (1978) Stimulation of contractile force and adenylate cyclase in the myocardium by histamine and dopamine: effects of adenosine. Pfluegers Arch [Suppl] 373:18
2. Baumann G, Rieß G, Erhardt WD, Felix SB, Blömer H (1980) Reduzierte β-adrenerge Ansprechbarkeit im nicht ischämischen Myokard nach experimentellem Herzinfarkt. Z Kardiol 69:209

3. Baumann G, Schrader J, Gerlach E (1981) Inhibitory action of adenosine on histamine- and dopamine-stimulated cardiac contractility and adenylate cyclase in guinea pigs. Circ Res 48:259
4. Baumann G, Rieß G, Erhardt WD, Felix SB, Ludwig L, Blümel G, Blömer H (1981) Impaired beta-adrenergic stimulation in the uninvolved ventricle post-acute myocardial infarction: reversible defect due to excessive circulating catecholamine-induced decline in number and affinity of beta-receptors. Am Heart J 101:569
5. Baumann G, Felix SB, Schrader J et al. (1981) Cardiac contractile and metabolic effects mediated via the myocardial H_2-receptor adenylate cyclase system: Characterization of two new specific H_2-receptor-agonists, impromidine and dimaprit, in the guinea pig and human myocardium. Res Exp Med Berl 171:81
6. Chidsey CA, Braunwald E, Morrow HG (1965) Catecholamine excretion and cardiac stores of norepinephrine in congestive heart failure. Am J Med 39:442
7. Delius W, Wirtzfeld A, Dominiak P, Sebening H, Blömer H, Grobecker H (1979) Effects of acute beta-adrenoceptor blockade (metoprolol i.v.) on plasma norepinephrine concentration and hemodynamics in postmyocardial infarction patients. Z Kardiol 68:441
8. Drummond GI, Duncan I (1970) Adenylate cyclase in cardiac tissues. J Biol Chem 245:976
9. Drummond GI, Severson DL (1979) Cyclic nucleotides and cardiac function. Circ Res 44:145
10. Gillespie TA, Ambos HD, Sobel BE, Robert R (1977) Effects of dobutamine in patients with acute myocardial infarction. Am J Cardiol 31:588
11. Gilman AG (1970) A protein-binding assay for adenosine 3′,5′-cyclic monophosphate. Proc Natl Acad Sci USA 67:303–331
12. Glick G, Parmley WW, Wechsler AS, Sonnenblick EH (1968) Glucagon: its enhancement of cardiac performance in the cat and dog and persistance of its inotropic action despite beta-receptor blockade with propranolol. Circ Res 22:789
13. Klein J, Levey GS (1971) Activation of myocardial adenyl cyclase by histamine in guinea pig, cat and human heart. J Clin Invest 50:1012
14. Mathes P, Gudbjarnason S (1971 a) Changes in norepinephrine stores in the canine heart following experimental myocardial infarction. Am Heart J 81:211
15. Mathes P, Cowan C, Gudbjarnason S (1971 b) Storage and metabolism of norepinephrine after experimental myocardial infarction. Am J Physiol 220:27
16. Neumann G, Erhardt W, Oberhuber B, Fritsch R, Blümel G (1980) A new highly potent and short-acting analgesic, Carfentanyl (R 33799) in combination with the hypnotic agent, Etomidat (R 26490), as a method of anaesthesia in guinea pigs. Res Exp Med 177: 135–143
17. Richardson JA (1963) Circulating levels of catecholamines in acute myocardial infarction and angina pectoris. Prog Cardiovasc Dis 6:56
18. Salomon Y, Londos C, Rodbell M (1974) A highly sensitive adenylate cyclase assay. Anal Biochem 58:541
19. Schrader J, Baumann G, Gerlach E (1977) Adenosine as inhibitor of myocardial effects of catecholamines. Pfluegers Arch 372:29
20. Sulakhe PV, Leung NL, St Louis PJ (1976) Stimulation of calcium accumulation in cardiac sarcolemma by protein kinase. Can J Biochem 54:438
21. Swedberg K, Waagstein F, Hjalmarson A, Wallentin I (1979) Prolongation of survival in congestive cardiomyopathy by β-receptor blockade. Lancet 1:1374
22. Thomas JA, Marks BH (1978) Plasma norepinephrine in congestive heart failure. Am J Cardiol 41:233
23. Tsien RW (1977) Cyclic AMP and contractile activity in heart. Adv Cyclic Nucleotide Res 8:363
24. Valori C, Thoms M, Shillingford JP (1967) Urinary excretion of free noradrenaline and adrenaline following acute myocardial infarction. Lancet 1:127
25. Wirtzfeld A, Klein G, Delius W, Sack D (1978) Treatment of acute myocardial infarction with metoprolol. Dtsch Med Wochenschr 103:566

Discussion

Szekeres, Szeged: There are data showing a loss of catecholamines in the non-ischemic myocardium after coronary occlusion. This could play a role in the reduced response to tyramine. How do you explain the decrease of isoproterenol action?

Baumann: The observed decrease of isoproterenol action in the non-ischemic myocardium is most likely due to conformational changes of receptor proteins. This probably leads to decreased affinity and number of β-receptors for their specific ligands. The same, of course, explains the reduced response of the heart to tyramine which acts through the release of endogeneous catecholamines from the terminal nerve endings.

Schümann, Essen: Have you measured the noradrenaline content and the binding capacity of left ventricular tissue after pretreatment with reserpine?

Baumann: After pretreatment with reserpine (5 mg/kg 24 and 4 h prior to the operation) the concentration of catecholamines in the myocardium is nearly zero! Binding capacity of left ventricular membrane preparations has not been measured.

Erdmann, Munich: I read from your experimental data that the number of ^{3}H-dihydroalprenolol binding sites in the nonischemic heart is decreased to 10% of control. On the other hand, the effect of catecholamines on myocardial contractility is diminished by only 50%. How do you explain this apparent dissociation?

Baumann: The decrease in contractile response of the heart to isoproterenol stimulation was nearly 60%. The dissociation observed may be due to a positive β-receptor-adenylate cyclase ratio, e.g. 2:1, which means that it is not necessary to occupy all β-receptors to get maximal stimulation of the adenylate cyclase. In addition, the isolated (isovolumetric) heart preparation may give better contractile results as compared to the physiological situation, in which hearts are really working with respect to increased pre- and afterload.

Herbaczynska-Cedro, Warsaw: While investigating the changes in blood catecholamine levels following acute myocardial infarction in dogs, we found that the main catecholamine released was adrenaline, which originated in the adrenal medulla. We have also looked at the relation between the activity of myocardial enzymes (SDH, ATPase) determined histochemically and the magnitude of blood adrenaline rise. In dogs with high circulating adrenaline the activity of these enzymes was low, also in the non-infarcted part of the heart. This might be relevant to your suggestion that high blood catecholamines in the early stage of infarction impair the sarcolemmal membrane and that this may lead to the decrease in the number of β-receptors which you have observed in your study.

Baumann: Blood concentrations of catecholamines alone are not representative for the concentration of catecholamines to which the myocardial cell is actually exposed. Especially in the early post-infarction period catecholamines are constantly released from the ischemic area. There is additional evidence indicating that catecholamines are released also from the uninvolved myocardium.

Starke, Freiburg: I have two points: 1. Could the loss of binding sites with little loss in sensitivity to isoproterenol be due to "spare receptors"?

2. Is a 90% loss in binding sites also demonstrable upon administration of exogenous β-agonists? 90% seems to be unusually high.

Baumann: 1. The term "spare receptor" is not well defined, and should therefore be avoided. If "spare receptors" are defined as preformed receptor structures present in the sarcolemmal membrane but not able to bind catecholamines for unknown reasons, I would probably agree.
2. Long-term infusion with dopamine (e.g. 200 μg/kg/h for 7 days) can induce similar changes as described here for endogenous catecholamines after myocardial infarction. On the average we observed a 60% loss in binding sites and in response. This discrepancy may indicate that the levels of catecholamines in the heart after myocardial infarction are even higher.

Erdmann, Munich: Dr. Starke has seduced you to accept his dangerous concept of "spare receptors" for β-agonists and β-antagonists. Please, could you tell me what "spare receptors" are and what is your evidence for the existence of "spare receptors"?

Baumann: I really don't want to favor the concept of "spare receptors" as outlined by Dr. Starke. The existence of "spare receptors" has never been proved as we all know. Nevertheless, this model may be helpful as a kind of working hypothesis. All the changes I talked about are reversible within 3 weeks, even if protein biosynthesis is pharmacologically blocked. This fact lends support to the hypothesis that performed receptor structures are present somewhere in the sarcolemma "in a masked form". But of course this is an unproved hypothesis.

Mattila, Helsinki: Metoprolol may have protected the cardiac tissue and its β-receptors against some toxic action, or it can have developed new active receptors from the spare pool. Which of these alternatives do you prefer?

Baumann: β-blockers certainly have a protective effect in so far as these compounds closely bind to the β-receptors and prevent prolonged exposure to excessively elevated plasma catecholamine concentrations. However, the other mechanism may also contribute to the effects observed.

Williams, Boston: To see a 90% loss of receptor number caused by down regulation due to chronic catecholamine exposure would be a unique finding. In most binding studies the decrement in β-receptor number is in the range of 40%–60%. Your finding of a 90% decrease in receptor number (making the Scatchard plot difficult to analyze) and a dramatic loss of receptor affinity raise the question of whether all of the endogenous catecholamine has been washed from your preparation at the time the binding studies were done.

Baumann: I cannot rule out this possibility completely, but to me it seems unlikely that endogenous catecholamines are still present in the membrane preparation, because the preparation procedure includes numerous washing and centrifugation steps. Binding of catecholamines to β-receptors is labile and repeated dilution results in rapid dissociation of the ligand-receptor complex. As I already pointed out, it should be considered that not all β-receptors linked to the adenylate cyclase system may be necessary for maximal activation.

Ziegelhöffer, Bratislava: I should like to support your idea. We have investigated the changes in myocardial cAMP level in dogs after acute coronary occlusion

(ACO). The results showed a rapid increase of cAMP in the non-infarcted parts of the left ventricle as a response to the release of endogenous catecholamines. A later slow and persistent increase which begins approximately 1 h after ACO, leads to a further elevation of cAMP. This increase was interpreted to be induced by exogenous plasma catecholamines liberated from the adrenal medulla. It is well known that cAMP induces a protein kinase-dependent phosphorylation of one or more regulative proteins in the membrane and via this enhances the Ca transport and the contractility of the heart. The phosphate bound is very stable and the phosphorylated state will be kept permanently near to maximum by high plasma levels of catecholamines. These effects together will explain the loss of heart adaptability as well as the beneficial effect of β-adrenoreceptor blocking agents, however, not the decrease in the number of receptors. This decrease measured by binding techniques may thus be regarded as an early result of changes in the reactivity, turnover and/or configuration of the catecholamine binding sites.

Baumann: I am somewhat happy to hear that you come to similar conclusions from your studies. The role of endogenous myocardial catecholamines should be further investigated, especially in the non-ischemic cardiac tissue after acute myocardial infarction. Besides catecholamines released from the adrenal medulla and the ischemic area most of the changes observed in our studies may be due to catecholamines directly released from the uninvolved myocardium, probably triggered by a myocardial reflex mechanism.

Heart Catecholamines in Genetic Hypertension

J. M. Saavedra, J. Fernandez-Pardal, P. Guicheney, J. B. Furness, R. McCarty, B. Rouot, F. M. A. Correa

Summary

Bilateral stellate ganglionectomy or peripheral administration of 6-hydroxydopamine resulted in an almost total loss of noradrenaline, but in only a partial decrease in dopamine and adrenaline content in the heart. Rats with bilateral adrenodemedullectomy showed only a slight decrease in the content of adrenaline in the heart. These observations suggest that most of the heart dopamine and adrenaline are formed in situ, outside of noradrenergic nerves.

Immobilization stress produced increased levels of dopamine and adrenaline, and a marked decrease in the heart noradrenaline content. The changes in catecholamine content after stress, as well as the basal values, were identical in spontaneously hypertensive rats (SHR) and their normotensive controls. An increase in all catecholamine levels was detected in the stellate ganglia of adult SHR.

Dahl, sodium sensitive (DS) rats fed a high salt diet showed decreased levels of noradrenaline and adrenaline in the heart, suggesting increased turnover and metabolism of the catecholamines in genetic, salt sensitive hypertension.

The number of β-adrenoceptors, and their affinity constants, were not changed in the hearts of genetically hypertensive rats.

No change was detected in the methylation of specific phospholipids by heart membranes after stress or after propranolol administration. The total methylation of chloroform-methanol extractable material, instead, was increased after immobilization stress or repeated administration of propranolol.

Our results suggest a role for dopamine and adrenaline in the heart and a possible relationship of these catecholamines with mechanisms operative in stress and genetic hypertension.

No correlation was found between the catecholamine turnover, the number of β-adrenoceptors and the methylation of specific endogenous phospholipids in heart membranes.

Introduction

Peripheral catecholamines play a role in the regulation of blood pressure and in the expression of some forms of hypertension [10, 23].

There are marked alterations in the metabolism of heart noradrenaline in at least one form of experimental hypertension, the DOCA-salt hypertensive rat. These alterations include a decrease in the content of noradrenaline in the heart and an increase in its turnover [9]. There is an inverse relationship between the degree of hypertension and the accumulation of exogenously administered noradrenaline in the heart of DOCA-salt rats [6]. In this model, the storage of heart noradrenaline is also related to the salt intake [8].

An involvement of heart catecholamines in the pathogenesis of genetic hypertension has also been postulated. In the spontaneously hypertensive rat, both catecholamine levels and the regulation of heart adenylate cyclase have been reported to be altered [1, 3, 20] but the molecular mechanisms involved in these alterations have not yet been clarified.

The stimulation of β-adrenoceptors by catecholamines, their linkage to adenylate cyclase, and the regulation of plasma membrane fluidity have been reported recently to be related to the degree of methylation of endogenous phospholipids in plasma membranes [11].

We have studied heart catecholamine levels, β-adrenoceptor number and affinity, and degree of phospholipid methylation in two models of genetic hypertension in the rat, the spontaneously (genetic) hypertensive rat (SHR) [27] and the genetic, salt sensitive hypertensive (*Dahl* salt sensitive, DS, and salt resistant, DR) rats, as well as the methylation of phospholipids in control rats, during stress and after chronic β-blockade. We report that changes in catecholamine metabolism occur in both models of hypertension, in the stellate ganglia in SHR and in the heart tissue itself in the Dahl salt sensitive hypertensive rats. No correlation exists, however, between the degree of hypertension, the level of heart catecholamines, the number and affinity of heart β-adrenoceptors, and the methylation of specific phospholipids by heart membranes.

Materials and Methods

Animals

Dahl Salt-Resistant (DR) and Salt-Sensitive (DS) Rats. Male Dahl salt-sensitive (DS) and salt-resistant (DR) rats [4] were obtained from Brookhaven National Laboratory, Upton, N.Y. Groups of 16 rats of each DS and DR lines were fed low sodium chloride chow (0.45% per dry weight) from weaning until 10 weeks of age. Similar groups of the DS and DR lines were fed high sodium chloride chow (8% per dry weight) (ICN Pharmaceutical, Cleveland, Ohio) over the same period. Tap water was provided for drinking ad lib.

Spontaneously Hypertensive Rats (SHR). Male SHR, 4 weeks and 14 weeks old, and age-matched normotensive rats of the Wistar-Kyoto substrain, from which SHR were derived [27] were maintained under identical conditions for one week after being received from Taconic Farms, Germantown, N.Y.

Control Rats. Sprague-Dawley male rats, weighing 200–250 g, were obtained from Taconic Farms, Germantown, N.Y. The animals were housed 6 per cage with alternate periods of light and dark (12 h light, 12 h dark, with lights on from 06.00 to 18.00) and were allowed free access to water and standard laboratory chow.

Pharmacological, Surgical and Physiological Manipulations

Surgical Procedures. Male Sprague-Dawley rats, weighing 300 g, were obtained from Zivic-Miller, Allison Park, Pa., after bilateral removal of the adrenal medulla or sham operation, and were kept under standard laboratory conditions for one week before killing. Male Sprague-Dawley rats, weighing 300 g, with bilateral stellate ganglionectomy and sham-operated controls, were obtained from Zivic-Miller, Laboratories, Allison Park, PA and were kept in the laboratory for two weeks following the heart denervation before killing.

Stress. Immobilization was used as the model of stress [16]. Rats were immobilized in a prone position by inserting their heads through steel wire loops fixed on a board and by fastening their limbs to 4 metal strips with adhesive tape. Acute stress was produced by immobilizing the animals for 150 min, once. Repeated stress was produced by immobilizing rats daily for 150 min with sacrifice on the seventh day, while still immobilized to the board. In each experiment, intact rats of the same age were used as controls. All animals were killed between 11 a.m. and 12 noon.

Administration of 6-Hydroxydopamine. Male Sprague-Dawley rats were injected intravenously with 200 mg/kg of 6-hydroxydopamine, two times, with a 3 day interval between injections, and were killed one week after the last injection.

Administration of Propranolol. Propranolol (1 mg/kg, i.p.) was administered twice daily at 12-h intervals for two weeks to a group of 16 Sprague-Dawley male rats. A similar group was used as control, and received an equivalent volume of saline (0.1 ml per 100 g weight, i.p.). After one week of treatment, both propranolol-treated and control groups were divided randomly into two groups of 8 animals each, and one of the subgroups was submitted to repeated immobilization stress, as described above.

Measurement of Blood Pressure. Systolic blood pressure was measured in unanesthetized rats by tail plethysmography, using a programmed electrosphygmomanometer (Narco Biosystems, Houston Texas, Model PE-500) as previously described [30].

Biochemical Determinations

Catecholamine Determinations. Rats were killed by decapitation at 9 a.m., and the stellate ganglia, and a 5 to 10 mg portion of the left atria and left ventricle were immediately removed and frozen on dry ice for catecholamine determinations. Tissue catecholamines were measured by a modification [29] of a radioenzymatic method [5], and corrected for the protein content of the tissues [22].

Determination of β-Adrenoceptor Binding in Heart Membranes. Hearts were removed immediately after decapitation, washed in ice-cold saline and the heart membranes prepared from the left and right ventricles according to the method of *Lefkowitz* and *Williams* (1977). β-Adrenoceptor binding sites and affinity constants were determined by incubating the membranes with [^{3}H]-dihydroalprenolol (DHA)

(New England Nuclear, Boston, Mass, specific activity 32.65 Ci/mmole) in absence and in presence of 10 μM non-radioactive alprenolol. The number and the affinity of specific [^{3}H]-DHA binding sites in myocardial membranes were calculated using Scatchard analysis [31] and the significance of the results was determined by the Wilcoxon paired test.

Determination of Chloroform-Methanol Extractable Methylated Products and Individual Methylated Phospholipids. The production of methylated phospholipids by incubation of a P_2 fraction of cardiac membranes in the presence of [^{3}H]-methyl-*S*-adenosyl-L-methionine (New England Nuclear, Boston, Mass.; specific radioactivity 12 Ci/mmol) was measured as reported by *Hirata* et al. [12] after isolation of the methylated phospholipids by thin layer chromatography. The total incorporation of [^{3}H]-methyl groups into chloroform-methanol extractable material was also studied [12].

Results

Effects of 6-Hydroxydopamine Administration, Surgical Denervation, and Adrenal Demedullation on Atrial Catecholamines

Dopamine, noradrenaline and adrenaline were measured in the left atria of Sprague-Dawley rats. Most of the catecholamines were present as noradrenaline, with dopamine and adrenaline representing not more than 2% of the total catecholamine content each (Fig. 1).

The administration of 6-hydroxydopamine resulted in an almost complete disappearance of noradrenaline in the atria, whereas a substantial amount of both dopamine and adrenaline remained present in the tissue ($P<0.01$, Student's *t*-test) (Fig. 1). A similar result was obtained after bilateral surgical removal of the stellate ganglia (Fig. 2) ($P<0.01$, Student's *t*-test). The removal of the adrenal medulla was followed by increased levels of atrial noradrenaline ($P<0.05$, Student's *t*-test), and only a partial decrease in atrial adrenaline ($P<0.05$, Student's *t*-test). Dopamine levels in the atria remained unchanged (Fig. 3).

Atrial Catecholamines and Heart β-Adrenoceptors in Spontaneously Hypertensive Rats

The systolic blood pressure as determined by the indirect method was: at 4 weeks of age, SHR 125 ± 5 vs WKY 110 ± 3 mmHg; at 14 weeks of age, SHR 180 ± 10 vs WKY 120 ± 10 mmHg. At both ages, the increased in blood pressure of SHR rats was statistically significant ($P<0.01$, Student's *t*-test).

Left atria catecholamine levels were examined in young (4 weeks old) and adult (14 weeks old) SHR and Wistar Kyoto normotensive controls. No differences in catecholamine content were noted either in young or adult hypertensive rats with respect to their age-matched normotensive controls (Table 1).

Acute immobilization stress produced significant increases in atrial dopamine and adrenaline, and a significant decrease in atrial noradrenaline. These changes were similar for both SHR and WKY adult rats (Fig. 4).

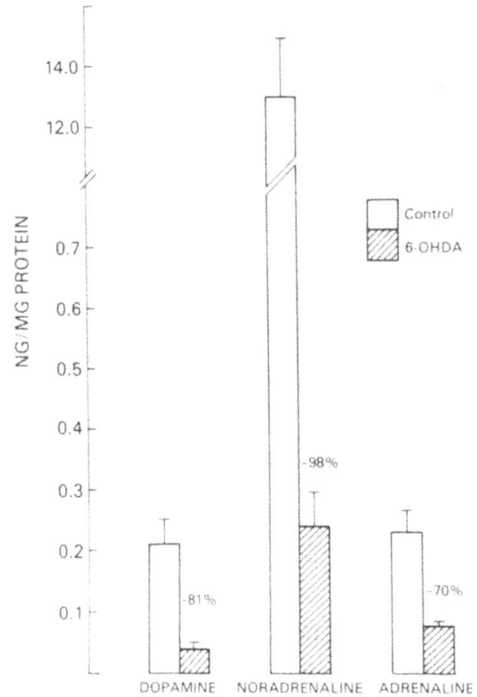

Fig. 1. Effect of 6-hydroxydopamine on left atria catecholamines in Sprague-Dawley rats

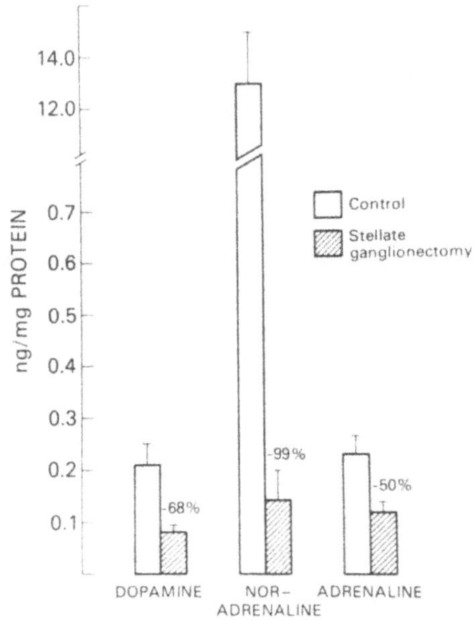

Fig. 2. Effect of surgical denervation on left atria catecholamines in Sprague-Dawley rats

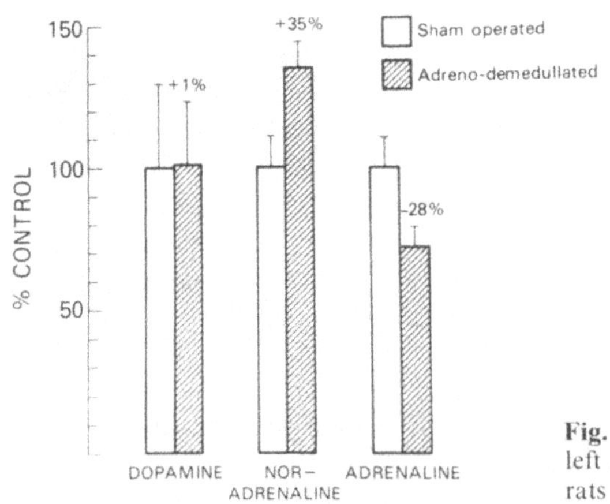

Fig. 3. Effect of adrenal medullectomy on left atria catecholamines in Sprague-Dawley rats

Table 1. Left heart atria catecholamines in SHR and WKY

Age = 4 weeks	Dopamine	Noradrenaline	Adrenaline
SHR	0.36 ± 0.03	16.79 ± 1.10	0.25 ± 0.01
WKY	0.27 ± 0.04	14.73 ± 1.31	0.21 ± 0.02
Age = 14 weeks			
SHR	0.20 ± 0.02	11.58 ± 0.50	0.12 ± 0.01
WKY	0.19 ± 0.01	12.08 ± 0.51	0.14 ± 0.01

Results are expressed as ng/mg protein, $\bar{x} \pm$ S.E.M., for groups of 11 rats, assayed individually.

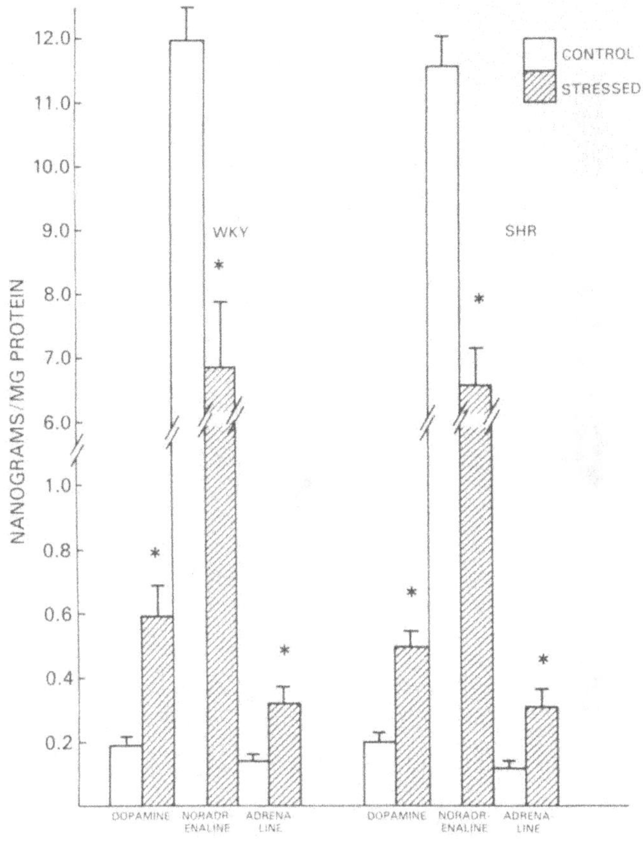

Fig. 4. Atrial catecholamines in hypertensive and stressed rats. * $P<0.05$; Student's t-test

Both the β-adrenoceptor number and affinity constants were not significantly different in the SHR when compared to their normotensive controls, either in young or in adult rats. Number of sites were for young, 4 weeks old rats: SHR 80.0 ± 10.1 vs WKY 68.0 ± 8.0 picomoles/g protein; for adult, 14 weeks old rats: SHR: 73.4 ± 5.0 vs WKY 64.0 ± 1.0 picomoles/g protein. The affinity constants (Kd) were for young rats: SHR 2.8 ± 0.4 vs WKY 2.7 ± 0.3 nM; for adult rats were: SHR 3.40 ± 0.4 vs WKY 3.0 ± 0.3 nM.

Stellate Ganglia Catecholamines in Spontaneously Hypertensive Rats

The content of dopamine, noradrenaline and adrenaline in the stellate ganglia of 14 weeks old SHR was increased with respect to that of age-matched Wistar-Kyoto normotensive controls (+100%, +75% and +120%, respectively) (Fig. 5).

Heart Catecholamines and β-Adrenoceptors in Dahl Salt-Resistant and Salt-Sensitive Rats

The content of catecholamines in the left atria and left ventricles, the number of β-adrenoceptors and their affinity constants (Kd) were determined in Dahl salt

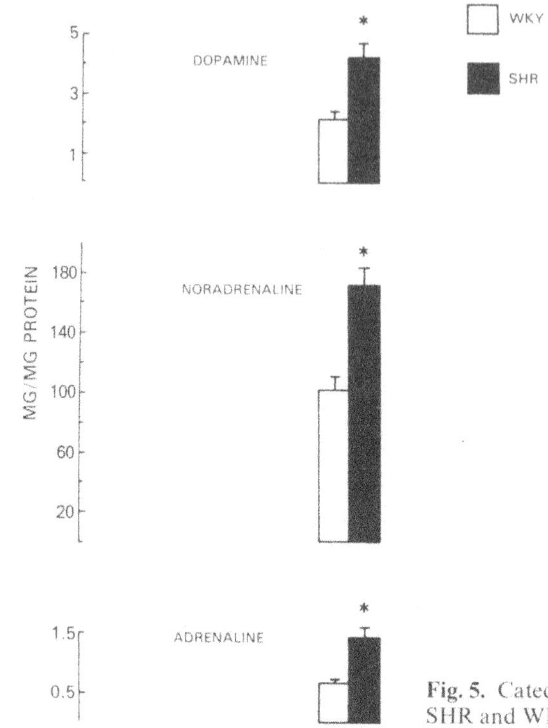

Fig. 5. Catecholamines in stellate ganglia of adult SHR and WKY rats. * $P < 0.01$; Student's t-test

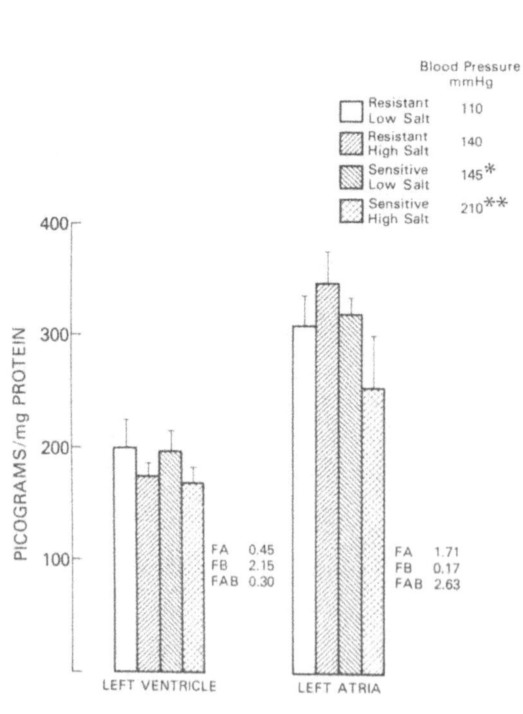

Fig. 6. Dopamine in the heart of Dahl rats. F_A = DR vs DS (lines); F_B = high salt vs low salt (treatment); F_{AB} = interaction; * $P < 0.05$, ** $P < 0.01$

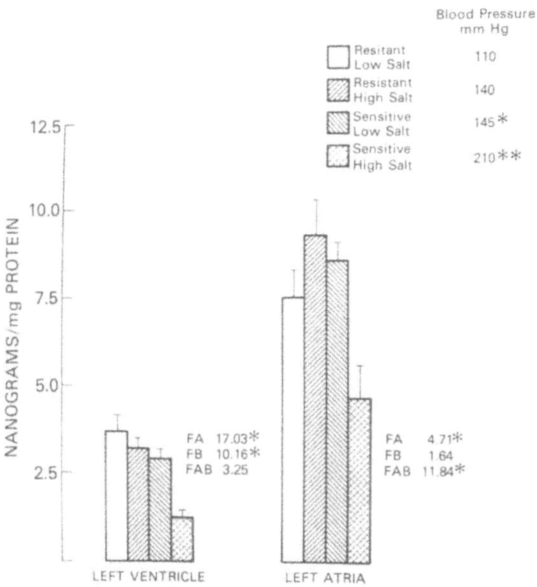

Fig. 7. Noradrenaline in the heart of Dahl rats. F_A = DR vs DS (lines); F_B = high salt vs low salt (treatment); F_{AB} = interaction; *P<0.05; ** P<0.01

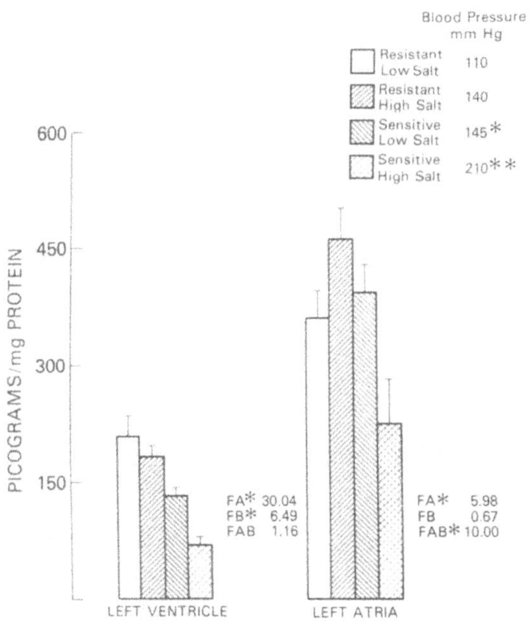

Fig. 8. Adrenaline in the heart of Dahl rats. F_A = DR vs DS (lines); F_B = high salt vs low salt (treatment); F_{AB} = interaction, *P<0.05; ** P<0.01

resistant (DR) and salt sensitive (DS) rats, maintained on a low (0.4%) or high (8%) salt diet. The blood pressures were as follows: DR, low salt; 110 mmHg; DR, high salt, 140 mmHg; DS, low salt, 145 mmHg; and DS, high salt, 210 mmHg. Statistical significance: DR low salt < DR high salt = DS low salt < DS high salt (P < 0.01). (Two-way ANOVA and Student's Newman-Keuls test for comparison of individual means) [32].

No significant differences were found for dopamine concentrations in either the left atria or left ventricle of Dahl rats (Fig. 6).

On the contrary, significant differences were detected both in the left atria and the left ventricles of Dahl rats, for noradrenaline and adrenaline (Figs. 7, 8).

In the left atria, a high salt diet affected the two lines differently, and there was evidence of a line × salt interaction (Figs. 6, 7). DS rats fed a high salt diet had a substantial decrease in their noradrenaline and adrenaline concentrations, whereas DR rats fed a high salt diet showed a tendency for an increase in their catecholamine levels (Figs. 7, 8).

Both the number of β-adrenoceptor sites and the affinity constants (Kd) for [^{3}H]-dihydroalprenolol binding were similar in the DR rats fed low or high salt diets, and in the hypertensive DS rats fed a high salt diet (number of sites, DR low salt, 74 ± 10; DR high salt, 72 ± 8; DS high salt, 72 ± 8, DS low salt, 59 ± 6 picomole/g protein; Kd: DR low salt 5.0 ± 1.0; DR high salt 4.96 ± 1.2; DS high salt 4.80 ± 0.9; DS low salt 3.4 ± 1.4 nM). Both the number of β-adrenoceptor sites and the affinity constants were lower in DS fed a low salt diet, but the results did not attain statistical significance.

Methyl Incorporation into Chloroform-Methanol Extractable Material from Heart Membranes

Both the administration of propranolol for 15 days, or the submission to immobilization stress, daily for one week, resulted in a marked increase in the incorporation of [^{3}H]-methyl groups from [^{3}H]-methyl-*S*-adenosyl-L-methionine into chloroform-methanol extractable material, after incubation with heart membranes (Fig. 9).

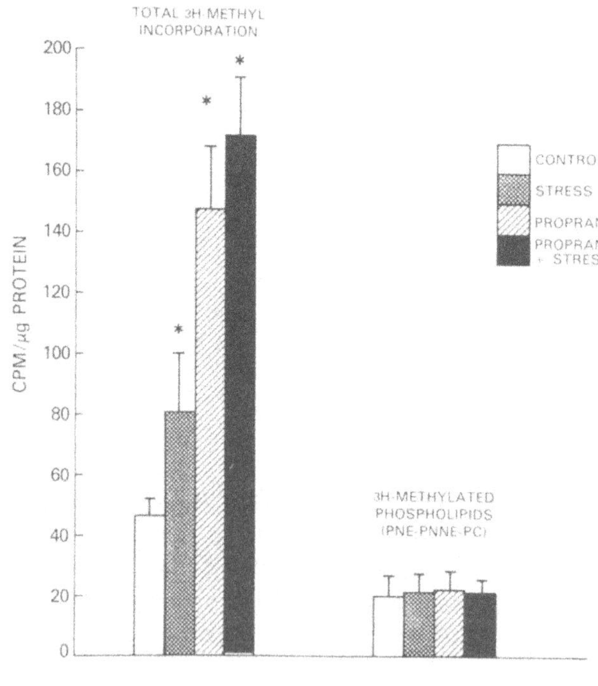

Fig. 9. Methylation of chloroform-extractable material by heart membranes after stress and propranolol administration, * $P < 0.05$

In contrast, none of the previously mentioned treatments resulted in a change in methylation of the specific phospholipids, *N*-methyl phosphatidylethanolamine, *N,N*-dimethylphosphatidylethanolamine or phosphatidylcholine (Fig. 9).

Discussion

Dopamine, noradrenaline and adrenaline were detected in the left heart atria of the rat. In this tissue, the relative concentrations of dopamine and adrenaline are small, not more than 4% of the total catecholamine content each.

Surgical denervation or peripheral administration of 6-hydroxydopamine, which resulted in an almost total disappearance of heart noradrenaline, only produced a partial decrease in dopamine and adrenaline, indicating that these two catecholamines are located mainly outside of sympathetic nerves. Removal of the source of peripheral adrenaline, the adrenal medulla, resulted in only a small decrease in heart adrenaline and no change in the dopamine content, indicating that adrenaline could also be formed in situ in the heart tissue, and not merely be taken up from the circulation. The presence in heart of the adrenaline-forming enzyme phenylethanolamine *N*-methyltransferase, PNMT (E.C. 2.1.1.28) [2] further supports this hypothesis.

As to the cellular localization of heart adrenaline, the amine could be located in the small intensive fluorescent (SIF) cells or chromaffin cells, which have been described to be present in clusters within the atrial tissue [14, 15, 28, 18, 33]. Alternatively, the small remaining amounts of adrenaline, and also of noradrenaline and dopamine present in the heart after denervation could be partially located in cardiac muscle cells and could represent a transient accumulation from the circulation, as it might be the case in the denervated intestine of the guinea-pig [13].

No change in left atria catecholamines was detected in either young or adult SHR when compared with their normotensive controls. These results are in agreement with prior observations [20] and contradict observations by other groups, which reported a decreased content of catecholamines in this model [26].

Both in hypertensive and normotensive rats, immobilization stress results in a substantial decrease in heart noradrenaline levels, indicating probable increased release and metabolism of this catecholamine, but in increased levels of dopamine and adrenaline. It is thus tempting to speculate that the synthesis of dopamine and adrenaline may increase after stress, and that these two catecholamines might have roles of their own in the heart under stress conditions.

Immobilization stress produced identical changes in heart catecholamines in the left atria of hypertensive and normotensive rats, indicating that the metabolism of heart catecholamines, both under basal conditions and under stress, may not be different in spontaneously hypertensive rats with respect to their normotensive controls.

In addition to the lack of changes in tissue levels of catecholamines, no statistically significant differences in number of β-adrenoceptor binding sites or in their affinity constants were detected in SHR. Our observations are in contradiction to those of *Limas* and *Limas* [19] who reported a decreased number of β-adrenoceptors in SHR, but are in agreement with recent results of *Mukherjee* et al. [23] in the same model.

The levels of stellate ganglia catecholamines were increased substantially in adult, hypertensive SHR. *Nakamura* and *Nakamura* have reported increased dopamine-β-hydroxylase activity in the stellate ganglia of young SHR [25]. Our observations indicate that there is also an increased synthesis of catecholamines in this ganglion during the phase of established hypertension in SHR. As to the source of dopamine and adrenaline in the ganglia, the most likely candidates are, as in atrial tissue, the SIF cells [15].

In contrast to the SHR, the genetic, salt sensitive hypertensive (Dahl, DS rats) showed marked changes in heart noradrenaline and adrenaline. A similar effect of high salt diet in heart catecholamines was noted before in DOCA-salt hypertensive rats [8]. Our results and those of *de Champlain* and coworkers demonstrate that the capacity of the sympathetic granules to bind and store noradrenaline can be influenced by the state of sodium balance. Our results further suggest that this regulatory property of the sodium balance could be extended to heart adrenaline, but not to heart dopamine. Furthermore, our data demonstrate the existence of a genetic component which determines the response of heart catecholamine metabolism to a salt load, since no decrease in heart catecholamines was seen in the salt-resistant, normotensive Dahl (DR) rats. Thus, in this model, the level of heart catecholamines can be correlated inversely with the level of blood pressure, as was the case in the DOCA-salt hypertensive rats [8].

Consistent with our findings in the SHR rat, no change in number of sites or affinity of β-adrenoceptors was detected in Dahl rats, indicating a lack of correlation between hypertension, catecholamine levels and β-adrenoceptor function in the heart.

No differences in the methylation of endogenous phospholipids by heart membranes could be detected after repeated stress, which results in a persistent, several-fold increase in plasma catecholamines, and in a substantial acceleration of noradrenaline turnover in the heart, or after propranolol administration, at a dose which results in blockade of β-adrenoceptors in the heart. Also, no differences in phospholipid methylation were found in heart membranes from SHR and Dahl salt-sensitive rats when compared to their corresponding controls (unpublished results). These results suggest that under the present experimental conditions, no correlation exists between catecholamine metabolism, β-adrenoceptor stimulation and phospholipid methylation in the heart. Notwithstanding, a substantial increase in total methyl group incorporation to chloroform-methanol extractable material was found after both repeated stress and propranolol administration. Some of the products formed under these circumstances can be temptingly identified as fatty acid methyl esters, which are also found in lung and other tissues in substantial amounts after incubation of plasma membranes with *S*-adenosyl-L-methionine [34]. The role of fatty acid methyl esters in the heart is presently not well understood.

Our results strongly suggest a role for endogenous dopamine and adrenaline, as well as noradrenaline in the rat heart. Our findings also demonstrate that heart catecholamines are involved in genetic hypertension and that their metabolism can be modulated by the amount of salt in the diet, provided the necessary genetic factors are present.

The absence of a significant correlation between the level of catecholamines, the number of β-adrenoceptors, and the methylation of endogenous phospholipids by heart membranes is not surprising, giving the heterogeneity and complexity of the different mechanisms studied. The simultaneous study of catecholamine turnover, the level of their metabolites, the subclasses of adrenoceptors, and the specific methylated lipids found could result in a better understanding of the molecular mechanisms involved in experimental and genetic models of hypertension.

Acknowledgments. The authors wish to thank Mr. Lester Carmon, Ms. Antonia del Carmine and Mrs. Dorothy Rutherford for their technical assistance and Roberta L. Holcomb and Patricia Ingalls for their editorial assistance. P. G. is indebted to the Hoffmann-La Roche Foundation, Nutley, New Jersey, for partial financial support. J. B. F. was supported by a Fulbright Fellowship.

References

1. Amer MS, Gomoll AW, Perhach JL, Ferguson HC, Mc Kinney GR (1974) Aberrations of cyclic nucleotide metabolism in the hearts and vessels of hypertensive rats. Proc Natl Acad Sci USA 71:4930–4934
2. Axelrod J (1962) Purification and properties of phenylethanolamine-N-methyltransferase. J Biol Chem 237:1657–1660
3. Chalmers JP, Howe PRC, Provis JC, West MJ (1979) Changes in cardiac norepinephrine in spontanteously hypertensive and stroke-prone rats. In: Usdin E, Kopin IJ, Barchas J (eds) Catecholamines: basic and clinical frontiers. Pergamon, New York, pp 776–778
4. Dahl LK, Heine M, Tassinari L (1962) Role of genetic factors in susceptibility to experimental hypertension due to chronic excess salt ingestion. Nature 194:480–482
5. Da Prada M, Zürcher B (1976) Simultaneous radioenzymatic determination of plasma and tissue adrenaline, noradrenaline and dopamine within the femtomole range. Life Sci 19:1161–1174
6. de Champlain J, Krakoff LR, Axelrod J (1966) A reduction in the accumulation of ^{3}H-norepinephrine in experimental hypertension. Life Sci 5:2283–2291
7. de Champlain J, Krakoff LR, Axelrod J (1967) Catecholamine metabolism in experimental hypertension in the rat. Circ Res 20:136–145
8. de Champlain J, Krakoff LR, Axelrod J (1968) Relationship between sodium intake and norepinephrine storage during the development of experimental hypertension. Circ Res 23:479–491
9. de Champlain J, Krakoff LR, Axelrod J (1969) Interrelationships of sodium intake, hypertension, and norepinephrine storage in the rat. Circ Res [Suppl I] 24/25:I/75–I/92
10. Grobecker H, Roizen MF, Weise V, Saavedra JM, Kopin KJ (1975) Sympathoadrenal medullary activity in young, spontaneously hypertensive rats. Nature 258:267–268
11. Hirata F, Axelrod J (1980) Phospholipid methylation and biological signal transmission. Science 209:1082–1090
12. Hirata F, Viveros OH, Diliberto EJ, Axelrod J (1978) Identification and properties of two methyltransferases in conversion of phosphatidylethanolamine to phosphatidylcholine. Proc Natl Acad Sci USA 75:1718–1721
13. Howe, PRC, Provis JC, Furness JB, Costa M, Chalmers JP (in press) Residual catecholamines in extrinsically denervated guinea-pig ileum. Clin Exp Physiol Pharmacol
14. Jacobowitz D (1976) Histochemical studies of the relationship of chromaffin cells and adrenergic nerve fibers to the cardiac ganglia of several species. J Pharmacol Exp Ther 158:227–240
15. Jacobowitz D (1970) Catecholamine fluorescence studies of adrenergic neurons and chromaffin cells in sympathetic ganglia. Fed Proc 29:1929–1944
16. Kvetnansky R, Mikulaj L (1970) Adrenal and urinary catecholamines in rats during adaptation to repeated immobilization stress. Endocrinology 87:738–743
17. Lefkowitz RJ, Williams LT (1977) Catecholamine binding to the beta-adrenergic receptor. Proc Natl Acad Sci USA 74:515–519
18. Libet B, Owmań Ch (1974) Concomitant changes in formaldehyde-induced fluorescence of dopamine interneurons and in slow inhibitory postsynaptic potentials of rabbit superior cervical ganglion, induced by stimulation of the preganglionic nerve or by a muscarinic agent. J Physiol (Lond), 237:635–662
19. Limas C, Limas CJ (1978) Reduced number of β-adrenergic receptors in the myocardium of spontaneously hypertensive rats. Biochem Biophys Res Commun 83:710–714

20. Louis WJ, Spector S, Tabei R, Sjoerdsma A (1969) Synthesis and turnover of norepinephrine in the heart of the spontaneously hypertensive rat. Circ Res 24:85–91
21. Louis WJ, Krauss KR, Kopin IJ, Sjoerdsma A (1970) Catecholamine metabolism in hypertensive rats. Circ Res 27:589–594
22. Lowry OH, Rosenbrough NJ, Farr AL, Randall RJ (1951) Protein measurement with the Folin phenol reagent. J Biol Chem 193:263–275
23. Mukherjee A, Graham RM, Sagalowsky AI, Pettinger W, McCoy K (in press) Myocardial beta-adrenergic receptors in the stroke-prone spontaneously hypertensive rat. J Mol Cell Cardiol
24. Nagatsu T, Kato T, Numata (Sudo) Y, Ikuta K, Umezawa H, Matsuzaki M, Takeuchi T (1974) Serum dopamine-β-hydroxylase activity in developing hypertensive rats. Nature 251:630–631
26. Nakamura K, Gerold M, Thoenen H (1971) Genetically hypertensive rats: relationship between the development of hypertension and the changes in norepinephrine turnover of peripheral and central adrenergic neurons. Naunyn-Schmiedebergs Arch Pharmacol 271: 157–169
25. Nakamura K, Nakamura K (1977) Selective activation of sympathetic ganglia in young spontaneously hypertensive rats. Nature 266:265–266
27. Okamoto K, Aoki K (1963) Development of a strain of spontaneously hypertensive rats. Jpn Circ J 27:282–293
28. Partanen S, Korkala O (1974) Catecholamines in human fetal heart. Experientia 30: 798–800
29. Saavedra JM (1979) Microquantitation of neurotransmitters in specific areas of the central nervous system. Int Rev Neurobiol 21:259–274
30. Saavedra JM, Grobecker H, Axelrod J (1978) Changes in central catecholaminergic neurons in the spontaneously (genetic) hypertensive rat. Circ Res 42:529–534
31. Scatchard B (1949) The attraction of proteins for small molecules and ions. Ann NY Acad Sci 51:660–672
32. Sokal RR, Rohlf FS (1969) *Biometry,* Freeman, San Francisco
33. Williams TH, Black AC Jr, Tanemichi C, Jew JY (1977) Species differences in mammalian SIF cells. Adv Biochem Psychopharmacol 16:505–511
34. Zatz M, Dudley P, Markey S (in press) Biosynthesis of methylated nonpolar lipids, including fatty acid methyl esters, by rat lung membranes. J Biol Chem

Discussion

Starke, Freiburg: What is the source of adrenaline in the heart? Could it be the system of chromaffin cells?

Saavedra: Our results suggest that an important part of cardiac adrenaline originates in the heart tissue itself, since adrenaline levels in the heart decrease very little after removal of the adrenal medulla. Denervation of the heart also results in only a partial decrease of adrenaline content. Moreover, the adrenaline-forming enzyme, phenylethanolamine N-methyltransferase (PNMT) is present in the heart. These data suggest that heart adrenaline mainly originates locally in places other than the sympathetic nerves.

The most likely candidates for adrenaline formation in the heart, as you suggested, are the chromaffin cells, or small intensive fluorescent (SIF) cells, which belong at least to two types. One type receives direct sympathetic innervation from the stellate ganglia; the other type forms clusters surrounding the small vessels in the atrial tissue, and may be of neurosecretory nature. Whether adrenaline could actually be released into the coronary circulation or could act in situ as a local hormone is not known. The precise localization of adrenaline in specific cells in the heart will require a combination of immunohistofluorescent techniques to establish the simultaneous presence of tyrosine hydroxylase, dopamine-β-hydroxylase and PNMT in the same cell.

Szekeres, Szeged: Which of your findings could be related to the pathomechanism of hypertension?

Saavedra: The changes in the metabolism of catecholamines in the heart of salt sensitive, hypertensive Dahl rats and the increased formation of catecholamines in the stellate ganglia of spontaneously hypertensive rats (SHR) are clearly related to the pathomechanism of hypertension. What is not known is whether these changes represent a causal mechanism or a secondary, adaptive phenomenon, and that is difficult to establish. To fully answer your question, a more detailed developmental study of these changes will have to be performed.

Schrader, Munich: We have recently carried out experiments in isolated hearts in which we have inhibited cellular transmethylation by increasing tissue content of S-adenosyl-homocysteine (infusion of adenosine and L-homocysteine). Under these conditions, the contractile response of the heart to catecholamines was essentially unaltered. In our opinion these findings indicate that there is no immediate influence of methylation on cardiac effects of catecholamines.

Saavedra: We could not find any changes of total ^{3}H-methyl group incorporation into chloroform-methanol extractable compounds after acute (4 h) immobilization stress, a situation which results in a several-fold increase in plasma catecholamine levels. These results agree well with yours, and suggest that there is no relationship between the acute effects of catecholamines and methylation in the heart. The changes which we report correspond to repeated (during 7 days) stress, and we could not show any changes in phospholipid methylation in heart membranes, but an increase in the methylation of compounds which we could now tentatively identify as fatty acid methyl esters. The role of this increased fatty acid methyl ester formation during stress is not known.

Palm, Frankfurt: What is the meaning of your results concerning chronic treatment with propranolol on total methylation of heart membranes? Obviously it is the same effect that is produced by chronic stress and I think you will get the same results when you treat your animals with diazepam.

Saavedra: We do not know the meaning of the changes observed after propranolol administration. We will have to identify all the compounds formed before advancing a hypothesis. Presently, we think that most if not all of the increase in methylation in heart membranes after propranolol or stress is due to increased fatty acid methyl ester formation.

We agree with you on the possibility of obtaining similar results after administration of diazepam. Preliminary experiments show that, at least in the adrenal cortex, administration of pentobarbital results in increased methyl incorporation into chloroform-methanol extractable material.

In any case, the results obtained, contrary to our original postulate, are clearly not related to β-adrenoceptor stimulation, and we have failed in our experimental model to demonstrate a correlation between catecholamine metabolism and release, β-adrenoceptor stimulation, and lipid methylation in the heart.

II. Biochemical and Physiological Implications of Adrenergic Receptor Stimulation

Principles of Signal Transformations Mediated by Subtypes of Adrenergic Receptors

G. Schultz, K. H. Jakobs

Summary

The current knowledge on the cellular signal transformations by subtypes of adrenergic receptors is presented. Stimulation of either β-adrenergic receptor, i.e., β_1 or β_2, causes an increase in the cellular concentration of cyclic AMP via stimulation of the adenylate cyclase. An opposite change in the cellular cyclic AMP concentration via inhibition of the adenylate cyclase appears to be the first intracellular signal caused by stimulation of α_2-adrenergic receptors. The coupling of β-adrenergic receptors to adenylate cyclase is mediated by a regulatory guanyl nucleotide-binding protein, whose interactions with the receptor and the enzyme are relatively well characterized. The molecular mechanisms of the α_2-adrenergic signal transformation, which also appears to be mediated by a guanyl nucleotide-binding component, are less clear and may involve an accelerated inactivation of the adenylate cyclase by an increased GTPase activity. Stimulation of α_1-adrenergic receptors causes several changes possibly involved in the final cellular response. Among these, the increase in the cytoplasmic calcium concentration is the most important signal, which may be triggered by an increased membrane phosphatidylinositol turnover, while the increase in cellular cyclic GMP levels appears to be a consequence of the increased calcium concentration and other cellular changes yet not fully understood. The relevance of these three main cellular signal-generating systems for the cellular regulation via other than adrenergic hormone and neurotransmitter receptors and receptor subtypes is discussed.

Introduction

Specialized cellular functions are precisely regulated by neuronal, endocrine and local control mechanisms, involving various intercellular signals. Regulations by rapidly acting hormones and neurotransmitters involve membrane-bound receptors. These cellular components recognize and reversibly bind the intercellular signals and induce the formation of intracellular signals mediating the cellular response to the extracellular stimulus. For several hormonal factors, the occurrance of receptor subtypes has recently been recognized, whose stimulations cause generations of different intracellular signals. The coupling of receptor subtypes to different cellular signal-generating components has most clearly been demonstrated for adrenergic receptors, which we like to discuss as examples of cellular signal-transforming components in the following.

Intracellular Signals Involved in the Mediation of the Effects of Different Adrenergic Receptor Subtypes

The physiologically occurring catecholamines, noradrenaline and adrenaline, can interact with four different receptor types, classified as α_1, α_2, β_1 and β_2, according

Table 1. Cellular changes induced by stimulation of different adrenergic receptor subtypes

	α_1	α_2	β
Cytoplasmic Ca^{2+} concentration	↑		
Phosphatidylinositol turnover	↑	–	–
Cyclic GMP concentration	↑	–	–
Cyclic AMP concentration	(↓)	↓	↑
Adenylate cyclase activity	–	↓	↑

to specificities and different affinities of various (synthetic) agonists and antagonists. Stimulation of either β-adrenergic receptor, i.e., β_1, occurring predominantly in the heart, and β_2, occurring predominantly in vascular, bronchial and uterine smooth muscle, causes an increase in the intracellular concentration of cyclic AMP (Table 1). An opposite change in the cyclic AMP concentration is the solely known intracellular signal caused by stimulation of α_2-adrenergic receptors, occurring presynaptically in adrenergic and cholinergic nerve endings and postsynaptically in various cell types, e.g., platelets and fat cells. Stimulation of α_1-adrenergic receptors, occurring in vascular smooth muscle, the myocardium and hepatocytes, causes several cellular changes which appear to be involved in the cellular regulation. The main intracellular signal is an increase in the cytoplasmic calcium concentration, which is accompanied by changes in phosphatidylinositol metabolism and a rise in the cyclic GMP concentration.

Stimulation of Cyclic AMP Formation by β-Adrenergic Agonists

The molecular mechanisms involved in stimulation of cyclic AMP formation by β-adrenergic agonists are relatively well characterized [6, 14, 20–22, 26, 27]. Studies in avian erythrocytes and several cultured cell lines have demonstrated that the hormone-sensitive adenylate cyclase is a multi-component system, which is composed of at least three membranous proteins: the β-adrenergic receptor, the

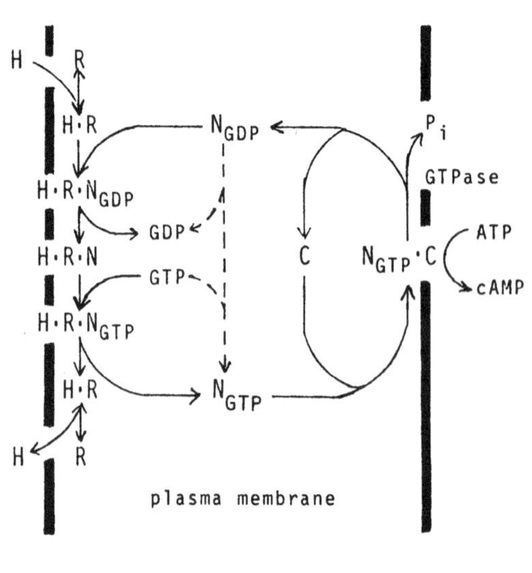

Fig. 1. Hormonal regulation of adenylate cyclase activity. *H*, hormonal agonist; *R*, hormone receptor (e.g., β-adrenergic receptor); *N*, guanyl nucleotide-binding regulatory protein; *C*, catalytic subunit (adenylate cyclase). For details see text

catalytic unit and a guanyl nucleotide-binding regulatory component. It appears that these proteins are separate entities floating within the bilayer of the plasma membrane with relative freedom and that collisions of β-adrenergic receptor, the regulatory component and the enzyme are the rate-limiting factor for the coupling process. Not very much is known about the stoichiometry of the individual components of the system in different cell types.

The sequence of events involved in stimulation of adenylate cyclase by β-adrenergic agonists appears to be as follows (Fig. 1): No evidence has as yet been presented for a direct interaction of hormone receptor and catalytic subunit. Instead, when agonist (H) is bound to the β-adrenergic receptor (R), the (low affinity) $H \cdot R$ complex first binds to the regulatory guanyl nucleotide-binding protein (N), and a ternary (high affinity) complex ($H \cdot R \cdot N$) is formed. In the absence of hormone, GDP is bound to N; the binding of $H \cdot R$ to N promotes a release of the bound GDP from N and a subsequent binding of GTP to N. The formation of N_{GTP} appears to induce the dissociation of $H \cdot R$, which is a low affinity hormone receptor complex. The GTP-loaded regulatory protein interacts with the catalytic subunit (C), i.e., the adenylate cyclase, which by itself is relatively inactive, to form the active adenylate cyclase complex ($N_{GTP} \cdot C$). The rapid decay of this enzymatically active complex involves the GTPase activity of the guanyl nucleotide-binding regulatory component. Hydrolysis of bound GTP to GDP and inorganic phosphate is followed by dissociation of N_{GDP} and inactive C. Thereby, the hormone-induced activation of the adenylate cyclase is terminated until N_{GTP} is regenerated by hormone-receptor interaction. When an antagonist is bound to the β-adrenergic receptor, the antagonist-receptor-complex does not interact with N and, therefore, also neither N_{GTP} nor the active $N_{GTP} \cdot C$ complex are formed.

This model of β-adrenergic agonist-induced activation of adenylate cyclase is based on and reflected in several typical findings in studies with membrane preparations. (a) GTP decreases the apparent affinity of the β-adrenergic receptor for agonists but not for antagonists. (b) For activation of the adenylate cyclase, the presence of GTP in addition to that of a hormone agonist is required. (c) When GTP analogues, stable to hydrolysis, such as guanylyl-5′-imidodiphosphate or guanosine-5′-O-(3-thiotriphosphate) are used, the hormone-stimulated adenylate cyclase is accumulated in its active form. (d) Pretreatment of the membrane preparation with cholera toxin, catalyzing ADP-ribosylation of the guanyl nucleotide-binding and -degrading component, largely reduces the GTPase activity and, thereby, also leads to the accumulation of the hormone-stimulated adenylate cyclase in its active form.

The regulatory protein was recently purified from rabbit liver to apparent homogeneity and shown to have a molecular weight of about 130 000 and to consist of three subunits of about 35 000, 45 000 and 52 000 daltons [25]. The 45 000- and 52 000-dalton proteins are ADP-ribosylated by cholera toxin. The relative contribution of the three polypeptides to the activity of the regulatory protein are not known. At least in some cellular systems, calmodulin appears to be involved as an additional component in hormonal regulation of adenylate cyclase activity. Calcium by interaction with calmodulin can enhance hormonal stimulation of adenylate cyclase in some neuronal tissues [5]. The interaction of calcium-calmodulin with the other components of the adenylate cyclase system is far form being clear.

Whereas in almost all cases the effects of β-adrenergic receptor activation are attributable to increased adenylate cyclase activity and, thereby, increased cellular cyclic AMP concentrations, in at least two cases no direct connection between increased cyclic AMP levels and cellular changes could be demonstrated. These are the inhibition of magnesium transport in S 49 lymphoma cells [23] and increased methylation of membrane phospholipids in several cell types [13] induced by β-adrenergic agonists. These responses may be mediated by a completely different effector pathway, which is independent of adenylate cyclase, but they may also be related to early steps in adenylate cyclase activation by β-adrenergic agonists. Similar to adenylate cyclase activation, these membrane effects are regulated by GTP and, as shown for the magnesium transport, involve the guanyl nucleotide-binding regulatory protein.

Inhibition of Cyclic AMP Formation by α_2-Adrenergic Agonists

As most other hormonal factors lowering intracellular cyclic AMP levels, α_2-adrenergic agonists have been shown in several cell types (human platelets, hamster fat cells, rat hepatocytes and others) to inhibit adenylate cyclase. During the last few years, hormonal inhibition of the enzyme in membrane preparations has intensively been characterized [15–18] (Table 2). Inhibition occurred without significant lag phase and was reversed by appropriate antagonists without substantial delay. Basal and hormonally stimulated forms of the enzyme were inhibited by maximally 50% to 80%. Maximally inhibitory effects of different hormonal factors were not additive. Similar to adenylate cyclase stimulation by β-adrenergic agonists, inhibition of the enzyme by α-adrenergic agonists and other hormonal factors depended on the presence of GTP; the concentrations of GTP required for inhibition were generally higher than those necessary for stimulatory hormonal effects. In addition to GTP, monovalent cations favored hormone-induced inhibition of adenylate cyclase. In some systems, e.g., adipocytes, the presence of monovalent cations was absolutely required. Na^+ was most potent, followed by $Li^+ > K^+ >$ choline. In the presence of GTP, which inhibits the enzyme in the

Table 2. Effects of β- and α_2-adrenergic agonists on adenylate cyclase activity and modulations of agonist-receptor binding

	Stimulatory hormone (β)	Inhibitory hormone (α_2)
Effect on adenylate cyclase		
Onset/reversal	Fast	Fast
Forms affected	Basal and hormone-inhibited	Basal and hormone-stimulated
Maximal effect	Several-fold	50%–80%
GTP	Required	Required (higher concentration)
Cholera toxin + GTP	Synergistic	Permit inhibition
Stable GTP analogues	Synergistic	Prevent or reverse inhibition
Optimal Na^+ concentration	Low	High (30–200 mM)
Agonist-receptor binding		
Modulation by GTP	+	+
Modulation by Na^+	(+)	+
Modulation by Mg^{2+}	+	+

adipocyte system, Na^+ caused a concentration-dependent adenylate cyclase activation which was reduced or prevented by α-adrenergic agonists and other antilipolytic hormonal factors. Whereas in other systems, e.g., myocardium, Na^+ amplified inhibitory effects, no clear effect of Na^+ was demonstrable in human platelet membranes. Similar to β-adrenergic receptors, binding of agonists but not antagonists to α_2-adrenergic receptors is modulated by GTP and other guanyl nucleotides. Additionally, the hormone-receptor binding is regulated by monovalent cations such as Na^+ and Li^+ [14].

The molecular mechanism, by which α-adrenergic agonists and other hormonal factors inhibit adenylate cyclase, is less clear than the mechanism of β-adrenergic stimulation of the enzyme. The turn-on reaction, i.e., the formation of the active form of the adenylate cyclase, was not retarded by inhibitory hormones. Several findings support the assumption that inhibitory hormonal factors accelerate the turn-off reaction, i.e., the rate of decay of the active adenylate cyclase to its inactive form [19]. Inhibition of the adenylate cyclase by hormones was prevented or reversed in the presence of a stable GTP analogue, which can cause persistent activation of the enzyme. Even at early time points when stable GTP analogues (like GTP) inhibit adipocyte adenylate cyclase, no hormonal inhibition was observed [19]. In contrast, the stimulatory effect of cholera toxin, causing ADP-ribosylation of the guanyl nucleotide-binding component and reduction of the GTPase activity and, thereby, accumulation of the adenylate cyclase in its active form, was counteracted by α-adrenergic agonists and other inhibitory hormonal factors. The concept that inhibitory hormonal factors promote the inactivation of the enzyme is supported by findings in hamster adipocytes, where Na^+ reduced the turn-off reaction and inhibitory hormonal factors accelerated this reaction [18]. Recent determinations of the specific GTPase apparently involved in the inactivation process indicate that inhibitory hormonal factors stimulate this enzyme in membrane preparations from hamster adipocytes [2] and human platelets [1].

Although various data support the concept that stimulatory hormonal factors promote the activation whereas inhibitory hormones promote the inactivation of adenylate cyclase, the basis of different coupling of β- and α_2-adrenergic receptors to the enzyme is not at all understood. If in either effect on adenylate cyclase activity the same regulatory guanyl nucleotide-binding component is involved, it is not clear why the active cyclase complex, $N_{GTP} \cdot C$, is more rapidly inactivated (by the GTPase activity of N) in the presence of an agonist bound to the α_2-adrenergic receptor than with an agonist bound to the β-adrenergic receptor. Thus, it is possible that different stimulatory and inhibitory regulatory components (N_s and N_i) [15, 26] are involved in coupling of β- and α_2-adrenergic receptors, respectively, to the adenylate cyclase. N_i could be composed in part of the regulatory protein involved in adenylate cyclase stimulation, which has been shown to be a multi-component system [25].

Cellular Signal Generation Stimulated by α_1-Adrenergic Agonists

Stimulation of α_1-adrenergic receptors promotes several changes that may serve as intracellular signals between membrane-bound receptors and the controlled cellular function(s). Based on studies in a variety of tissues, it is now generally accepted that an increase in the cytoplasmic calcium concentration represents the main intracel-

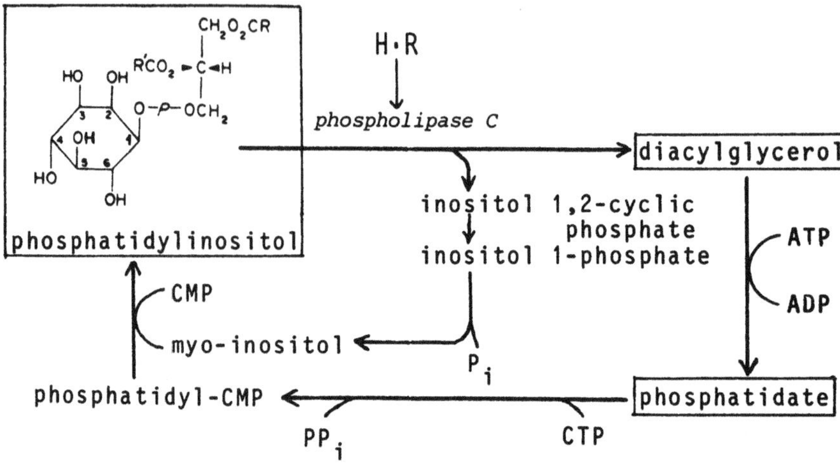

Fig. 2. Phosphatidylinositol metabolism: Stimulation by α_1-adrenergic agonists and other hormonal factors

lular signal [3, 4, 9, 24]. Several inactive calcium pools may contribute to the influx of this ion into the cytosol in a tissue-dependent manner. Besides the extracellular space, intracellular storage sites such as mitochondria, sarcoplasmic reticulum, various vesicles and membrane phospholipids are possible sources of the hormone-induced increase in cytoplasmic calcium. The molecular mechanisms involved in signal transformation by α_1-adrenergic receptors are not clear.

The α_1-adrenergic receptors are apparently not coupled to the regulatory guanyl nucleotide-binding system involved in the coupling of β- and α_2-adrenergic receptors to other membrane components; binding of agonists to α_1-adrenergic receptors is generally not affected by guanyl nucleotides [14]. On the other hand, stimulation of α_1-adrenergic receptors causes rapid breakdown and subsequent resynthesis of membrane phosphatidylinositol (PI) (Fig. 2). It is not clear how phospholipase C, an apparently soluble enzyme, catalyzing PI breakdown, is stimulated when α_1-adrenergic receptors become occupied by agonists. The changes in PI metabolism have been shown in several systems to be independent of the presence of extracellular calcium. This correlation between ligand-induced calcium mobilization and PI breakdown lead Michell and associates [24] to propose that this PI response is a signal primary to and causal for calcium gating. However, there are still large gaps in the understanding of the proposed mechanism, e.g., how the agonist-stimulated PI breakdown induces the calcium gating in the plasma membrane and, especially, how calcium is mobilized from internal storage sites by changes in plasma membrane PI metabolism. Furthermore, it has not been ruled out that the PI response is a signal parallel to increased calcium influx; in some systems, the changes in PI metabolism may even be an event secondary to the calcium signal [24].

Another possible intracellular signal observed in cellular responses involving increased cytoplasmic calcium concentrations is cyclic GMP. An increase in the cellular cyclic GMP concentration has been shown with α-adrenergic agonists in the rat ductus deferens [29], whose α-adrenergic receptors belong to the α_1-type. The effects of α-adrenergic agonists and other hormonal factors are restricted to intact cells or tissues and appear to be the result of increased cyclic GMP formation by one or the other form of the guanylate cyclase. Hormonal effects on cyclic GMP formation are thought to be the result of a chain of receptor-triggered reactions. As α-adrenergic and other hormonal effects on cyclic GMP levels are generally reduced or abolished when the extracellular calcium concentration is reduced, the hormone-induced increase in the cytoplasmic calcium concentration is assumed to be a primary and necessary step in coupling of receptors to guanylate cyclase [28]. Oxygen is additionally required as shown for hormone and calcium effects on cyclic GMP levels in two tissues, umbilical arteries [7] and kidney slices [8]. α-Adrenergic, cholinergic and calcium effects on cyclic GMP levels in the ductus deferens were reduced by mepacrine, an inhibitor of phospholipases, and by eicosatetraynoic acid, an arachidonate analogue, inhibiting unsaturated fatty acid peroxidation by lipoxigenases and cyclooxigenase [30]. On the basis of these findings it has been proposed that hormone-induced stimulation of cyclic GMP formation involves a calcium-induced release of arachidonic acid and possibly other unsaturated fatty acids and their peroxidation [30]. This assumption is supported by findings in crude or partially purified preparations of soluble guanylate cyclase, which were stimulated by arachidonic acid and fatty acid peroxides such as prostaglandin G_2 and various eicosatetraenoic acids [12].

Soluble guanylate cyclase has recently been purified from bovine lung to apparent homogeneity, retaining important regulatory functions of the crude enzyme [11]. The purified enzyme exhibited the absorption spectrum of a hemoprotein; Fe and additionally Cu were detected at concentrations of each about one mole per mole of holoenzyme [10]. Heme and Cu may be involved in the regulation of soluble guanylate cyclase activity by fatty acids and oxygen (or by fatty acid peroxides) and by other factors. The exact roles of heme and Cu and the molecular mechanism of the regulation(s) of this enzyme's activity need to be elucidated. In addition, hormone receptors may be coupled to the membrane-bound form of guanylate cyclase, whose control is not at all understood, and this coupling process may involve alternative mechanisms.

Signal Transformations by Hormone Receptor Subtypes

Adrenergic receptor subclassification has lead to the concept that the β-, α_2- and α_1-receptor subtypes are coupled to different cellular signals (Fig. 3). For hormonal factors other than catecholamines, the involvements of different intracellular signals in the cellular responses have been demonstrated [4] (Table 3). However, differentiated signal transformation by subtypes of receptors is less clear, as receptor subtypes have been established only for some hormonal factors.

Cellular reactions specific for α_1-adrenergic stimulation, such as PI response, involvement of calcium and increase in the cyclic GMP level, and for an α_2-adrenergic stimulation, such as inhibition of cyclic AMP formation and GTP modulation of agonist binding to the receptor, have also been observed with muscarinic cholinergic agonists and angiotensin II. No receptor subtypes have been described, mediating the two types of cellular responses. With histamine, serotonin and vasopressin (antidiuretic hormone), cellular responses of the α_1-adrenergic type and of the β-adrenergic type (increased cyclic AMP formation, GTP modulation of agonist binding) have been observed. In contrast, opiates, adenosine and prosta-

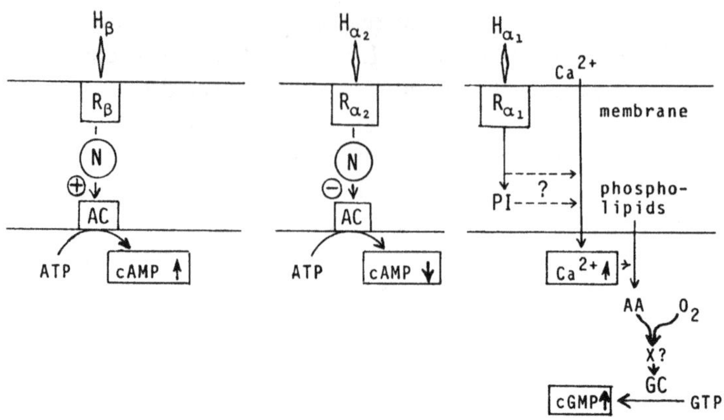

Fig. 3. Signal-generating systems coupled to subtypes of adrenergic receptors. *H,* adrenergic agonist; *R,* adrenergic receptor; *N,* guanyl nucleotide-binding regulatory protein; *AC,* adenylate cyclase (catalytic subunit); *PI,* changes in phosphatidylinositol metabolism (PI response); *GC,* guanylate cyclase; *AA,* arachidonic acid

Table 3. Cellular changes induced by hormonal stimuli

Hormonal factor	Tissue/cell type	AC ↑	GTP	AC ↓	GTP	Ca^{2+} ↑	PI ↑	cGMP ↑
Adrenergic β	Various	+	+					
α_2	Platelets, N×G, adipocytes, liver			+	+			
α_1	smooth muscle, liver					+	+	+
Cholinergic (muscarinic)	Myocardium			+	+			+
	Smooth muscle					+	+	+
Angiotensin II	Liver			+	+			
	Liver, smooth muscle					+	+	
Histamine H_2	Stomach, myocardium	+	+					
H_1	Smooth muscle					+	+	+
Serotonin	Neuronal cells	+	+					
	Smooth muscle					+	+	+
Vasopressin	Kidney	+	+					
	Smooth muscle, liver, platelets					+	+	
Opiates	Rabbit adipocytes	+						
	N×G, neuronal cells			+	+			
PGE_1, PGE_2	Platelets, frog erythrocytes	+	+					
	Adipocytes			+	+			
Adenosine	Platelets	+						
	Adipocytes			+	+			
ADP	Platelets			+		+	+	+

AC, adenylate cyclase activity; GTP, modulation of agonist-receptor binding by guanyl nucleotides; PI, stimulation of phosphatidylinositol metabolism (PI response); N×G, neuroblastoma×glioma hybrid cells glandin E_1 and E_2 are capable of inducing cellular reactions of the β- and the α_2-adrenergic types.

In some cases, different cellular response types are elicited by one agonist in one particular cell type. For instance, angiotensin II and ADP are capable of inducing responses of the α_1- and α_2-adrenergic types in rat hepatocytes and human platelets, respectively. It is not clear whether in these cases one type of receptor can be coupled to two signal-generating systems or whether different receptor subtypes will be differentiated. It is anticipated that more sophisticated studies on agonist- and antagonist-receptor interactions will discriminate receptor subclasses for hormonal factors other than catecholamines, supporting the concept that one type of receptor can be coupled only to one particular signal-generating component.

Acknowledgements. The authors' studies reported herein were supported by grants from the Deutsche Forschungsgemeinschaft.

References

1. Aktories K, Jakobs KH (1981) Epinephrine inhibits adenylate cyclase and stimulates a GTPase in human platelet membranes via α-adrenoceptors. FEBS Lett 130:235–238
2. Aktories K, Schultz G, Jakobs KH (in press) Stimulation of a low K_m GTPase by inhibitors of adipocyte adenylate cyclase. Mol Pharmacol
3. Berridge MJ (1975) The interaction of cyclic nucleotides and calcium in the control of cellular activity. Adv Cyclic Nucleotide Res 6:1–98
4. Berridge MJ (1980) Receptors and calcium signalling. Trends Pharmacol Sci 1:419–424
5. Brostrom MA, Brostrom CO, Wolff PJ (1979) Calcium dependence of hormone-stimulated cAMP accumulation in intact glial tumor cells. J Biol Chem 254:7548–7557
6. Cassel D, Selinger Z (1978) Mechanism of adenylate cyclase activation through the β-adrenergic receptor: catecholamine-induced displacement of bound GDP by GTP. Proc Natl Acad Sci USA 75:4155–4159
7. Clyman RI, Blacksin AS, Manganiello VC, Vaughan M (1975) Oxygen and cyclic nucleotides in human umbilical artery. Proc Natl Acad Sci USA 72:3883–3887
8. DeRubertis FR, Craven PA (1978) Alterations in rat renal cortical and medullary guanosine 3′,5′-monophosphate accumulation by oxygen- and calcium-dependent and -independent mechanisms: evidence for a calcium-independent action of oxygen in rat inner medulla. metabolism 27:855–868
9. Exton JH (1981) Molecular mechanisms involved in α-adrenergic responses. Mol Cell Endocrinol 23:233–264
10. Gerzer R, Böhme E, Hofmann F, Schultz G (1981) Soluble guanylate cyclase purified from bovine lung contains heme and copper. FEBS Lett 132:71–74
11. Gerzer R, Hofmann F, Schultz G (1981) Purification of a soluble, sodium-nitroprusside-stimulated guanylate cyclase from bovine lung. Eur J Biochem 116:479–486
12. Graff G, Stephenson JH, Glass DB, Haddox MK, Goldberg ND (1978) Activation of soluble splenic cell guanylate cyclase by prostaglandin endoperoxides and fatty acid hydroperoxides. J Biol Chem 253:7662–7676
13. Hirata F, Axelrod J (1980) Phospholipid methylation and biological signal transmission. Science 209:1082–1090
14. Hoffman BB, Lefkowitz RJ (1980) Radioligand binding studies of adrenergic receptors: New insights into molecular and physiological regulation. Ann Rev Pharmacol Toxicol 20:581–608
15. Jakobs KH (1979) Inhibition of adenylate cyclase by hormones and neurotransmitters. Mol Cell Endocrinol 16:147–156
16. Jakobs KH, Schultz G (1980) Actions of hormones and neurotransmitters at the plasma membrane: inhibition of adenylate cyclase. Trends Pharmacol Sci 1:331–333

17. Jakobs KH, Aktories K, Lasch P, Saur W, Schultz G (1980) Hormonal inhibition of adenylate cyclase. In: Dumont J, Nunez J (eds) Hormones and cell regulation, vol 4. Elsevier/North-Holland, Amsterdam, pp 89–106
18. Jakobs KH, Aktories K, Schultz G (1981) Inhibition of adenylate cyclase by hormones and neurotransmitters. Adv Cyclic Nucleotide Res 14: 173–187
19. Jakobs KH, Aktories K, Schultz G (submitted) Acceleration of the adipocyte adenylate cyclase turn-off reaction by inhibitory hormonal factors.
20. Lefkowitz RJ, Hoffman BB (1980) New directions in adrenergic receptor research. Trends Pharmacol Sci 1:314–318
21. Levitzki A (1981) The beta-adrenergic receptor and its mode to coupling to adenylate cyclase. CRC Crit Rev Biochem 10:81–112
22. Levitzki A, Helmreich EJM (1979) Hormone-receptor-adenylate cyclase interactions. FEBS Lett 101:213–219
23. Maguire ME, Erdos JJ (1980) Inhibition of magnesium uptake by β-adrenergic agonists and prostaglandin E_1 is not mediated by cyclic AMP. J Biol Chem 255: 1030–1035
24. Michell RH, Kirk CJ (1981) Why is phosphatidylinositol degraded in response to stimulation of certain receptors? Trends Pharmacol Sci 2:86–89
25. Northup JK, Sternweis PC, Smigel MD, Schleifer LS, Ross EM, Gilman AG (1980) Purification of the regulatory component of adenylate cyclase. Proc Natl Acad Sci USA 77:6516–6520
26. Rodbell M (1980) The role of hormone receptors and GTP-regulatory proteins in membrane transduction. Nature 284: 17–22
27. Ross EM, Gilman AG (1980) Biochemical properties of hormone-sensitive adenylate cyclase. Ann Rev Biochem 49:533–564
28. Schultz G, Hardman JG, Schultz K, Baird CE, Sutherland EW (1973) The importance of calcium ions for the regulation of guanosine 3′,5′-cyclic monophosphate levels. Proc Natl Acad Sci USA 70:3889–3893
29. Schultz G, Schultz K, Hardman JG (1975) Effects of norepinephrine on cyclic nucleotide levels in the ductus deferens on the rat. Metabolism 24:429–437
30. Spies C, Schultz K-D, Schultz G (1980) Inhibitory effects of mepacrine and eicosytetraynoic acid on cyclic GMP elevations caused by calcium and hormonal factors in rat ductus deferens. Naunyn-Schmiedebergs Arch Pharmacol 311:71–77

Discussion

Krawietz, Munich: In your schema you showed the GTPase unit at a different place in the membrane and not as a part of the nucleotide binding protein. From the experiments of Pfeuffer and Gilman, I would say, there is more evidence that the GTPase is part of the nucleotide binding protein.

Schultz: I did not intend to separate the GTPase from the guanyl nucleotide-binding, regulatory component; it is an inherent activity of this system. However, it is not yet known which one of the three constituents apparently forming the regulatory component possesses GTPase activity, since this activity is lost in the purified system.

Krawietz: You postulated an inhibitory nucleotide binding protein. Could it not be that merely a different mode of aggregation of the individual subunits of the nucleotide binding protein results in a stimulatory or inhibitory effect on adenylate cyclase activity?

Schultz: It has been proposed by Rodbell and associates that aggregation and disaggregation of components of the adenylate cyclase system are involved in the regulation of enzyme activity. This is a likely possibility and one which could be involved in addition to the mechanisms of regulation I presented.

Palm, Frankfurt/M: Is it possible to distinguish α_1- and α_2-adrenoceptors by their effects on the enhancement of intracellular Ca^{2+}, i.e. do α_1- or α_2-adrenoceptors differ with respect to an enhanced Ca^{2+}-influx from the extracellular space or an enhanced release of Ca^{2+} from intracellular stores?

Schultz: Van Zwieten has recently presented data indicating α_2-promoted influx of calcium in vascular smooth muscle. As most data were obtained from in vivo systems, I am not convinced of a coupling of α_2-adrenoceptors to the Ca^{2+} system in a manner typical for α_1-adrenoceptors.

Opie, Cape Town: I get the impression from the literature that the effects of α-stimulation on the heart are very controversial – for example, does cyclic GMP consistantly rise?

Schultz: α_1-Adrenergic effects on the myocardium are not well understood. To my knowledge, no positive inotropic response and no rise of cyclic GMP levels have been reported although increased calcium influx appears to be involved. However, dissociations between calcium and cyclic GMP elevation have been reported for other tissues, too.

Starke, Freiburg: There seems to be one exceptional stimulatory effect mediated by α_2-adrenoceptors, namely contraction of vascular smooth muscle. This requires more calcium at the contractile machinery. May an exceptional α_2-receptor be coupled to some calcium channel or calcium transfer mechanism?

Schultz: Again, I am not convinced of the existence of direct coupling of α_2-adrenoceptors to calcium influx. However, an α_2-induced fall in the cyclic AMP level could affect calcium efficacy and distribution by reducing myosin light chain kinase phosphorylation and calcium sequestration, respectively.

Occurrence and Functional Significance of Cardiac Muscle Protein Kinases

R. Mewes, F. Hofmann

Summary

Phosphorylation of contractile and membrane proteins has been implicated in the control of cardiac muscle contractility by intracellular signals such as cAMP, calcium, or cGMP, although the precise functional role of some of these phosphorylation/dephosphorylation reactions has not been elucidated. Some insight into the possible function of these reactions has come from studies designed to characterize in vitro the components of these phosphorylation systems. Cardiac muscle contains at least three different signal-regulated protein kinases, namely calcium/calmodulin stimulated myosin light chain kinase (MLCK), cAMP-dependent protein kinase and cGMP-dependent protein kinase. All three enzymes have been purified to homogeneity and characterized. Each enzyme has a high specific activity (in the range of 5–20 μmoles of phosphate transferred per min and mg pure enzyme), if the appropriate substrate is used. cAMP-dependent protein kinase has in vitro a rather broad substrate specificity which does also exist in vivo. cGMP-dependent protein kinase again has in vitro a rather broad substrate specificity. However, an in vivo substrate has not been detected, so far. In contrast, cardiac MLCK phosphorylates with high activity only cardiac myosin light chain-2.

In further experiments the effects of the purified protein kinases on highly enriched cardiac sarcolemmal vesicles were investigated. Addition of calcium and calmodulin in the presence or absence of MLCK or phosphorylase kinase increased slightly the phosphorylation of a M_r 24 000 protein. In contrast, cAMP-dependent protein kinase phosphorylated several protein bands, a M_r 22 000 protein being the major substrate. Similar results as with cAMP-dependent protein kinase were obtained by addition of cGMP-dependent protein kinase. The M_r 22 000 protein could be converted to a M_r 11 000 protein by adequate denaturation conditions. These results suggest that cardiac sarcolemma does contain only a substrate for cAMP-dependent protein kinase which may be involved in the regulation of calcium fluxes.

Introduction

It is now generally accepted that activation of specific protein kinases by intracellular generated signals – such as calcium; cAMP, cGMP and other less well defined small molecules – and concomitant phosphorylation of key regulatory proteins or enzymes is one important mechanism by which many hormones, neurotransmitters and autacoids (locally generated and acting hormones) regulate cellular functions (for an introduction see Ref. 11). This mechanism has been found to operate in many different tissues including cardiac tissue. From the results so far reported it is quite clear that cardiac muscle contraction is basically controlled by the intracellular concentration of calcium and the interaction of calcium with the troponin-tropomyosin system. This basic control mechanism does not involve covalent modification of proteins. However, the in vivo phosphorylation of a number of cardiac proteins has been observed in connection with the addition of catecholamines (see Ref. [1, 5, 15, 31, 32]). In most cases a positive correlation between this modification and the increase in cAMP-levels and the positive inotropic effect of catecholamines has been found. Therefore, it has been suggested that the modulatory effect of some hormones (catecholamines, acetylcholin etc.) on cardiac contractility might work through activation of specific protein kinases and concomitant phosphorylation of regulatory proteins. The following article will briefly summarize the occurrence and functional significance of signal regulated, cardiac muscle protein kinases.

Occurrence and Properties of Signal-Regulated Protein Kinases

The occurrence of signal-regulated protein kinases has been studied using bovine cardiac muscle. As shown in Fig. 1 DEAE-cellulose chromatography of crude extracts of this tissue revealed the presence of at least three different protein kinases, namely cAMP-dependent protein kinase type II, cGMP-dependent protein kinase and a calcium/calmodulin-dependent protein kinase. The latter enzyme phosphorylates cardiac myosin light chains and is identical with myosin light

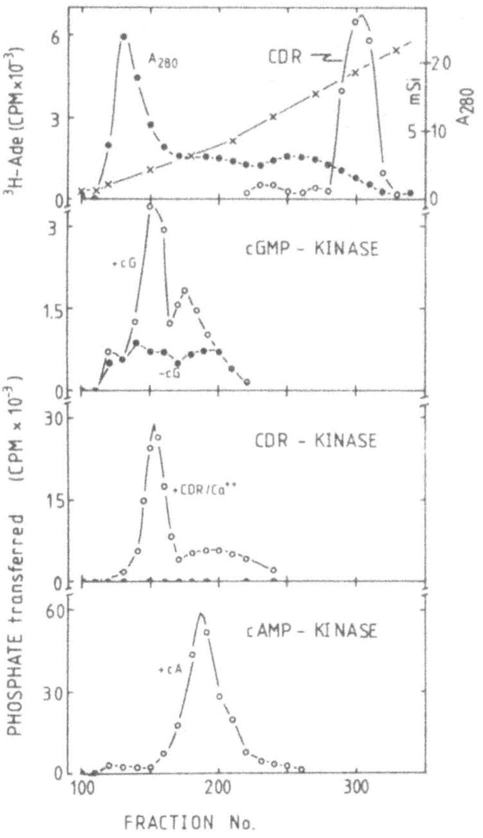

Fig. 1. DEAE-cellulose column profiles of bovine cardiac muscle kinases. Extraction of minced bovine cardiac muscle and chromatography on DEAE-cellulose was carried out as described [6]. The activity of cGMP-, cAMP- and calcium/calmodulin-dependent protein kinases was determined in 10 μl aliquots as described in refs. 6, 9 and 34, respectively. Calmodulin (CDR) was determined as described in [34] using a preparation of calmodulin-deficient phosphodiesterase. *Ade*, adenosine

Table 1. Molecular weight, subunit composition and active species of purified cardiac muscle protein kinases

Protein Kinase	Activator	Number of subunits	Minimal M_r	Composition of enzyme species	
				inactive	active
cAMP-dependent	cAMP	4	56 000 (R) 40 000 (C)	Tetramer (R_2C_2)	Monomer C
cGMP-dependent	cGMP	2	75 000	Dimer $(RC)_2$	Ternary complex $cGMP_2 \cdot (RC)_2$
Myosin light chain kinase	Calcium/ CaM	1	94 000	Monomer (C)	Ternary complex $Ca_4^{2+} \cdot CaM \cdot C$

The values shown have been taken from ref. 6, 9, 34, 35. R, regulatory subunit; C, catalytic subunit; RC, subunit containing a catalytic and binding domain; CaM, calmodulin

chain kinase. Calmodulin, the activator of a number of calcium-dependent enzymes, is eluted separately from its target enzymes at rather high salt concentrations. Each of the shown protein kinases has been purified to apparent homogeneity [6, 9, 34]. As expected, the purified enzyme differ in a number of properties including their specific activator molecule, molecular weights, subunit composition and composition of the active species (Table 1). cAMP-dependent protein kinase is active only in its dissociated form, whereas cGMP-dependent and myosin light chain kinase are only active upon formation of a ternary complex containing the inactive enzyme and its activator(s).

On the other hand, the three enzymes have in common a rather high specific activity, if suitable substrates are added in vitro (Table 2). Myosin light chain kinase and cAMP-dependent protein kinase phosphorlyate their preferred substrates with velocities above 20 μmoles phosphate transfered per min and mg enzyme. Lower specific activities are observed with cGMP-dependent protein kinase. This lower reaction velocity is presumably not a property of the enzyme but of the substrates. These substrates are not phosphorylated in vivo by cGMP-dependent protein kinase [2]. In addition histones inactivate slowly the enzyme [33] suggesting that a higher velocity will be obtained if a physiological substrate is used.

Table 2. Specific activities of cardiac muscle protein kinases

Protein kinase	Specific activity [μmol/min/mg]	Substrate
cAMP-dependent	20–40	Kemptide
cGMP-dependent	2– 5	Histone IIb
	1	Troponin-I (cardiac)
Myosin light chain kinase	18–30	Cardiac myosin light chain 2

Specific activities have been determined using pure preparations of protein kinases. For assay conditions see ref. 6, 10, 34. Kemptide is the peptide which contains the phosphorylation site of pyruvate kinase (L-type) [16]

Functional Significance of Cardiac Protein Kinases

The occurrence of the signal regulated protein kinases in cardiac muscle and the rather high velocity with which they catalyze the phosphate transfer to proteins, suggest that in vivo, hormone-induced alterations in the levels of calcium, cAMP or cGMP are expressed through a change in protein kinase activity and that these enzymes are able to alter the state of phosphorylation of one or several cardiac proteins within a short time. This interpretation has been confirmed by a number of laboratories [5, 15, 31, 32], which showed that in vivo the phosphate content of a number of proteins or enzymes increases rapidly when the concentration of cAMP is elevated by the addition of isoproterenol. Among the phosphorylated proteins are the inhibitory subunit of troponin [25], the C-protein of myofibrillar origin [13], phosphorylase kinase [23], and "phospholamban", a protein of the sarcoplasmic reticulum [21]. As summarized in Table 3 the functional change induced by phosphorylation has been elucidated for most of these proteins. It has been shown that phosphorylation of phosphorylase kinase by cAMP-dependent protein kinase increases its affinity for calcium [3, 4], and facilitates activation of the enzyme at low calcium concentration. The active enzyme in turn phosphorylates and activates phosphorylase leading to a breakdown of glycogen. In contrast to this example phosphorylation of the inhibitory subunit of troponin decreases the affinity of the troponin complex for calcium [25, 29]. This decreased affinity is presumably responsible for the decrease in relaxation time observed during the positiv inotropic effect of catecholamines. Finally, phosphorylation of phospholamban increases the rate of calcium transport into the sarcoplasmic reticulum [15] decreasing thereby the free calcium concentration of the cytoplasma. In connection with the phosphorylation of troponin-I this latter phosphorylation will abbreviate primarily the relaxation time. However, since more calcium is pumped into the sarcoplasmic reticulum, it is also feasible that more calcium will be released from this storage site during following contraction cycles. This latter possibility has been proposed to explain the positiv inotropic effects of catecholamines since it would result in an elevated intracellular concentration of free calcium [15]. The functional role of the rapid phosphorylation of the C-protein is not known presently.

In contrast to the cAMP-system much less solid data are available about the function of the cardiac myosin light chain phosphorylation. Several laboratories

Table 3. Possible physiological relevance of cardiac protein kinase

Kinase	Substrate	Functional change
cAMP-dependent	Phosphorylasekinase	Increase in glycogenolysis
	Troponin-I	Decrease in relaxation time
	Sarcoplasmic reticulum	Increase in calcium uptake; Decrease in relaxation time
	C-protein (myofibril)	Unknown
Calcium-dependent	Myosin light chain 2	Unknown
	Sarcoplasmic reticulum	Increase in calcium uptake; Decrease in relaxation time
cGMP-dependent	Unknown	Unknown

For references see text

have reported that myosin light chain-2 is phosphorylated in vivo [8, 14]. However, with one exception [17] no functional significance could be attributed to this phosphorylation [8, 14]. Other studies have shown that compared to its protein substrate far less than stoichiometric concentrations of myosin light chain kinase are present in cardiac muscle and that calmodulin activates the enzyme by increasing its binding affinity for the substrate [12]. Taken together both observations suggest that only catalytic amounts of the enzyme are present in cardiac tissue and that phosphorylation of a significant part of the light chains is a rather slow process, which can not be directly involved in the control of cardiac contractility. On the other hand, myosin light chain kinase is a very specific enzyme which does not phosphorylate other proteins suggesting that some, jet to be elucidated, function is controlled by this enzyme. The enzyme phosphorylating "phospholamban" in a calcium/calmodulin-dependent reaction has not been identified so far [20] but is not identical with the myosin light chain kinase. This latter reaction is thought to introduce the similar changes in calcium transport shown previously for the cAMP-dependent phosphorylation of the sarcoplasmic reticulum. Neither a physiological substrate for cGMP-dependent protein kinase, nor a general agreement on a possible correlation between cGMP levels and cardiac function have been reported. The previously observed correlation [7] between the negative inotropic effect of acetylcholin and the increase in cGMP concentration has been questioned recently [22].

Significance of Phosphorylation of Sarcolemmal Proteins

The results summarized so far are compatible with the idea that certain parts of the positive inotropic effect of β-adrenergic agents are mediated by cAMP and protein phosphorylation. However, these data do not provide unequivocal evidence for the mechanism by which β-adrenergic agents increase the force of contraction. It is widely accepted that this change in contractile properties is mediated by an increase in the myoplasma concentration of calcium. As discussed above one hypothesis is that cAMP-dependent phosphorylation of the sarcoplasmic reticulum results in the elevated calcium levels. On the other hand, electrophysiological studies suggested that an increased influx of calcium from the extracellular space or release from plasma membrane stores is the primary cause of the elevated calcium concentration during positiv inotropic stimulation. Therefore it has been proposed that cAMP-dependent phosphorylation of sarcolemmal proteins occurs during β-adrenergic stimulation of the heart, and that this reaction modifies the voltage-dependent slow calcium channel in such a way that more calcium is delivered to the myoplasma during each twitch [27, 30]. So far, the biochemical evidence for this mechanism is circumstantial. Several groups have shown that in vitro a sarcolemmal protein (app. M_r 11 000) is phosphorylated by a cAMP-dependent reaction (for a review see [5 and 36]). However, it has been difficult to correlate this phosphorylation with a change in calcium transport or flux. An additional setback of these earlier studies was, that heterogeneous plasma membrane fractions and only partially purified enzyme preparations were used. In order to clarify the questions posed, we have started to reinvestigate this problem using highly purified preparations of sarcolemma and preparations of homogeneous kinases.

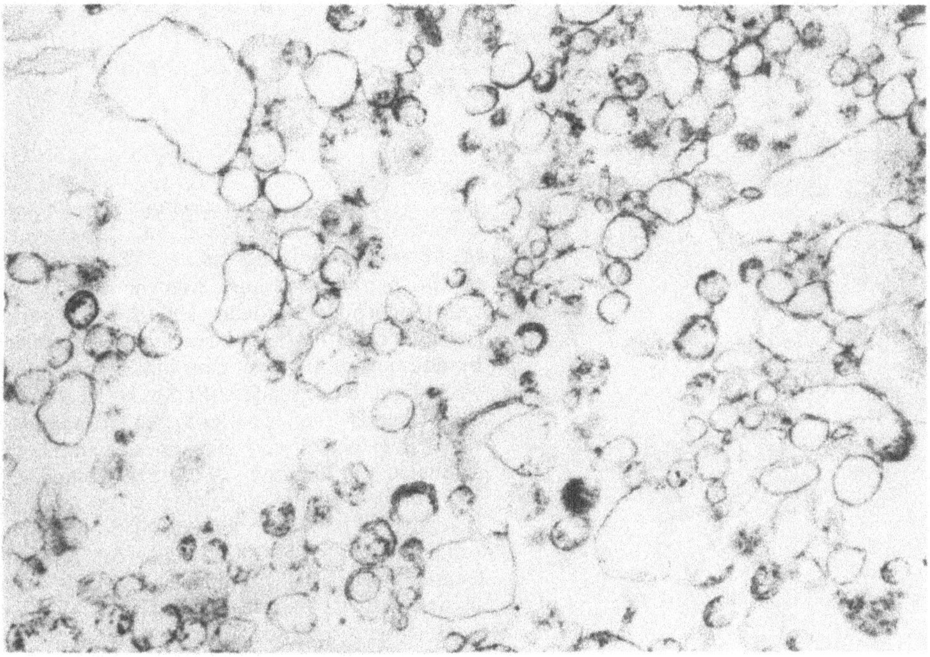

Fig. 2. Electron micrograph of bovine cardiac muscle sarcolemma. The preparation of sarcolemmal membranes was carried out according to [26] including the discontinuous sucrose gradient centrifugation step. Aliquots were fixed in a double volume of 6% glutaraldehyde solution, washed in 0.1 *M* cacodylate buffer, pH 7.3, encoated with 1% agar. After postfixation in osmium tetroxyde, the agar-encoated fractions were dehydrated with a increasing series of alcohol, past through propylene oxide and embedded in epon. Ultrathin sections were stained with uranyl acetate and lead citrate and examined in a Philips 301 electron microscope [24]. The shown section is representative for three different preparations examined. X = 15 000

As shown by electronmicroscopy (Fig. 2) a homogeneous population of small vesicles was obtained by standard procedures [26]. The majority of membranes present in this preparation was of sarcolemmal origin as suggested by the high specific binding capacity of ouabain (range for 4 preparations 50–80 pmoles ouabain bound/mg proteins). Further characterization of this preparation showed that the staining pattern for proteins in sodium dodecyl sulfate (SDS) polyacrylamide gels depended on the denaturation temperature (Fig. 3). Denaturation of sarcolemnal proteins in the presence of SDS for 5 min at 100 °C resulted in the disappearence of two Coomassie blue stained bands (M_r 90 000 and 100 000) which were present when denaturation was carried out for 30 min at 30 °C. The disappearance of these stained bands could not be prevented by the addition of EDTA (2 m*M*), benzamidine (50 m*M*), PMSF (0.5 m*M*) aprotinin (1.1 mg/ml), soja bean trypsin inhibitor (0.4 mg/ml) or leupeptin (0.1 mg/ml) – applied singly or together – to samples before they were boiled. This inability of several known protease inhibitors to prevent these change in staining pattern makes it very unlikely that the observed change is caused by a proteolytic process.

The membrane preparation contains a low amount of an apparently cAMP-dependent protein kinase, which phosphorylated in the presence of cAMP and MgATP several distinct bands to a low extent. Addition of pure catalytic subunit

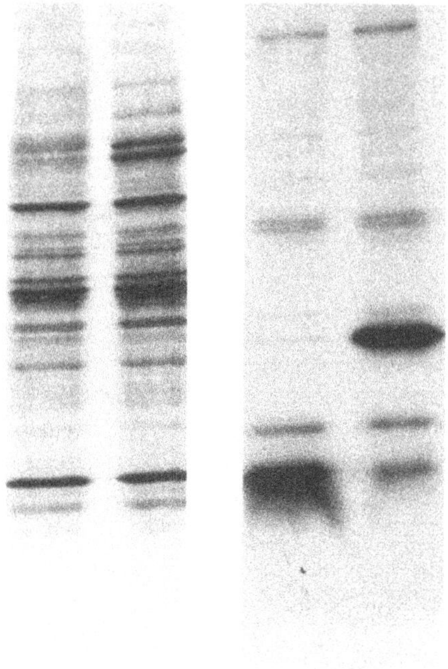

1 2 3 4

Fig. 3. Polyacrylamide gel electrophoresis of sarcolemmal proteins. Aliquots of sarcolemmal membranes were incubated at 30 °C for 60 sec in the presence of pure catalytic subunit of cAMP-dependent protein kinase (1 μg) and [γ-^{32}P] ATP (1 m*M*). The reaction was stopped by addition of 10 μl running buffer containing 5% sodium dodecylsulfate (SDS) and 5% 2-mercaptoethanol. Immediately thereafter, samples were either boiled for 5 min (tracks 1 and 3) or incubated for 30 min at 30 °C (tracks 2 and 4). Denatured samples containing 40 μg protein were layered on a discontinuous SDS-polyacrylamid gradient (10%–20% acrylamide) gel and focused through a 4% sample gel [18]. Gels were stained with Coomassie blue (tracks 1 and 2) and thereafter exposed for a week to an X-ray film (Kodak X-Omat S) (tracks 3 and 4). Gels were run from top to bottom

(2.2 μg) to the membrane fraction increased the total amount of ^{32}P incorporated within 75 sec from 110 pmole/mg protein to 450 pmole/mg protein. Separation of the phosphorylated membrane proteins on SDS polyacrylamid gels (Fig. 3) indicated that the majority of radioactive phosphate was associated with a band of M_r 22 000 when denaturation was carried out at 30 °C. In addition several other radioactive band were detected (approx M_r's 150 000; 72 000; 59 000; 45 000; 42 000; 14 000 and 12 000). The radioactivity associated with this latter bands increased when the phosphorylation reaction was carried out for one hour. Denaturation of phosphorylated membranes for 5 min at 100 °C resulted in a shift of the major radioactive band from M_r 22 000 to a M_r of 12 000 (Fig. 3). This shift was not prevented when the same protease inhibitor as mentioned above were added before the denaturation step. This shift in the apparent molecular weight was also found when samples were phosphorylated for 1 h at 30 °C. No difference in phosphorylation pattern was observed when homogeneous cGMP-dependent protein kinase (4.5 μg) was used instead of the catalytic subunit.

Effect of Calcium/Calmodulin on the Phosphorylation

It had been reported that addition of calcium and calmodulin to crude particulate fractions of cardiac muscle increase the phosphorylation of a 54 000 dalton protein [28]. In addition, the calcium/calmodulin dependent phosphorylation of "phospholamban" has been reported [20]. Therefore, the effect of calcium/calmodulin was tested. Addition of calcium (0.6 m*M*) and calmodulin (1 μg) applied alone or in combination did not increase the phosphorylation of the membrane proteins significantly compared with control samples incubated in the presence of 0.6 m*M* EGTA. Occasionally a slightly phosphorylated band (app. M_r 24 000) was observed in the presence of calcium/calmodulin. This occasionally observed band could be due to a minor contamination of the membrane preparation by sarcoplasmic reticulum since the calcium dependent phosphorylation of "phospholamban" has been reported.

It was possible that the inability of calcium/calmodulin to stimulate the phosphorylation of membrane proteins was due to the lack of a calcium responsive protein kinase. Therefore, in another series of experiments phosphorylation of membrane proteins was carried out in the presence of pure preparations of two calcium/calmodulin dependent protein kinases, namely cardiac muscle myosin light chain kinase (1.5 μg) and phosphorylase kinase (5 μg). Incubations were carried out in the presence and absence of calcium/calmodulin for 1 min and 1 h and denaturation of phosphorylated membranes was done at 30 °C and at 100 °C. Autoradiogramms of the respective gels showed no major change in the distribution or intensity of the radioactive bands when the enzymes were added. In control samples in which the enzymes were incubated in the absence of the membrane fraction autophosphorylation of the enzymes was evident.

The experiments described are in agreement with earlier reports [5, 36] that cardiac sarcolemma contains a good substrate for cAMP-dependent protein kinase, which has a minimal molecular weight of 11 000 dalton. In contrast to previous reports the M_r of this protein depends on the particular denaturation condition applied. The data shown suggest that it consists of at least two subunits. Similar observations have been reported [19]. This protein is not phosphorylated in a calcium/calmodulin dependent reaction suggesting that it is not identical to the sarcoplasmic reticulum protein "phospholamban". In addition neither this nor another sarcolemmal protein is phosphorylated by phosphorylase kinase or myosin light chain kinase. This may indicate that a calcium dependent phosphorylation of cardiac muscle plasma-membrane proteins is not an important reaction controlling the flux or transport of itself or of other ions. The relationship of the cAMP-dependent phosphorylation of the low molecular peptide with calcium transport or flux processes is not known and remains to be elucidated.

Acknowledgement. We would like to thank Dr. P. Cohen, University of Dundee for his generous gift of phosphorylase kinase; Mrs P. Janko and H. Vogt for expert technical support, Mrs. E. Messmer for typing the manuscript and Dr. W. Minuth, Universität Heidelberg für die electron-microscopical figures. Part of this work was supported by a grant from the Deutsche Forschungsgemeinschaft and by financial support of Fonds der Chemie.

References

1. Adelstein RS, Eisenberg E (1980): Regulation and kinetics of the actin-myosin-ATP interaction. Ann Rev Biochem 49:921–956
2. Blumenthal DK, Stull JT, Gill GN (1978) Phosphorylation of cardiac troponin by guanosine-3',5'-monophosphate-dependent protein kinase. J Biol Chem 253:334–336
3. Brostrom CD, Hunkeler FC, Krebs EG (1971) The regulation of skeletal muscle phosphorylase kinase by Ca^{2+}. J Biol Chem 246:1961–1967

4. Cohen P (1980) The role of calcium ions, calmodulin and troponin in the regulation of phosphorylase kinase from rabbit skeletal muscle. Eur J Biochem 111:563–574
5. England PJ (1980) Protein phosphorylation in the regulation of muscle contraction. In: Cohen P (ed) Molecular aspects of cellular regulation. vol I. Elsevier/North-Holland Biochem Press, p 153–173
6. Flokerzi V, Speichermann N, Hofmann F (1978): A guanosine-3′,5′-monophosphate-dependent protein kinase from bovine heart muscle. J Biol Chem 253:3395–3399
7. George WJ, Polson JB, O'Toole HG, Goldberg ND (1970) Elevation of guanosine-3′,5′-monophosphate in rat heart after perfusion with acetylcholine. Proc Natl Acad Sci USA 66:398–401
8. High CW, Stull JT (1980) Phosphorylation of myosin in perfused rabbit and rat hearts. Am J Physiol 239:H756–H764
9. Hofmann F (1980): Apparent constants for the interaction of regulatory and catalytic submit of cAMP-dependent protein kinase I and II. J Biol Chem 255:1559–1564
10. Hofmann F, Schultz G (1980) Wirkungsprinzipien von Hormonen und Neurotransmittern. Drug Res 30(II):1991–1995
11. Hofmann F, Beavo JA, Bechtel PJ, Krebs EG (1975): Comparison of adenosin-3′,5′-monophosphate-dependent protein kinases from rabbit skeletal and bovine heart muscle. J Biol Chem 250:7795–7801
12. Jeacocke SA, England PJ (1980) Phosphorylation of a myofibrilla protein of M_r 150 000 in perfused rat heart and the tentative identification of this as C-protein. FEBS Lett 122: 129–132
13. Jeacocke SA, England PJ (1980) Phosphorylation of myosin light chains in perfused rat heart. Biochem J 188:763–768
14. Janko P, Wolf H, Hofmann F (submitted) Calmodulin activates bovine cardiac muscle light chain kinase by increasing the affinity for myosin light chains
15. Katz AM (1979) Role of the contractile proteins and sarcoplasmic-reticulum in the response of the heart to catecholamines: an historical review. Adv Cyclic Nucleotide Res 11:303–343
16. Kemp BE, Graves DJ, Benjamini E, Krebs EG (1977) Role of multiple basic residues in determining the substrate specificity of cyclic AMP-dependent protein kinase. J Biol Chem 252:4888–4894
17. Kopp SJ, Barany M (1979) Phosphorylation of the 19 000-dalton light chain of myosin in perfused rat heart under the influence of negative and positive inotropic agents. J Biol Chem 254:12007–12012
18. Laemmli UK (1970) Cleavage of structural proteins during the assembly of the head of bacteriophage T 4. Nature 227:680–685
19. Lamers JMJ, Stinis JT (1980) Phosphorylation of low molecular weight proteins in purified preparations of rat heart sarcolemma and sarcoplasmic reticulum. Biochem Biophys Acta 624:443–459
20. Le Peuch CJ, Guilleux JC, Demaille JG (1980) Concerted regulation of cardiac sarcoplasmic reticulum calcium transport by cyclic adenosine monophosphate-dependent and calcium-dependent phosphorylations. Biochemistry 18:5150–5157
21. Le Peuch CJ, Guilleux JC, Demaille JG (1980) Phospholamban phosphorylation in the perfused rat heart is not solely dependent on β-adrenergic stimulation. FEBS Lett 114:165–168
22. Linden J, Brooker G (1979) The questionable role of cyclic guanosine 3′,5′-monophosphate in heart. Biochem Pharmacol 28:3351–3360
23. McCullough TE, Walsh DA (1979) Phosphorylation and dephosphorylation of phosphorylase kinase in the perfused rat heart. J Biol Chem 254:7345–7352
24. Minuth WW, Kriz W (1980) Newly synthesized proteins in rat glomerulosa fraction. renal Physiol 3:88–96
25. Ray KP, England PJ (1976) Phosphorylation of the inhibitory submit of troponin and its effect on the calcium dependence of cardiac myofibrill adenosine triphosphatase. FEBS Lett 70:11–16
26. Reeves JP, Sutko JL (1979) Sodium-calcium ion exchange in cardiac muscle membrane vesicles. Proc Natl Acad Sci USA 76:590–594
27. Reuter H, Scholz H (1977): The regulation of the calcium conductance of cardiac muscle by adrenaline. J Physiol 264:49–62

28. Schulman H, Greengard P (1978) Ca^{2+}-dependent protein phosphorylation system in membranes from various tissues and its activation by "calcium-dependent regulator". Proc Natl Acad Sci USA 75:5432–5436
29. Solaro PJ, Holroyde MJ, Crouch TH, Robertson S, Johnson JD, Potter JD (1980) Myofilament protein phosphorylation in striated muscles. 8. Cold Spring Harbor Conference on Protein Phosphorylation (Abstr) p64
30. Sperelakis N, Schneider JA (1976) A metabolic control mechanism for calcium ion flux that may protect the ventricular myocardial cell. Am J Cardiol 37:1079–1085
31. Stull JT (1980) Phosphorylation of contractile proteins in relation to muscle function. Adv Cyclic Nucleotide Res 11:p39–93
32. Stull JT, Blumenthal DK, Cooke R (1980) Regulation of contraction by myosin phosphorylation: a comparison between smooth and skeletal muscles. Biochem Pharmacol 29:2537–2543
33. Walton GM, Gill GN (1981) Regulation of cyclic nucleotide-dependent protein kinase activity by histones and poly(C-argine). J Biol Chem 256:1681–1688
34. Wolf H, Hofmann F (1980) Purification of myosin light chain kinase from bovine cardiac muscle. Proc Natl Acad Sci USA 77:5852–5855
35. Wolf H, Hofmann F (1981) Basic properties of myosin light chain kinase from bovine cardiac muscle. In: Rosen and Krebs (ed) Cold Spring Harbor Conference on Protein Phosphorylation. Vol 59, p301–304
36. Wollenberger A, Will H (1978) Protein kinase-catalysed membrane phosphorylation and its possible relationship to the role of calcium in the adrenergic regulation of cardiac contraction. Life Sci 22:1159–1178

Discussion

Trautwein, Homburg/Saar: You mentioned that there was no evidence of a functional role of cGMP in the heart. We have injected cGMP into isolated ventricular cells and found a depression of the plateau of the action potential (an effect opposite to that of cAMP injction) indicating perhaps a depression of the slow inward current.

Hofmann: No comment.

Cyclic AMP-Dependent Activation of Protein Kinases in the Myocardium

J. G. Dobson, Jr.

Summary

Upon catecholamine stimulation of the heart, the cyclic AMP-dependent activation of protein kinases appears to play an important role in the enhancement of contractility and glycogenolysis presumably by promoting the phosphorylation of myocardial proteins. In the intact rat heart, epinephrine produced a rapid elevation of myocardial cyclic AMP and activation of both cyclic AMP-dependent protein kinase and phosphorylase *b* kinase within 5–10 sec. This preceded the conversion of phosphorylase *b* to phosphorylase *a* and the augmentation of contractile state that occurred within 12–18 sec. The β-adrenergic blocking agent, practolol, had no effect on the basal levels of the above parameters, but did prevent the epinephrine produced increases. The importance of catecholamine activation of the kinases in catalyzing protein phosphorylation was investigated in isolated myocardial preparations previously exposed to ^{32}P to label intracellular ATP. In ventricular muscle slices, isoproterenol increased the ^{32}P incorporation into trichloroacetic acid insoluble protein as well as the phosphate content of this protein. In isolated perfused hearts, the catecholamine produced an increase in ^{32}P incorporation into a $29\,000 \times g$ (membrane and mitochondria) and $100\,000 \times g$ (sarcoplasmic reticulum) pellet fractions and a $100\,000 \times g$ soluble fraction within 30 sec. While the β-adrenergic blocking agent, propranolol, did not by itself affect ^{32}P incorporation, it prevented the isoproterenol-induced incorporation of ^{32}P into the pellet and soluble fractions and the acid-insoluble protein. Removal of isoproterenol from the isolated preparations eliminated the differences in ^{32}P incorporation indicating that the effects of the catecholamines were reversible. Sodium dodecyl sulfate gel electrophoresis of ventricular muscle homogenates of isolated perfused hearts revealed that isoproterenol enhanced the ^{32}P incorporation into several myocardial proteins having molecular weights of 155 000, 92 000, 30 000, 16 000 and 13 000 within 30 sec. These findings are consistent with the notion that cyclic AMP-dependent activation of protein kinases by catecholamine stimulation enhances the phosphorylation of myocardial proteins which in turn may play an important role in the augmentation of contractility and glycogenolysis.

Introduction

β-adrenergic stimulation of the heart resulting in an enhancement of contractile state and glycogenolysis involves at least in part adenosine 3′,5′-monophosphate (cyclic AMP) [6, 18] and cyclic AMP-dependent protein kinase [1, 3, 11]. The inotropic response caused by the catecholamines has been suggested not only to involve cyclic AMP-dependent protein kinase and phosphorylase *b* kinase, but also possibly cyclic AMP-independent protein kinases and myosin light chain kinase that catalyze the phosphorylation of sarcoplasmic reticulum [12, 14, 19, 23, 26], sarcolemmal membrane [22, 25] and proteins associated with contractile filaments

[7–9, 13, 21, 27]. The glycogenolytic response caused by catecholamine activation of glycogen phosphorylase (i.e. conversion of the *b* form of the enzyme to the *a* form by phosphorylation) is thought to involve cyclic AMP-dependent protein kinase catalyzed phosphorylation of phosphorylase *b* kinase, thereby transforming the latter enzyme to its activated form [5, 16, 24].

The purpose of this report was to examine the relationship between the catecholamine-induced cyclic AMP-dependent activation of protein kinases, phosphorylation of myocardial proteins and the elevation of both contractile state and phosphorylase activity in the myocardium subjected to catecholamine stimulation.

Methods and Materials

Preparations

Three rat heart preparations were used in the present studies. An open-chest rat preparation was used when the in situ heart was studied. For these studies rats were anesthetized with sodium pentobarbital (25 mg/kg, i.p.) and α-chloralose (50 mg/kg, i.p.). In addition the rats received atropine sulfate (1 mg/kg, i.p.) to suppress vagal responses in the heart. Acute bilateral adrenalectomy was performed in all rats to eliminate the influence of adrenal catecholamines. Body temperature was maintained at 37 °C and the jugular vein was cannulated for the administration of drugs into the superior vena cava. The chest of the rat was opened and the preparation was ventilated with room air via a tracheal cannula at 80/min. The rate of left ventricular pressure development (dP/dt) was obtained from a cannula inserted through the apex of the heart into the left ventricle as described previously [5]. Epinephrine was administered as a bolus via the jugular cannula. At various times after the bolus injection the hearts were frozen in situ by compression with liquid nitrogen cooled clamps. In some preparations practolol, a β-adrenergic blocking agent, was administered via the jugular cannula 5 min prior to the epinephrine injection.

In a second preparation ventricular muscle slices were prepared from hearts excised from anesthetized rats. The myocardium from both the left and right ventricles was cut into slices 0.5 mm thick with a Stadie-Riggs tissue slicer. The slices were washed several times in oxygenated bathing solution containing in mmol/l: 118.4 NaCl, 4.69 KCl, 2.52 $CaCl_2$, 25.0 $NaHCO_3$ and 1.18 $MgSO_4$ pH 7.4 as described previously [2]. The slices were finally incubated in oxygenated bathing solution containing glucose (10 m*M*) and ^{32}P (60–300 mCi/mmol, 0.2–1.0 m*M* total phosphate for 40 min at 37 °C to partially label intracellular ATP pools. During the last 10 min of the incubation isoproterenol, epinephrine, norepinephrine, or propranolol was added to the bathing solution as indicated in the Results. The incubation was terminated by freezing the slices as mentioned above and homogenizing the frozen slices in a buffer containing ethylenediamine-tetraacetic acid (EDTA, 10 m*M*), KF (20 m*M*), KH_2PO_4 (20 m*M*) and sucrose (0.25 *M*) pH 6.8 which inhibited the activity of endogenous protein kinases and phosphatases.

The third preparation was an isolated rat heart perfused (constant flow, 10 ml/min) with oxygenated physiological saline solution (PSS) containing 30–60 mCi ^{32}P/mmol (0.1 m*M* total phosphate) for 60 min at 37 °C. Some hearts had a balloon cannula inserted via the left atrium down through the mitral valve into the lumen of the left ventricle for recording the rate of left ventricle, pressure development as described above. Isoproterenol was administered directly into the aortic perfusion cannula via an infusion pump (0.2 ml/min) and the hearts were frozen as described above while still attached to the perfusion cannula. The heart wafers were pulverized in liquid nitrogen [3] and homogenized as described above.

Some homogenates obtained from ventricular slices and perfused hearts were subjected to differential centrifugation to obtain a $1000 \times g$, $29\,000 \times g$ and $100\,000 \times g$ pellet and $100\,000 \times g$ supernatant fractions. Using adenylate cyclase as a membrane marker, cytochrome C reductase as a mitochondrial marker and Ca-Mg-ATPase as a sarcoplasmic reticulum marker the $29\,000 \times g$ pellet fraction appeared to be primarily composed of sarcolemmal membrane and mitochondria whereas the $100\,000 \times g$ pellet and $100\,000 \times g$ supernatant fractions seemed to be composed primarily of sarcoplasmic reticulum and soluble cytoplasm, respectively. Some of the homogenates were treated with an equal volume of 20% trichloroacetic acid, centrifuged, dissolved in 0.1 *M* NaOH (0 °C) and reprecipitated with the acid to obtain an acid-insoluble fraction.

Biochemical Determinations and Procedures

Myocardial cyclic AMP concentration was determined utilizing an assay based on the activation of skeletal muscle protein kinase [6]. Cyclic AMP-dependent protein kinase activity was measured by determining the ability of the enzyme to phosphorylate histone in the absence and presence of 2 μ*M* cyclic AMP [3]. Protein kinase is expressed as the ratio of activity as determined in the absence to that in the presence of cyclic AMP. An increase in the protein kinase activity ratio is assumed to reflect enzyme activation by an increase in the degree of dissociation of the holoenzyme into its regulatory and catalytic subunits. Phosphorylase *b* kinase activity was determined by measuring the ability of the enzyme to catalyze the formation of phosphorylase *a* from crystalline skeletal muscle phosphorylase *b* [5]. The results are expressed as the activity ratio at pH 6.8 to that measured at pH 8.2. An increase in the activity ratio of phosphorylase *b* kinase indicates transformation to the activated (phosphorylated) form of the enzyme. Glycogen phosphorylase activity was measured by the production of glucose-1-phosphate in the absence and presence of 2 m*M* 5′-AMP [5]. The results are expressed as the ratio of phosphorylase activity without 5′-AMP to phosphorylase activity with 5′-AMP. An increase in the activity ratio indicates an increase in the conversion of phosphorylase *b* to phosphorylase *a* which is physiologically more active. The specific radioactivity of formed myocardial [γ-^{32}P]ATP, substrate for protein phosphorylation, was determined utilizing an assay system that measures the incorporation of ^{32}P into casein from [γ-^{32}P]ATP in the presence of cyclic AMP-dependent protein kinase [2]. Within any given set of experiments the ^{32}P incorporation data is corrected for any differences that may have existed in the specific radioactivity of [γ-^{32}P]ATP. Inorganic phosphate was determined by a colorimetric procedure of *Itaya* and *Ui* [10] utilizing malachite green. Protein was determined as described previously using bovine serum as a standard [3]. Sodium dodecyl sulfate (SDS) slab 10%–20% gradient polyacrylamide gel electrophoresis was performed as described previously [4].

Materials

Male 200–250 g Sprague Dawley rats were obtained from Charles River. All salts, dextrose, EDTA, acids, solvents were certified grade from local suppliers. ^{32}P was obtained from New England Nuclear. Epinephrine, norepinephrine and isoproterenol were supplied by Sigma in the chloride form and prepared fresh before use in 0.1% sodium metabisulfite. Propranolol was from Sigma. Practolol was generously supplied by Dr. D. J. Marshall of Ayerst Laboratories. The gel electrophoresis reagents were obtained from Bio Rad.

Statistical Methods

Analysis of variance was performed for paired and unpaired observations [20]. A probability of <0.05 was accepted as indicating a statistically significant difference.

Results

Time-course of Protein Kinase Activation and the Elevation of Both Contractile State and Phosphorylase Activity Induced by Catecholamine Stimulation

In the in situ working rat heart preparation bolus injection of 1 µg/kg epinephrine produced a rapid and maximal elevation in myocardial cyclic AMP concentration within 5–7 sec (Fig. 1). Cyclic AMP-dependent protein kinase and phosphorylase *b* kinase were maximally activated within a similar time period. However, the increase in the rate of ventricular pressure development (dP/dt, an index of contractility) and activation of phosphorylase occurred slower and reached maximum levels later in time (approximately 15 sec) after the bolus injection of epinephrine. All five parameters returned towards control levels 30 sec after the catecholamine injection. Practolol, a β-adrenergic blocking agent, at 5 mg/kg prevented all of the above epinephrine-induced responses. Thus, upon β-adrenergic stimulation of the heart there is a rapid and concomitant increase in cyclic AMP, cyclic AMP-dependent protein kinase and phosphorylase *b* kinase that is followed in time by an elevation in contractile state and phosphorylase activation.

Relationship Between Both Cyclic AMP-Dependent Protein Kinase and Phosphorylase *b* Kinase and Catecholamine-Elicited Changes in Myocardial Cyclic AMP

The activation of cardiac cyclic AMP-dependent protein kinase and phosphorylase *b* kinase was investigated with respect to the myocardial concentration of cyclic AMP in the in situ rat heart preparation. The activity ratios of cyclic AMP-dependent protein kinase and phosphorylase *b* kinase increased in a parallel manner as the concentration of cyclic AMP increased in response to epinephrine stimulation (Fig. 2). The data indicate that the activation of cyclic AMP-dependent protein kinase and phosphorylase *b* kinase are linearly related to myocardial cyclic AMP content at the epinephrine doses employed.

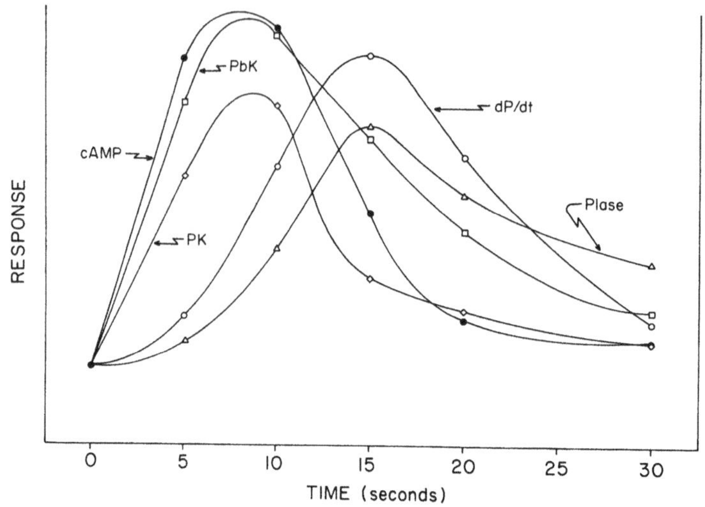

Fig. 1. Time-course of the epinephrine (bolus injection of 1 μg/kg)-induced elevation in the concentration of cyclic AMP (*cAMP*); activity ratios of cyclic AMP-dependent protein kinase (*PK*), phosphorylase *b* kinase (*PbK*) and phosphorylase (*Plase*); and the rate of left ventricular pressure development (*dP/dt*) in the in situ rat heart. The recorded maximal changes in the five above parameters were from: 0.5 to 1.5 (μmol/kg tissue, wet weight) for cAMP, 0.2 to 0.7 (−cAMP/+cAMP) for PK, 0.04 to 0.14 (pH 6.8/8.2) for PbK, 0.15–0.72 (−5′-AMP/+5′-AMP) for Plase and 3200 to 7100 (mmHg/sec) for dP/dt. Each time point represents the mean of 5 hearts. The increases in cAMP, PK and PbK were significantly different from zero time within 5 sec, whereas the increases in Plase and dP/dt did not become significantly different from zero time until 10 sec after the epinephrine injection

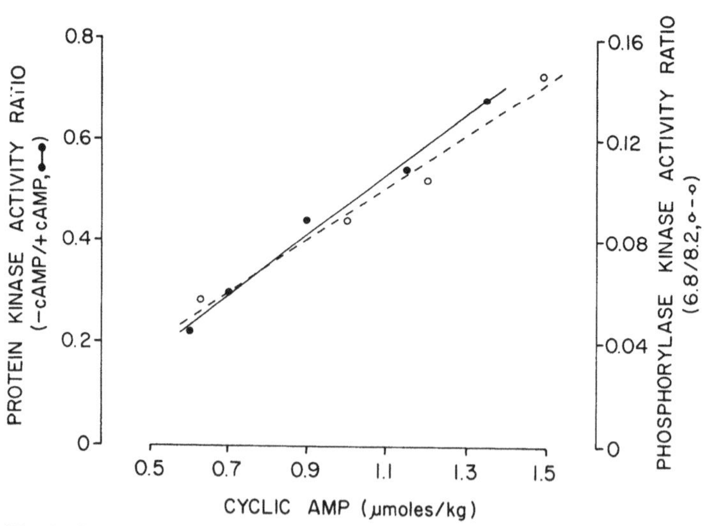

Fig. 2. Relationship between epinephrine-induced increases in myocardial cyclic AMP concentration and the elevation in the activity ratios of cyclic AMP-dependent protein kinase and phosphorylase *b* kinase. The data points were obtained from hearts frozen in situ 10 sec after bolus injection of epinephrine at 0.1–1.0 μg/kg into the open-chest rat heart preparation. Each point represents the mean of 5 to 6 experiments

Incorporation of ^{32}P into Cardiac Muscle Elicited by Catecholamine Stimulation

The possibility that the phosphorylation of myocardial proteins could be observed upon catecholamine-induced activation of cyclic AMP-dependent protein kinase and phosphorylase *b* kinase was investigated using ventricular muscle slices. Incubating ventricular slices in the presence of ^{32}P for 50 min and 2×10^{-5} *M* isoproterenol for the last 10 min did not significantly increase ^{32}P content (based on inorganic phosphate) in ventricular slice homogenates (Fig. 3). Isoproterenol failed to produce a significant increase in ^{32}P content in ventricular slices even if the ^{32}P incorporation was based on homogenate protein. Therefore, to determine if the large amount of phosphate in the slice homogenate was possibly obscuring a catecholamine-induced increase in ^{32}P incorporation, the homogenate was centrifuged as described in the Methods to yield a $1000 \times g$, $29\,000 \times g$ and $100\,000 \times g$ pellet and $100\,000 \times g$ supernatant fractions.

A significant increase in ^{32}P incorporation into the $100\,000 \times g$ pellet (primary sarcoplasmic reticulum) and $100\,000 \times g$ supernatant (essentially soluble cytoplasmic constituents) fractions was observed with a 10 min exposure of the ventricular slices to isoproterenol (Fig. 3). A significant increase in ^{32}P incorporation into the $100\,000 \times g$ pellet and $100\,000 \times g$ supernatant fractions was observed as early as 2 min after addition of isoproterenol to the bathing solutions of the ventricular slices. While propranolol, a β-adrenergic blocking agent, had little effect alone on

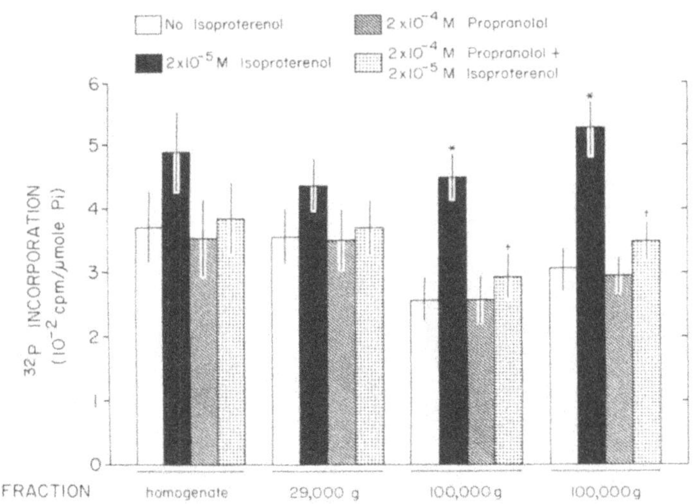

Fig. 3. Effect of isoproterenol on the incorporation of ^{32}P into cardiac muscle protein based on total inorganic phosphate. Rat ventricular muscle slices were incubated in the presence of ^{32}P for 50 min in a bathing solution containing no isoproterenol, isoproterenol, propranolol or propranolol plus isoproterenol for the last 10 min as indicated. In experiments where propranolol was used, the agent was present in the bathing solution for the last 15 min of the incubation. After the incubation period slices were frozen, homogenized and the homogenate, pellet and supernatant fractions prepared as outlined in the Methods. Each value represents the mean ± 1 SE for 6 experiments. *Asterisks* denote a significant difference from the appropriate control (no isoproterenol). *Daggers* denote a significant difference from the appropriate isoproterenol value

the incorporation of ^{32}P into ventricular muscle slices, the blocking agent significantly prevented the isoproterenol-induced ^{32}P incorporation in the $100\,000 \times g$ pellet and $100\,000 \times g$ supernatant fractions. Norepinephrine and epinephrine at 10^{-5} *M* had the same effect as isoproterenol in the above experiments and their effects were blocked by propranolol. The contribution of ^{32}P-phospholipids in the observed ^{32}P incorporation into ventricular muscle slices was found to be insignificant. The results indicate that β-adrenergic stimulation is capable of producing an increase in ^{32}P incorporation into ventricular muscle.

^{32}P Incorporation into Ventricular Muscle Acid-Insoluble Protein

To determine if isoproterenol enhanced ^{32}P incorporation into ventricular muscle ester phosphates which are presumably associated with serine and threonine residues, ventricular slices were incubated in either the absence or presence of isoproterenol in the bathing solution and the resulting homogenate was treated to isolate acid-insoluble protein as described in the Methods. When based on protein the degree of ^{32}P incorporation into acid-insoluble protein of such homogenates was only 3%–7% of that observed in the untreated homogenates. Stimulation of ventricular muscle slices with isoproterenol produced an 86% increase in the ^{32}P incorporation into homogenate acid-insoluble protein that was reversed upon removal of the catecholamine from the bathing solution (Fig. 4) indicating that the

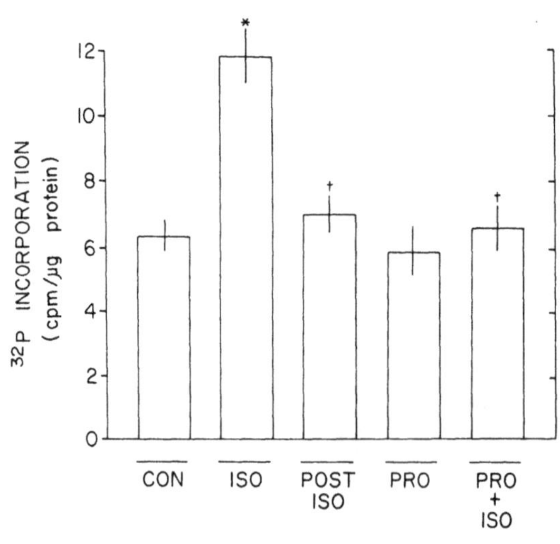

Fig. 4. Effect of isoproterenol on the incorporation of ^{32}P into cardiac muscle acid-insoluble protein. Rat ventricular slices were incubated in bathing solution containing ^{32}P for 50 min (*CON:* control). Isoproterenol (10^{-5} *M, ISO*), propranolol (10^{-4} *M, PRO*) or both *PRO* +*ISO*) were present in the bathing solution for the last 10 min as indicated. When propranolol was used it was present in the bathing solution for the last 15 min. In some instances isoproterenol was present in the bathing solution from time 30 to 40 min, then removed by placing the slices in bathing solution containing no isoproterenol but the same specific radioactivity of ^{32}P for the last 10 min (*POST ISO*). After the incubation the slices were frozen, homogenized and the acid-insoluble proteins obtained as described in the Methods. Each value represents the mean ± 1 SE for 5 experiments. *Asterisks* denote significant difference from control. *Daggers* denote a significant difference from the isoproterenol value

enhanced phosphorylation elicited by the catecholamine is indeed reversible. The measured inorganic phosphate increased significantly from 27.0±1.2 to 34.3±1.6 nmol/mg protein. Propranolol, while having no effect alone, prevented the isoproterenol-induced increase in ^{32}P incorporation into ventricular slice acid-insoluble protein. Isoproterenol at 10^{-5} *M* in the bathing solution increased significantly the ^{32}P incorporation into ventricular muscle slice acid-insoluble protein within as little time as 1 min, indicating that the incorporation of ^{32}P into acid-insoluble protein was rapid. Treatment of acid-insoluble fractions with 0.2 *M* NaOH at 95 °C eliminated the isoproterenol-produced increase in ^{32}P incorporation into acid-insoluble protein and decreased the phosphate content by approximately 85%.

Catecholamine-Induced ^{32}P Incorporation in the Myocardium of the Intact Heart

Since the specific radioactivity of formed [γ-^{32}P]ATP is generally greater in isolated hearts perfused with ^{32}P as compared to ventricular muscle slices [2], isolated rat hearts were perfused with PSS containing ^{32}P by recirculation to more fully label intracellular ATP. In theory, if the specific radioactivity of intracellular [γ-^{32}P]ATP is enhanced then the ability to detect catecholamine-induced phosphorylations should be facilitated. Exposure of isolated perfused hearts to 10^{-5} *M* isoproterenol for 2 min produced an increase in ^{32}P incorporation into a 100 000×*g* pellet and 100 000×*g* supernatant fractions and acid-insoluble protein obtained from the heart homogenates (Fig. 5). These catecholamine-induced incorporations of ^{32}P

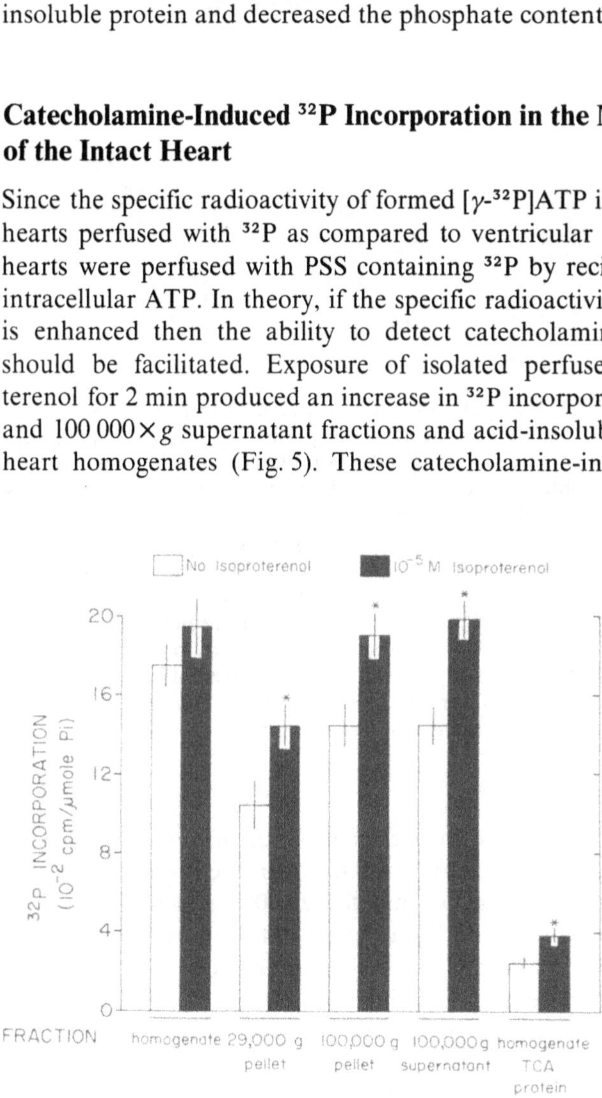

Fig. 5. Effect of isoproterenol on the incorporation of ^{32}P into the intact heart. Isolated rat hearts were perfused (10 ml/min) with PSS containing ^{32}P in a recirculating system for 60 min. Some hearts were exposed to 10^{-5} *M* isoproterenol during the last 2 min of perfusion. After the perfusion period all hearts were frozen, homogenized and homogenate, pellet, supernatant and acid-insoluble (homogenate TCA) protein prepared as described in the Methods. Each value represents the mean ±1 SE for 4 experiments. *Asterisks* denote a significant different from the value in the absence of isoproterenol

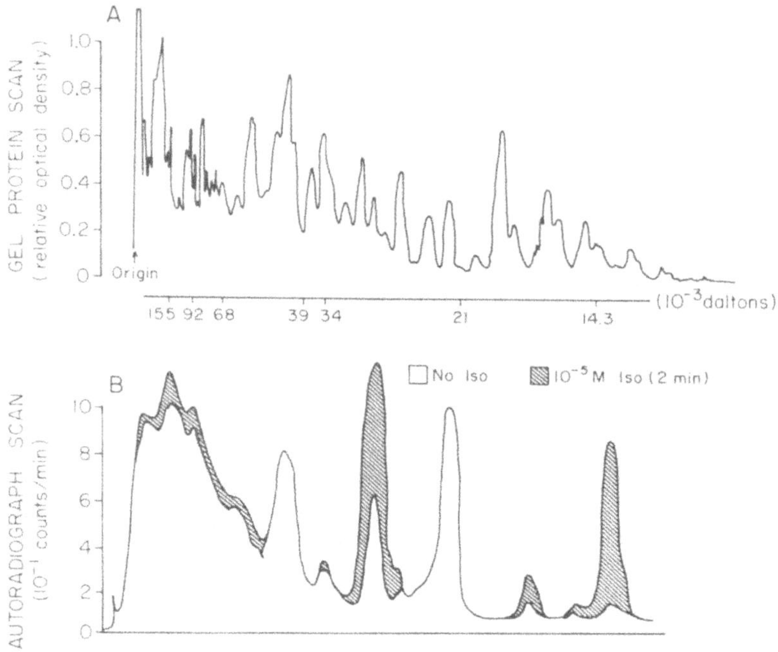

Fig. 6. Effect of isoproterenol on ^{32}P incorporation into cardiac muscle proteins separated with SDS-slab gel electrophoresis. Isolated rat hearts were perfused with oxygenated PSS containing ^{32}P in the absence and presence of 10^{-5} *M* isoproterenol as described in the legend of Fig. 5. The homogenates obtained were solubilized and subjected to electrophoresis on 10–20% gradient polyacrylamide SDS gels. *Panel A* is the absorbance tracing of Coomassie blue-stained proteins obtained from applying 75 μg protein of the homogenate protein at the origin. *Panel B* is a densitometer tracing of x-ray film (Kodak X-R5) exposed to gel tracks containing homogenates from a control and isoproterenol stimulated heart. The *cross-hatched area* is the change in x-ray film development caused by isoproterenol. Under the conditions employed the density of the developed film was linearly related to the radioactivity (up to 200 cpm) on the gel. The molecular weight standard used were RNA-polymerase (155 000, β-subunit), phosphorylase (92 000), serum albumin (68 000), core enzyme (39 000 α-subunit), tropomysin (34 000), trypsin inhibitor (21 500) and lysozyme (14 300)

were prevented by propranolol and are in agreement with that observed above using ventricular muscle slices. However, in the isolated heart the isoproterenol also caused a 39% increase in ^{32}P incorporation into the 29 000×*g* pellet (primarily sarcolemmal membranes and mitochondria) fraction obtained from the heart homogenates. A significant incorporation of ^{32}P in these four fractions was detected with only a 30 sec exposure of the isolated hearts to isoproterenol.

^{32}P Incorporation into Myocardial Proteins from the Intact Heart Separated by SDS Gel Electrophoresis

Isoproterenol (10^{-5} *M*) stimulation of isolated perfused hearts for 2 min produced an increase in the ^{32}P incorporation into various myocardial proteins as judged by radiography of slab gel electrophoresis of perfused heart homogenates (Fig. 6). A significant increase in ^{32}P incorporation was observed in proteins having molecular

Table 1. Effect of 10^{-5} *M* isoproterenol on contractility and incorporation of ^{32}p into myocardial proteins in the isolated rat heart

Time of isoproterenol perfusion (min)	dP/dt (mmHg/sec)	^{32}p incorporation (cpm/100 μg protein) into proteins having molecular weighs of:				
		155 000	92 000	30 000	16 000	13 000
0	2710	296	214	193	36	40
0.5	4730	601	466	527	68	217
1	5860	594	375	614	70	210
2	5520	–	344	–	54	205
4	5340	606	447	625	53	118

Isolated rat hearts were perfused for 60 min with PSS containing ^{32}p. 10^{-5} *M* isoproterenol was present in the perfusing fluid for the last 0–4 min as indicated. Hearts were frozen, homogenized and solubilized homogenate protein subjected to gel electrophoresis and radiography as described in the legend to Fig. 6. Each value represent the mean of 2 to 4 experiments. All values for dP/dt and ^{32}p incorporation in the presence of isoproterenol were significantly different from the zero time values.

weights of 155 000, 92 000, 30 000, 16 000 and 13 000. The increase in the incorporation of ^{32}P into these proteins was observed as early as 30 sec after the initiation of isoproterenol stimulation of the isolated hearts (Table 1). The enhanced ^{32}P incorporation paralleled the increase in dP/dt. There were other groups of proteins which appeared to show an enhanced isoproterenol-induced ^{32}P incorporation but the differences were not found to be significantly different with the present methods employed.

Discussion

Bolus injection of epinephrine produced a transient elevation in cyclic AMP concentration and an activation of both cyclic AMP-dependent protein kinase and phosphorylase *b* kinase that occurred simultaneously and preceded in time the transitory increase in contractility and activation of phosphorylase in the intact working heart. It is also important to note that the responses are reversible as reflected by their transient nature after bolus epinephrine injection. This probably indicates removal of the catecholamine from the coronary circulation by wash out. The *β*-adrenergic blocking agent, practolol, prevented the increase in all five parameters measured.

The degree of cyclic AMP-dependent protein kinase and phosphorylase *b* kinase activation was linearly related to the concentration of cyclic AMP in the intact working heart. The striking correlation between epinephrine-induced increases in cyclic AMP-dependent protein kinase, phosphorylase *b* kinase and cyclic AMP suggests that the actions of epinephrine on these protein kinases are mediated solely by an increase in cyclic AMP. The data are consistent with the hypothesis that upon *β*-adrenergic stimulation, cyclic AMP, cyclic AMP-dependent protein kinase and phosphorylase *b* kinase not only initiate the sequence of events that leads to glycogenolysis via phosphorylase activation but may also be involved in the enhancement of myocardial contractile state as well.

β-Adrenergic stimulation of ventricular muscle slices and isolated perfused hearts, previously exposed to ^{32}P to partially label intracellular ATP, did not significantly increase ^{32}P incorporation into homogenate protein obtained from the two cardiac preparations. However, an enhanced accumulation of ^{32}P into protein of various subcellular fractions prepared from the homogenates was observed. In particular, isoproterenol caused an increase in ^{32}P incorporation into $100\,000 \times g$ pellet and $100\,000 \times g$ supernatant fractions from slice homogenates. In addition the catecholamine increased ^{32}P incorporation into a $29\,000 \times g$ pellet fraction from perfused heart homogenates. The $29\,000 \times g$ and $100\,000 \times g$ pellet fractions represented mainly sarcolemmal membranes plus mitochondria material and sarcoplasmic reticulum, respectively, whereas the $100\,000 \times g$ supernatant contained primarily soluble cytoplasmic protein. Moreover, the ^{32}P incorporation into the fractions occurred within 0.5–2 min and was found to be reversible and prevented by the β-adrenergic blocking agent, propranolol.

The incorporation of ^{32}P into acid-insoluble protein obtained from trichloroacetic acid treated homogenates of both ventricular muscle slices and isolated perfused hearts exposed to isoproterenol was elevated within 0.5–1 min and found to be labile in 95 °C NaOH. This indicates that the incorporation resulted in the formation of ester phosphates in serine and/or threonine residues of protein [15]. The elevation in ^{32}P incorporation into the prepared fractions and acid-insoluble protein from isoproterenol-stimulated isolated cardiac preparations was found whether based on protein or inorganic phosphate. Isoproterenol also caused an increase in inorganic phosphate concentration (based on protein) in acid-insoluble protein. This would suggest that the ^{32}P incorporation probably occurred as a result of protein phosphorylation rather than simply phosphate exchange.

While there was not a significant increase in ^{32}P incorporation into homogenates obtained from isolated perfused hearts exposed to isoproterenol, there was an enhanced incorporation of ^{32}P into certain homogenate proteins separated by SDS-polyacrylamide gel electrophoresis. This was presumably due to the fact that electrophoresis removed the high interfering levels of ^{32}P present in the homogenates. The increase in ^{32}P occurred in proteins having molecular weights of 155 000, 92 000, 30 000, 16 000 and 13 000. The 155 000 dalton protein band may contain the C-protein of myosin [17]. The 92 000 dalton protein may be phosphorylase since it migrated with a phosphorylase standard. The exact identity of the labeled proteins remains to be determined. However, the ^{32}P labeling of the five proteins or protein groups occurred within 30 sec after isoproterenol stimulation of isolated perfused hearts. Thus, the ^{32}P incorporation of cardiac muscle protein caused by isoproterenol occurred within a reasonable time frame with respect to the catecholamine mediated enhancement of myocardial contractility and phosphorylase activation. This would suggest that protein phosphorylation may indeed play an important role in the catecholamine-elicited contractile and glycogenolytic responses in the heart.

Acknowledgements. The author wishes to thank Sherri Bale, Allison Gergely and Delia Mulvaney for their excellent technical assistance. The findings reported herein were supported by Public Health Service Grants 18280 and 22828 and by Research Career Development Award HL-00602 (J.G.D., Jr.) from the National Heart, Lung and Blood Institute.

References

1. Brown DF, Honeyman TW, Dobson JG Jr (1978) Properties of epinephrine-induced activation of cardiac adenosine 3′,5′-monophosphate-dependent protein kinase. Biochim Biophys Acta 544:462–473
2. Cogoli JM, Dobson JG Jr (1981) An easy and rapid method for the measurement of [γ-^{32}P]ATP specific radioactivity in tissue extracts obtained from *in vitro* rat heart preparations labeled with ^{32}Pi. Anal Biochem 110:331–337
3. Dobson JG Jr (1978) Protein kinase regulation of cardiac phosphorylase activity and contractility. Am J Physiol 234:H638–H645
4. Dobson JG Jr (1981) Catecholamine-induced phosphorylation of cardiac muscle proteins. Biochim Biophys Acta 675:123–131
5. Dobson JG Jr, Mayer SE (1973) Mechanisms of activation of cardiac glycogen phosphorylase in ischemia. Circ Res 33:412–420
6. Dobson JG Jr, Ross J Jr, Mayer SE (1976) The role of adenosine 3′,5′-monophosphate and calcium in the regulation of contractility and glycogen phosphorylase activity in guinea pig papillary muscle. Circ Res 39:388–395
7. England PJ (1976) Studies on the phosphorylation of the inhibitory subunit of troponin during modification of contraction in perfused rat heart. Biochem J 160:295–304
8. Frearson N, Solaro RJ, Perry SV (1976) Changes in phosphorylation of P light chain of myosin in perfused rabbit heart. Nature 264:801–802
9. Holroyde MJ, Howe E, Solaro RJ (1979) Modification of calcium requirements for activation of cardiac myofibrillar ATPase by cyclic AMP dependent phosphorylation. Biochim Biophys Acta 586:63–69
10. Itaya K, Ui M (1966) A new micromethod for the colorimetric determinations of inorganic phosphate. Clin Chim Acta 14:361–366
11. Keely SL, Corbin JD (1977) Involvement of cAMP-dependent protein kinase in the regulation of heart contractile force. Am J Physiol 233:H269–H275
12. Kirchberger MA, Chu G (1976) Correlation between protein kinase-mediated stimulation of calcium transport by cardiac sarcoplasmic reticulum and phosphorylation of a 22 000 dalton protein. Biochim Biophys Acta 419:559–562
13. Kopp SJ, Barany M (1979) Phosphorylation of the 19 000-dalton light chain of myosin in perfused rat heart under the influence of negative and positive inotropic agents. J Biol Chem 254:12007–12012
14. LaRaia PJ, Morkin E (1974) Adenosine 3′,5′-monophosphate-dependent membrane phosphorylation. A possible mechanism for the control of microsomal calcium transport in heart muscle. Circ Res 35:298–306
15. Manning DR, DiSalvo J, Stull JT (1980) Protein phosphorylation: Quantitative analysis *in vivo* and in intact cell systems. Mol Cell Endocrinol 19:1–19
16. Namm DH, Mayer SE (1968) Effects of epinephrine on cardiac cyclic 3′,5′-AMP, phosphorylase kinase and phosphorylase. Mol Pharmacol 4:61–69
17. Offer G, Moos C, Starr R (1973) A new protein of the thick filament of vertebrate skeletal myofibrils. Extraction, purification and characterization. J Mol Biol 74:653–676
18. Robison GA, Butcher RW, Oye I, Morgan HE, Sutherland EW (1965) The effect of epinephrine on adenosine 3′,5′-phosphate levels in the isolated perfused rat heart. Mol Pharmacol 1:168–177
19. Schwartz A, Entman ML, Kanicki K, Lane LK, van Winkle WB, Bornet EP (1976) The rate of calcium uptake into sarcoplasmic reticulum of cardiac muscle and skeletal muscle. Effects of cyclic AMP-dependent protein kinase and phosphorylase *b* kinase. Biochim Biophys Acta 426:57–72
20. Sokal RR, Rohlf FJ (1969) Biometry. Freeman, San Francisco, pp 175–203
21. Solaro RJ, Moir AJG, Perry SV (1976) Phosphorylation of troponin I and the inotropic effect of adrenaline in the perfused rabbit heart. Nature 262:615–617
22. Sulakhe PV, St Louis PJ (1976) Membrane phosphorylation and calcium transport in cardiac and skeletal muscle membranes. Gen Pharmacol 7:313–319
23. Tada M, Kirchberger MA, Katz AM (1975) Phosphorylation of a 22 000-dalton component of the cardiac sarcoplasmic reticulum by adenosine 3′,5′-monophosphate-dependent protein kinase. J Biol Chem 250:2640–2647

24. Walsh DA, Perkins JP, Brostrom CO, Ho ES, Krebs EG (1971) Catalysis of the phosphorylase kinase activation reaction. J Biol Chem 246:1968–1976
25. Walsh DA, Clippinger MS, Sivaramakrishnan S, McCullough TE (1979) Cyclic adenosine monophosphate dependent and independent phosphorylation of sarcolemma membrane proteins in perfused rat heart. Biochemistry 18:871–877
26. Wray HL, Gray RR (1977) Cyclic AMP stimulation of membrane phosphorylation and Ca^{2+}-activated, Mg^{2+}-dependent ATPase in cardiac sarcoplasmic reticulum. Biochim Biophys Acta 461:441–459
27. Wyborny LE, Reddy YS (1978) Phosphorylated cardiac myofibrils and their effect on ATPase activity. Biochem Biophys Res Commun 81:1175–1179

Discussion

Opie, Cape Town: Do changes in glycogen levels or turnover really play a role in the contractile response to catecholamine stimulation?

Dobson: This is a question people have asked for years, but I believe that the answer is "no" for 2 reasons:

1. There are inotropic interventions other than with catecholamines that enhance contractile state without influencing myocardial glycogen levels.
2. Isolated perfused hearts in which we previously depleted most of the endogenous glycogen by anoxic insult show a normal mechanical response to catecholamine stimulation when the hearts are perfused with oxygenated physiological saline solution containing glucose. In such hearts we have not determined glycogen turnover, but glycogen content does not change particularly when insulin is not present in the perfusion fluid.

Kukovetz, Graz: You have demonstrated that quite a number of proteins have been phosphorylated and I wonder whether you have tried to identify them further.

Dobson: The 92 000 dalton protein in which we see an increase in ^{32}P incorporation comigrates with phosphorylase on SDS gels. The 155 000 dalton protein may be the C-protein of myosin. Regarding the ^{32}P incorporation into other proteins, we have several ideas but I think it is rather premature for me to speculate as to their identity.

Gerlach, Munich: Why did you use rather high concentrations of isoproterenol?

Dobson: I agree that the concentrations of isoproterenol used in the present studies are "rather" high. We employed such concentrations in these initial studies simply to determine whether we could see an increase in ^{32}P incorporation into myocardial protein. In more recent studies (not reported here) we have found that 10^{-7} *M* isoproterenol is capable of producing a significant increase in ^{32}P incorporation into myocardial protein.

Berne, Charlottesville: When was the propranolol administered?

Dobson: The propranolol (10^{-4} *M*) was administered into the perfusion fluid of isolated hearts during the last 5 minutes of the perfusion period. In other words, hearts were exposed to propranolol for 3 min before and then during the 2 min of isoproterenol infusion.

Antiadrenergic Action of Adenosine in the Heart: Possible Physiological Significance

J. Schrader, G. Baumann[1], *E. Gerlach*

Summary

Adenosine was recently demonstrated by us to attenuate the inotropic and glycogenolytic effects of catecholamines in the guines pig heart [24]. In an attempt to further characterize this effect, the influence of adenosine on cardiac contractility and adenylate cyclase activity stimulated by isoproterenol (ISO), histamine (H), and dopamine (DOP) was studied. Furthermore, the possible role of endogenously formed adenosine in modulating the response of the heart to β-adrenergic stimulation was investigated.

Adenosine applied to the isolated perfused guinea pig heart produced a dose-dependent inhibition of contractile force development elicited by bolus injections of ISO, H, and DOP. The action of adenosine was rapid in onset and fully reversible. In a particulate membrane preparation of ventricular tissue the adenylate cyclase activity stimulated by ISO, H, and DOP was inhibited by adenosine and this effect could be reversed by theophilline.

Reduction of perfusion pressure from 60 to 40 and 20 cm H_2O diminished the contractile responsiveness of the heart to isoproterenol. This condition was associated with a progressive increase in the concentration of adenosine in the coronary effluent perfusate. Time course of release of adenosine correlated well with the diminished inotropic response, and the concentrations of adenosine were in the catecholamine-antagonistic range. Similar results were obtained when hearts were perfused with a hypoxic medium.

Our findings suggest that adenosine formed by the heart may act as an important negative feedback-inhibitor of the actions of catecholamines, thus preventing sympathetic overstimulation of the heart.

Introduction

Different sites of action of adenosine in the heart have been reported: (a) adenosine dilates coronary resistance vessels [12] by interacting with specific receptor sites on the outer membrane of coronary myocytes [25]. Several lines of evidence suggest that adenosine formed and released by the heart may be involved in the regulation of coronary blood flow [5]; (b) adenosine depresses impulse formation and conduction in the sinus and AV-node, respectively [12]. This action of adenosine has recently been implicated in causing AV-conduction disturbances associated

1 Present address: Department of Internal Medicine, Technical University of Munich, Ismaninger Straße 22, D-8000 Munich 80 FRG

with myocardial hypoxia [3]; (c) adenosine decreases contractile force in atrial myocardium [12] most likely by inhibiting transmembranal Ca^{2+}-influx [2, 23].

Besides its well recognized actions on the coronary vessels, adenosine has more recently been reported to have additional effects on the ventricular myocardium. We have shown adenosine to antagonize the contractile and some metabolic effects of catecholamines in the guinea pig heart [24] and similar actions have been demonstrated for the rat heart [10, 11]. Furthermore, adenosine was reported to inhibit adrenergic transmission at a presynaptic site [14] and to decrease myocardial oxygen consumption [16].

It was the purpose of the present study to further define the antiadrenergic action of adenosine. Furthermore, the possible role of endogenously formed adenosine in modulating the response of the heart to β-adrenergic stimulation was investigated.

Methods

Isolated hearts from guinea pigs (320–380 g) were perfused according to the Langendorff technique either at constant pressure (60 cm H_2O) or at constant flow (12 ml/min) by use of a roller pump. The perfusate consisted of a modified Krebs-Henseleit solution gassed with 95% O_2–5% CO_2 as recently described [22]. All substances to be tested were either infused (< 80 µl/min) for the time indicated in figures, or injected as a bolus (10 µl) directly into the perfusion stream about 2 cm above the orifice of the coronary arteries.

Coronary flow was monitored with a Statham M-4001 electromagnetic flowmeter and an IVM flow probe built into the aortic cannula. Perfusion pressure was measured with a Statham pressure transducer connected to a small rubber balloon which had been inserted into the left ventricle. Balloon pressure was electronically differentiated to yield heart rate and dP/dt. The hemodynamic parameters were displayed on a Beckman dynograph (type R-411).

For quantitation of cAMP, isolated hearts still attached to the perfusion system were frozen between aluminium blocks precooled in liquid nitrogen. Tissue extraction with 0.5 *N* percloric acid, neutralization of the extracts, and determination of cAMP by a radioisotope dilution assay were performed as described [24].

A sarcolemma preparation of guinea pig ventricles was prepared [24] and adenylate cyclase activity was determined by measuring the conversion of [^{14}C] ATP to [^{14}C]cAMP, employing thin layer chromatography [1]. The assay system (final volume: 150 µl) contained Tris-HCl, 40 m*M*, pH 7.5; $MgSO_4$, 15 m*M*; EGTA, 0.1 m*M*; [8-^{14}C]ATP, 0.4 m*M*; papaverine hydrochloride, 1 m*M*; creatine phosphate, 10 m*M*; and creatine kinase (3 U/150 µl). Hormones were added at final concentrations as indicated in Results.

Results

Haemodynamic Actions of Adenosine

As illustrated in Fig. 1, adenosine is a potent inhibitor of the inotropic actions of isoproterenol. At a concentration of $8 \cdot 10^{-6}$ *M*, adenosine not only dilated the coronary vessels, but also completely abolished the contractile effect of β-

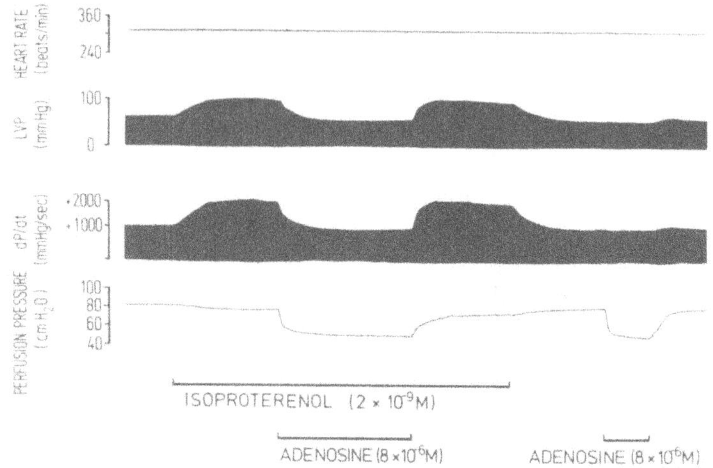

Fig. 1. Effects of adenosine on the isoproterenol-induced increase in ventricular contractile force in the isolated guinea pig heart. Heart was electrically paced and perfused at constant flow (12 ml/min)

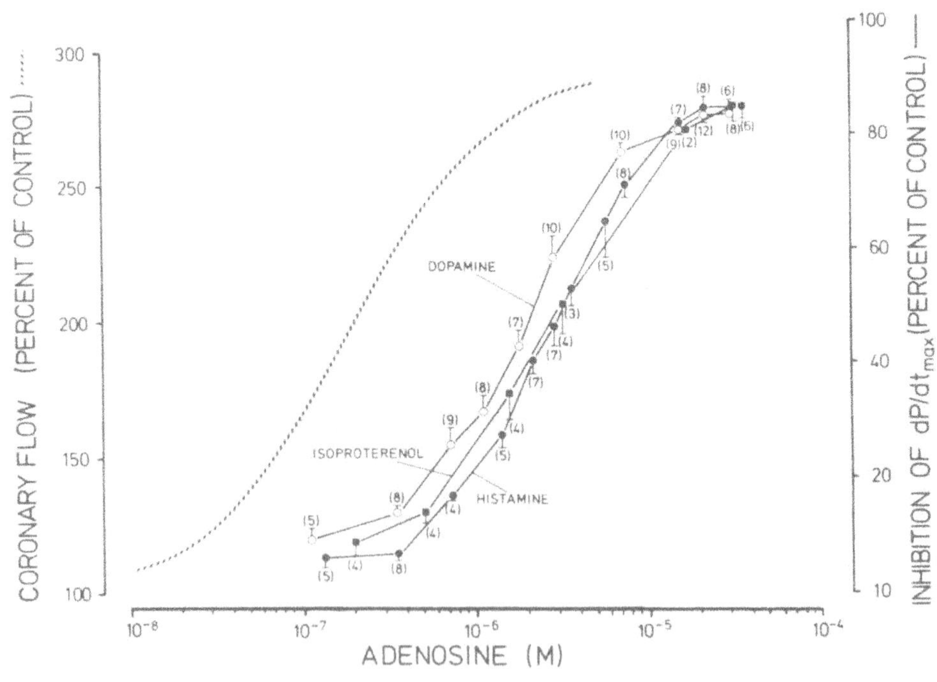

Fig. 2. Dose-response curves of the effect of adenosine on coronary flow and contractile force development of hearts stimulated by isoproterenol (40 pmol), histamine (7.5 nmol) and dopamine (15 nmol). The contractile responses to inotropic agents obtained in electrically paced hearts were quantitated by measuring from the original tracing, the total area of the hormone-induced increase of dP/dt_{max}. $\bar{x} \pm$ S.E.M.

adrenergic stimulation. The action of adenosine was rapid in onset and fully reversible within 1 minute after cessation of adenosine infusion. Interestingly, application of adenosine to the nonstimulated heart did not significantly influence contractile force development (Fig. 1).

Similar experiments as with isoproterenol were carried out with histamine and dopamine, which are known to act via different sarcolemma receptor sites [17]. From our data summarized in Fig. 2 it is obvious that the dose response curves for the different hormones proved to be very similar. Inhibition of increases in dP/dt_{max} started at about 10^{-7} *M* adenosine and reached maximal values at about $2 \cdot 10^{-5}$ *M*. For comparison, the dose-response curve for the coronary dilating action of adenosine [5] is included in Fig. 2. As can be seen, the coronary vessels are about 10 times more sensitive to adenosine, compared with the contractile effect of this nucleoside.

Inhibition of Adenylate Cyclase by Adenosine

We have previously shown that adenosine inhibits cAMP accumulation induced by isoproterenol [24]. These findings already suggested that adenosine may interfere with the adenylate cyclase-cAMP system in the heart. We have therefore investigated the influence of adenosine on adenylate cyclase activity of a crude particulate membrane preparation from guinea pig ventricles. From our data summarized in Fig. 3 it can be seen that the increase in adenylate cyclase activity brought about by equieffective doses of isoproterenol, histamine and dopamine was inhibited by adenosine in all cases. This inhibitory effect was sepecific for adenosine, since degradative products of adenosine, namely inosine and hypoxanthine were found to be ineffective. Interestingly, theophilline inhibited the action of adenosine on adenylate cyclase, irrespective of the hormone used for stimulation.

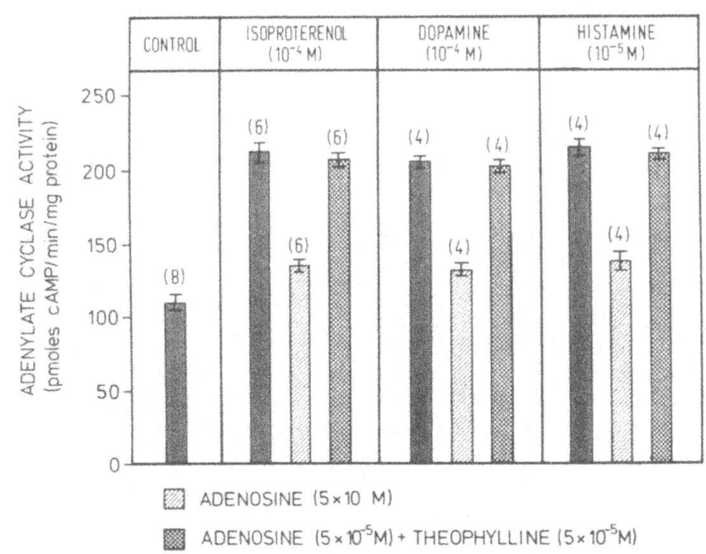

Fig. 3. Influence of adenosine and theophilline on isoproterenol-, dopamine- and histamine-stimulated adenylate cyclase activity in a particulate membrane preparation of guinea pig ventricles. $\bar{x} \pm$ S.E.M.

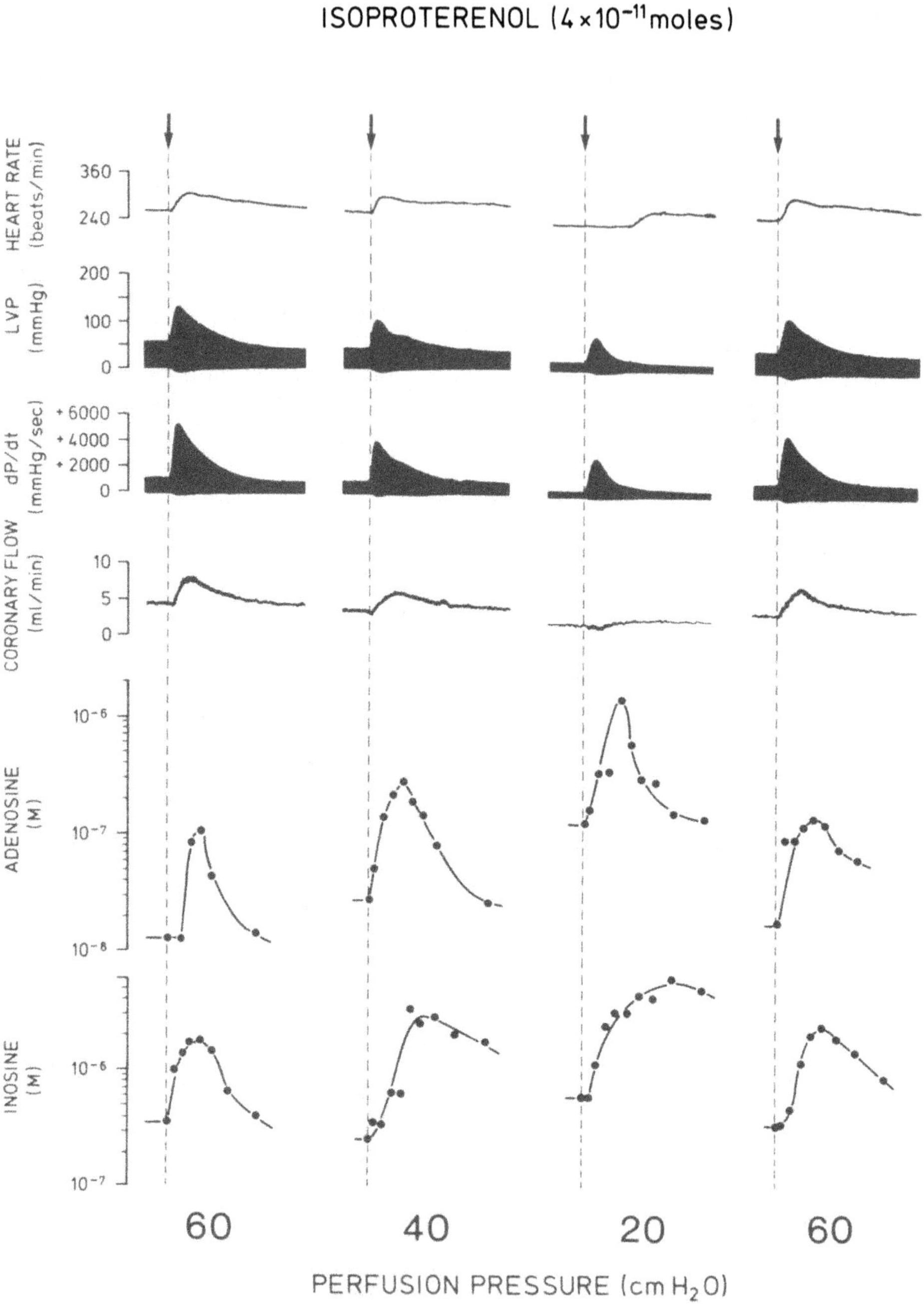

Fig. 4. Effects of bolus injections of isoproterenol during perfusion of an isolated guinea pig heart at different perfusion pressures. Haemodynamic parameters and concentrations of adenosine and inosine are from the same experiment

Relationship Between Myocardial Adenosine Formation and Response to Catecholamines

In order to ascribe a physiological role to adenosine in the heart, it must be demonstrated that this nucleoside is formed and released in amounts which are sufficient to mimic the biological response.

As shown in Fig. 4, a reduction of perfusion pressure from 60 to 40 and 20 cm H_2O, respectively, considerably diminished the isoproterenol-induced increase of left ventricular pressure and dP/dt_{max} in a reversible manner. This hemodynamic condition was associated with considerable changes in the release of adenosine and its degradative product inosine into the effluent perfusate. Reduction in perfusion pressure per se already led to a stepwise increase in perfusate levels of adenosine. Upon bolus injection of isoproterenol there was an additional rapid increase of adenosine. The peak concentration values for adenosine were in the catecholamine-antagonistic range.

Similar observations were made when hearts were rendered hypoxic by reducing the oxygen-content of the equilibrating gas-mixture from normal 95% O_2 to 30% O_2 (Fig. 5). Hypoxic perfusion decreased the inotropic response to catecholamines and this effect was again associated with an augmented release of adenosine.

Provided our assumption is valid, that endogenously formed adenosine inhibits adenylate cyclase in the intact heart, one could expect that the decrease in mechanical responsiveness to catecholamines is paralleled by an attenuated

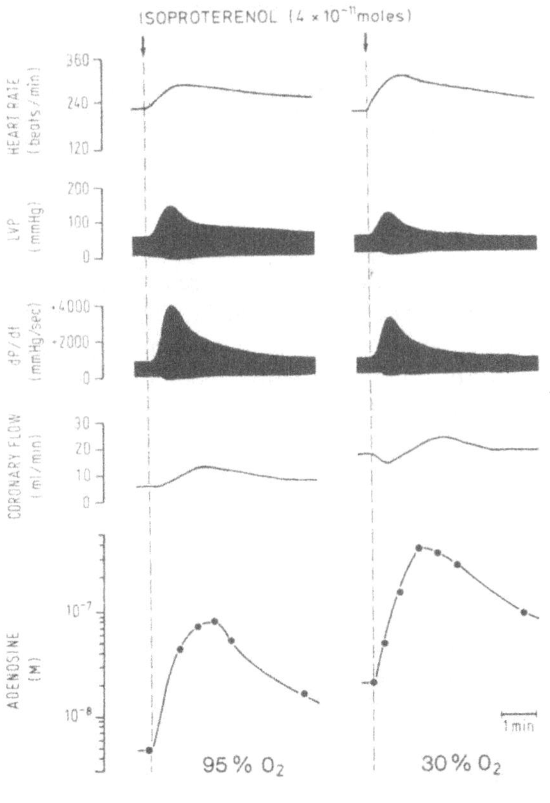

Fig. 5. Influence of hypoxic perfusion on the haemodynamic response to bolus injections of isoproterenol as well as perfusate concentrations of adenosine

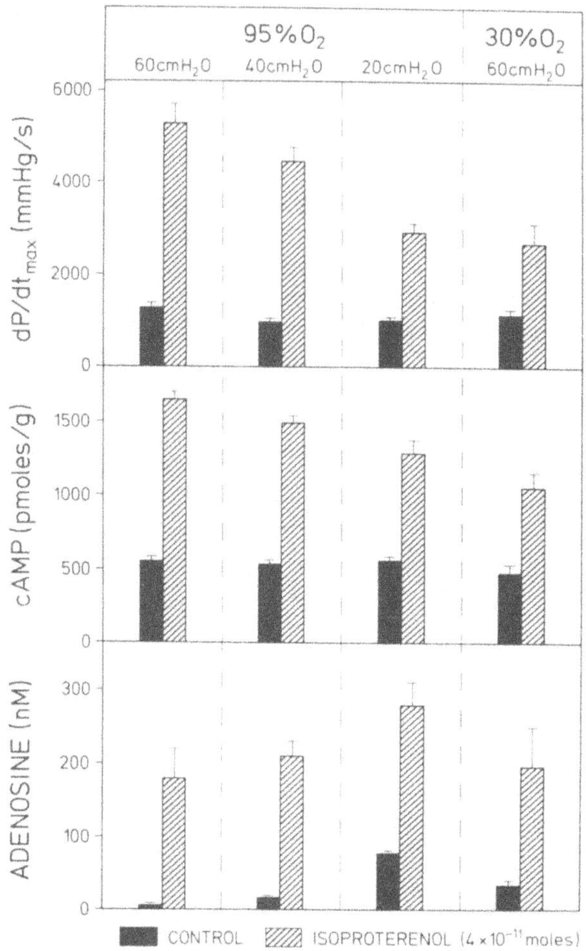

Fig. 6. Influence of changes in perfusion pressure and O_2-content in perfusion medium on contractile response, tissue content of cyclic 3′,5′-AMP and effluate concentration of adenosine in the absence and presence of isoproterenol. Experiments were similar to those shown in Figs. 4 and 5. Hearts were freeze-clamped after application of isoproterenol when maximal contractile responses were reached (15–20 s); $n=5$–7; $\bar{x}\pm$S.E.M.

increase in the tissue levels of cAMP. We have therefore measured cAMP under the two experimental conditions described above. As shown in Fig. 6, hypoxia alone – either elicited by reduction in perfusion pressure or by lowering the oxygen content in the perfusion medium – did not significantly alter tissue cAMP-levels. However, the more adenosine was produced by the hypoxic heart, the less cAMP accumulated in the myocardium.

Discussion

The present study demonstrates that adenosine is a potent inhibitor of the inotropic and metabolic effects of catecholamines. This inhibitory action of adenosine is not limited to β-adrenergic stimulation but could also be demonstrated in hearts which were stimulated by histamine and dopamine. Adenosine, however, did not influence basal contractile force nor did it alter the increase in contractile force brought about by elevation of extracellular $[Ca^{2+}]$ [1]. Adenosine was reported also not to antagonize the positive inotropic action mediated by α-adrenoreceptors of the heart [13]. It thus appears that the inhibitory action of adenosine can only be demonstrated when hormonal stimuli are involved which act via the adenylate cyclase-cAMP system of the heart.

Stimulation of cardiac β-adrenoreceptors causes an enhanced Ca^{2+}-influx by increasing the slow inward current [21]. It has been suggested that it is the number of functional conductance channels for Ca^{2+} which are increased by catecholamines [21]. In atrial myocardium from guinea pigs we and others have reported adenosine to depress the catecholamine-induced, Ca^{2+}-mediated slow action potentials [2, 23]. Recent evidence also indicates that adenosine rather selectively can inhibit the catecholamine-induced ventricular slow action potentials in embryonic chick hearts [4]. Based on these findings the mechanical data reported in this study may be explained by an adenosine mediated reduction of the number of newly formed Ca^{2+}-channels.

Methylxanthines are known to inhibit phosphodiesterase activity at high concentrations (10^{-4} to 10^{-2} *M*) [7]. In the range of $1\text{–}5 \cdot 10^{-5}$ *M*, however, these compounds specifically antagonize the action of adenosine [3, 6, 7, 14]. It was the latter concentration range in which theophilline inhibited the antiadrenergic effects of adenosine on contractile force and adenylate cyclase activity. Although the precise mechanism of theophilline action is not fully understood it may involve competition with adenosine receptor binding at the sarcolemma membrane [28].

Adenosine was proposed to modulate adenylate cyclase activity by two different sites, designated P- and R-site [18]. From our data and studies with different adenosine derivatives [1, 24, 29] it appears that the effect of adenosine on cardiac contractility is mediated by the R-site, located on the external cell surface. Inhibition of cardiac adenylate cyclase by adenosine most likely also explains the observed reduction in cAMP levels of hormonally stimulated hearts. Similar to our observation in guinea pig hearts, adenosine was recently reported to also depress the isoproterenol-induced changes of cAMP in cultured heart cells [15] as well as cAMP-dependent protein kinase and glycogen phosphorylase in perfused rat heart [11]. It is interesting to note in this respect that in frog ventricles adenosine lowers cAMP tissue levels, while cGMP increased concomitantly [27].

It is a well documented observation that the inotropic effects of catecholamines are greatly attenuated in the hypoxic heart [8, 9, 19]. The mechanism underlying this effect, however, is only poorly understood. The findings reported in this study provide first evidence that endogenously formed adenosine may play an important role. Not only did the formation of adenosine rapidly increase with isoproterenol, also the concentration values of adenosine were in the catecholamine-antagonistic range. These findings strongly suggest that an adenosine-mediated mechanism may be operative in the normal heart which participates in the attenuation and limitation of β-adrenergic stimulation.

Our view concerning the possible interaction between adenosine formation and catecholamine action at the molecular level is summarized in Fig. 7.

Adenosine in the myocardial cell can be formed by degradation of cardiac adenine nucleotides, and also from S-adenosylhomocysteine [26] which is an obligatory product of S-adenosylmethionine-dependent methylation reactions. Whenever adenosine is produced at an enhanced rate it diffuses into the extracellular space and reaches its two principal sites of action: the coronary resistance vessels and the sarcolemma membrane. Since coronary smooth muscle

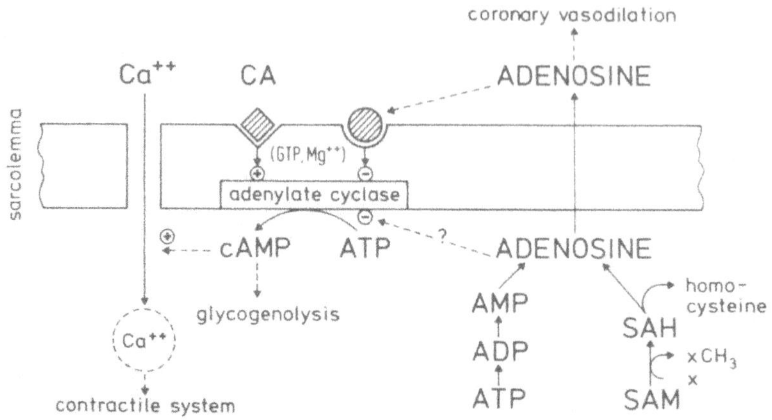

Fig. 7. Possible interrelationships of the site(s) of adenosine production and the site(s) of adenosine action in the myocardium

cells are more sensitive to adenosine than ventricular myocardium (Fig. 2) the dilatory action of adenosine is likely to be prominent at low rates of adenosine production. However, when adenosine production is further enhanced by catecholamines, the antiadrenergic action of adenosine becomes operative.

Separate receptors for adenosine and catecholamines are located on the outer surface of the sarcolemma membrane and are linked to the adenylate cyclase which requires GTP and Mg^{+}. It is possible that a second inhibitory site of adenosine may be located on the inner surface of the membrane in close association with the catalytic unit of this enzyme [18]. Adenylate cyclase is viewed to integrate stimulatory and inhibitory signals, thereby setting the rate of intracellular cAMP production. cAMP in turn modulates the density of the calcium channels in the sarcolemma membrane possibly by phosphorylation of specific membrane proteins [20]. Since catecholamines not only stimulate cardiac contractility but also enhance production of adenosine, this nucleoside may act as an important negative feed-back inhibitor of the actions of catecholamines. The physiological significance of this mechanism may be the following: whenever the stimulatory effects of catecholamines can not be matched by an adequate O_2-supply, adenosine is formed at a greatly accelerated rate. Adenosine may thus prevent additional sympathetic stimulation and further deterioration of cardiac energy metabolism.

References

1. Baumann G, Schrader J, Gerlach E (1981) Inhibitory action of adenosine on histamine- and dopamine-stimulated cardiac contractilit and adenylate cyclase in guinea pigs. Circ Res 48:259–266
2. Belardinelli L, Rubio R, Berne RM (1979) Blockade of Ca^{2+}-dependent rat atrial slow action potentials by adenosine and lanthanium. Pfluegers Arch 380:19–27
3. Belardinelli L, Belloni FL, Rubio R, Berne RM (1980) Atrioventricular conduction disturbances during hypoxia. Possible role of adenosine in rabbit and guinea pig heart. Circ Res 47:684–691
4. Belardinelli L, Vogel S, Berne RM (1981) Antiadrenergic effect of adenosine on ventricular slow action potentials in embrionic chick hearts. Fed Proc 40:561
5. Berne RM, Rubio R (1979) Coronary Circulation. In: Berne RM, Sperelakis N (eds) Handbook of Physiology – The Cardiovascular System I. pp 873–952

6. Bünger R, Haddy FJ, Gerlach E (1975) Coronary responses to dilating substances and competitive inhibition by theophylline in the isolated perfused guinea pig heart. Pfluegers Arch 358:213–224
7. Chasin M, Harris DN (1976) Inhibitors and activators of cyclic nucleotide phosphodiesterase. Adv Cyclic Nucleotide Res 7:225–264
8. Daniell HB, Bagwell EE, Walton RP (1967) Limitation of myocardial function by reduced coronary blood flow during isoproterenol action. Circ Res 21:85–98
9. Davidson S, Maroko PR, Braunwald W (1974) Effects of isoproterenol on contractile function of the ischemic and anoxic heart. Am J Physiol 227:439–443
10. Dobson JG (1978) Reduction by adenosine of the isoproterenol-induced increase in cyclic 3′,5′-monophosphate formation and glycogen phsophorylase activity in rat heart muscle. Circ Res 43:785–793
11. Dobson JG (1981) Properties of the adenosine inhibition of isoproterenol elicited inotropic and glycogenolytic responses in rat heart. Fed Proc 40:616
12. Drury A, Szent-Gyögyi A (1929) The physiological activity of adenine compounds with special reference to their action upon mammalian heart. J Physiol (Lond) 68:213–226
13. Endoh M, Yamashita S (1980) Adenosine antagonizes the positive inotropic action mediated via β- but not α-adrenoreceptors in the rabbit papillary muscle. Eur J Pharmacol 65:445–448
14. Fredholm BB, Hedqvist (1980) Modulation of neurotransmission by purine nucleotides and nucleosides. Biochem Pharmacol 29:1635–1643
15. Hazeki O, Michio U (1981) Modification by islet-activating protein of receptor-mediated regulation of cyclic AMP accumulation in isolated rat heart cells. J Biol Chem 256:2856–2862
16. Jesmok GJ, Gross GJ, Hardman HF (1978) The effect of adenosine on myocardial metabolism and oxygen consumption in the isolated dog heart preparation. J Mol Cell Cardiol 10:249–261
17. Klein J, Levey GS (1971) Activation of myocardial adenylcyclase by histamine in guinea pig, cat and human heart. J Clin Invest 50:1012–1015
18. Londos C, Wolff J (1977) Two distinct adenosine-sensitive sites on adenylate cyclase. Proc Natl Acad Sci USA 74:5482–5486
19. Mahler JT, Manchanda SC, Cymeran A, Wolfe DL, Hartley LH (1975) Cardiovascular responsiveness to β-adrenergic stimulation and blockade in chronic hypoxia. Am J Physiol 228:477–481
20. Reuter H (1979) Properties of two inward membrane currents in the heart. Ann Rev Physiol 41:413–424
21. Reuter H, Scholz H (1977) The regulation of the calcium conductance of cardiac muscle by adrenaline. J Physiol (Lond) 264:49–62
22. Schrader J, Gerlach E (1976) Compartmentation of cardiac adenine nucleotides and formation of adenosine. Pfluegers Arch 367:129–135
23. Schrader J, Rubio R, Berne RM (1975) Inhibition of slow action potentials of guinea pig atrial muscle by adenosine: a possible effect on Ca^{2+} influx. J Mol Cell Cardiol 7:427–433
24. Schrader J, Baumann G, Gerlach E (1977) Adenosine as inhibitor of myocardial effects of catecholamines. Pfluegers Arch 372:29–35
25. Schrader J, Nees S, Gerlach E (1977) Evidence for a cell surface adenosine receptor on coronary myocytes and atrial muscle cells. Pfluegers Arch 369:251–257
26. Schrader J, Schütz W, Bardenheuer H (1981) Role of S-adenosylhomocysteine hydrolase in adenosine metabolism in mammalian heart. Biochem J 196:65–70
27. Singh J, Flitney FW (1980) Adenosine depresses contractility and stimulates 3′,5′ cyclic nucleotide metabolism in the isolated frog ventricle. J Mol Cell Cardiol 12:285–297
28. Trost T, Schwabe U (1981) Adenosine receptors in fat cells. Identification by (–)-N^6-[^{3}H] Phenylisopropyladenosine binding. Mol Pharmacol 19:228–235
29. Webster S, Olsson RA (1981) Adenosine regulation of canine cardiac adenylate cyclase. Biochem Pharmacol 30:369–373

Discussion

Berne, Charlottesville: Observations in our laboratory by Dr. Belardinelli on right ventricular strips of embryonic (10–15 days) chick hearts treated with TTX indicate that field stimulation in the presence of isoproterenol induces an increase in maximum rate of rise of the action potential ($\dot{V}_{max}$). This effect is decreased in a dose dependent manner by adenosine, but never below basal levels. My question is, why cannot the level of contractility and $\dot{V}_{max}$, that is increased by catecholamines, be depressed *below* basal levels by adenosine?

Schrader: There are reasons to assume that catecholamines increase the number of functional conduction channels for calcium (*Reuter, Scholz,* J. Physiol. (1977) *264:*49–62). It appears that adenosine only affects those newly formed calcium channels and does not influence normal calcium conductance. Consistant with this view is the observation that adenosine only antagonizes the catecholamine stimulated contractility, but not basal contractile force.

Opie, Cape Town: 1. Have you tested the effects of adenosine alone on the rate of formation of cyclic AMP?

2. Have you examined the effects of adenosine on the action of the naturally occurring catecholamines, adrenaline and noradrenaline?

Schrader: 1. Adenosine in the concentration range used in our study does not influence cAMP levels of the heart. In the millimolar range, however, we found cAMP to be elevated. This latter effect could be due to an increased substrate concentration for the adenylate cyclase. In fact, adenosine was recently shown to considerably increase myocardial ATP levels (*Isselhard* et al. J. Mol. Cell. Cardiol. *12:*619–634, 1980).

2. There was no principal difference in the action of adenosine in hearts stimulated by isoproterenol, adrenaline or noradrenaline.

Kukovetz, Graz: We have found in isolated guinea pig hearts that ^{3}H-adenosine is taken up very rapidly from the perfusate. Therefore, I wonder if it is safe to conclude from your data that adenosine exerts its inhibitory effect on adenylate cyclase activity by an action at a surface site of the sarcolemma? Furthermore, was the action of your protein-coupled adenosine you referred to in your discussion slide at coronary smooth muscle membranes or at myocardial cell membranes?

Schrader: Several lines of evidence suggest a surface located site of adenosine action: 1. Phenyl-isopropyl-adenosine, an adenosine derivative which is not metabolized, inhibited catecholamine action in the heart at concentrations starting even at 10^{-8} *M*.

2. Adenosine covalently linked to a macromolecular support (Pflügers Arch. *372:*29–35, 1977) which cannot penetrate across the myocardial cell membrane, inhibited the hemodynamic actions of catecholamines and also induced coronary vasodilation.

Mattila, Helsinki: How does theophylline work in your experimental system when it is applied (a) in moderate doses not modifying phosphodiesterase (PDE) and (b) in higher doeses when it behaves as a PDE inhibitor?

Schrader: In concentrations up to 5×10^{-5} *M*, theophylline only antagonizes the action of adenosine and does not interfere with phosphodiesterease-activity. In this concentration range there is also no influence on hemodynamic parameters of the heart. Only in the millimolar range does theophilline inhibit phosphodiesterase and this is associated with increases in contractile force and coronary flow.

Dobson, Worcester: Your very interesting data neatly support our findings suggesting that adenosine may be serving as a negative-feedback modulator of catecholamine stimulation, particularly when the myocardial adenosine concentrations are elevated.
Not only is this of physiological importance, but it may have clinical relevance, especially in regard to protecting the ischemic myocardium (where adenosine concentrations are markedly elevated) against excessive catecholamine stimulation.

Krawietz, Munich: Levitzki showed a stimulatory effect of adenosine on adenylate cyclase activity of turkey erythrocyte membranes in the absence of L-isoprenaline. Can you explain this different effect of adenosine?

Schrader: Whether adenosine stimulates or inhibits may be critically dependent on the Mg^{2+}-concentration in the assay system. It has recently been shown that at low Mg^{2+} a stimulatory effect of adenosine can be elicited also on myocardial adenylate cyclase (*Webster, Olsson,* Biochem. Pharmacol. (1981) *30:*369–373).

Schultz, Heidelberg: At previous conferences, you showed P-site inhibitory effects of adenosine on myocardial adenylate cyclase. Now you propose the occurrence of an R-site adenosine effect, i.e. the involvement of typical purinergic receptors (for definition of P- and R-site cf. *Londos, Wolff,* Proc. Natl. Acad. Sci. USA (1977) *74:*5482–5486). Could you summarize the evidence for the involvement of purinergic receptors in your adenosine effects on adenylate cyclase?

Schrader: We found phenyl-isopropyl-adenosine to be the most potent inhibitor of the effects of isoproterenol on contractile parameters. This compound, however, only slightly influenced adenylate cyclase. On the other hand, some purine modified adenosine derivatives also inhibited contractile force and this effect could be antagonized by theophylline. Thus, the distinction between R- and P-site may not be as clear cut in the intact heart as in the broken cell preparation. The reason why there are no apparent P-site receptors in the membrane preparation is unclear at present.

Effects of Ionic Composition on Cardiac Sarcolemmal Adenylate Cyclase

L. Leclerc, P. Harris

Summary

Adenylate cyclase activity has been estimated in a myocardial sarcolemmal preparation. The "basal" activity was increased by chloride ions and, to a lesser extent, by acetate ions. Sodium and potassium ions appeared to have no effect. GTP stimulated basal activity. The combination of GTP and Cl^- had an effect which was more than would be expected from the simple addition of the individual effects. Isoprenaline stimulated basal activity and, to a greater extent, the activity in the presence of GTP. The presence of chloride or acetate reduced the stimulatory effect of isoprenaline.

Introduction

For some years, there has been an interest in the effects of cations and anions on adenylate cyclase activity from cardiac and other tissues. The present communication reports part of a group of studies intended to investigate the action of physiological cations and anions in the light of the new knowledge concerning the components of the adrenergic receptor/modulator/cyclase system and their control.

Methods

A sarcolemmal fraction has been prepared from the rabbit heart by the method of *Mas Oliva* et al. [4].

The volume of the assay system was 250 µl. It consisted of: 25 m*M* Tris acetate buffer (pH 7.4); Mg acetate 15 m*M*; Tris-ATP 5 m*M*; EDTA 0.4 m*M*; dithiothreitol 0.4 m*M*; theophylline 10 m*M*; creatine phosphate 5 m*M*; creatine kinase 2 mg/ml; plasmalemmal protein 0.1 mg. The reaction was started by the addition of ATP and proceeded for 20 min at 30 °C. The reaction was stopped by the addition of 500 µl of acetate buffer (pH 4; 1 m*M* EDTA). After adjustment of the pH with Tris, aliquots of the supernatant were taken for cyclic AMP estimation. The estimation was carried out by means of the Amersham International kit, based on competitive binding to a specific binding protein.

When GTP was added, the assay mixture was pre-incubated for 30 min. When l-isoprenaline was added, the mixture contained 1 m*M* ascorbic acid.

Results

To compare the effects of 100 m*M* K^+, Cl^- and CH_3COO^-, a sarcolemmal preparation was incubated with 100 m*M* KCl, 100 m*M* $CH_3 \cdot COO \cdot K$, 100 m*M* HCl adjusted to pH 7.4 with Tris and 100 m*M* acetic acid adjusted to pH 7.4 with Tris.

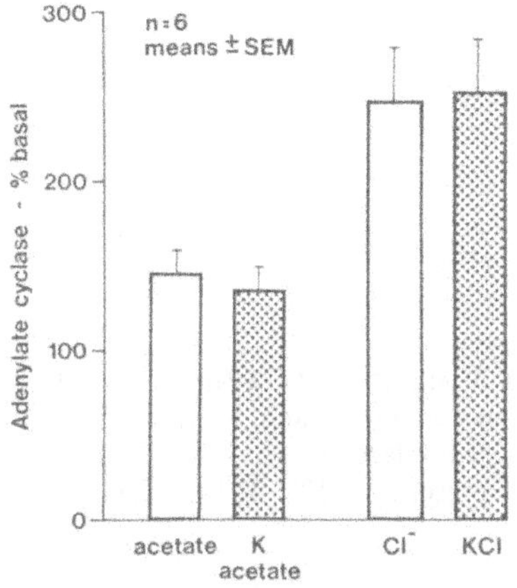

Fig. 1. Effects of chloride, acetate and potassium ions on adenylate cyclase activity of cardiac sarcolemma

Figure 1 shows the results of six such investigations in which the activities are expressed as a percentage of the "basal" activity of the enzyme. The results show clearly that chloride ions stimulated basal activity much more than acetate ions, while the presence of K^+ had no effect.

Figure 2 compares the effects of 10^{-4} *M* GTP, 100 m*M* KCl and the combination of the two on basal activity and on activity in the presence of 10^{-4} *M* *l*-isoprenaline. Both GTP and KCl, given separately, stimulated activity in the absence of isoprenaline. When GTP and KCl were given together the resultant activity was significantly more than would have been expected from the summation of the two separate effects. Isoprenaline stimulated cyclase activity both under basal conditions and in the presence of GTP alone. The absolute effect of isoprenaline was greater in the presence of GTP but the proportional effect was similar. KCl alone also increased basal cyclase activity, but, in the presence of KCl, the stimulatory effect of isoprenaline disappeared. Similarly the effect of isoprenaline was reduced in the presence of GTP and KCl given together.

Figure 3 makes a similar comparison between the effects of 10^{-4} *M* GTP, 100 m*M* CH_3COOK and the combination of the two in the absence and presence of 10^{-4} *M* *l*-isoprenaline. The stimulatory effect of acetate alone was less than that of chloride. When GTP and acetate were given together, the resultant activity was a

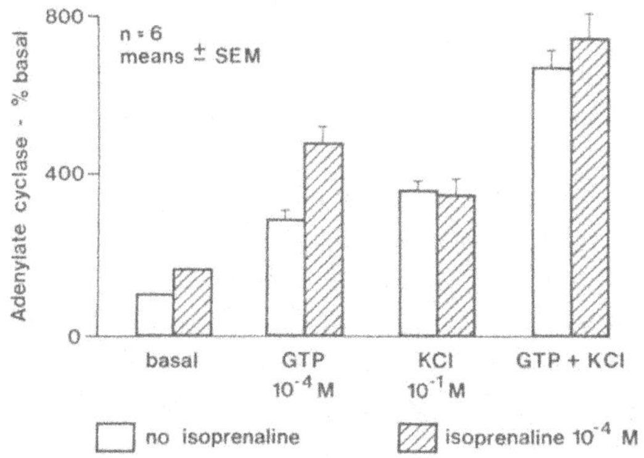

Fig. 2. Effects of GTP and KCl given separately and together on adenylate cyclase activity in the absence and presence of 10^{-4} *M* *l*-isoprenaline

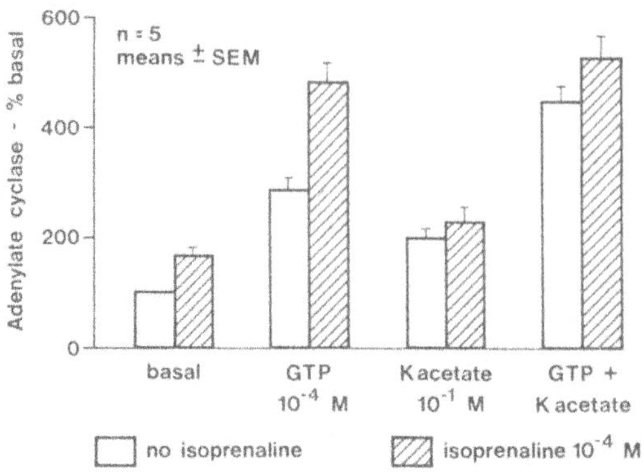

Fig. 3. Effects of GTP and CH_3COOK given separately and together on adenylate cyclase activity in the absence and presence of 10^{-4} *M* *l*-isoprenaline

little greater than would have been expected from a summation of the two, but not significantly so. The stimulatory effect of isoprenaline was reduced in the presence of acetate.

Discussion

The influence of anions on basal adenylate cyclase activity has been demonstrated in a number of tissues, including the liver [1, 2, 3], the kidney [5], the pancreas [6] and fat cells [2]. *Tada* et al. [7] showed a stimulation of basal activity of cardiac adenylate cyclase by KCl and NaCl. They established that K^+ and Na^+ were equivalent and interchangeable, but did not attribute the effect to the anion. Our studies show that, at 100 m*M*, the effects of KCl and CH_3COOK are due to the anions. *Roy* et al. [5] found no great difference between the effects of sodium and potassium salts on renal adenylate cyclase activity, although Li^+ inhibited. *Katz* et al. [3] found differences in the effects of cations on hepatic adenylate cyclase activity, but only at concentrations considerably above 100 m*M*.

The more-than-additive effects of chloride and GTP given together have previously been noted by *Svoboda* and *Christophe* (1978) in the pancreas and *Katz* et al. [3] in the liver.

The influence of anions on hormone-stimulation has been less extensively studied. In the liver, *Johnson* et al. [1] noted that 200 m*M* NaCl prevented the stimulation of adenylate cyclase by glucagon, while *Katz* et al. [3] found that NaCl increased the effect of glucagon up to a concentration of 200 m*M*, after which NaCl became inhibitory. Our results show that the effect of isoprenaline on cardiac adenylate cyclase activity was reduced in the presence of 100 m*M* KCl and CH_3COOK.

The cyclase enzyme is presumed to be located on the cytoplasmic surface of the plasmalemma, while the adrenergic receptor is located on the external surface. In the living cell, these molecules are, therefore, bathed in an entirely different ionic environment. The concentration of chloride surrounding the cyclase enzyme will normally be relatively low. Movements of chloride from the exterior of the cell may, therefore, have an important influence on the rate of formation of cyclic AMP.

References

1. Johnson RA, Pilkis SJ, Hamet P (1975) Liver membrane adenylate cyclase. Synergistic effects of anions on fluoride, glucagon, and guanyl nucleotide stimulation. J Biol Chem 250:6599–6607
2. Kalish MI, Piñeyro MA, Cooper B, Gregerman RI (1974) Adenyl cyclase activation by halide anions other than fluoride. Biochem Biophys Res Commun 61:781–788
3. Katz MS, Kelly TM, Piñeyro MA, Gregerman RI (1980) Anions and cations as stimulators of liver adenylate cyclase. Biochim Biophys Acta 632:11–25
4. Mas-Oliva J, Williams A, Nayler WG (1979) ATP-induced stimulation of calcium binding to cardiac sarcolemma. Biochem Biophys Res Commun 87:441–447
5. Roy C, Le Bars N-C, Jard S (1977) Vasopressin-sensitive kidney adenylate cyclase. Differential effects of manovalent ions on stimulation by fluoride, vasopressin and guanylyl 5′-imidodiphosphate. Eur J Biochem 78:325–332
6. Svoboda M, Christophe J (1978) The monovalent anions chloride and azide as potent activators of NaF^- and p(NH)ppG-stimulation of pancreatic adenylate cyclase. FEBS Lett 86:230–234
7. Tada M, Kirchberger MA, Iorio J-AM, Katz AM (1975) Control of cardiac sarcolemmal adenylate cyclase and sodium, potassium-activated adenosinetriphosphatase activities. Circ Res 36:8–17

Discussion

Krawietz, Munich: I would like to pose three questions: 1. Does Mg^{2+} or F^- have a stimulatory effect on your membrane preparation?
2. Have you examined the stimulatory effect of Cl^- in the presence of different nucleotides (GMP, GDPβS, GppNHp) bound to the nucleotide binding site?
3. To check, it there is any effect of Cl^- on the catalytic unit, it would be necessary to show its effect in CYC^- S49 lymphoma cells which lack the nucleotide binding protein.

Harris: 1. We have examined the effect of different Mg^{2+} concentrations which appear to stimulate up to a maximum at a concentration of 15 m*M*. The action of Mg^{2+} is complicated because Mg · ATP is the substrate for the enzyme and free ATP may be inhibitory. Thus the degree of association of Mg^{2+} and ATP is important and the specific influence of free Mg^{2+} is difficult to distinguish.
2. We are in the process of carrying out these experiments. Cl^- is stimulatory in the presence of GppNHp.
3. That is a very interesting suggestion. We have not carried out such experiments.

Berne, Charlottesville: What is the mechanism of acetate in affecting GTP stimulation of cAMP with isoprenaline?

Harris: We believe its action to be similar to, but less potent than that of cloride. However, we have no knowledge of its molecular site of action.

Dienstl, Innsbruck: Did you also perform similar experiments with lactate?

Harris: No, we have made no observations on lactate.

Stiles, Durham: It has been demonstrated in some systems that washing membranes with EGTA solutions can remove bound GDP from the guanine nucleotide regularoty protein. How do you think the alterations in ion composition from this procedure may work and how would this phenomenon fit into your schema for activation of adenylate cyclase?

Harris: Our assay system routinely includes EGTA. We are at present investigating the effects of complete removal of guanine nucleotides using pre-incubation with isoprenaline and GMP (Method of *Lad* et al.).

Ziegelhöffer, Bratislava: When investigating the influence of electrical stimulation and hormonal stimulation on the activities of sarcolemmal ATPases we have found a less pronounced stimulation in the presence of nitrates. Do you have some experience with nitrate anions?

Harris: No, unfortunately, we have no studies with nitrate ions.

Catecholamines and Cardiac Adenine Nucleotide Metabolism [1]

H.-G. Zimmer, H. Ibel

Summary

Time course studies revealed that in rat hearts in vivo two phases in response to application of isoproterenol can be distinguished: In the initial phase (during the first 5 h), the enhancement of adenine nucleotide biosynthesis coincided with the glycogenolytic effect characterized by an elevation of cAMP and glucose-6-phosphate levels. In the second phase (after 12 h) in which glycogenolysis had become normalized, there was an increase in the activity of glucose-6-phosphate dehydrogenase, the first and rate-limiting enzyme of the hexose monophosphate shunt. This enhanced enzyme activity was associated with the acceleration of adenine nucleotide biosynthesis though maintained on a somewhat lower level. In both phases, the available pool of 5-phosphoribosyl-1-pyrophosphate proved to be elevated. The increase in the activity of glucose-6-phosphate dehydrogenase observed during the second phase could be abolished by inhibitors of protein synthesis (actinomycin D and cycloheximide). Continuous i.v. infusion of dopamine and dobutamine for 48 h did also elicit enhancement of cardiac glucose-6-phosphate dehydrogenase activity and elevation of the available pool of 5-phosphoribosyl-1-pyrophosphate. It thus appears that catecholamines, apart from their well-known hemodynamic and acute glycogenolytic effects, exert a long-term stimulatory influence on the myocardial hexose monophosphate shunt by affecting the synthesis of its first and rate-limiting enzyme. In this way, they may contribute to the maintenance of the enhanced adenine nucleotide biosynthesis.

Introduction

Previous studies have revealed that catecholamines, especially isoproterenol, stimulate adenine nucleotide biosynthesis in rat hearts in vivo in a dose-dependent manner [20]. Basically, there are two mechanisms by which this effect may have been elicited.

1. Isoproterenol induces the well-known Ca^{2+}-dependent positive inotropic effect which leads to the diminution of cardiac adenine nucleotide levels, in particular of ATP [5, 20]. This diminution of ATP causes release of feedback inhibition of the first and rate-regulating enzyme of the biosynthesis of adenine nucleotides, 5-phosphoribosyl-1-pyrophosphate amidotransferase (EC 2.4.2.14). In this way, the de novo formation of adenine nucleotides can be stimulated.

2. Isoproterenol stimulates adenylate cyclase (EC 4.6.1.1) resulting in the increase of 3′,5′-cyclic AMP (cAMP) and ultimately in the stimulation of glycogenolysis. Provided the hexose monophosphate shunt is also enhanced under these conditions, the available pool of 5-phosphoribosyl-1-pyrophosphate (PRPP) would be elevated which is an essential precursor for adenine nucleotide biosynthesis [20].

---

1 Supported by the Deutsche Forschungsgemeinschaft (Zi 199/1–3)

Application of the β-adrenergic blocker propranolol and of the Ca^{2+}-antagonistic compound D 600 made it possible to distinguish between these two possible mechanisms. Propranolol prevented both the positive inotropic and the metabolic effect. In contrast, compound D 600 only abolished the positive inotropic effect, whereas cAMP was still elevated and glycogenolysis as well as adenine nucleotide biosynthesis were accelerated [20, 25]. From these studies it was concluded that, apart from the feedback mechanism exerted by ATP on the PRPP amidotransferase, there must be an additional mode of regulation of adenine nucleotide biosynthesis in the myocardium.

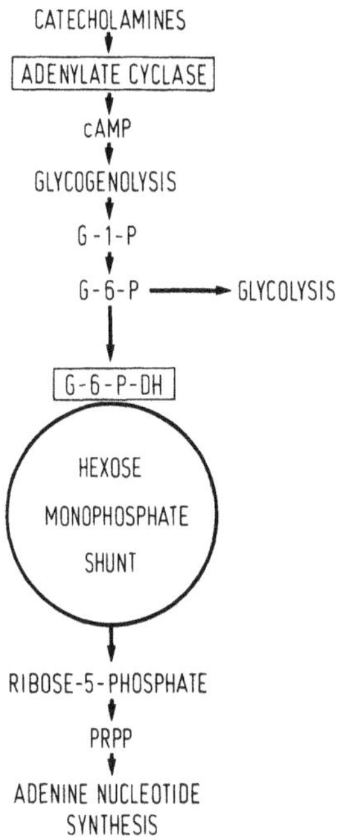

Fig. 1. Schematic representation of the sequence of metabolic events occurring in the myocardium subsequent to β-adrenergic receptor stimulation with catecholamines

The possible metabolic pathway involved in bringing about the stimulation of adenine nucleotide biosynthesis is schematically depicted in Fig. 1. As is well known, catecholamines induce stimulation of glycogenolysis which leads to production of glucose-6-phosphate [3, 10, 15]. Glucose-6-phosphate can be further metabolized either via the glycolytic pathway [13, 18] or via the hexose monophosphate shunt. One of the endproducts of the hexose monophosphate shunt, ribose-5-phosphate, is converted to PRPP, which can then be utilized in the de novo synthesis of adenine nucleotides. To elucidate the possible involvement of the hexose monophosphate shunt in the catecholamine-induced stimulation of cardiac adenine nucleotide biosynthesis, studies were performed, in which the activity of the hexose monophosphate shunt was related to the stimulation of glycogenolysis on the one hand and to the enhancement of adenine nucleotide biosynthesis on the other hand. In addition to isoproterenol, dopamine and dobutamine were examined, since these substances have recently become of clinical interest in the treatment of certain kinds of heart failure [9, 16, 19].

Materials and Methods

All experiments were done on female Sprague-Dawley rats (220–240 g body weight) fed a diet of Altromin with free access to water. Isoproterenol, actinomycin D and cycloheximide were purchased from Sigma Chemie München. Dopamine (kindly provided by Giulini Pharma Hannover) and dobutamine (which was a gift from Lilly GmbH Giessen) were applied as continuous i.v. infusion over 48 h. To do that, a steriflex catheter (Vygon Company Aachen) with an outside diameter of 1 mm was inserted into the right jugular vein of rats anesthetized with ether. It was subcutaneously directed toward the neck of the animals, fixed there and connected to a 20-ml syringe which was placed in an infusion pump (Type 5003, Infors AG Hofstetten). After the operation, the animals could move around freely in individual cages. The infusion rate was 5 ml/kg/h. To prepare the hearts for measuring glucose-6-phosphate dehydrogenase activity, a cannula was placed in the ascending aorta after thoracotomy and tightly fixed. The hearts were rapidly excised, and the coronary arteries were perfused via the cannula with an ice-cold KCl-solution (0.15 *M* containing 8 ml of 0.02 *M* $KHCO_3$/liter) in order to remove blood and to stop beating. In experiments in which the levels of cAMP and glucose-6-phosphate were to be determined, the animals were anesthetized with ether at different times after administration of isoproterenol, the hearts were rapidly removed after thoracotomy and immersed in Freon at a temperature of −156 °C. When the available pool of PRPP and the biosynthesis of adenine nucleotides should be determined, the hearts were excised after thoracotomy, they were blotted on filter paper and immersed in liquid nitrogen.

Levels of myocardial cAMP were determined according to the method of *Gilman* [6]. The content of glucose-6-phosphate was measured using the method of *Lang* and *Michal* [12]. The hexose monophosphate shunt was assessed by measuring the activity of glucose-6-phosphate dehydrogenase, the first and rate-limiting enzyme of this shunt [4] and the availability of PRPP. Measurements of glucose-6-phosphate dehydrogenase activity (EC 1.1.1.49) were done according to the methods of *Glock* and *McLean* [7, 8]. Enzyme activity was determined in the supernatant after homogenization of the hearts in the perfusion medium, centrifugation and dialysis. Protein concentration in the dialysate was determined using the modified biuret reaction [14]. The specific activity of glucose-6-phosphate dehydrogenase was expressed as units/g protein. The available pool of PRPP was assessed from the radioactivity of adenine nucleotides after 5, 10 and 15 min of in vivo exposure of the animals to i.v. injected ^{14}C-adenine (8-^{14}C-adenine, specific activity: 54.2 mCi/mmole, and U-^{14}C-adenine, specific activity: 282 mCi/mmole, both obtained from Amersham Buchler Braunschweig). Since PRPP is consumed in the reaction by which adenine is converted to AMP catalyzed by adenine phosphoribosyltransferase (EC 2.4.2.7) and since AMP is in rapid equilibrium with ADP and ATP via the myokinase reaction (EC 2.7.4.3), the total radioactivity of adenine nucleotides is an indirect measure of the available pool of PRPP [21]. Rates of adenine nucleotide biosynthesis were determined on the basis of the incorporation of 1-^{14}C-glycine into myocardial adenine nucleotides. The radioactivity of adenine nucleotides was related to the specific activity of the tissue precursor pool of glycine thus giving actual rates of de novo formation of adenine nucleotides [24].

Results

The effects of isoproterenol applied as a single high dose of 25 mg/kg subcutaneously on the level of cAMP, glucose-6-phosphate, the activity of glucose-6-phosphate dehydrogenase and on the rates of adenine nucleotide biosynthesis are shown in Fig. 2. There was an immediate increase in glycogenolysis indicated by the elevation of cAMP and of glucose-6-phosphate levels. This stimulation of glycogenolysis was associated with the immediate enhancement of adenine nucleotide biosynthesis. However, 12 h after application of isoproterenol, glycogenolysis had become normalized, since the levels of cAMP and glucose-6-phosphate had returned to their respective base line values. However, adenine nucleotide biosynthesis was still increased, though on a lower level. Exactly at that time the activity of glucose-6-phosphate dehydrogenase started to be enhanced reaching a maximum after 48 h.

Further studies were carried out to evaluate whether the available pool of PRPP might be altered after 5 h, when glycogenolysis was stimulated, but glucose-6-phosphate dehydrogenase activity remained unaltered, as well as after 24 h, when glycogenolysis had become normalized, but glucose-6-phosphate dehydrogenase activity was stimulated. As is evident from the data given in Fig. 3, the incorporation of ^{14}C-adenine into cardiac adenine nucleotides was increased both after 5 and after 24 h, indicating an elevation of the available pool of PRPP. It thus appears that the pool of PRPP and the biosynthesis of adenine nucleotides are increased at all times subsequent to β-adrenergic stimulation with isoproterenol.

From these results it is obvious, that isoproterenol stimulates the hexose monophosphate shunt, since the activity of glucose-6-phosphate dehydrogenase and the available pool of PRPP were found to be increased after 24 h. The question therefore arose whether other catecholamines might also have such a long-term effect on the hexose monophosphate shunt. In this respect, dopamine and dobutamine appeared to be of particular interest, since they are used clinically [9, 16, 19]. These catecholamines were administered as continuous i.v. infusion in unanesthetized and unrestrained rats. The data presented in Fig. 4 show that 48 h after continuous application of dopamine, the activity of glucose-6-phosphate dehydrogenase was enhanced in a dose-dependent manner. Furthermore, the available pool of PRPP also turned out to be elevated when dopamine was applied in a dose of 10 mg/kg/h. Essentially similar results were obtained after 48 h of continuous i.v. infusion of dobutamine, except that the increase in glucose-6-phosphate dehydrogenase activity was not as pronounced as observed with dopamine (Fig. 5).

Additional investigations were performed to elucidate the possible mechanism responsible for the enhancement of enzyme activity. Since it required at least 12 h for glucose-6-phosphate dehydrogenase activity to become enhanced, one possibility may be that protein synthesis is involved. To test this, inhibitors of protein synthesis were applied together with isoproterenol, and glucose-6-phosphate dehydrogenase activity was then measured after 24 h. The results given in Fig. 6 demonstrate that treatment with actinomycin D or with cycloheximide abolished the isoproterenol-induced enhancement of glucose-6-phosphate dehydrogenase activity. This effect was specific for this enzyme, since the activity of 6-phosphogluconate dehydrogenase (EC 1.1.1.44), the second enzyme of the hexose monophosphate shunt, was not affected at all. From these results it appears that the increase

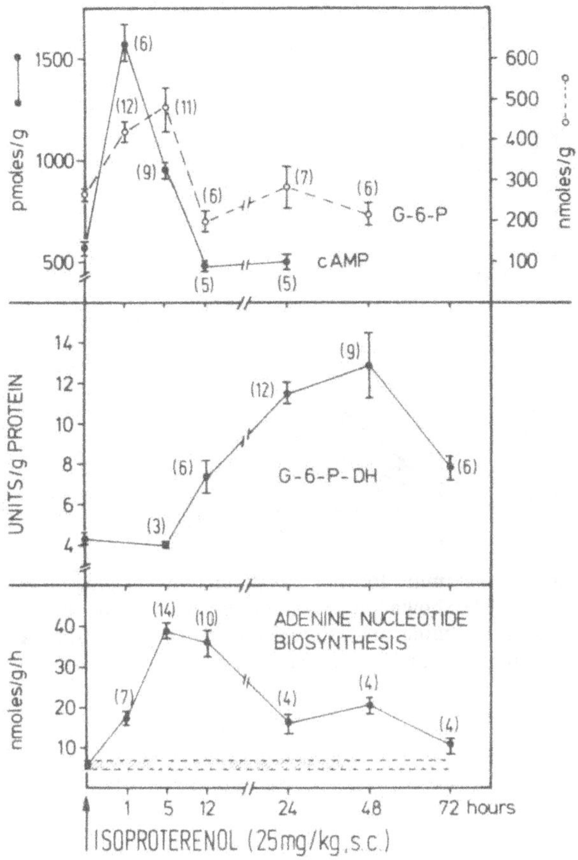

Fig. 2. Time course of changes in the levels of 3′,5′-cyclic AMP (cAMP) and glucose-6-phosphate (G-6-P), in the activity of glucose-6-phosphate dehydrogenase (G-6-P-DH) and in adenine nucleotide biosynthesis in rat hearts in vivo subsequent to s.c. application of isoproterenol

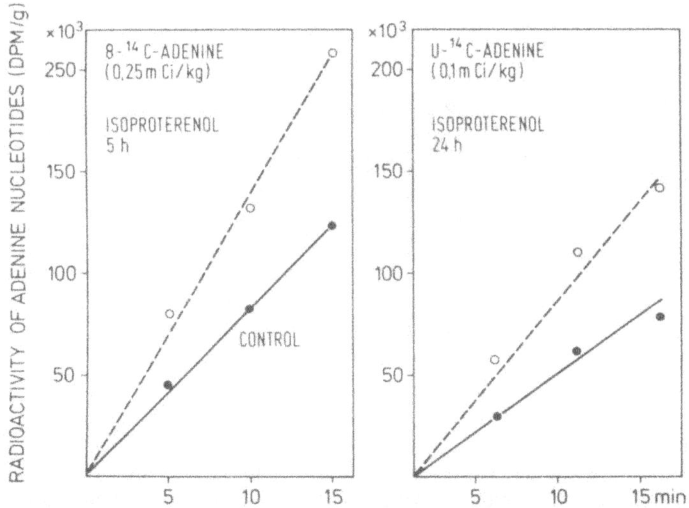

Fig. 3. Incorporation of ^{14}C-adenine into cardiac adenine nucleotides in rat hearts in vivo under control conditions as well as 5 and 24 h after s.c. application of 25 mg isoproterenol/kg. Each point represents mean values of two experiments

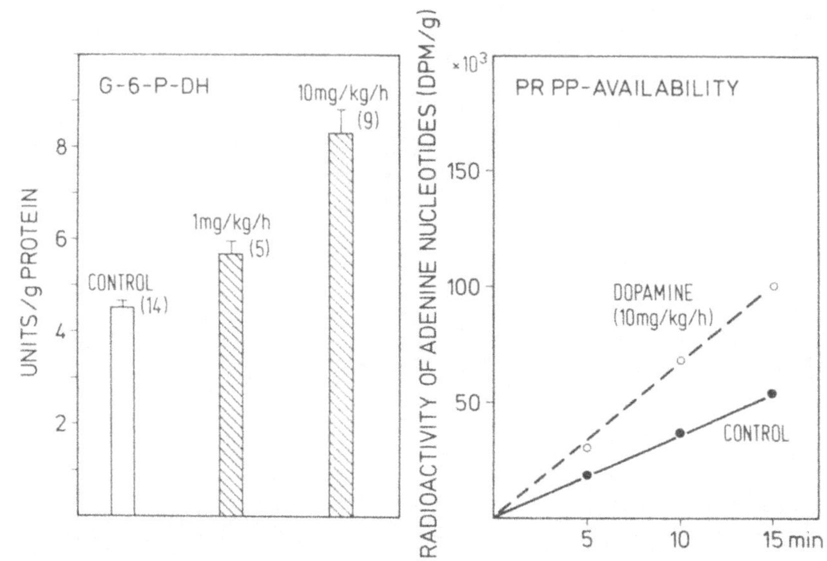

Fig. 4. Effects of continuous i.v. infusion of dopamine for 48 h on the activity of glucose-6-phosphate dehydrogenase activity (G-6-P-DH) shown on the *left-hand panel* and on the incorporation of ^{14}C-adenine into myocardial adenine nucleotides (PRPP-availability on the *right-hand panel*). Mean values ± SEM, number of experiments in parentheses. The data on the *right-hand panel* represent mean values of two experiments

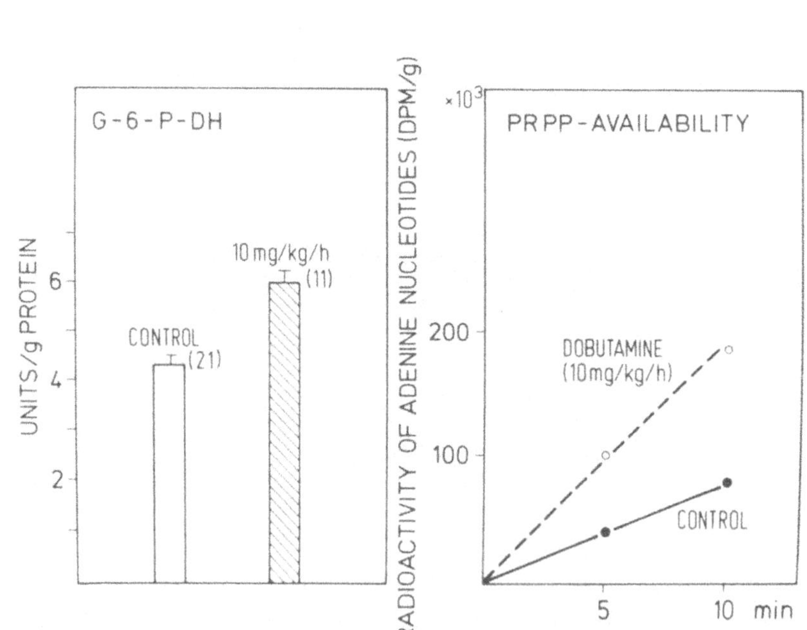

Fig. 5. Changes in cardiac glucose-6-phosphate dehydrogenase activity and in PRPP-availability under the influence of dobutamine 48 h after continuous i.v. administration. Mean values ± SEM, number of experiments in parentheses. The data on the *right-hand panel* represent mean values of two experiments

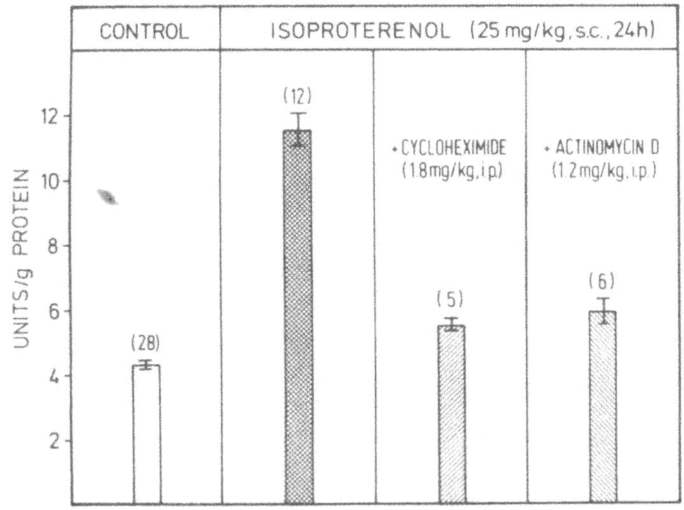

Fig. 6. Effects of cycloheximide and actinomycin D on the isoproterenol-induced stimulation of cardiac glucose-6-phosphate dehydrogenase activity. Mean values ± SEM, number of experiments in parentheses

in protein synthesis is necessary for the stimulation of glucose-6-phosphate dehydrogenase activity elicited by isoproterenol.

Discussion

It has been shown previously that two mechanisms are involved in the regulation of cardiac adenine nucleotide biosynthesis:

1. Feedback control exerted by ATP at PRPP amidotransferase, the first and rate-limiting enzyme of adenine nucleotide biosynthesis.

2. Availability of PRPP, which is an essential precursor for adenine nucleotide synthesis and which is formed from ribose-5-phosphate originating in the hexose monophosphate shunt [21]. The results obtained in the present study provide additional evidence for the involvement of PRPP in the regulation of myocardial adenine nucleotide biosynthesis.

Catecholamines were utilized as experimental tools to stimulate cardiac adenine nucleotide biosynthesis. As could be demonstrated, there is a good correlation between the elevation of the PRPP pool and the enhancement of adenine nucleotide synthesis (Figs. 2, 3) under the influence of isoproterenol. However, there appear to exist two different mechanisms by which the available pool of PRPP can be enhanced, since two phases of metabolic alterations in response to isoproterenol can be distinguished: The first phase (during the first 5 h) is characterized by the enhancement of glycogenolysis, indicated by the elevation of the cAMP and glucose-6-phosphate levels, the increase in the PRPP pool and the immediate acceleration of adenine nucleotide biosynthesis. In the second phase (after 12 h) glycogenolysis has become normalized, but glucose-6-phosphate dehydrogenase activity is enhanced together with the available pool of PRPP and adenine nucleotide biosynthesis. Thus, the increase in the available pool of PRPP and in adenine nucleotide synthesis coincides with the stimulation of glycogenolysis during the first phase and with the enhancement of glucose-6-phosphate dehydrogenase activity during the second phase.

From the increased activity of glucose-6-phosphate dehydrogenase and the elevation of the PRPP pool observed in the second phase, one can conclude that the flow through the oxidative branch of the hexose monophosphate shunt is stimulated. Since, however, glucose-6-phosphate dehydrogenase activity is not enhanced during the first phase, other metabolic reactions seem to be involved in bringing about the elevation of the PRPP pool. Possible alternatives may be (a) the increase in glucose-6-phosphate, which may be metabolized by the oxidative part of the hexose monophosphate shunt without an apparent enhancement of glucose-6-phosphate dehydrogenase activity, (b) an alteration in the $NADP^+/NADPH$ ratio which has been shown to influence glucose-6-phosphate dehydrogenase activity [4], and (c) the reactions within the non-oxidative part of the hexose monophosphate shunt. Which of these possibilities may actually play a role cannot be decided from the results obtained in this study.

A partial explanation for the fact that glucose-6-phosphate dehydrogenase activity is not enhanced during the first phase after isoproterenol application is provided by the studies with inhibitors of protein synthesis. Since the isoproterenol-induced stimulation of glucose-6-phosphate dehydrogenase activity is delayed and can almost entirely be prevented by cycloheximide [17] or actinomycin D (Fig. 6), it appears that protein synthesis is required for the enhancement of glucose-6-phosphate dehydrogenase activity to occur. In addition, time course studies have revealed that the rise in cAMP level is the first metabolic response to isoproterenol (Fig. 2) followed by an increase in protein synthesis. When protein synthesis is maximally enhanced, glucose-6-phosphate dehydrogenase activity starts to increase [23]. Furthermore, prevention of cAMP elevation by β-receptor blockade with atenolol did also abolish the isoproterenol-induced enhancement of glucose-6-phosphate dehydrogenase activity. Thus, the increase in glucose-6-phosphate dehydrogenase activity seems to be mediated via cardiac β-adrenergic receptors, and cAMP seems to be involved in bringing about the increase in the synthesis of proteins such as glucose-6-phosphate dehydrogenase. Although the underlying mechanisms are not known exactly, it is tempting to assume that catecholamines may affect RNA synthesis via stimulation of nuclear protein phosphorylation through the cAMP-dependent protein kinase system [11], thus inducing an increase in the synthesis of proteins and also of enzymes like glucose-6-phosphate dehydrogenase.

Since glucose-6-phosphate dehydrogenase activity is increased in response to β-adrenergic stimulation with isoproterenol, the question arises whether it is this enzyme which actually regulates the cardiac hexose monophosphate shunt [4]. In this context it is interesting to note that only the activity of this first enzyme of the shunt was enhanced, but not the activity of the second enzyme 6-phosphogluconate dehydrogenase. Another question concerns the localization of the hexose monophosphate shunt within the myocardium. Since isoproterenol in the dose applied induces focal myocardial cell lesions [5], the increase in glucose-6-phosphate dehydrogenase activity could well have occurred in inflammatory or connective tissue cells. However, when the isoproterenol-induced focal myocardial injury was prevented by simultaneous administration of compound D 600 [5], glucose-6-phosphate dehydrogenase activity was still enhanced [22]. Furthermore, application of very small doses of isoproterenol which are unlikely to induce cardiac cell lesions, did elicit stimulation of enzyme activity [24]. Finally, studies on isolated cardiac myocytes obtained from rat hearts also support the notion that the hexose monophosphate shunt is active and that it can be stimulated in myocardial cells [1].

A further question concerns the problem whether catecholamines other than isoproterenol or positive inotropic agents not acting via cardiac β-adrenergic receptors have also a stimulatory effect on the hexose monophosphate shunt. In this regard, dopamine and dobutamine deserve special attention because of their clinical importance [9, 16, 19]. Long-term i.v. administration of these catecholamines has been shown in this investigation to stimulate the cardiac hexose monophosphate shunt to about the same extent as does isoproterenol. However, other agents which influence cardiac hemodynamics without affecting β-adrenergic receptors such as ouabain and triiodothyronine [26] do not alter glucose-6-phosphate dehydrogenase activity. Thus, the stimulation of this enzyme activity is a specific response to catecholamines. Another experimental condition, in which the myocardial hexose monophosphate shunt is also activated is hypertrophy of the heart induced by aortic constriction [2, 17, 26]. Since in all these situations, the biosynthesis of adenine nucleotides has been shown to be also accelerated, these processes may represent adaptive and compensatory alterations to meet the increased energy demand of the heart.

References

1. Burns AH, Reddy WJ (1977) Hexose monophosphate shunt in isolated cardiac myocytes from normal rats. Am J Physiol 232(6): E570–E573
2. Clark MG, Williams JF, Kolos G, Hickie JB (1972) The role of the pentose phosphate pathway in myocardial hypertrophy. Int J Biochem 3:629–636
3. Drummond GI, Duncan L, Hertzman E (1966) Effect of epinephrine on phosphorylase b kinase in perfused rat hearts. J Biol Chem 241:5899–5903
4. Eggleston LV, Krebs HA (1974) Regulation of the pentose phosphate cycle. Biochem J 138:425–435
5. Fleckenstein A (1971) Pathophysiologische Kausalfaktoren bei Myokardnekrose und Infarkt. Wien Z Inn Med 52:133–143
6. Gilman AG (1970) A protein binding assay for adenosine 3′,5′-cyclic monophosphate. Proc Natl Acad Sci 67:305–312
7. Glock GE, McLean P (1953) Further studies on the properties and assay of glucose-6-phosphate dehydrogenase and 6-phosphogluconate dehydrogenase of rat liver. Biochem J 55:400–408
8. Glock GE, McLean P (1954) Levels of enzymes of the direct oxidative pathway of carbohydrate metabolism in mammalian tissues and tumours. Biochem J 56:171–175
9. Goldberg LI (1972) Cardiovascular and renal actions of dopamine: potential clinical applications. Pharmacol Rev 24:1–29
10. Hammermeister KE, Yunis AA, Krebs EG (1965) Studies on phosphorylase activation in the heart. J Biol Chem 240:986–991
11. Jungman RA, Russell DH (1977) Cyclic AMP, cyclic AMP-dependent protein kinase, and the regulation of gene expression. Life Sci 20:1787–1798
12. Lang G, Michal G (1974) D-Glucose-6-phosphat and D-Fructose-6-phosphat. In: Bergmeyer HU (ed) Methoden der enzymatischen Analyse. Chemie, Weinheim, S 1283–1287
13. Mansour TE (1963) Studies on heart phosphofructokinase: purification, inhibition and activation. J Biol Chem 238:2285–2292
14. Robinson HW, Hodgen CG (1940) The biuret reaction in the determination of serum proteins. J Biol Chem 135:727–731

15. Robison GA, Butcher RW, Øye I, Morgan HE, Sutherland EW (1965) The effect of epinephrine on adenosine 3',5'-phosphate levels in the isolated perfused rat heart. Mol Pharmacol 1:168–177
16. Tuttle RR, Mills J (1975) Dobutamine. Development of a new catecholamine to selectively increase cardiac contractility. Circ Res 36:185–196
17. Valadares JRE, Singhal RL, Parulekar MR, Beznak M (1969) Influence of aortic coarctation on myocardial glucose-6-phosphate dehydrogenase. Can J Physiol Pharmacol 47:388–391
18. Williamson JR (1964) Metabolic effects of epinephrine in the isolated, perfused rat heart. J Biol Chem 239:2721–2729
19. Wirtzfeld A, Klein G, Delius W, Himmler C, Volger E, Davidson J (1978) Dopamin and Dobutamin in der Behandlung der schweren Herzinsuffizienz. Dtsch Med Wochenschr 103:1915–1921
20. Zimmer H-G, Gerlach E (1974) Effect of beta-adrenergic stimulation on myocardial adenine nucleotide metabolism. Circ Res 35:536–543
21. Zimmer H-G, Gerlach E (1978) Stimulation of myocardial adenine nucleotide biosynthesis by pentoses and pentitols. Pfluegers Arch 376:223–227
22. Zimmer H-G, Ibel H (1979) Effects of isoproterenol and dopamine on the myocardial hexose monophosphate shunt. Experientia 35:510–511
23. Zimmer H-G, Ibel H (1979) Studies on the mechanism for the isoproterenol-induced stimulation of cardiac glucose-6-phosphate dehydrogenase. FEBS Lett 106:335–337
24. Zimmer H-G, Trendelenburg C, Kammermeier H, Gerlach E (1973) De novo synthesis of myocardial adenine nucleotides in the rat. Acceleration during recovery from oxygen deficiency. Circ Res 32:635–642
25. Zimmer H-G, Steinkopff G, Gerlach E (1974) Studien über die Regulation der de novo-Synthese von Adenin-Nucleotiden im Herzen. Verh Dtsch Ges. Kreislaufforsch 40: 348–352
26. Zimmer H-G, Ibel H, Gerlach E (1980) Significance of the hexose monophosphate shunt in experimentally induced cardiac hypertrophy. Basic Res Cardiol 75:207–213

Discussion

Kübler, Heidelberg: In your excellent study you showed an increase in adenine nucleotide synthesis by about 100% after β-stimulation. Since myocardial adenine nucleotide levels are unchanged or even falling after application of isoproterenol, an increased adenine nucleotide turnover has to be assumed, which probably should be regulated by catabolic processes. Do you have any further information regarding the catabolic regulation?

Zimmer: The isoproterenol-induced increase in myocardial adenine nucleotide biosynthesis is clearly time-dependent, amounting to about 500% after 5 h. There is always a diminution of cardiac adenine nucleotide levels subsequent to application of isoproterenol. Thus, the enhancement of adenine nucleotide synthesis appears to be an adaptation, partly caused by the diminution of adenine nucleotide levels as a result of catabolic processes.

Schwarzmeier, Vienna: Do you have data on whether, and if, how phosphodiesterase (PDE) inhibitors influence hexose monophosphate shunt activity under your experimental conditions?

Zimmer: We have not examined the effect of phosphodiesterase inhibition on myocardial glucose-6-phosphate dehydrogenase activities. However, application of dibutyryl-cAMP and of aminophylline have been shown in previous studies to increase the rate of cardiac adenine nucleotide biosynthesis. Since according to our studies cAMP is involved in bringing about the enhancement of glucose-6-phosphate dehydrogenase activity one can expect that this enzyme also becomes activated in the presence of PDE inhibitors.

Harris, London: 1. We found an increased activity of cardiac ornithine decarboxylase activity within 30 min after the injection of isoprenaline in rats, and presumed that it represented an increased synthesis of enzyme protein.
2. Would you care to relate the increased activity of the monophosphate shunt to the timing of the increased formation of RNA?

Zimmer: 1. The increased activity in cardiac ornithine decarboxylase activity observed 30 min after isoproterenol injection may well be due to conformational changes of this particular enzyme, since protein synthesis has not been found to be enhanced so soon.
2. The earliest enhancement of RNA synthesis has been observed 3 h after application of isoproterenol (*Wood* et al., J Mol Cell Cardiol (1971) *3,* 127). This occurs well before the increase in protein synthesis and in glucose-6-phosphate dehydrogenase activity.

Gerlach, Munich: Would you care to speculate on the cause of the delayed onset of the isoproterenol-induced elevation in G-6-PDH synthesis compared to its prompt other metabolic actions. Can this late éffect be actually attributed to a direct action of catecholamines?

Zimmer: The sequence of metabolic events leading to the enhanced synthesis of cardiac glucose-6-phosphate dehydrogenase under the influence of isoproterenol involves an increase in cAMP level and in protein synthesis. Since the processes involved in protein synthesis require some time, hours at the least, it is understandable that the increase in glucose-6-phosphate dehydrogenase activity is delayed. As far as the mechanism for the enhancement of protein synthesis is concerned, one may assume that phosphorylation of nuclear protein through the cAMP-dependent protein kinase system leads to an increase in RNA synthesis which is a prerequisite for any enhancement of protein synthesis.

Hjalmarson, Göteborg: I have a comment to your interesting presentation. Some years back we studied the effects of physical training on the development of hypertrophy of rat heart and skeletal muscles. We found activation of hexose monophosphate shunt enzymes as you have reported. There might be similar effects of catecholamines and work load.

Zimmer: I can confirm your results in regard to the stimulation of the cardiac hexose monophosphate shunt during development of hypertrophy. After aortic constriction in rats, the activity of glucose-6-phosphate dehydrogenase and the available pool of 5-phosphoribosyl-1-pyrophosphate turned out to be enhanced, indicating an increased flow through the hexose monophosphate shunt under these conditions.

Beta-Adrenoceptor Stimulation and Membrane Calcium Permeability [1]

H. Reuter

Summary

β-Adrenoceptor occupancy enhances the calcium-dependent slow inward current in cardiac cell membranes. This catecholamine effect is mediated by cyclic AMP and seems to be due to an increase in the number of functional calcium channels. The regulation of calcium channels by catecholamines is of major functional significance for the positive inotropic and chronotropic effects of these drugs.

Introduction

Postsynaptic membranes recognize neurotransmitter molecules by specific receptors. The neurotransmitter-receptor binding reaction is transduced into a response that often involves membrane permeability changes. In the case of β-sympathomimetic actions on the heart, β-adrenoceptors located on or within the surface membrane are the recognition sites of the natural neurotransmitter, noradrenaline, or other β-adrenoceptor agonists. The transduction of the signal is complex and involves activation of the enzyme adenylate cyclase via a GTP-binding protein (N-protein, see [14]). Activation of adenylate cyclase by the β-adrenoceptor-agonist reaction results in a transient increase in cyclic AMP within the cell which in turn may regulate a variety, if not all, biochemical and biophysical β-adrenergic responses. A prominent β-adrenergic response is a permeability change of the surface membrane to calcium (Ca) ions. In the following, evidence will be summarized (a) that the β-adrenergic effect on the Ca permeability of cardiac cell membranes involves regulation of the number of functional Ca channels in the membranes, (b) that this effect is mediated via cyclic AMP, and (c) that it is of major functional significance for regulation of the heart beat. Since these effects have been reviewed several times [4, 11, 16] only the main results will be illustrated and summarized briefly.

The Effect of Catecholamines on the Ca Permeability of Cardiac Cell Membranes

In the heart two distinct ionic inward currents carry the charge required for depolarization of the cardiac cell membrane during excitation, an initial fast inward sodium current and a secondary slow inward current largely carried by Ca ions (for review see [11]). The slow inward current, responsible for the plateau phase of the cardiac action potential, is modulated by β-adrenergic agonists. This is illustrated in Fig. 1 where the results of a voltage clamp experiment performed in a cow ventricular trabecula are plotted. The Ca-dependent slow inward current has been measured at two different extracellular Ca-concentrations (0.45 and 1.8 mM,

---

1 Supported by the Swiss National Science Foundation

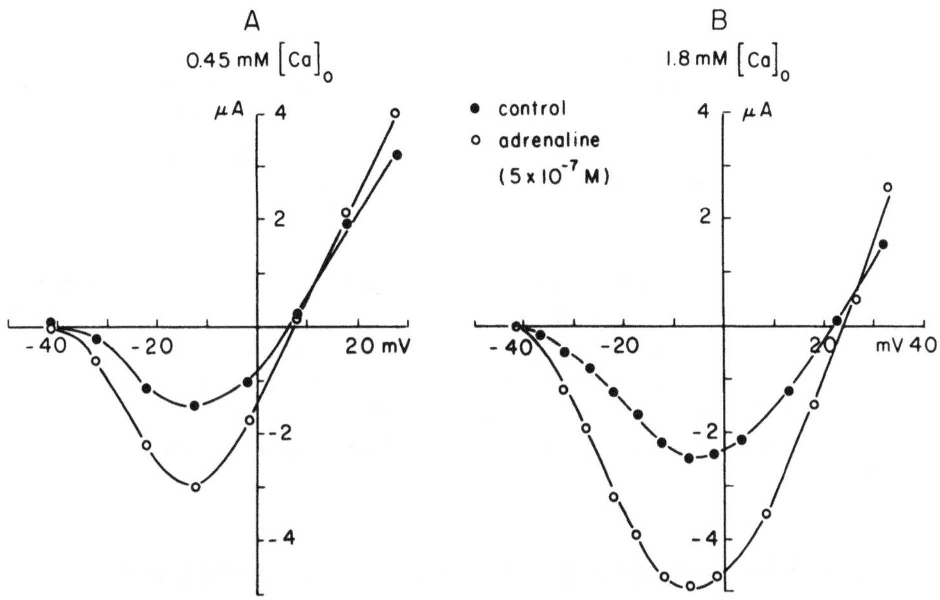

Fig. 1 A, B. Current-voltage relationships obtained, under voltage clamp conditions, from a cow ventricular trabecula at two different Ca-concentrations (**A** 0.45 m*M*; **B** 1.8 m*M*) in Tyrode solution. The currents represented by *closed circles* were obtained under control conditions, those plotted as *open circles* in the presence of adrenaline (5×10^{-7} *M*). Note twofold increase in peak slow inward current by adrenaline in **A** and **B**

Figs. 1 A and 1 B respectively). The current is clearly larger at the higher Ca concentration and peak inward current is reached around −10 mV. In both cases the size of peak inward current is doubled by the addition of adrenaline (5×10^{-7} *M)* to the bathing solution. This effect of adrenaline on the Ca-dependent slow inward current has been extensively analyzed by *Reuter* [10] and *Reuter* and *Scholz* [13]. The main result of the analysis is that neither the kinetics of the slow inward current, nor its driving force are altered by β-adrenergic catecholamines, but that the limiting Ca conductance of the membrane is increased. These results have been confirmed by *Pappano* and *Carmeliet* [9] in sheep heart Purkinje fibres, and by *Noma* et al. [8] in the rabbit sinoatrial node. The increase in the limiting Ca conductance by catecholamines in the heart has been explained by *Reuter* and *Scholz* [13] by an increase in the number of functional Ca channels in the membrane (for extensive review see *Reuter* [12]).

Evidence Concerning Regulation of the Ca Permeability of the Cardiac Cell Membrane by Cyclic AMP

Since the limiting Ca conductance is supposed to reflect the number of functional Ca channels in the membrane, it is of interest to ask the question: what regulates this conductance? Early experiments by *Tsien* et al. [17] in cardiac Purkinje fibres have indicated that monobutyryl cyclic AMP could increase the Ca-dependent slow inward current responsible for the plateau phase of the action potential. These results have been confirmed and extended by *Reuter* [10] who demonstrated an increase in the slow inward current by dibutyryl- and monobutyrylcyclic AMP in cat papillary muscles under voltage clamp conditions. As with catecholamines, the cyclic AMP derivates increase the limiting Ca conductance of the cell membrane. Similar effects have been observed with the phosphodiesterase inhibitor theophylline at low concentrations (< 1 m*M*; [15, 17]). Moreover, acetylcholine which reduces cyclic AMP in the heart [see 2], also reduces the Ca-dependent slow inward current [3]. This occurs again without a noticeable change in the kinetics of the slow inward current, and is due to a marked reduction in the limiting Ca conductance [5].

All these results provide important evidence that cyclic AMP is involved in the regulation of the Ca conductance of cardiac cell membranes. Cyclic AMP could exert its effects on the availability of functional Ca channels via a cyclic AMP-dependent, proteinkinase catalyzed phosphorylation reaction of Ca channels within the membrane [11, 12]. The final steps in the reaction sequence leading to an increase of the Ca permeability of the cardiac cell membrane are, however, entirely hypothetical.

Functional Significance of the Increase in Slow Inward Current

Is the increase in Ca-dependent slow inward current of cardiac cell membranes by β-adrenergic catecholamines related to the most prominent effects of these drugs on the heart, the positive inotropic and chronotropic actions? Figure 2 illustrates an experiment performed in a cat papillary muscle. In this experiment mechanical

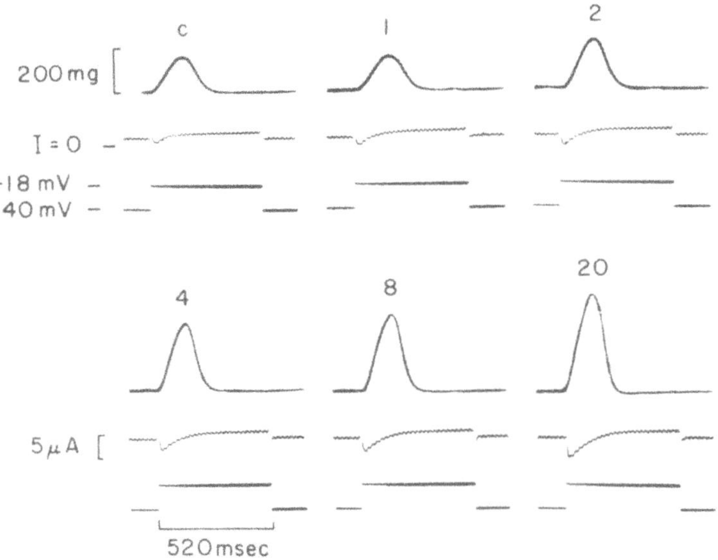

Fig. 2. Voltage clamp experiment in a cat papillary muscle showing the development of noradrenaline (10^{-6} *M*) effect on force of contraction (*upper traces*) and slow inward (calcium) current (*middle traces*). Voltage clamp pulses were applied from a holding potential of –40 mV to a clamp potential of –18 mV at a frequency of 0.3 Hz. The panels show control records (C) and records during 1st, 2nd, 4th, 8th and 20th depolarization after addition of noradrenaline to the bathing solution. Slow inward current is indicated by the initial downward deflection of the current traces during voltage clamp steps. (*Reuter,* 1974)

force and Ca-dependent slow inward current have been measured simultaneously under voltage clamp conditions. The muscle was depolarized by voltage clamp steps of equal size (lower traces) at a rate of 0.3 Hz. After a control record (C), noradrenaline was applied into the bath. The next depolarization (1) already indicated a slight increase in slow inward current, while tension was still unaffected. During the next depolarizations (2–20), inward current and tension increased simultaneously, and a steady state was reached with the twentieth depolarization. This result indicates that the Ca-dependent slow inward current is increased before a noticeable increase in tension. The extra Ca flowing into the fibre during depolarization in the presence of catecholamines may replenish intracellular Ca stores from where it can be released during subsequent depolarizations. The larger amount of Ca released may cause a larger increase in force of contraction [10]. In frog heart, where the main Ca store, the sarcoplasmic reticulum, is much more sparse than in mammalian heart muscle, a more direct relationship exists between increased Ca influx and increase in force by adrenaline [6, 7]. These results suggest that the increase in Ca-dependent slow inward current during the cardiac action potential by catecholamines is an essential step in the positive inotropic effect of these drugs.

The positive chronotropic effect is also closely related to the increase in slow inward current. This is illustrated in Fig. 3. In this experiment *Brown* and *Di Francesco* [1] have measured action potentials (Fig. 3A) and membrane current (Fig. 3B) in a sino-atrial preparation of the rabbit heart. Adrenaline (5×10^{-8} *M*) increased the rate of spontaneous activity and the amplitude and upstroke velocity

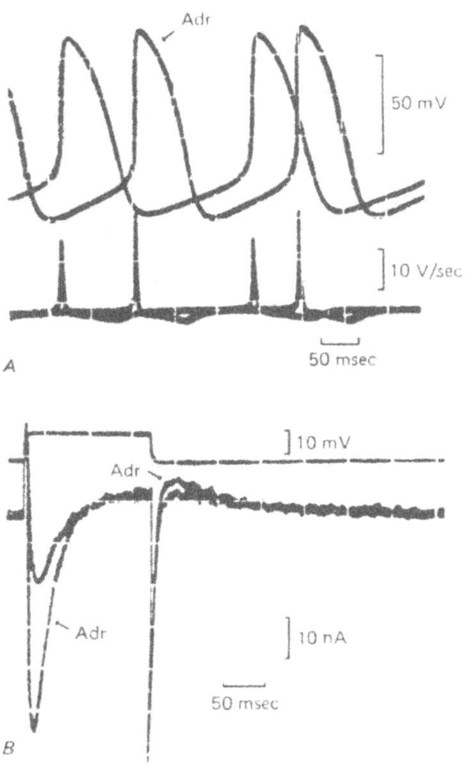

Fig. 3A+B. A Spontaneous pacemaker activity and action potentials recorded from a rabbit sino-atrial node preparation under control conditions and in presence of adrenaline (5×10^{-8} *M*); *lower traces* differentiated action potentials showing increased upstroke velocities of action potentials in presence of adrenaline. **B** membrane current (*lower traces*) measured in another sino-atrial node preparation during a 10 mV depolarizing voltage clamp pulse from a holding potential of –45 mV (*upper trace*) in absence and presence of adrenaline. Note large increase by adrenaline in slow inward current (downward deflection of current traces) during clamp pulse and slight increase in outward current after termination of clamp pulse. Results were obtained in presence of TTX (10^{-7} g/ml). (*Brown* and *Di Francesco,* 1980)

of action potentials. Voltage clamp analysis of these effects showed a large increase in slow inward current during a depolarizing voltage pulse, and a much smaller increase in outward current upon repolarization. A detailed analysis of these effects by *Noma* et al. [8] showed that the increase in slow inward current can be explained by the same mechanism like that suggested by *Reuter* and *Scholz* [13], namely an increase in the limiting conductance. They also concluded from their study that the chronotropic effect of adrenaline results largely from the increase in slow inward current, while *Brown* and *Di Francesco* [1] suggest that another pacemaker current might be important for this effect.

In conclusion, the increase in Ca-dependent slow inward current, though not the only membrane conductance system that is affected by catecholamines in cardiac cell membranes, is of major functional significance for the primary β-adrenergic actions on the heart, the positive inotropy (Fig. 2) and chronotropy (Fig. 3).

References

1. Brown H, Di Francesco D (1980). Voltage-clamp investigation of membrane currents underlying pace-maker activity in rabbit sino-atrial node. J Physiol 308:331–351
2. Drummond GI, Severson DL (1979) Cyclic nucleotides and cardiac function. Circ Res 44:145–153
3. Giles W, Noble SJ (1976) Changes in membrane currents in bullfrog atrium produced by acetylcholine. J Physiol 261:103–123
4. Hagiwara S, Byerly L (1981) Calcium channel. Ann Rev Neurosci 4:69–125
5. Hino N, Ochi R (1980) Effect of acetylcholine on membrane currents in guinea pig papillary muscle. J Physiol 307:183–197
6. Morad M, Sanders C, Weiss J (1981) The inotropic actions of adrenaline on frog ventricular muscle: relaxing versus potentiating effects. J Physiol 311:585–604
7. Niedergerke R, Page S (1977) Analysis of catecholamine effects in single atrial trabeculae of the frog heart. Proc R Soc Lond [Biol] 197:333–362
8. Noma A, Kotake H, Irisawa H (1980) Slow inward current and its role mediating the chronotropic effect of epinephrine in the rabbit sinoatrial node. Pfluegers Arch 388:1–9
9. Pappano AJ, Carmeliet EE (1979) Epinephrine and the pacemaking mechanism at the plateau potentials in sheep cardiac Purkinje fibers. Pfluegers Arch 382:17–26
10. Reuter H (1974) Localization of *beta* adrenergic receptors and effects of noradrenaline and cyclic nucleotides on action potentials, ionic currents and tension in mammalian cardiac muscle. J Physiol 242:429–451
11. Reuter H (1979) Properties of two inward membrane currents in the heart. Ann Rev Physiol 41:413–424
12. Reuter H (1980) Effects of neurotransmitters on the slow inward current. In: Zipes DP, Bailey JC, Elharrar V (eds) The slow inward current and cardiac arrhythmias. Nijhoff The Hague Boston London Developments in Cardiovascular Med 7, 205–219 (1980)
13. Reuter H, Scholz H (1977) The regulation of the calcium conductance of cardiac muscle by adrenaline. J Physiol 264:49–62
14. Rodbell M (1980) The role of hormone receptors and GTP-regulatory proteins in membrane transduction. Nature 284:17–22
15. Scholz H, Reuter H (1976) Effect of theophylline on membrane current in mammalian cardiac muscle. Naunyn-Schmiedebergs Arch Pharmacol 293:R19
16. Tsien RW (1977) Cyclic AMP and contractile activity in heart. Adv Cyclic Nucleotide Res 8:363–420
17. Tsien RW, Giles W, Greengard P (1972) Cyclic AMP mediates the effects of adrenaline on cardiac Purkinje fibres. Nature 240:181–183

Discussion

Parratt, Glasgow: 1. Is the shift in plateau height by noradrenaline a β- or an α-adrenoceptor mediated effect? I thought that Gouvier and also Giotti and Ledda had demonstrated that action potential prolongation was an α-mediated effect.
2. Might not action potential prolongation induced by noradrenaline under certain circumstances be an antiarrhythmic intervention? I will present evidence this afternoon that noradrenaline under certain circumstances (e.g. early in myocardial ischaemia) can reduce ventricular ectopic activity.

Reuter: 1. The shift of the plateau height is a purely β-adrenergic effect. It can be inhibited by β-adrenoceptor antagonists. The prolongation depends, as I mentioned in my talk, very much on the experimental conditions, notably stimulation rate. It could also occur with α-adrenoceptor agonists in some cardiac tissues. 2. Action potential prolongation could, in principle, have an antiarrhythmic effect. However, this effect may not be predictable with β-adrenoceptor agonists, since it depends on various factors, for example heart rate, as I mentioned before.

Lichtlen, Hannover: By which mechanisms is Ca transported out of the cell?

Reuter: There are two transport systems involved in Ca extrusion: (a) the Na-Ca exchange system which extrudes Ca by exchanging internal Ca for external Na, and (b) an ATP-driven Ca pump which may be modulated by calmodulin. Theoretically, this second system could be affected by β-adrenergic catecholamines via cAMP.

Lichtlen, Hannover: Are there any drugs which act on the outward movement of Ca from the cells, especially on the ATP-driven Ca-pump?

Reuter: Glycosides may act via the Na-Ca exchange system by inhibiting the Na pump. This would cause a rise in $[Na^+]_i$ and, subsequently, of $[Ca^{2+}]_i$. The ATP-driven Ca pump could perhaps be modulated by calmodulin, just like the corresponding pump in the sarcoplasmic reticulum is influenced by phospholamban.

Trautwein, Homburg/Saar: 1. Have you any suggestion why the larger Ca inflow induced by adrenaline prolongs the action potential as you have shown in your slide, whereas injection of Ca^{2+} into cells strongly shortens the action potential?
2. Are the properties of the catecholamine-induced Ca channels identical to the "control" Ca channels? In isolated ventricular cells epinephrine produces spontaneity with slow diastolic depolarization at negative membrane potentials where slow inward channels should not be activated.

Reuter: 1. The increased efficiency of the sarcoplasmic reticulum Ca pump may keep the Ca concentration sufficiently low near the membrane, thus preventing Ca from having an effect on K^+ conductance. This, of course, would not hold true for an increase in Ca influx by increasing external $[Ca^{2+}]$, where one finds indeed a pronounced shortening of the action potential.
2. Judging from the study of the ion selectivity of these channels they seem to be identical to the "control" channels.

Kukovetz, Graz: Are similar changes in Ca influx as under epinephrine observed with other stimulators of adenylate cyclase, in particular with histamine or glucagon?

Reuter: Indeed, we and others have seen very similar effects with a variety of drugs, all of which increase cAMP. In guinea pig atria we have seen that only occupancy of histamine H_2-receptors, but not of H_1-receptors, causes effects similar to β-adrenoceptor occupancy. Only occupancy of H_2-receptors results in a stimulation of the adenylate cyclase in this tissue.

Szekeres, Szeged: 1. We have shown in 1970 that adrenoceptor agonists which act on both α- and β-receptors have a biphasic action on diastolic and fibrillation thresholds. The first effect (lowering of the threshold) is brought about by stimulation of β-receptors, the mechanism responsible for the second effect (increase in threshold) *could* not be identified. Did you find such a time dependent change in Ca currents in the isolated heart?
2. Do you conclude that the action of adrenergic agonists on Ca flux is brought about mainly by increasing the *number* of available Ca channels?

Reuter: 1. I cannot really answer your first question, since we have not investigated threshold changes due to β- or α-adrenergic effects. However, excitability threshold is a very complicated matter and involves *many* parameters.
2. Yes, I think so.

The Problem of Catecholamine Oxygen Wasting in the Heart

E. O. Feigl, G. A. Rooke

Summary

The question of myocardial catecholamine oxygen wasting was reexamined in a two-part study. In the first phase, the effects of independently varying systolic blood pressure, heart rate, and stroke volume were observed in closed-chest, anesthetized dogs. This was achieved by cardiac pacing, phenylephrine infusion, controlled hemorrhage, and adjustment of arteriovenous shunts. Pretreatment with propranolol was used to hold contractility constant. From these results, where myocardial oxygen consumption was varied by changing preload and afterload, a new pressure-work index was devised and fitting parameters for other indexes of myocardial oxygen consumption were determined.

These indexes were prospectively applied to a second set of experiments in closed-chest, anesthetized dogs given norepinephrine, isoproterenol, dobutamine, nembutal, and propranolol to alter myocardial contractility. Inotropic "oxygen wasting" was observed with the tension-time and other indexes, but not with the pressure-work or systolic pressure-rate indexes.

It is concluded that whether or not "catecholamine oxygen wasting" is observed depends on the index that is used to evaluate cardiac performance. Using the systolic pressure-rate or pressure-work indexes, the same cardiac performance can be achieved by changes in preload or by catecholamines without postulating "oxygen wasting" due to augmented contractility.

Introduction

Myocardial "catecholamine oxygen wasting" is the concept that cardiac performance requires additional oxygen consumption when the inotropic state of the heart is augmented. That is, myocardial oxygen consumption will be greater for a given level of cardiac performance if it is achieved with catecholamines than with manipulation of the preload and afterload. In other words, there is an additional cost in myocardial oxygen consumption associated with increased contractility. Braunwald and co-workers related the additional oxygen consumption induced by catecholamines to the velocity of cardiac contraction, as indicated by ejection velocity or dP/dt [2, 3, 10, 18]. *Bretschneider* and co-workers have also incorporated contractility (dP/dt) terms into an equation for prediction of myocardial oxygen consumption [1, 4].

The crux of the problem is defining cardiac performance in relation to oxygen consumption. It has long been recognized that external cardiac work alone is poorly correlated with myocardial oxygen consumption, and that work at a high aortic pressure (pressure work) requires more oxygen than the same amount of work at a low pressure (flow work) [5, 6, 8]. This led to the development of indexes of myocardial oxygen consumption based on pressure development without a stroke volume or work term, such as the tension-time index [16], the product of systolic blood pressure and heart rate, the product of mean blood pressure and heart rate [9, 11], and estimated wall tension [15]. Similarly, the integral of wall tension (force) over time correlates well with myocardial oxygen consumption [20].

On the other hand, a Fenn effect (additional energy cost of shortening or external work) has been demonstrated for cardiac muscle [7], and indexes that incorporate a work term also correlate well with myocardial oxygen consumption [5, 6, 12, 19].

The question of catecholamine oxygen wasting was reexamined in a two-part study. In the first phase, the effects of independently varying systolic blood pressure, heart rate, and stroke volume were observed in closed-chest, anesthetized dogs. From these results, where myocardial oxygen consumption was varied by changing preload and afterload, a new pressure-work index was devised and fitting parameters for other indexes of myocardial oxygen consumption were determined.

These indexes were prospectively applied to a second set of experiments in closed-chest, anesthetized dogs given norepinephrine, isoproterenol, dobutamine, nembutal, and propranolol to alter myocardial contractility. Inotropic "oxygen wasting" was observed with the tension-time and other indexes, but not with the pressure-work or systolic pressure-rate indexes.

Methods

Dogs had an initial surgery where an electromagnetic flow transducer was implanted around the ascending aorta. In phase I, but not in phase II, complete atrioventricular heart block was produced by cauterization of the bundle of His. Following a 2-week surgical recovery period, the animals were premedicated with morphine and anesthetized with chloralose. Circumflex coronary artery blood flow was measured with a cannula-tip flowmeter [17], and coronary sinus blood was sampled for calculation of myocardial oxygen consumption without opening the thorax. Left ventricular pressure and its derivative (dP/dt) were measured with a catheter tip manometer.

In phase I experiments with atrioventricular block, the heart rate was paced, and propranolol was administered to prevent contractility changes. Phenylephrine infusion, controlled hemorrhage, and adjustment of arteriovenous shunts were used to manipulate stroke volume, afterload, and heart rate in an organized array of data points. Data from phase I experiments were used to obtain the fitting parameters for previously published indexes of myocardial oxygen consumption, and for a new pressure-work index.

Phase II animals with chronically implanted aortic electromagnetic flow transducers, but without heart block, propranolol, or arteriovenous shunts, were studied closed-chest as in phase I. Inotropic changes were achieved by infusions of norepinephrine, isoproterenol, dobutamine, nembutal, and, at the end of the experiment, propranolol. The formulas and fitting parameters obtained from phase I were *prospectively* applied *without change* to the phase II data. This tested the predictive ability of the various indexes when contractility was changed.

Results

Phase I: When heart rate and systolic blood pressure were held constant, an increase in stroke volume always resulted in an increase in myocardial oxygen consumption. This effect was more pronounced at high systolic pressures than at low pressures, suggesting that the augmented myocardial oxygen consumption is related to external cardiac work.

A new index that combined the systolic pressure rate product and external work was devised to fit the data.

$$M\dot{V}O_2 = K_1\ (SBP \times HR) + K_2 \left(\frac{(0.8\ SBP + 0.2\ DBP) \times HR \times SV)}{BW} \right) + 1.43$$

Where:

$M\dot{V}O_2$ = Left ventricular myocardial oxygen consumption (ml O_2/min/100 g)
SBP = Systolic blood pressure (mmHg)
DBP = Diastolic blood pressure (mmHg)
HR = Heart rate (beats/min)
SV = Stroke volume (ml)
BW = Body weight (kg)
K_1 = 4.08×10^{-4}
K_2 = 3.25×10^{-4}

The phase I data, where contractility changes were prevented by propranolol pretreatment, were *retrospectively* fit with various indexes of myocardial oxygen consumption. When applied in this manner most indexes had a good correlation with changes in myocardial oxygen consumption (range: 0.867–0.985), as shown in Table 1.

In phase II, where contractility was varied by catecholamine infusion, the indexes were applied *prospectively* using the fitting constants exactly as they were determined from the phase I data. When contractility varied, the correlation between indexes and changes in myocardial oxygen consumption was variable. Some indexes had a good correlation, while others did very poorly (range: 0.562–0.978), as shown in Table 2.

More importantly, the observed and predicted oxygen consumptions diverged as a function of contractility (maximum dP/dt) for some indexes but not for others. The higher the contractility, the greater predictive error became. Thus, additional oxygen consumption above the value predicted by the index, i.e., "oxygen wasting",

Table 1. Correlation between the percent change in oxygen consumption and percent change in index without changes in contractility

Formula	Correlation coefficient
Pressure-work index	0.985
Systolic pressure-rate	0.924
Estimated wall tension	0.955
External work	0.939
Mean pressure-rate	0.867
E_t (Bretschneider)	0.978
Tension-time index	0.935

Table 2. Correlation between the percent change in oxygen consumption and percent change in index with changes in contractility

Formula	Correlation coefficient
Pressure-work index	0.978
Systolic pressure-rate	0.947
Estimated wall tension	0.902
External work	0.892
Mean pressure-rate	0.821
E_t (Bretschneider)	0.946
Tension-time index	0.562

was observed at elevated contractility with some indexes but not with others. The estimated wall tension, external work, mean pressure-rate, and, particularly, the tension-time index showed this "oxygen wasting" effect of augmented contractility. The pressure-work and systolic pressure-rate indexes did not show oxygen wasting.

Discussion

The results from the phase I experiments demonstrate that at a given systolic pressure-heart rate product an increase in stroke volume is associated with an augmented myocardial oxygen consumption. This effect was achieved by altering the cardiac preload in the presence of propranolol and is unlikely to be secondary to contractility changes. The data were best fit by a new index that incorporates both a systolic pressure-heart rate product term and an external cardiac work term. However, several other indexes also fit the data well.

The concept of catecholamine "oxygen wasting" is based on having some index of myocardial oxygen consumption to determine what the oxygen consumption ought to be for a given level of cardiac performance in the absence of catecholamines. Without an index value separate from an actual oxygen consumption measurement, there is no reference point to which "oxygen wasting" can be related.

Raab [14], in reviewing the literature, observed that catecholamines augmented myocardial oxygen consumption proportionally more than the increase in external cardiac work. This led to the concept of myocardial "catecholamine oxygen wasting" using external cardiac work as the reference index.

Klocke and co-workers [13] observed little increase in myocardial oxygen consumption in potassium-arrested hearts, while *Sonnenblick* et al. [18] found that the tension-time index grossly underestimated myocardial oxygen consumption during catecholamine infusion and proposed that the oxygen wasting effect was due to contraction velocity. That is, an additional contractility term needed to be added to the tension-time index to account for oxygen consumption. This led to the concept that augmented contractility carries an additional oxygen cost, or is "oxygen wasting" [3].

The present results confirm previous studies in that a large and significant additional oxygen consumption is observed during catecholamine infusion *if* the tension-time index is used as a reference index. This effect could be dealt with by adding an index term directly related to contractility, such as maximum dP/dt, similar to *Bretschneider's* approach [1, 4].

The rationale of the present experiments was to retrospectively determine the index fitting constants and equations from a data set where contractility changes were prevented with propranolol (phase I) and then prospectively apply these indexes to a different data set where contractility was varied by catecholamine infusion.

The phase I data were best fit with a newly developed pressure-work index. When the pressure-work index or the simple systolic pressure-rate product index were prospectively applied to the catecholamine infusion (phase II) data, no evidence for additional oxygen consumption attributable to contractility was found during catecholamine infusion. That is, indexes and fitting constants developed from data without contractility changes predicted the catecholamine data without an "oxygen wasting" effect.

It is concluded that whether or not "catecholamine oxygen wasting" is observed depends on the index that is used to evaluate cardiac performance. Using the systolic pressure-rate or pressure-work indexes, the same cardiac performance can be achieved by changes in preload or by catecholamines without postulating "oxygen wasting" due to augmented contractility.

References

1. Baller D, Bretschneider HJ, Hellige G (1979) Validity of myocardial oxygen consumption parameters. Clin Cardiol 2:317–327
2. Boerth RC, Covell JW, Pool PE, Ross J Jr (1969) Increased myocardial oxygen consumption and contractile state associated with increased heart rate in dogs. Circ Res 24: 725–734
3. Braunwald E (1971) Control of myocardial oxygen consumption. Physiologic and clinical considerations. Am J Cardiol 27:416–432
4. Bretschneider HJ (1972) Die hämodynamischen Determinanten des myokardialen Sauerstoffverbrauches. In: Dengler (ed) Die therapeutische Anwendung β-sympathikolytischer Stoffe. Stuttgart New York, Schattauer, pp45–60
5. Britman NA, Levine HJ (1964) Contractile element work: a major determinant of myocardial oxygen consumption. J Clin Invest 43:1397–1408
6. Burns JW, Covell JW (1972) Myocardial oxygen consumption during isotonic and isovolumic contractions in the intact heart. Am J Physiol 223:1491–1497
7. Coleman HN, Sonnenblick EH, Braunwald E (1969) Myocardial oxygen consumption associated with external work: the Fenn effect. Am J Physiol 217:291–296
8. Evans CL, Matsuoka Y (1914/15) The effect of various mechanical conditions on the gaseous metabolism and efficiency of the mammalian heart. J Physiol 49:378–405
9. Feinberg H, Katz LN, Boyd E (1962) Determinants of coronary flow and myocardial oxygen consumption. Am J Physiol 202:45–52
10. Graham TP, Covell JW, Sonnenblick EH, Ross J Jr, Braunwald E (1968) Control of myocardial oxygen consumption: relative influence of contractile state and tension development. J Clin Invest 47:375–385
11. Katz LN, Feinberg H (1958) The relation of cardiac effort to myocardial oxygen consumption and coronary flow. Circ Res 6:656–669
12. Khalafbeigui F, Suga H, Sagawa K (1979) Left ventricular systolic pressure-volume area correlates with oxygen consumption. Am J Physiol 237:H566–H569
13. Klocke FJ, Kaiser GA, Ross J Jr, Braunwald E (1965) Mechanism of increase of myocardial oxygen uptake produced by catecholamines. Am J Physiol 209:913–918
14. Raab W (1956) The adrenergic-cholinergic control of cardiac metabolism and function (physio-pathological and clinical aspects). Adv Cardiol 1:65–152
15. Rodbard S, Williams CB, Rodbard D, Berglund E (1964) Myocardial tension and oxygen uptake. Circ Res 14:139–149

16. Sarnoff SJ, Braunwald E, Welch GH Jr, Case RB, Stainsby WN, Macruz R (1958) Hemodynamic determinants of oxygen consumption of the heart with special reference to the tension-time index. Am J Physiol 192: 148–156
17. Smith FD, D'Alecy LG, Feigl EO (1974) Cannula-tip coronary blood flow transducer for use in closed-chest animals. J Appl Physiol 37:592–595
18. Sonnenblick EH, Ross J Jr, Covell JW, Kaiser GA, Braunwald E (1965) Velocity of contraction as a determinant of myocardial oxygen consumption. Am J Physiol 209:919–927
19. Suga H (1979) Total mechanical energy of a ventricle model and cardiac oxygen consumption. Am J Physiol 236: H498–H505
20. Weber KT, Janicki JS (1977) Myocardial oxygen consumption: the role of wall force and shortening. Am J Physiol 233: H421–H430

Discussion

Bassenge, Freiburg: Recently Bretschneider's group presented a paper demonstrating that ventricular pacing increased $M\dot{V}O_2$ by up to 30% (at the same level of heart work), thereby substantially decreasing the efficiency of the heart. If this effect plays a significant role, it should be reflected in your analysis when pacing is applied, especially at high heart rates. How do you interprete your analysis on the basis of these observations?

Feigl: This is another example of the "oxygen wasting" problem. The Bretschneider observation is dependent on the Bretschneider index. It is possible that the effect will not be observed with other indices. Our experiments did not directly test the question of ventricular pacing, so a specific answer is not available.

Spieckermann, Göttingen: You tested your new formula in a very limited range of O_2-consumption. Energy demand may increase by a factor of up to 8 or 10. Does your formula work in such an expanded range?

Feigl: We deliberately avoided high, and thus potentially toxic levels of catecholamine. These data include myocardial oxygen consumption values from 7 to 32 ml O_2/min per 100 g myocardium. The important conclusion here is not our new pressure-work index, but the demonstration that at a constant systolic pressure-heart rate product an increase in stroke volume augments myocardial oxygen consumption, and that "catecholamine oxygen wasting" is entirely dependent on the index employed.

Bünger, Bethesda: In hemodynamically stable, catecholamine-stimulated isolated working guinea pig heart preparations there is an essentially linear relationship between myocardial oxygen consumption and the product [heart rate]×[peak systolic pressure – enddiastolic pressure]. Extrapolation to zero function (product is zero) yields a "resting" consumption, which is approximately 0.5–1.2 μmol $O_2 \times min^{-1} \times g^{-1}$ (wet weight). This is compatible with the fact that e.g. potassium arrested hearts still take up oxygen.

Feigl: We also observe a positive oxygen consumption for the arrested heart by extrapolation, as given by the C value (1.43 ml O_2/min per 100 g) in our pressure-work index. In the interest of time I only presented changes in oxygen consumption so the intercept value was not apparent.

Koch, Berlin: You very nicely analysed the effect of variations in stroke volume on myocardial oxygen uptake. I assume that you also looked into the influence of changes in ventricular filling pressures on myocardial oxygen consumption. Would you please summarize your findings concerning the effect of filling pressures on myocardial oxygen uptake in comparison to the effects of changes in stroke volume. Presumably changes in stroke volume are much more important in this respect.

Feigl: We were able to independently control stroke volume and systolic blood pressure in closed chest dogs in these experiments. Our aim was to evaluate stroke volume rather than left ventricular diastolic pressure, because work (stroke volume × mean ejection pressure) seems a more fundamental thermodynamic variable than filling pressure.

Schrader, Munich: From your experiments it certainly appears that oxygen wastage of the myocardium cannot be derived from hemodynamic parameters. On the other hand, it appears conceivable that there may be oxygen wastage at a biochemical level. This would apply to situations in which cellular oxygenases, besides the respiratory chain enzymes, are stimulated. Are there conditions known where this might occur in the heart?

Feigl: The physiological function of the heart is mechanical – that is to generate pressure and cardiac output. Thus it is rational to evaluate cardiac function in mechanical terms. These data do not indicate myocardial "catecholamine oxygen wasting" when evaluated mechanically. There may be a biochemical oxygen wasting, but I would guess this may be significant only at high and perhaps toxic levels of catecholamines.

Effects of Adrenergic Stimulation and Inhibition on Infarcting Myocardium

L. H. Opie

Summary

High doses of catecholamines can initiate a sequence which involves: excess circulating hormone; reaction with the β_1-receptor; excess formation of the messenger, cyclic AMP; and unwanted effects such as a fall of the ventricular fibrillation threshold, impaired inotropic response, oxygen wastage and enzyme release. Of the harmful effects, the arrhythmogenic effect and increased enzyme release are mediated by cyclic AMP or associated with cyclic AMP activity; evidence for such a role of cyclic AMP in the impaired inotropic response and in oxygen wastage remains to be proven.

Therapeutic procedures, such as Ca^{2+} antagonism (verapamil and/or decreased external Ca^{2+} concentration), β-blockade and (for enzyme release) inhibition of cardiac lipolysis, may inhibit the unwanted effects of excess β_1-stimulation and produce a beneficial therapeutic effect. In the case of the infarcting myocardium, the complex metabolic patterns mean that therapy will be correspondingly complex. Both β-blockade and Ca^{2+}-antagonism are therapies of potential in decreasing the metabolic and the arrhythmogenic effects of catecholamine-stimulation found in the acutely infarcting myocardium.

Catecholamine-Mediated Myocardial Damage

Acute myocardial infarction in man is accompanied by both general metabolic response and a local metabolic response in the ischaemic tissue. This article emphasizes the effects of adrenergic stimulation and inhibition on the infarcting myocardium with particular reference to β-agonists and antagonist effects.

Direct Myocardial Damage Caused by Catecholamines

To understand the effects of β-agonists and β-blockade on the infarcting myocardium requires an appreciation of the harmful effects of excess catecholamine stimulation on the non-ischaemic heart (Fig. 1). In the intact animal, both histochemical [9] and histological [1] evidence suggests that damage to the non-ischaemic zone of the heart with infarction can be catecholamine-mediated. Direct evidence that it is the catecholamines that are cardiotoxic stems from experiments in vivo [36] and with isolated hearts. Thus both adrenaline and noradrenaline have a concentration-related effect in causing loss of enzymes from the non-ligated rat heart [12, 52]. A dose of adrenaline 10^{-6} mol/liter, for example, causes a peak loss

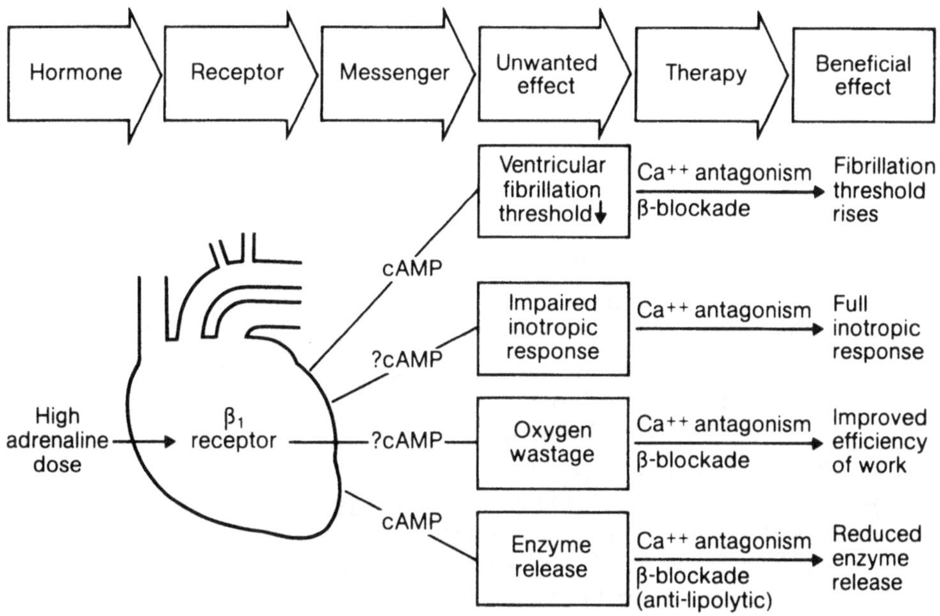

Fig. 1. Excess β_1-stimulation of the heart can give rise to a series of unwanted effects: a fall in ventricular fibrillation threshold and impaired inotropic response, oxygen wastage and increased enzyme release. The first and the last of these effects are associated with an increased intracellular cyclic AMP which may, therefore, be the second messenger. In every case except the impaired inotropic response, therapy can be either by calcium antagonist or by β-blockade. Part of the effects of β-blockade may be mediated by antilipolytic agents such as nicotinic acid. These data suggest that harmful effects of cyclic AMP are mediated via an accumulation of intracellular calcium. For details see *Opie* [23]

of lactate dehydrogenase at a rate of about 700 mU/g/min or 42/U/g/h, which is comparable to a peak rate reached in coronary-ligated hearts perfused with unphysiologically high fatty acid solutions [6]. In man, phaeochromocytomas are associated with a focal myocarditis [48] similar to the lesions found in rat hearts perfused with noradrenaline [52].

The mechanism of such catecholamine effects relevant to the processes whereby excess catecholamine stimulation can harm the infarcting myocardium. The hypothesis of Fleckenstein is that catecholamines, by permitting excess entry of calcium ions, allow a state of 'excess' coupling of extraction-contraction with breakdown of high-energy phosphate compounds [8]. We found a major protective effect of calcium antagonism and an added effect of the antilipolytic agent, nicotinic acid, against the harmful mechanical and metabolic effects of excess perfusate adrenaline [28]. That the calcium ion was involved in the deleterious effects of catecholamines on the normal perfused heart was shown specifically by the protective effects of the calcium-antagonist procedures such as verapamil, halving the perfusate Ca^{2+}, or doubling the Mg^{2+} value, in decreasing the release of enzymes [12, 28]. That the β-receptor was involved was shown by the protection afforded by β-blockade and by the relation between enzyme release and the increase of tissue cyclic AMP [12].

Thus a catecholamine-mediated increase of cyclic AMP could lead to an increased entry of Ca^{2+} ions which, in general, promotes cell damage [8]. Additionally, as is suggested by the beneficial effects of nicotinic acid, cyclic AMP could stimulate tissue lipases with breakdown of cellular phospholipids such as lysolecithin and (possibly and speculatively) accumulation of harmful lipid metabolites such as free fatty acids, acyl CoA, acyl carnitine and lysophosphoglyceride.

Further evidence that cyclic AMP can mediate harmful effects of β-stimulation is provided by experiments in which the heart was perfused with dibutyryl cyclic AMP, which also caused loss of enzyme even in the presence of β-blockade [12].

Catecholamines and Arrhythmias

We have proposed a hypothesis that adenosine 3',5'-cyclic monophosphate (cyclic AMP) may be the second messenger of one serious pathological effect of beta-adrenergic stimulation, namely the provocation of cardiac arrhythmias such as fatal ventricular fibrillation (Fig. 2, [17]).

In 1978 *Podzuweit* et al. [32] reported studies on a baboon model in which the left anterior descending coronary artery was ligated and sudden death due to ventricular fibrillation followed about 35 min later. Cyclic AMP values in minidrill biopsies repetitively taken from the ischaemic tissue consistently showed a rise starting 5–10 min before fibrillation. A similar association between increased tissue cyclic AMP and ventricular fibrillation was found in a cat model [2]. We have proposed that the increased tissue cyclic AMP may be either the direct cause of ventricular fibrillation or at the least an important precipitating factor [26, 27, 31]. In a pig heart model of acute myocardial infarction, the relation between beta-adrenergic stimulation, tissue cyclic AMP, and ventricular arrhythmias has been explored by my colleague, Cecilia Muller [21].

Beta-Adrenergic Stimulation, Tissue Cyclic AMP Levels, and Ventricular Arrhythmias in the Pig

The pig model used is described in detail in her thesis. Cyclic AMP in minibiopsies taken from the non-ischaemic and mid-ischaemic zones at regular intervals after coronary ligation was determined by the method of *Tovey* et al. [45]. Arrhythmias were quantified using an arbitrary scoring system. Ligation of a lateral branch of the anterior descending coronary artery provided a suitable model with only minor rhythm disturbances and minimal changes in tissue cyclic AMP after ligation.

To study the effects of beta-adrenergic stimulation on tissue levels of cyclic AMP and the incidence of ventricular arrhythmias after coronary artery ligation, isoproterenol (infused at 0.5 μg/kg/min) was started 3 min after ligation and continued up to the end of the experiment (7 pigs). A regular continuous ventricular tachycardia not exceeding a rate of 210 per min was precipitated in 6 out of 7 pigs. Seventeen episodes of ventricular fibrillation occurred and only 2 animals survived the full duration of the experiment. Arrhythmia scores were much higher ($P < 0.001$) than in the control group although infarct size was similar. Cyclic AMP in the non-ischaemic zone rapidly increased for the duration of the experiment ($P < 0.001$ for areas under the curves when compared with values in ischaemic zone). But changes in tissue levels of ATP, phosphocreatine, lactate, and potassium could not be linked to the development of these serious arrhythmias [21].

The striking effects of isoproterenol infusion was to greatly increase the incidence and severity of ventricular arrhythmias, and also to increase the tissue cyclic AMP in both ischaemic and non-ischaemic zones of cyclic AMP infused hearts, but more

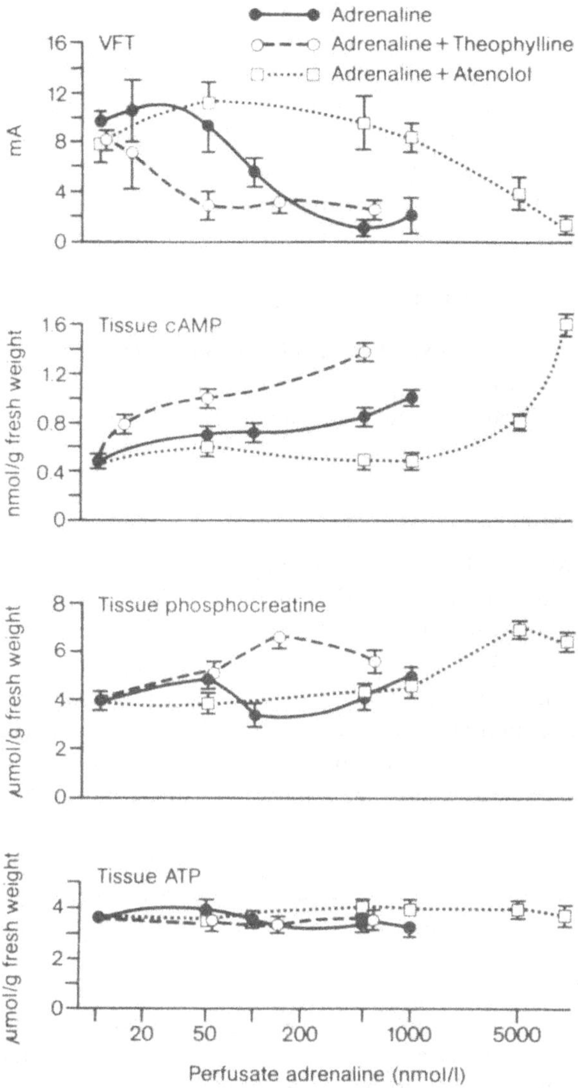

Fig. 2. As the perfusate adrenaline of the isolated perfused rat heart is increased, there is a dose-related response curve and fall in the ventricular fibrillation threshold (VFT). At the same time the tissue cyclic AMP rises. In the presence of added theophylline (10^{-3} mol/liter) the ventricular fibrillation threshold falls more rapidly and the tissue cyclic AMP rises more rapidly. In the presence of β_1-blockade by atenolol (3.8×10^{-5} mol/liter) the dose-response curves are markedly shifted to the right, so that the rise in tissue cyclic AMP is much delayed, as is the fall in the ventricular fibrillation threshold. Note inconstant changes in tissue phosphocreatine and tissue ATP, showing that the hypothesis of *Fleckenstein* [8] whereby catecholamine stimulation breaks down ATP does not operate. Published by kind permission of Journal of Clinical Investigation. (*Lubbe* et al. (1978) [17])

so in the non-ischaemic zone (Fig. 1). The values of cyclic AMP in the non-ischaemic zone were comparable to those that reduced the fibrillation threshold in the rat model [17].

Inequalities of cyclic AMP concentrations are expected to contribute to different electrophysiological effects in non-ischaemic and in ischaemic zones, thereby causing electrical inhomogeneity and setting the stage for re-entrant arrhythmias. *Podzuweit* et al. found that direct infusion of dibutyryl cyclic AMP near the infarct edge, but not into the central infarct zone, provoked arrhythmias [33].

Proposed Mechanisms of Cyclic-AMP-Induced Arrhythmias

Two major proposals have been made to explain the mechanism of cyclic-AMP-induced arrhythmias: first, cyclic AMP can precipitate slow responses in K^+-blocked fibres; secondly, cyclic AMP can promote after-potentials.

Cyclic AMP and Slow Responses

The intracellular content of cAMP has been shown to play an important role in determining permeability of cell membranes to Ca^{2+} ions [55]; a calcium-dependent current fosters the establishment of slow responses [41, 53] which favour the development of re-entry mechanisms [4]. *Reuter* compared the effects of noradrenaline (10^{-6} *M*) with DbcAMP (1 m*M*) on the normal cardiac action potential [35]. Noradrenaline increased both the action potential duration and the force of cardiac contraction of the ventricular muscle in the calf. DbcAMP increased the duration of the action potential more markedly and also has a prolonged effect on tension development.

At first sight, these changes, being similar to those produced by class III anti-arrhythmic agents such as amiodarone [42], should be anti-arrhythmic rather than arrhythmogenic. However, the arrhythmic role of cAMP is probably not directly related to its role in increasing the action potential duration, but to a closely related role in promoting the slow response.

Schneider and *Sperelakis* [41] have studied the *slow response* elicited in an isolated perfused guinea-pig heart with a floating micro-electrode. The normal action potential of the isolated guinea-pig heart had a peak rate of depolarization (dv/dt) of 86 V/s in a medium with K^+ 2.7 m*M*. In a high-potassium medium (K^+ 27 m*M*) there was decreased polarization in the resting state, a total absence of the action potential and a zero value for dv/dt. When catecholamines, methyl-xanthines, histamine, angiotensin II, cAMP or DbcAMP were introduced to the depolarized fibres, a slow response developed with a very much lower value for dv/dt, only 15 V/s. Thus, the agents used are thought either to increase cardiac cAMP or to imitate the effects of cAMP, and all could produce the slow response. Direct evidence for the role of cyclic AMP in causing the slow response has been obtained by microiontophoresis of cyclic AMP into cardiac Purkinje fibres superfused with a high external K^+ [49]. According to the concepts of *Cranefield* [4] such a slow response predisposes to re-entrant arrhythmias in the presence of regional ischaemia.

However, there are *two criticisms* of this hypothesis: First, some energy is required to elicit the slow response. If the ATP level is severely lowered by

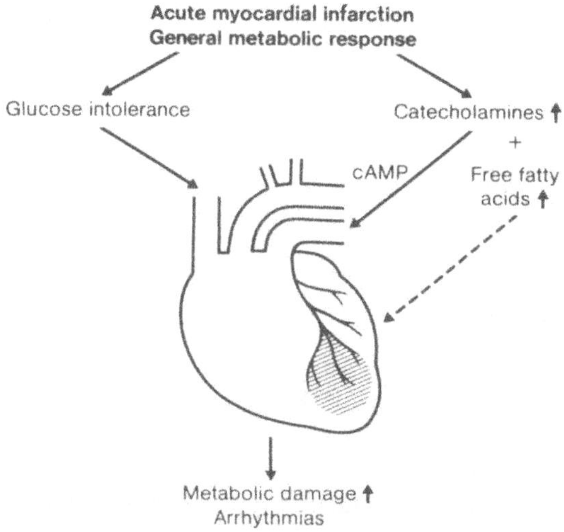

Fig. 3. As part of the general metabolic response to acute myocardial infarction there is glucose intolerance, an elevation of circulating and urinary catecholamines, and increased blood free fatty acids. These factors predispose to metabolic damage and may provoke arrhythmias. In the case of increased catecholamine stimulation, the mediator of metabolic damage and of arrhythmias is probably cyclic AMP. For references, see text

metabolic inhibition (severe hypoxia, metabolic poisons) then the slow response can no longer be obtained [40]. Regional ischaemia is characteristically associated with ATP depletion, which should therefore limit slow responses. A possible answer to this criticism is that where ischaemic and non-ischaemic tissue meet, the extracellular K^+ derived from adjacent ischaemic cells could act on non-ischaemic cells which have a relative normal ATP content.

Secondly, it should be noted that in developing regional ischaemia, the extracellular potassium levels are much lower than 27 m*M*. Thus in a model of regional ischaemia produced by coronary artery ligation, the local venous potassium values were, at the most, increased by 2 m*M* [30]. The use of a special K^+-sensitive electrode in the interstitial space shows that the interstitial K^+ can rise to values of 10–14 m*M* during acute ischaemia in the dog [10], and to nearly 18 m*M* in the pig [11]. These much smaller changes in potassium do indeed alter the action potential pattern, also producing slow responses [27]. Our preliminary data (*Nathan* and *Opie,* unpublished) show that cAMP, when added to these partially K^+-inhibited fibres, has a more marked effect in prolonging the action potential duration than in normal fibres. Thus it is possible that cAMP could exaggerate the potassium-dependent tendency towards the slow response formation.

Our hypothesis does not exclude the role of *metabolic factors other than cAMP* (Fig. 3). Shortening of the action potential duration may theoretically predispose to the development of arrhythmias, for example in the presence of ischaemia. The fuel supply (substrate) to the heart may influence the development of arrhythmias in the presence of ischaemia [27]. We have studied the effect of changing from glucose to lactate and back to glucose, on the action potential duration and on force development of the guinea-pig papillary muscle [27]. When glucose was replaced by lactate, the action potential duration fell significantly, as did the tension development. These data could also be explained by postulating that glycolytic ATP, produced only by glucose and not by lactate, is required for the maintenance of the normal action potential duration. *Cowan* and *Vaughan Williams* [3] found that during hypoxia the addition of free fatty acid to the glucose-containing perfusate depressed the action potential duration. Upon reoxygenation, fatty acids no longer decreased the action potential duration. This effect of non-glucose fuels in decreasing the action duration was more marked in the presence of hypoxia.

Thus maintenance of the action potential duration is thought to be an anti-arrhythmic process and may be substrate-dependent, especially in conditions of ischaemia. This part of our hypothesis would explain the anti-arrhythmic properties of glucose in acute experimental myocardial infarction [34, 37]; a relative increase in glucose extraction is achieved by β-blockade in dogs with coronary artery ligation [24].

Other Electrophysiological Properties of Cyclic AMP

Our recent experimental data studied simulated ischaemia (hypoxia, high K^+, high lactate, dibutyryl cyclic AMP 10^{-3} and 2.5×10^{-3} *M*) and simulated 'reperfusion' (return to normal conditions). On return to normal medium, runs of extrasystoles and early after-depolarizations were observed. Extrasystoles in the form of completely 'normal' action potentials were possibly the result of after-potentials reaching threshold in cells elsewhere in the preparation with subsequent propagation to the site of the recording electrode, giving an apparently normal action potential. In the absence of dibutyryl cyclic AMP, automaticity did not occur ($P < 0.02$). These data suggest a mechanism for reperfusion arrhythmias which are generally observed following reinstatement of coronary flow to ischaemic tissue [29].

Cyclic AMP can also be arrhythmogenic in non-ischaemic tissue by promoting phase 4 depolarization in Purkinje fibres [14]. In otherwise non-automatic cells adrenergic stimulation or phosphodiesterase inhibition can provoke after-potentials and thereby cause automaticity [16].

Another interesting possibility for cyclic AMP action is by mediating electrical uncoupling and contractures in hypoxic muscle, possibly as a result of increased Ca^{2+} entry [55]. Finally, it must be kept in mind that cyclic AMP might not always be the second messenger of all beta-adrenergic electrophysiological effects. Thus, epinephrine-induced automaticity and the increase of cyclic AMP content of Purkinje fibres can in some circumstances be dissociated [5].

Calcium and Arrhythmias

Recently, *Thandroyen* [44] has found that a variety of calcium-antagonist drugs (verapamil, nifedipine, diltiazem) all have a concentration-dependent effect in increasing the ventricular fibrillation threshold after coronary ligation in the isolated rat heart. He could not totally exclude an effect of these agents in lowering tissue cyclic AMP values (possibly by a non-specific 'anti-ischaemic' effect). However, *Muller* [21] in her pig preparation showed that infusion of verapamil could inhibit isoproterenol-induced arrhythmias without altering tissue cyclic AMP values. Thus it seems that the hypothesis that cyclic AMP is arrhythmogenic by promoting influx of Ca^{2+} ions, is further supported.

Revised Hypothesis (Fig. 4)

The original hypothesis [31] stressed the arrhythmogenic properties of cAMP itself. Our revised hypothesis [26] allows for a more complex genesis of ventricular ar-

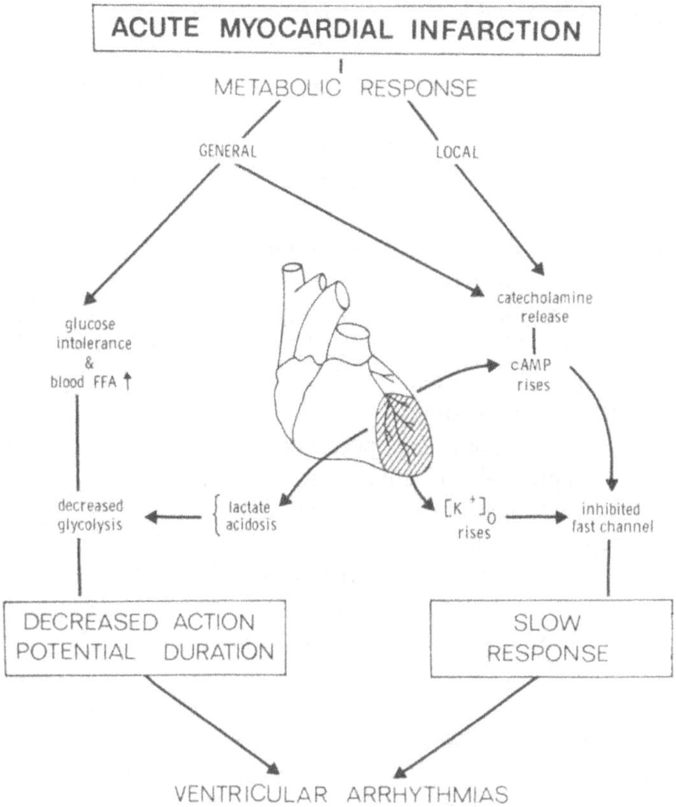

Fig. 4. The genesis of ventricular arrhythmias in acute myocardial infarction is complex. Two of the possible mechanisms are shown in this scheme. First, an elevation of cyclic AMP in the ischaemic zone, could in the presence of increased external potassium, provoke the slow response. Secondly, a decreased action potential duration has been linked to conditions decreasing glycolysis such as accumulation of lactate and protons. These local metabolic changes may be related to the general metabolic response which includes catecholamine release, glucose intolerance and increased blood free fatty acids. For further details, see *Opie, Muller* and *Lubbe,* Lancet II:921, 1978. By permission of the Editors of The Lancet

rhythmias in acute myocardial infarction (Fig. 4). The three basic mechanisms of production of ventricular arrhythmias are held to be: a) decreased action potential duration; b) generation of the slow response; c) formation of afterpotentials.

The hypothesis can be divided into factual aspects and inferred events. The facts are that: a) cAMP rises after coronary artery ligation in a variety of models (pigs, baboon and cat); b) the rise of cAMP coincides with the development of ventricular arrhythmias; c) an increase of extracellular potassium inhibits the fast channel; d) venous and extracellular potassium rise after coronary artery occlusion; e) in certain circumscribed experimental conditions elevations of circulating free fatty acids, lactate and protons can decrease glycolysis, and can decrease the action potential duration; and f) procedures increasing tissue cyclic AMP can cause automaticity in papillary muscle preparations. Inferred events are that: a) the extracellular potassium rises sufficiently to inhibit the fast channel; b) K^+ inhibition can, in the presence of an elevated tissue cAMP, produce the slow response; c) cyclic AMP can cause ventricular ectopic beats and arrhythmias; and d) the changes in substrate metabolism early in regional ischaemia are able to shorten the action potential duration. However, tight links between decreased glycolysis and decreased action potential duration remain to be shown except in conditions of experimental hypoxia or ischaemia; hence critical aspects of the hypothesis have only been inferred on the basis of indirect evidence.

Implications and Criticisms of the cAMP Hypothesis

The hypothesis has implications for the development of new antiarrhythmic drugs and predicts that all agents decreasing tissue cAMP will be anti-arrhythmic [27]. The hypothesis also predicts that calcium antagonist agents will become agents used against ventricular arrhythmias, provided that side effects on the AV-node as well as depression of contraction can be avoided [18].

Another implication of the cAMP hypothesis is that the production of cyclic guanosine monophosphate, commonly held to oppose the effects of cAMP, should be anti-arrhythmic. Thus in general, vagal activity should decrease arrhythmias (with the exception of bradyarrhythmias precipitated by a slow heart rate). Most present information is in agreement with that concept. When vagal tone was increased in acute myocardial infarction by treatment with edrophonium hydrochloride (Tensilon), there was enhanced electrical stability [19]. However, the mixed effects of heart rate and of acetylcholine stimulation on the arrhythmias allow for alternate interpretations, and evidence for cyclic GMP as mediator of vagal effects is not too strong.

The major criticism of the cAMP hypothesis is that it appears to stress the role of cAMP itself as the mechanism whereby catecholamine activity can produce arrhythmias. In reality, cAMP itself is only likely to be one of a long chain of molecular events (such as protein kinase activation with entry of Ca^{2+}) in the production of arrhythmias. Furthermore, cAMP is but one of several arrhythmogenic factors (K^+, lactate, diminished glycolysis) at work and there are probably additional mechanisms of arrhythmia production, not dependent on slow-channel activity or on action potential duration shortening, which include the development if tachyarrhythmias mediated by the fast channel [39]; thus lignocaine, the classic anti-arrhythmic agent used in the management of ischaemic arrhythmias in man, has its major site of action on the fast channel. There are also marked differences in the response to infused adrenaline at various times after coronary occlusion in the dog. We must still show that such a time-dependent response to adrenaline relates to differences in the tissue cAMP response to adrenaline. Furthermore the role of cyclic AMP must be differentiated from that of a low pH [39].

Beta-Blockade in the Treatment of Acute Myocardial Infarction

Catecholamines are secreted in excess amounts for many hours after the onset of chest pain but serious arrhythmias only result when catecholamine stimulation is combined with ischaemia, i.e. at the very onset of symptoms or when the infarction process extends. Because of evidence showing that infarction commonly extends in the days after the onset, catecholamine-induced arrhythmias could occur later during hospitalization, with the danger of ventricular fibrillation if there is also myocardial ischaemia.

General Measures

General therapeutic measures can help to reduce catecholamine drive. Such measures include calmness, reassurance and sympathy on the part of the attending doctor, as well as the use of sedation and morphine. Morphine has an anti-arrhythmic effect in animals because it lessens sympathetic reflexes [7]. In addition, the possible use of β-blockade should be evaluated in the therapy of acute myocardial infarction, with special reference to the contraindications and the possible benefits.

Indications for Beta-Blockade for Arrhythmias of Acute Myocardial Infarction

Beta-blockade is seldom used as the therapy of choice in such arrhythmias because supraventricular arrhythmias usually respond better to verapamil which has little inotropic effect in therapeutic concentrations, and it is usually held that ventricular arrhythmias respond better to lignocaine or disopyramide. However, *Jewitt* and *Croxson* [13] treated 40 patients with acute myocardial infarction and ventricular arrhythmias with practolol 5–25 mg intravenously, with a beneficial effect in about half. *Taylor* and *Burley* [43] used propranolol or oxprenolol (2–16 mg) or practolol 5–40 mg (all given intravenously) and found that in half to two-thirds of the treated patients the tachyarrhythmias reverted to sinus rhythm. They conclude that β-blockade is most effective in the therapy of myocardial infarction complicated by sinus tachycardia, ventricular premature systoles and paroxysmal atrial tachycardia, but least effective for atrial fibrillation. Doses for ventricular extrasystoles were less than for supraventricular tachycardias. It is important to stress that patients with congestive heart failure, pulmonary oedema and heart block were all excluded, as were those with severe obstructive airways disease.

Beta-Blockade and Reduction of Ischaemic Damage (Fig. 5)

The possible benefits of β-blockade are not limited to the therapy of ventricular arrhythmias, and must be seen alongside the theoretically desirable possibility of

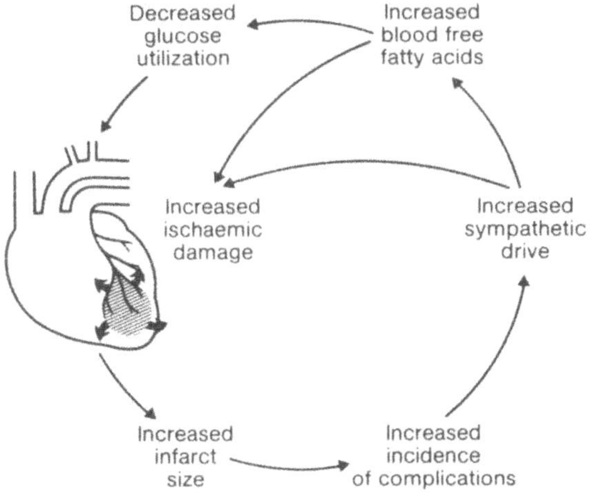

Fig. 5. In acute myocardial infarction, patients with larger infarcts have an increased incidence of complications and increased sympathetic drive. Such increased sympathetic drive will further increase the extent of ischaemic damage and a vicious circle mechanism will operate. The increased sympathetic drive may act either directly, or by elevating blood free fatty acids, and/or by decreasing glucose utilization. For full explanation of hypothesis and reservations involved, see [25]

reducing the severity of ischaemic damage and of ultimate 'infarct size'. Thus a definitive study would compare the effects of β-blockade with those of other anti-arrhythmic agents (lignocaine, disopyramide, procainamide) both on arrhythmias and on 'infarct size'. Not only benefits but also side-effects would have to be compared. In the meantime, β-blockade therapy is most suited for those patients in whom sinus tachycardia occurs in the absence of recognizable causes and is, therefore, 'inappropriate', provided that contraindications do not exist, and only if careful monitoring of the effects of β-blockade is possible. In such highly selected and carefully monitored patients β-blockade may be the anti-arrhythmic therapy of choice to minimize ischaemic damage and arrhythmias. In a recent trial with intravenous atenolol, *Yusuf* et al. [56] found beneficial effects: reduced enzyme release, less fall of R-wave, less heart failure and (in a preliminary study) decreased mortality.

Beta-Blockade Versus Calcium-Antagonists

Both these categories of agents are being evaluated for their possible protective effects on the ischaemic myocardium. Experimentally (Table 1) the calcium-antagonist diltiazem is able to give an equal degree of reduction of enzyme release to metoprolol, with the added advantage of increasing power production. Thus with diltiazem more cardiac work is performed even though enzyme release is decreased. The presumed mechanism is through increased coronary flow. In these experiments, the concentration of metoprolol was high and properties other than pure beta-blockade might have been evident. Such animal data cannot be extrapolated to patients, but do suggest that careful clinical comparison of calcium antagonists in acute myocardial infarction is warranted, especially because the protective effect is found in concentrations much lower than those of β-antagonists (Table 1).

Table 1. Comparative effects of metoprolol (10^{-4} *M*) and diltiazem ($4 \cdot 10^{-7}$ *M*) on release of lactate dehydrogenase and mechanical performance of isolated coronary-ligated working rat heart (values given as percentages of control, with *P* values in comparison with control) (Hamm, Thandroyen and Opie, unpublished data)

	Metoprolol	Diltiazem
Coronary flow (ischaemic and non-ischaemic)	91% (NS)	121% ($P<0.05$)
Oxygen uptake (ischaemic and non-ischaemic)	92% (NS)	122% ($P<0.01$)
Power production	117% (NS)	179% ($P<0.01$)
Release of lactate dehydrogenase (0–60 min after coronary ligation)	55% ($P<0.001$)	53% ($P<0.001$)

Isolated rat hearts perfused for 60 min after coronary artery ligation (see de Leiris and Opie, 1978) with glucose 11 m*M* as substrate. Hearts paced via right atrium at 230 beats/min. Concentration of metoprolol determined as (a) that concentration giving least release of enzyme in dose response curve, using 10^{-5} *M*, 5×10^{-5} *M* and 1×10^{-4} *M*); (b) that concentration avoiding tachycardia in response to isoprenaline 10^{-6} *M*. For both these requirements, metoprolol 10^{-4} *M* was required. Concentration of diltiazem determined from dose response curve using 10^{-7} *M*, 1×10^{-7} *M*, and 4×10^{-7} *M;* higher concentrations caused complete heart block.

Beta-Blockade and Pain

Waagstein and *Hjalmarson* [51] have widened the indications for β-blockade to include pain relief (always in patients without contraindications, which in their study were: no clinical heart failure; systolic blood pressure above 100 mmHg; no severe bradycardia; no atrioventricular block; no QRS duration above 100 ms). Practolol (mean dose 8 mg) was given intravenously on 126 occasions to patients with acute myocardial infarction, and pain relief was obtained on 108 occasions. In another study, pain relief was found in all patients with transmural infarctions, but in only half of those with non-transmural infarctions [52]. Similar effects were found with a variety of β-blocking agents. *Waagstein* and *Hjalmarson* [52] propose that pain relief by β-blockade may be the result of decreased myocardial ischaemia.

It may be concluded that the indications, contraindications, benefits and risks of β-blockade in acute myocardial infarction remain incompletely understood. We stress that our cAMP hypothesis does not necessarily provide firm evidence for the benefit of β-blockade in the therapy of acute myocardial infarction.

If cAMP Is Arrhythmogenic, Why Is Beta-Blockade Not More Effective?

As pointed out, β-blockade is not a proven agent of choice in the therapy of ischaemic ventricular arrhythmias. If, indeed, cAMP is an important arrhythmogenic mechanism, and if an early effect of coronary occlusion is release of catecholamines, then a logical question is why β-blockade is not more effective in preventing ischaemic arrhythmias. There are several possible explanations, but the three major arguments from experimental data are that the rise in cAMP is only short-lived after experimental coronary artery ligation in situ, that cAMP is only one of several factors producing ischaemic arrhythmias, and that the rise of cAMP is probably multifactorial and hence cannot be completely blocked. The last argument is most important. Although it would be natural to suppose that the rise of tissue cAMP in ischaemia is mediated by β_1-adrenoceptor stimulation, other factors are probably at work, possibly including inhibition of phosphodiesterase activity. Rational therapy could only be based on a clear understanding of the cause of the rise of cAMP in ischaemic tissue. It is important to note that in the pig with coronary ligation, rise of tissue cyclic AMP is *not* stopped by β-antagonist agents, even in maximal doses such as metoprolol 10 mg/kg [21].

Protective Effect of Beta-Blockade

Further studies are now required to assess whether patients who develop myocardial infarction while taking β-adrenergic blocking agents are protected from the development of ventricular fibrillation. The published observations of such a protective effect obtained in animals [15] and in man [54] deserve further investigation by carefully controlled clinical trials, in which the possible protective action of these β-adrenergic blocking agents is balanced against the negative inotropic and chronotropic effects of these agents and their propensity to induce or aggravate cardiac failure. The recent large Norwegian trial on timolol supports those who believe that prophylactic β-blockade after myocardial infarction will improve mortality and outlook [22]. A possible mechanism is that pretreatment with β-blockade could reduce formation of cyclic AMP during an acute ischaemic episode.

Acknowledgements. The original work here reported was supported by the Medical Research Council of South Africa and the Chris Barnard Fund who are both thanked. The cooperation of my colleagues Dr. F. Thandroyen, Ms. Cecilia Muller and Prof. W. F. Lubbe has been invaluable.

References

1. Baroldi G, Silver MD, Lixfeld W, McGregor DC (1977) Irreversible myocardial damage resembling catecholamine necrosis secondary to acute coronary occlusion in dogs: its prevention by propranolol. J Mol Cell Cardiol 9:687–691
2. Corr PB, Witkowski FX, Sobel BE (1978) Mechanisms contributing to malignant dysrhythmias induced by ischemia in the cat. J Clin Invest 61:109–119
3. Cowan JC, Vaughan Williams EM (1977) The effects of palmitate on intracellular potentials recorded from Langendorff-perfused guinea-pig hearts in normoxia and hypoxia, and during perfusion at reduced rate of flow. J Mol Cell Cardiol 9:327–342
4. Cranefield P (1975) The conduction of the cardiac impulse. Futura, New York
5. Danilo P Jr, Vulliemox Y, Verosky M, Rosen MR (1978) Epinephrine-induced automaticity of canine cardiac Purkinje fibers and its relationship to the adenylate cyclase-adenosine 3',5'-monophosphate system. J Pharmacol Exp Ther 205:175–182
6. De Leiris J, Opie LH (1978) Effect of substrates and of coronary artery ligation on mechanical performance and on release of lactate dehydrogenase and creatine phosphokinase in isolated working rat hearts. Cardiovasc Res 10:585–596
7. De Silva RA, Verrier RL, Lown B (1978) Protective effect of the vagotonic action of morphine sulphate on ventricular vulnerability. Cardiovasc Res 12:167–172
8. Fleckenstein A (1975) Metabolic factors in the development of myocardial necrosis and micro-infarcts. Triangle 14:27–36
9. Herbaczyńska-Cedro K (1970) The influence of adrenaline secretion on the enzymes in heart muscle after acute coronary occlusion in dogs. Cardiovasc Res 4:168–175
10. Hill JL, Gettes LS (1977) Ischemia-induced changes in interstitial potassium in *in situ* myocardium. (Abstr) Circulation 56:III/108
11. Hirche HJ, Franz Chr, Bös L, Bissig R, Lang R, Schramm M (1980) Myocardial extracellular K^+ and H^+ increase and noradrenaline release possible cause of early arrhythmias following acute coronary artery occlusion in pigs. J Molec Cell Cardiol 12:579–593
12. Horak AF, Pudzuweit T, Opie LH (1980) Cyclic AMP as mediator of catecholamine-induced enzyme release from isolated perfused working rat heart. Ad Myocardiol I: 367–373
13. Jewitt DE, Croxson R (1971) Practolol in the management of cardiac arrhythmias following myocardial infarction and cardiac surgery. Postgrad Med J [Suppl] 47:25–29
14. Kentera D, Zdravkovic M, Varagić V (1978) Induction of automatism of Purkinje fibers with dibutryl 3',5'-cyclic AMP. Res Commun Chem Pathol Pharmacol 21:157–160
15. Khan MI, Hamilton JT, Manning GW (1972) Protective effect of beta-adrenoceptor blockade in experimental coronary occlusion in conscious dogs. Am J Cardiol 30:832–837
16. Lazzara R, Hope RR, Yeh BK (1978) Implication of cAMP and calcium as mediators of automaticity induced in working myocardium (Abstr). Am J Cardiol 41:417
17. Lubbe WF, Podzuweit T, Daries PS, Opie LH (1978) The role of cyclic adenosine monophosphate in adrenergic effects on ventricular vulnerability to fibrillation in the isolated perfused rat heart. J Clin Invest 61:1260–1269
18. Lubbe WF, McFayden ML, Muller CA, Worthington M, Opie LH (1979) Protective action of amiodarone against ventricular fibrillation in isolated perfused rat heart. Am J Cardiol 43:533–546
19. Millard RW (1980) Changes in cardiac mechanics and coronary blood-flow of regionally ischemic procine myocardium induced by dilitiazem. Chest 78 (1): 193–199
20. Miller RR, Olson HG, Vera Z, De Maria AN, Amsterdam EA, Mason DT (1977) Clinical evaluation of the enhancement of vagal tone in acute myocardial infarction by edrophonium hydrochloride: effects on ventricular arrhythmias, His bundle electrography and left ventricular function. Am Heart J 93:222–228

21. Muller CA (submitted) Ventricular arrhythmogenesis in developing myocardial infarction in the pig: With special reference to the role of cyclic AMP. Ph D Thesis, University of Cape Town
22. Norwegian Multi-Center Study Group (1981) Timolol-induced reduction in mortality and reinfarction in patients surviving acute myocardial infarction. N Engl J Med 304:801–807
23. Opie LH (1980) Beta-blockade: metabolic effects in experimental myocardial infarction. Triangle 19:11–18
24. Opie LH, Thomas M (1976) Propranolol and experimental myocardial infarction: substrate effects. Postgrad Med J [Suppl 4] 52:124–132
25. Opie LH, Tansey MJ, Kennelly BM (1977) Proposed metabolic vicious circle in patients with large myocardial infarcts and high plasma free fatty acid concentrations. Lancet 2:890–892
26. Opie LH, Muller CA, Lubbe WF (1978) Cyclic-AMP and arrhythmias revisited. Lancet 2:921–923.
27. Opie LH, Nathan D, Lubbe WF (1979) Biochemical aspects of arrhythmogenesis and ventricular fibrillation. Am J Cardiol 43:131–148
28. Opie LH, Thandroyen FT, Muller C, Bricknell PL (1979) Adrenaline-induced "oxygen-wastage" and enzyme release from working rat heart. Effects of calcium antagonism, beta-blockade, nicotinic acid and coronary artery ligation. J Mol Cell Cardiol 11:1073–1094
29. Opie LH, Muller C, Nathan D, Daries P, Lubbe WF (1980) Evidence for role of cyclic AMP as second messenger of arrhythmogenic effects of beta-stimulation. Adv Cyclic Nucleotide Res 12:63–69
30. Owen P, Thomas M, Young V, Opie LH (1970) Comparison between metabolic changes in local venous and coronary sinus blood after acute experimental coronary arterial occlusion. Am J Cardiol 25:562–570
31. Podzuweit T, Lubbe WF, Opie LH (1976) Cyclic adenosine monophosphate, ventricular fibrillation and antiarrhythmic drugs. Lancet 1:341–342
32. Podzuweit T, Dalby AJ, Cherry GW, Opie LH (1978) Cyclic AMP levels in ischaemic and non-ischaemic myocardium following coronary artery ligation: Relation to ventricular fibrillation. J Molec Cell Cardiol 10:81–94
33. Podzuweit T, Els DJ, Shanley BC (1978) Catecholamine-cAMP induced arrhythmias in the infarcting pig heart. J Mol Cell Cardiol 10 [Suppl 1]:77
34. Regan TJ, Harman MA, Lehan PH, Burke WM, Oldewurtel HA (1967) Ventricular arrhythmias and K^+ transfer during myocardial ischemia and intervention with procaine amide, insulin, or glucose solution. J Clin Invest 46:1657–1668
35. Reuter H (1974) Localization of beta-receptors, and effects of noradrenaline and cyclic nucleotides on action potentials, ionic currents and tension in mammalian cardiac muscle. J Physiol (Lond) 242:429–451
36. Rona G, Chappel CI, Balazs T, Gaundry R (1959) An infarct-like myocardial lesion and other toxic manifestations produced by isoproterenol in the rat. Arch Pathol 67:443–455
37. Russell DC, Oliver MF (1979) Effect of glucose on ventricular vulnerability during acute myocardial ischaemia in the dog. J Mol Cell Cardiol 11:31–44
38. Sano T, Sawanobori T, Hiraoka M (1977) Re-evaluation of arrhythmogenetic role of slow responses (Abstr). J Mol Cell Cardiol 9 (11) [Suppl]: 18
39. Sano T, Sawanobori T, Adaniya H (1978) Generation mechanism of arrhythmia in low pH. J Mol Cell Cardiol 10 [Suppl 1]:92
40. Schneider JA, Sperelakis N (1974) The demonstration of energy dependence of the isoproterenol-induced transcellular Ca^{2+} current in isolated perfused guinea pig hearts – an explanation for mechanical failure of ischemic myocardium. J Surg Res 16:389–403
41. Schneider JA, Sperelakis N (1975) Slow Ca^{2+} and Na^+ responses induced by isoproterenol and methylxanthines in isolated perfused guinea pig hearts exposed to elevated K^+. J Mol Cell Cardiol 7:249–273
42. Singh BN, Vaughan Williams EM (1970) The effect of amiodarone, a new anti-anginal drug on cardiac muscle. Br J Pharmacol 39:657–667
43. Taylor SH, Burley DM (1977) Beta-adrenoceptor blocking drugs in postmyocardial infarction dysrhythmias. Proc R Soc Med 70 [Suppl 11]: 11–15
44. Thandroyen FT (in press) Protective effect of calcium channel antagonist agents against ventricular fibrillation. J Mol Cell Cardiol

45. Tovey KC, Oldham KG, Whelan JAM (1974) A simple direct assay for cyclic AMP in plasma and other biological samples using an improved competitive protein binding technique. Clin Chim Acta 56:221–234
46. Tsien RW (1973) Adrenaline-like effects on intracellular iontophoresis of cyclic AMP in cardiac Purkinje fibres. Nature 245:120–122
47. Tsien RW (1977) Cyclic-AMP and contractile activity in heart in ADV. P Greengard, GA Robinson (eds) Raven Press, NY. Cy NU R 8:363–420
48. Van Vliet PD, Burchell HB, Titus JL (1966) Focal myocarditis associated with pheochromocytoma. N Engl J Med 274:1102–1108
49. Vogel S, Sperelakis N (1981) Induction of slow action potentials by micro-iontophoresis of cyclic AMP into heart cells. J Mol Cell Cardiol 13:51–64
50. Waagstein F, Hjalmarson AC (1976) Effect of cardioselective beta-blockade on heart function and chest pain in acute myocardial infarction. Acta Medica Scand [Suppl] 587: 193–200
51. Waagstein F, Hjalmasson AC (1976) Double-blind study of the effect of cardioselective beta-blockade on chest pain in acute myocardial infarction. Acta Med Scand [Suppl] 587:201–208
52. Waldenström AP, Hjalmarson AC, Thornell LT (1978) A possible role of noradrenaline in the development of myocardial infarction. An experimental study in the isolated rat heart. Am Heart J 95:43–51
53. Watanabe AM, Besch HR Jr (1974) Cyclic adenosine monophosphate modulation of slow calcium influx channels in guinea pig hearts. Circ Res 35:316–324
54. Wilhelmsson C, Wilhelmsen L, Vedin JA, Wilhelmsen L, Tibblin G, Werkö L (1974) Reduction of sudden deaths after myocardial infarction by treatment with alprenolol. Lancet 2:1157–1160
55. Wojtcazk J (1979) Cyclic AMP-mediated potentiation of the electrical uncoupling and contractures in hypoxic cardiac muscle: an explanation of the relation of cAMP to cardiac arrhythmias? J Mol Cell Cardiol 11 [Suppl 2]:68
56. Yusuf S, Ramsdale D, Peto R, Fursc L, Bennett D, Bray C, Sleight P (1980) Early intravenous atenolol in suspected acute myocardial infarction. Lancet 1:273–276

Discussion

Poole-Wilson, London: The result with diltiazem is intriguing. At what time were the drugs given? If the drugs were given after the decline of contractility did the drugs reduce enzyme release? In these concentrations did the drugs alter contractility or coronary flow in the absence of coronary ligation, or is the preparation maximally vasodilated under control conditions?

Opie: These data have been published in abstract form (Am J Cardiol 1981). The drugs were given before coronary ligation in the isolated working rat heart model (*Kannengieser* et al., J Mol Cell Cardiol (1975) *7,* 135). The preparation was not maximally vasodilated and calcium antagonists increased the coronary flow. We presume that diltiazem increased coronary flow in the ischemic zone [19].

Sleight, Oxford: What was the effect of pacing plus a Ca blocker on enzyme release compared with Ca-blockade without pacing?

Opie: Calcium blocking drugs decreased enzyme release in the coronary ligated heart; when paced to the rate of control hearts, there was still reduced enzyme release, and in the case of diltiazem the cardiac output actually increased (*Hamm* and *Opie,* Circulation Research, Supplement, in press); see also Table 1.

Tritthart, Graz: My question concerns early dysrhythmias in cardiac infarction. The present state of discussion is that reentry plays a key role in these dysrhythmias, whereas triggered activity or increased automaticity induced by cAMP increase is of little significance. Could you briefly comment as to the applicability of your experimental model to the clinical situation.

Opie: First, calcium-mediated slow responses could arise in areas of the ischemic heart where cyclic AMP rises. Such slow responses could introduce conduction delays. Similarly, the increase of intercellular resistance reported by *Wojtczak* in Poland [55] could also predispose to impaired conduction, and hence to re-entry arrhythmias.

Hjalmarson, Göteborg: Using the isolated rat heart preparation with global ischaemia (ball-valve technique) we have not been able to demonstrate any effect of α-receptor stimulation or blockade. All effects of catecholamines could be blocked by β_1-blockade (metoprolol) with no augmentation by additional α-blockade. This is in contrast to the interpretation of your labetalol–metoprolol comparison. You have, however, used very high dosages of both blockers and perhaps that is why there is no effect of pacing. Could it be non-specific effects of labetalol?

Opie: Our preparation used in the studies showing that labetalol was better than an equivalent β-blocking dose of metoprolol in decreasing enzyme release, was the isolated rat heart with coronary ligation (*Hamm* and *Opie,* submitted for publication). We used a variety of doses of metoprolol, ranging from 10^{-5} *M*, the concentration which you used, to 10^{-4} *M*. For labetalol we used 10^{-7} *M* to 10^{-5} *M*. The "equivalent" β-blocking doses in response to isoproterenol were metoprolol 10^{-4} *M* and labetalol 10^{-5} *M*. We cannot be sure that labetalol had no non-specific effects, and we did not add another α-blocking drug to metoprolol.

Palm, Frankfurt/M.: In your experiments metoprolol was used in concentrations as high as 10^{-4} *M* and labetalol at 10^{-5} *M*. I doubt that the effects at these concentrations are receptor-specific. In my opinion these are membrane stabilizing concentrations and possibly are not stereo-selective. 15 years ago a Swedish group has shown that inhibition of enzyme release from isolated hearts induced by anoxia is inhibited by β-adrenoceptor blocking drugs only in very high concentrations; inhibition was not stereo-specific, and thus non-specific, i.e. induced by membrane stabilization.

Opie: Thank you for an important point. We determined a dose response curve for both, metoprolol and labetalol (see answer to *Hjalmarson*). Thus we selected concentrations that completely stopped the tachycardia found in response to increasing doses of isoprenaline. Those are the concentrations compared. Of course, we cannot exclude additional non-specific effects.

Gerlach, Munich: It should be possible to decide whether your observed effects of β-blockers in high doses on lactate dehydrogenase (LDH) release are of an unspecific nature as Dr. Palm assumes just by determining the total activity of LDH remaining in the myocardial tissue at the end of perfusion. Did you perform such measurements?

Opie: Non-specific effects of agents structurally related to propranolol, like 3-methyl-propranolol, can decrease "infarct size". We did not measure tissue LDH because only a very small percentage is lost over the first hour of regional ischemia. However, in the rat heart in situ it is known that both β-blockers and calcium-antagonists can decrease pathologically measured infarct size (*de Leiris,* personal communication). Thus decreased enzyme loss is in general associated with decreased "infarct size".

Catecholamines in Myocardial Hypoxia and Ischemia[1]

R. Dietz, A. Schömig, R. Strasser, W. Kübler

Summary

In the isolated perfused rat heart noradrenaline release – independent of central sympathetic discharge – during myocardial ischemia was quantitatively assessed. An increased noradrenaline net release from the myocardium of the rat was detected after ischemic periods of more than 10 min; after 60 min of ischemia about 1×10^{-9} mol/heart was found in the effluent. This amount can be calculated to correspond to a noradrenaline concentration in the extracellular space of the ischemic myocardium of about $5–10 \times 10^{-6}$ *M*. During myocardial ischemia neuronal reuptake of noradrenaline is progressively impaired. The amount of neurotransmitter released from the ischemic myocardium is directly related to the impairment of neuronal uptake. This may indicate a central role of this mechanism for noradrenaline net release in the ischemic heart.

Introduction

Ischemic heart disease is frequently associated with malignant arrhythmias and with damage of cardiac tissue. Both, the occurrence of arrhythmias and the necrosis of cardiac cells have been related to high local concentrations of noradrenaline at cardiac adrenergic receptors [1, 11]. Increased local concentrations of noradrenaline after inadequate perfusion of the myocardium may be the consequence of two different mechanisms:

1. Stimulation of sympathetic activity via a central reflex in response to pain, fear and probably a fall in blood pressure.
2. A direct alteration of the sympathetic nerve ending leading to stimulation of sympathetic activity either as a consequence of an increased local discharge and/or of an impaired inactivation of noradrenaline.

The aim of the present study was to investigate the effects of inadequate myocardial perfusion on cardiac noradrenaline storage, uptake and release independently of centrally mediated sympathetic discharge.

Materials and Methods

Isolated Perfused Rat Heart Preparation

Female Wistar rats weighing 150–200 g were kept on a standard pellet diet and water ad libitum. Rats were anesthetized with Inactin (50 mg/kg b.w. intraperito-

1 These investigations were supported by a grant from the Deutsche Forschungsgemeinschaft (Sonderforschungsbereich 90 – Kardiovaskuläres System)

neally) after previous heparinization (1000 I.U./kg b.w.). Hearts were then rapidly excised and dropped into a beaker containing ice-cold saline. The aorta was cannulated and the coronary system was perfused retrogradely (Langendorff technique) with a modified Krebs-Henseleit bicarbonate solution (pH 7.4) containing in addition pyruvate (3×10^{-3} *M*) and glucose (1.1×10^{-2} *M*) at a constant flow rate of 3.0 ml/min/g tissue without recirculation. The solution was equilibrated with 95% O_2 and 5% CO_2 for all normoxic perfusions, and the temperature was adjusted to 37 °C. The hearts were kept in humid chambers which were gassed with air for experiments under normoxic conditions and with nitrogen when the effects of ischemia or hypoxia were studied. The time between opening of the thorax and start of the retrograde perfusion did not exceed 1 min. With respect to heart rate and tension developed this preparation was constant at least for 180 min after the begin of the retrograde perfusion (for more details see [5]).

Uptake and Inactivation of Noradrenaline by the Myocardium

^{3}H-noradrenaline[2] (specific activity 2.7 Ci/10^{-6} mol) and unlabelled noradrenaline were added to the perfusate yielding an activity of 10^4 cpm/ml and a final concentration of 10^{-7} *M*.

Arterial perfusate and venous effluents were collected during various intervals and aliquots were used for direct measurement of total radioactivity by liquid scintillation counting (betascint BF800, Berthold and Friesecke, Wildbad) and for the determination of the deaminated metabolites 3,4-dihydroxyphenylglycol (DOPEG) and 3,4-dihydroxymandelic acid (DOMA), as well as the O-methylated metabolites normetanephrine (NMN), 3-methoxy-4-hydroxy mandelic acid (VMA) and 3-methoxy-4-hydroxy phenylglycol (MOPEG) according to the method of *Graefe* (1973, 1976). The values were corrected for quench, recovery (for noradrenaline and metabolites more than 90%) and for crosscontamination (only relevant for DOMA). Noradrenaline uptake and its rate of metabolism were calculated from the arteriovenous differences under steady state conditions.

Neuronal uptake was determined after inhibition with cocaine (final concentration in the perfusate 3×10^{-5} *M*), extraneuronal uptake after additional inhibition with corticosterone (final concentration 10^{-4} *M*) by measuring the arterio-coronary differences of noradrenaline and its metabolites.

Release of Noradrenaline from the Myocardium

During a loading period of 40 min the hearts were perfused with the modified Krebs-Henseleit solution containing labelled and unlabelled noradrenaline in the above mentioned concentrations. During the following wash-out period of 30 min no labelled or unlabelled noradrenaline were added to the perfusate. Thereafter noradrenaline release was determined by measuring the radioactivity in the effluent. In another series of experiments the release of endogenous noradrenaline without any preceding loading period, was determined by direct measurement of the noradrenaline concentration in the perfusate using the radioenzymatic assay of *Da Prada* and *Zürcher* [2].

---

2 (7-^{3}H(N))-L-noradrenaline (NET-377) was purchased from New England Nuclear, Dreieichenhain, W. Germany

Interventions

For hypoxic perfusion the modified Krebs-Henseleit solution was equilibrated with nitrogen (99.9% purity), pH was adjusted to 7.4 by addition of CO_2. Complete or stop flow ischemia was achieved by cessation of coronary flow, incomplete or low flow ischemia was induced by reduction of flow to approximaely 0.05 ml/min/g. Hyperkalemic perfusates were produced by stepwise replacement of sodium by potassium up to a final concentration of 1.2×10^{-1} *M*.

Results

Release of Endogenous Noradrenaline from Isolated Rat Hearts Perfused Under Normoxic Conditions

Immediately after instalation of retrograde perfusion endogenous noradrenaline release from the isolated hearts amounts to 3×10^{-12} mol/min, this value being increased to about 4×10^{-12} mol/min after addition of cocaine. With continuation of the aerobic perfusion endogenous noradrenaline net release decreases exponentially. After 54–60 min of perfusion the release of endogenous cardiac noradrenaline has declined to about 1×10^{-13} mol/min (Fig. 1).

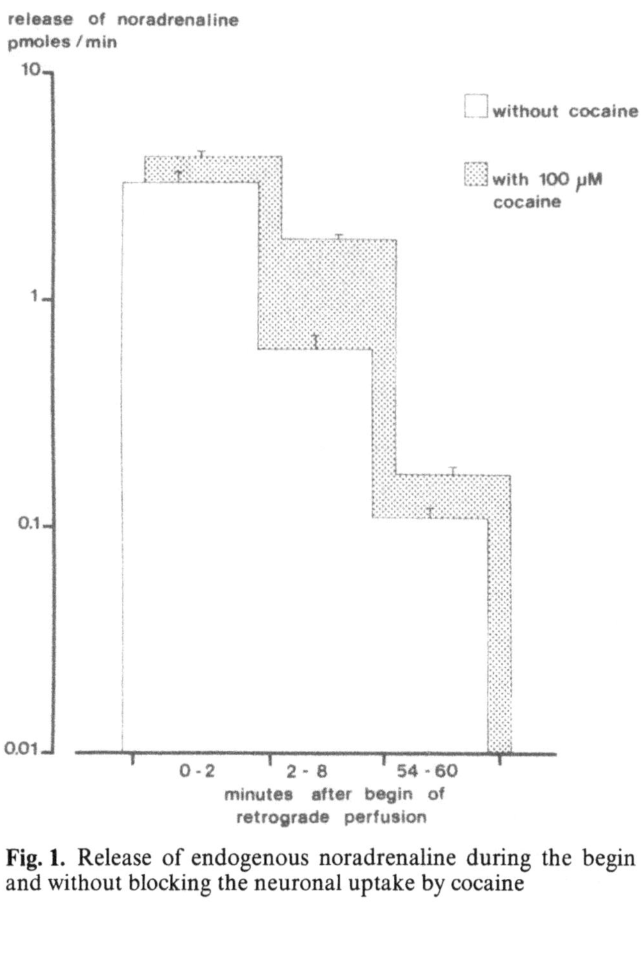

Fig. 1. Release of endogenous noradrenaline during the begin of retrograde perfusion with and without blocking the neuronal uptake by cocaine

Noradrenaline Release from the Ischemic Myocardium

After total ischemia for 60 min, the bulge of previously stored ^{3}H-noradrenaline is detected in the venous effluent during the first few minutes of reperfusion. The amount of ^{3}H-noradrenaline released decreases with time exponentially indicating simple wash-out kinetics. Endogenous noradrenaline washed out after 60 min of complete ischemia amounts to about 1×10^{-9} mol/rat heart.

The loss of cardiac noradrenaline depends on the duration of ischemia. No increase in noradrenaline net release can be detected in periods of ischemia ranging up to 7 min, either by determination of ^{3}H-noradrenaline or by direct measurement of the noradrenaline concentration in the effluent. After prolongation of the ischemic period for 10 min and more increasing amounts of noradrenaline are released.

During complete or stop flow ischemia, noradrenaline release is about 100 times higher than during incomplete or low flow ischemia. Blockade of the neuronal uptake by cocaine induces a threefold increase in cardiac noradrenaline release during incomplete ischemia, whereas noradrenaline release after complete ischemia (Fig. 2) is not significantly affected by cocaine.

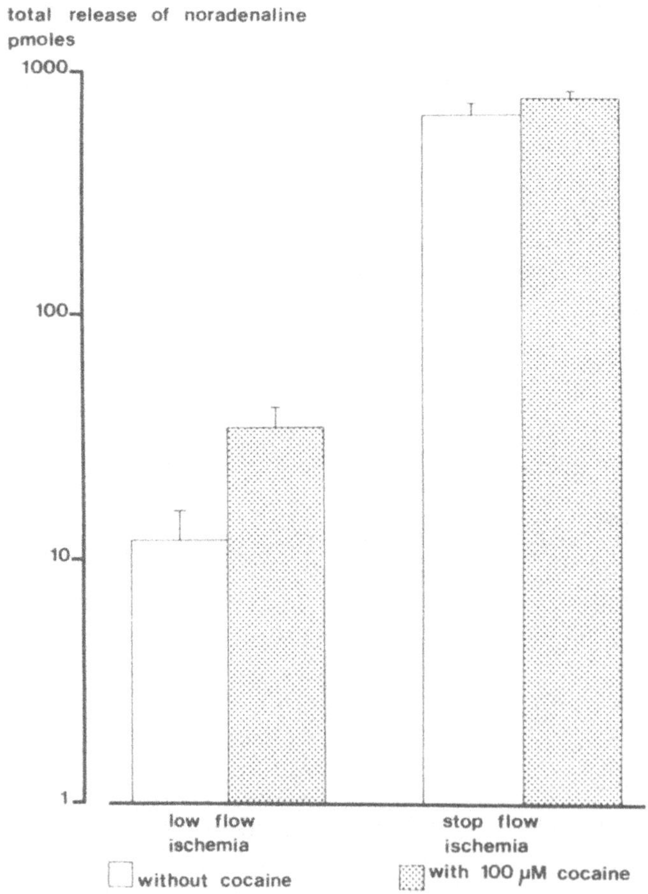

Fig. 2. Comparison of total net release of endogenous noradrenaline during low flow vs. stop flow ischemia with and without addition of cocaine

Cardiac Noradrenaline Release During Hypoxia, Hyperkalemia and Acidosis

During hypoxic perfusion for up to 30 min cardiac net release of ^{3}H-noradrenaline is not significantly changed when compared with aerobic conditions. Prolongation of the hypoxic perfusion for a further 30 min results in a doubling of the cumulative ^{3}H-noradrenaline release as compared with the control values obtained under normoxic conditions. Total cardiac release of ^{3}H-noradrenaline during hypoxic perfusion for 60 min, however, corresponds to only 1/10 of the amount released during an ischemic period of the same duration.

Potassium concentrations of more than 2×10^{-2} M in the perfusate induce a dose dependent increase in ^{3}H-noradrenaline net release. The cumulative amount of endogenous noradrenaline released during hyperkalemia is increased for all potassium concentrations tested when cocaine is added to the perfusate (Fig. 3). Changing the pH of the perfusate from 7.4 to 6.6 does not result in a significant increase in the net release of ^{3}H-noradrenaline from the hearts.

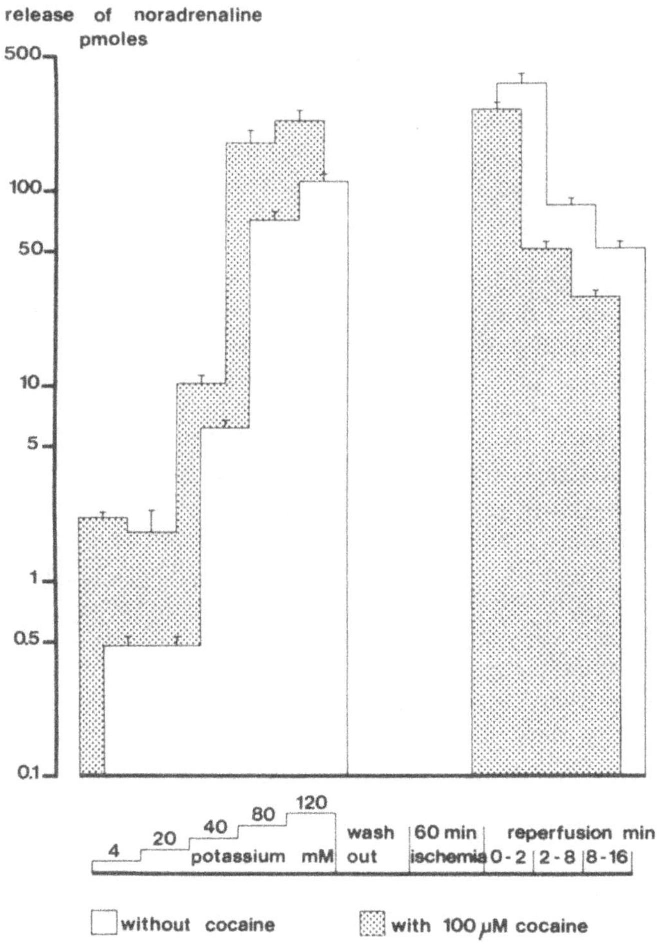

Fig. 3. Release of endogenous noradrenaline during stepwise increases in potassium concentration and following 60 minutes of stop flow ischemia with and without addition of cocaine

Cardiac Noradrenaline Uptake Under Conditions of Ischemia, Hypoxia, Hyperkalemia and Acidosis

The rate of cardiac noradrenaline uptake decreases with increasing duration of the ischemic period. Ischemic periods of less than 10 min do not significantly reduce the rate of uptake. A marked diminution, however, is found after longer ischemic periods. After 60 min of complete ischemia noradrenaline uptake is virtually zero.

In contrast to the conditions of complete ischemia, no significant deterioration of noradrenaline uptake is observed during and following a 60 min period of hypoxic perfusion.

Hyperkalemia of more than 2×10^{-2} *M* causes a dose dependent reduction in noradrenaline uptake. At 4×10^{-2} *M* potassium uptake is reduced to 40% and at 8×10^{-2} *M* it is almost completely abolished.

Noradrenaline uptake is progressively impaired by acidosis. The effect, however, is not pronounced: at pH 6.6 in the perfusate the rate of uptake is still 60% of the control value (pH 7.4).

Discussion

Local metabolic changes occurring in the heart itself are of great importance for the noradrenaline net release from the underperfused myocardium. Centrally mediated effects on sympathetic discharge during myocardial ischemia may play a role at the very onset of acute ischemia, with prolongation of the ischemic period, however, the local factors regulating net-noradrenaline release in the heart itself are probably predominant.

In the present study cardiac noradrenaline net release under the conditions of complete ischemia could only be observed after ischemic periods longer than 10 min. *Shahab* et al. [10] found a rather early net release of noradrenaline from the isolated perfused heart of the dog and of the rabbit – as early as after 3 min of complete ischemia. *McGrath* et al. [8] reported no increase in coronary sinus concentration of noradrenaline after 8 min of ischemia due to ligation of the left anterior descending branch of the left coronary artery in the dog. Similar results were obtained by *Riemersma* et al. (peronal communication) after 10 min of coronary artery occlusion in the dog. Our results are in agreement with these studies of *McGrath* et al. [8] and *Riemersma* et al. (personal communication) but not with the data reported by *Shahab* et al. [10] and *Wollenberger* et al. [12]. This discrepancy can hardly be explained by differences in the experimental procedure; it might, however, be due to methodological problems related to the determination of noradrenaline in the past (fluorometric assay).

The total amount of noradrenaline released after a 60-min period of ischemia yields about 1×10^{-9} mol/heart. Assuming a homogeneous distribution within the extracellular space of the rat heart a mean extracellular concentration of $5\text{–}10 \times 10^{-6}$ *M* can be expected. If administered exogenously to an isolated heart these high concentrations of the neurotransmitter seem to be sufficient to cause ultrastructural damage and leakage of intracellular enzymes [4, 8, 11]. The enzyme release can be markedly reduced after catecholamine depletion by reserpine pretreatment [3, 9].

The present study demonstrates a close relationship between the impairment of the neuronal uptake mechanism and the amount of noradrenaline released into the venous effluent: the more the rate of uptake decreases the more noradrenaline is found in the venous effluent. The parallel time course of impairment of neuronal uptake and of noradrenaline net release from the ischemic myocardium suggests that impaired neuronal uptake may play a central role for the increased net release of noradrenaline from the ischemic heart. As this mechanism is a main regulator of the noradrenaline concentration within the synaptic cleft and at the receptor site, any disturbance of neuronal uptake will induce marked changes in adrenergically mediated functions. Alterations in the neuronal uptake mechanism will also greatly influence the portion of noradrenaline that is spilled over into the plasma after its discharge from the sympathetic nerve ending into the synaptic cleft.

Of the different metabolic changes occurring during ischemia, hypoxia alone does not seem to be of greater significance for cardiac noradrenaline net-release. A rise in potassium concentration above 4×10^{-2} *M*, however, a concentration which most likely is reached in the extracellular space during ischemia, leads to an impairment of neuronal noradrenaline uptake and an increased noradrenaline net release. The effects induced by high potassium concentrations as well as by acidosis, however, are quantitatively not sufficient to explain the observed release of noradrenaline during ischemia.

In conclusion: the present experiments demonstrate the significance of local metabolic changes occurring at the sympathetic nerve endings for the increased net release of noradrenaline during myocardial ischemia. The results are compatible with the hypothesis that myocardial ischemia for periods longer than 10 min increases noradrenaline concentrations at myocardial adrenergic receptors sufficient to induce additional myocardial damage.

References

1. Corr PB, Witkowski FX, Sobel BE (1977) Increased adrenergic tone in ischemic myocardium underlying ventricular fibrillation. Clin Res 25:454 A
2. Da Prada M, Zürcher G (1976) Simultaneous radioenzymatic determination of plasma and tissue adrenaline, noradrenaline and dopamine within femtomole range. Life Sci 19:1161
3. De Leiris J, Feuvray D, Come D (1972) Acetylcholine induced release of lactatedehydrogenase from isolated rat heart. J Mol Cell Cardiol 4:357
4. Ganotte CE, Seabra Gomez R, Nayler WG, Jenning RB (1975) Irreversible myocardial injury in anoxic perfused rat hearts. Am J Pathol 80:419
5. Graefe KF (1976) Aufnahme und Metabolisierung von ^{3}H-(–)-noradrenalin im isoliert durchströmten Katzen- und Kaninchenherzen. Habilitationsschrift, Universität Würzburg
6. Graefe KF, Stefano FJE, Langer SZ (1973) Preferential metabolism of ^{3}H-norepinephrine through the deaminated glycol in the rat vas deferens. Biochem Pharmacol 22:1147
7. Hearse DJ, Humphrey SM, Chain EB (1973) Abrupt reoxygenation of the anoxic potassium arrested perfused rat heart: a study of myocardial enzyme release. J Mol Cell Cardiol 5:395
8. Hearse DJ, Garlick PB, Shilling JP (1975) Effect of beta-adrenergic agonists on enzyme release from hypoxic isolated perfused rat heart. Biochem Soc T 3(3):422–425
9. McGrath BP, Lim SP, Leversha L, Shanahan A (1981) Myocardial and peripheral catecholamine responses to acute coronary artery constriction before and after propranolol treatment in the anesthetised dog. Cardiovasc Res 15:28
10. Opie LH, Nathan D, Lubbe WF (1979) Biochemical aspects of arrhythmogenesis and ventricular fibrillation. Am J Cardiol 43:131–148
11. Sakai K, Spiekerman PG (1975) Effects of reserpine and propranolol on anoxia induced enzyme release from the isolated perfused guinea pig heart. Naunyn Schmiedebergs Arch Pharmacol 291:123

12. Shahab L, Wollenberger A, Haase M, Schiller U (1969) Noradrenalinabgabe aus dem Hundeherzen nach vorübergehender Okklusion einer Koronararterie. Acta Biol Med Germ 22:135
13. Waldenström AP, Hjalmarsson AC, Thormall L (1978) A possible role of noradrenaline in the development of myocardial infarction. Experimental study in isolated rat heart. Am Heart J 95:43
14. Wollenberger A, Shahab L (1965) Anoxia-induced release of noradrenaline from the isolated perfused heart. Nature 207:88

Discussion

Starke, Freiburg: 1. Wollenberger and Shahab reported that, e.g. in rabbit hearts, noradrenaline release is massively increased already after approx 3 min of anoxia. Did you make any analogous observations? We ourselves did not find any comparable release from rabbit hearts during anoxia.
2. Do you think that inhibition of noradrenaline re-uptake is the major factor causing the release? In my opinion, inhibition of re-uptake alone would not nearly lead to such high rates of overflow.

Dietz: 1. The results of Wollenberger and Shahab are, in fact, contradictory to ours. More recent investigations by Riemersma, McGrath and Waldenström, however, support our data on the time course of noradrenaline release under ischemia: no net release of noradrenaline can be detected after ischemic periods of less than 10 min.
2. All those interventions we studied which resulted in an increase in noradrenaline net efflux from the hearts had also an effect on the direct release of noradrenaline from its stores. The amount of noradrenaline found in the effluent, however, is greatly modulated by the simultaneous changes in uptake activity.

Gerlach, Munich: In addition, you have of course to consider species differences. Guinea pig hearts, for instance, do not respond to ischemia with a measurable release of catecholamines.

Dietz: There are certainly species differences. The fact that the time course of noradrenaline release during ischemia in a whole animal preparation (dog) closely follows the time course we observed in the isolated perfused rat heart strongly suggests that some species independent mechanisms are acting in ischemia.

Opie, Cape Town: Wollenberger and Shahab presented their data in Milan in 1966. They found an increased coronary venous concentration of catecholamine after two minutes of ischaemia, but only when the coronary ligation was released and there was re-perfusion.

Schrader, Munich: You interpreted your results in terms of changes in uptake and release of catecholamines, tacitly assuming that net synthesis of catecholamines does not play a significant role. Can one really do that, and do you have any data on the turnover of myocardial catecholamines?

Dietz: The importance of an impaired neuronal uptake of noradrenaline in cardiac diseases is evident from studies on heart failure, where a decreased tissue content is associated with increased noradrenaline turnover and concentration in plasma. A reduced rate of synthesis can not explain the discrepancy, whereas a reduced rate of re-uptake nicely explains these findings. In myocardial ischaemia, de novo synthesis of noradrenaline presumably is of neglible significance, since it is oxygen dependent.

Dobson, Worchester: Have you measured myocardial cyclic AMP in your hearts? Such information may provide a clue to whether the noradrenaline release is of physiological importance.

Dietz: We did not measure concentrations of cAMP ourselves. From data in the literature, however [10] it is evident that increases in cAMP after myocardial ischemia are closely related to the changes observed in noradrenaline release.
From recent experiments reported by *Meesmann* (1981) it appears that a preceding depletion of the necrotic zone from catecholamines largely prevents the occurrence of post-ischemic arrhythmias.

Spieckermann, Göttingen: Can you give us a rough measure of how many cells are irreversibly destroyed after 20 or 60 min of ischemia under your experimental conditions?

Dietz: Although it is well known that myocardial muscle cells are completely destroyed after a period of 60 min of ischemia, this does not hold true for sympathetic nerve fibres. These have a much longer survival time and it was their function we were interested in. Sympathetic nerve terminals may even be investigated in tissue which was excised about 2 h post mortem.

Hirche, Cologne: You had to use very high K^+ concentrations in your isolated hearts to induce noradrenaline (NA) release. When we investigate myocardial NA release in pig hearts in vivo we observe a reduction of catecholamine fluorescence – indicating a release of catecholamines from adrenergic nerve fibers – after only 3–4 min of severe ischemia. At this time the extracellular K^+-concentration has increased only to 10 mM. In pig hearts which develop ventricular fibrillation already 3–5 min after the onset of LAD occlusion NA release can be detected in the coronary sinus blood. How would you explain the fact that you need much higher K^+ concentrations to induce NA release?

Dietz: Do you really know the concentration of K^+ at the adrenergic terminals in your model?

Hirche: The extracellular K^+ concentration ($[K^+]e$) has been measured using ion-selective polyvinylchloride (PVC) liquid membrane electrodes placed on the epicardium of the ischemic region of the left ventricle in anesthetized pigs. However, *Hill* and *Gettes* [Circulation (1980) 61:768] have shown that $[K^+]e$ increases to comparably high values in deeper layers of the ischemic left ventricle of the pig heart.

III. Pathophysiological, Clinical and Therapeutic Aspects of Catecholamines and Adrenergic Receptors

Myocardial Catecholamines and Biosynthetic Enzymes in Various Heart Diseases of Man [1]

O. Penttilä, E. Merikallio, E. Klinge, P. Siltanen, J. Pispa, K. Kyösola

Summary

In a comparative study samples excised from a standard site of right auricular tissue during open heart surgery were assessed for their noradrenaline (NA) content as well as for their tyrosine hydroxylase (TH) and dopamine-β-hydroxylase (DBH) activities. In a total of 55 adult patients representing four different groups of heart diseases the NA contents were as follows: in ischaemic heart disease (IHD) 2.51 ± 0.71 µg/g wet tissue (mean $\pm$ SD, $n = 19$), in left-sided valvular heart disease (VHD) 1.78 ± 0.79 ($n = 10$), in uncomplicated atrial septal defect (ASD) 1.33 ± 0.70 ($n = 14$), and in congestive heart failure (CHF) 0.64 ± 0.34 ($n = 12$). The NA content was significantly higher in the IHD group than in the three other groups. The results of a corresponding fluorescence histochemical study supported the above findings. Both the TH and DBH activities were highest in the IHD group (TH 0.48 ± 0.16 and DBH 2.33 ± 0.92 nmoles $\times$ $g^{-1} \times min^{-1}$) although the differences were signifiant only when compared with the ASD (TH 0.30 ± 0.15, DBH 1.46 ± 0.90) and the CHF (TH 0.19 ± 0.11, DBH 0.90 ± 0.58) group. In the whole material there was a significant positive correlation between the NA content and the TH and the DBH activities. The exact cause of the NA depletion in the failing heart is not clear. A number of factors may be involved including depressed synthesis and impaired neuronal uptake and storage. In IHD the myocardial NA content and the activities of the biosynthetic enzymes are relatively high at least when compared with those in ASD and CHF. This possibly reflects an accentuated adrenergic tone and NA turnover in IHD.

Introduction

The myocardial noradrenaline (NA) content is unknown in healthy normal subjects but in various heart diseases of man it has been studied biochemically by several authors [2, 6, 8, 11, 21, 24, 31] and also histochemical demonstrations of the adrenergic structures have been performed [2, 13, 22]. The NA content, which generally reflects the density of the adrenergic innervation, is highest in the atria and lowest in the left ventricle and a positive correlation exists between the NA contents of these tissues [8, 24]. The cardiac NA stores have repeatedly been verified to decrease in myocardial failure [3, 6] and a concomitant reduction in the activity of tyrosine hydroxylase (TH), the rate limiting enzyme in NA synthesis [14], has been reported [11]. The NA depletion, which according to experimental studies may be reversible [33], appears to be confined to the myocardial adrenergic neurones since other tissues are not affected [3]. Regardless of the underlying heart disease only

1 From the Departments of Pharmacology, Pharmacy, Medical Chemistry and Anatomy, University of Helsinki, and the Department of Thoracic Surgery and the Cardiovascular Laboratory. First Department of Medicine, University Central Hospital, Helsinki, Finland

small amounts of dopamine and adrenaline are found in the myocardium [2, 21, 24].

The aim of this paper is to review our studies on the NA content and the activities of two biosynthetic enzymes, i.e. of TH and DBH (dopamine-β-hydroxylase), as well as our attempts to elucidate histochemically the adrenergic structures of human right atrium in different cardiac disorders. Special emphasis will be put on ischaemic heart disease (IHD) and on congestive heart failure (CHF). For obvious reasons no relevant normal control material has been available.

Patients and Methods

Classification of Patients

Atrial biopsies were excised from a standard site during open heart surgery from 55 adult patients. The first group included 19 patients suffering from IHD. They were selected for operation on the basis of anamnestic, clinical and coronary angiographic data and all of them had a two- or three-vessel disease and their exercise tolerance was low (NYHA III–IV). The second group consisted of 10 patients with left-sided valvular heart disease (VHD) without clinically overt congestive myocardial failure. Several of them had a multivalvular disease but the predominant lesions were as follows: mitral stenosis in 3, mitral regurgitation in 2, aortic stenosis in 1 and aortic regurgitation in 4 patients. The third group consisted of 14 patients with uncomplicated atrial septal defect (ASD) of secundum type. The fourth group consisted of 12 patients, who had experienced congestive heart failure (CHF) prior

Table 1. Clinical and laboratory data and atrial noradrenaline (NA) content and tyrosine hydroxylase (TH) and dopamine-β-hydroxylase (DBH) activities in the four patient groups (mean ± SD). (After 23)

	N	Age (yrs)	Heart volume (cm^3/m^2)	BP (mm Hg)	NA (μg/g)	TH (nmoles $\times g^{-1} \times min^{-1}$)	DBH (nmoles $\times g^{-1} \times min^{-1}$)
IHD	19	44.3	462	141/92	2.51	0.48	2.33
		7.8	72	16 10	0.71	0.16	0.92
VHD	10	46.9	787[a]	133/72[b]	1.78[c]	0.41	1.93
		10.4	198	13 19	0.79	0.21	0.96
ASD	14	41.6	567[a]	134/82[b]	1.33[c]	0.30[d]	1.46[e]
		13.1	73	24 13	0.70	0.15	0.90
CHF	12	48.7	970[a]	130/73[b]	0.64[c]	0.19[d]	0.90[e]
		7.0	207	28 18	0.34	0.11	0.58

[a] Significantly higher than in the IHD group ($P < 0.001$)

[b] Significantly lower than diastolic BP in the IHD group (VHD/IHD and CHF/IHD $P < 0.001$, ASD/IHD $P < 0.02$)

[c] Significantly lower than in the IHD group (VHD/IHD $P < 0.02$, ASD/IHD and CHF/IHD $P < 0.001$)

[d] Significantly lower than in the IHD group (ASD/IHD $P < 0.01$, CHF/IHD $P < 0.001$)

[e] Significantly lower than in the IHD group (ASD/IHD $P < 0.02$, CHF/IHD $P < 0.001$)

IHD, ischaemic heart disease; VHD, valvular heart disease; ASD, atrial septal defect; CHF, congestive heart failure to surgery based either on ASD in 2 or on VDH in 10 subjects. The patients underwent operation in a clinically stable state. The degree of failure had been assessed by means of clinical and laboratory examinations and cardiac catheterization. All the patients of the fourth group belonged to the functional classes III or IV by the NYHA criteria.

The heart volume and systemic blood pressure were recorded during the week preceding the operation. A summary of some clinical and laboratory data on the different patient groups is presented in Table 1.

Drug Treatment and Anaesthesia

The usual medication with digitalis, diuretics, antiarrhythmic and antihypertensive drugs was continued up to the operation with the exception of beta-adrenergic blocking drugs, which were withdrawn seven days before. No drugs known to interfere with the tissue NA content were used. Standard premedication (atropine + pethidine) and anaesthesia (thiopentone sodium, oxygen + nitrous oxide, pethidine or anileridine, alcuronium or pancuronium) were employed.

Excision of Biopsies

In order to insert the venous perfusion cannula for cardiopulmonary bypass, 150–350 mg of the right auricular appendage had to be removed and were used for the studies described below. The biopsies were excised before institution of cardiopulmonary bypass and operative correction.

Determination of Myocardial NA Content and Enzyme Activities

For the quantitative determination of NA a spectrophotofluorimetric method [1] was used slightly modified [21]. For the determination of enzyme activities samples weighing 100–200 mg were put immediately after biopsy into liquid nitrogen and stored therein until the whole material was collected. The TH activity was assayed according to *Waymire* et al. [35] and the DBH activity according to *Molinoff* et al. [18]. For further details see *Penttilä* et al. [23].

Histochemical Demonstration of Adrenergic Nerve Fibres

From a comparable group of 22 adult patients (mean age $\pm$ SD 49 $\pm$ 12 years, 8 females and 14 males) suffering from IHD ($n = 7$), VHD ($n = 5$), ASD ($n = 5$) and CHF ($n = 5$) identical samples of the right atrial appendage were excised during open heart surgery and immediately frozen in liquid nitrogen and stored until the whole material was collected. The glyoxylic acid fluorescence histochemical method [15] was used slightly modified [13]. From each specimen a large number of stretch preparations was made. All the preparations were analyzed in series, independently by three investigators without any knowledge of the code. In the qualitative and semiquantitative analysis of the fluorescence histochemical characteristics particular emphasis was laid on (a) the intensity of the specific fluorescence of the adrenergic nerve fibres and the varicosities (b) the density of the visualized adrenergic nerve network and (c) the average size and number of the visualized varicosities per given length of axon and the proportional share of the largest varicosities. Also (d) the ratio between the densities of the perivascular and the three-dimensional myocardial loose-meshed nerve networks was evaluated as well as (e) the occurrence of droplet fibres (see [22]).

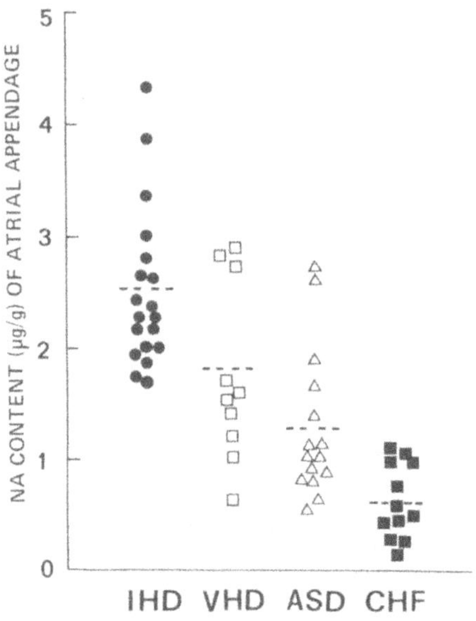

Fig. 1. Noradrenaline (*NA*) content in biopsy specimens of right atrial appendage from 19 patients with ischaemic heart disease (*IHD*), 10 with left-sided uni- or multivalvular heart disease (*VHD*), 14 with uncomplicated atrial septal defect (*ASD*) of secundum type, and from 12 patients who had experienced congestive heart failure (CHF) prior to surgery based on either ASD or VHD

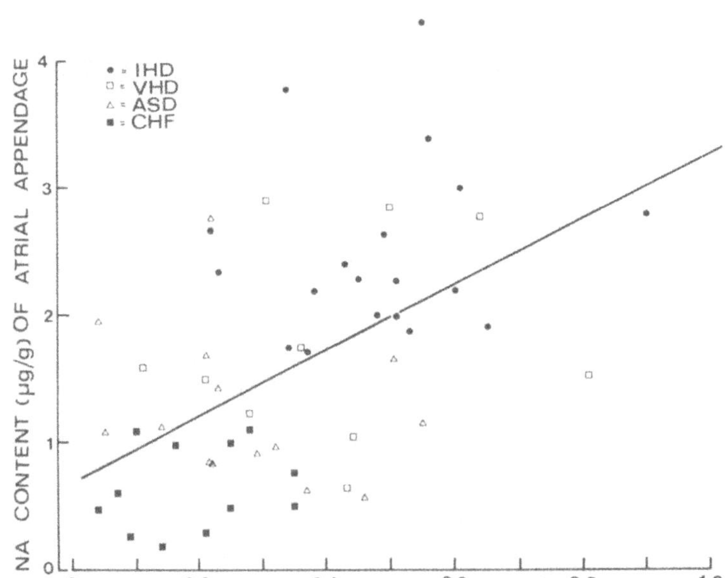

Fig. 2. Relationship between tyrosine hydroxylase (*TH*) activity and noradrenaline (*NA*) content in right auricular samples excised from patients with ischaemic heart disease (*IHD*, $n = 19$), valvular heart disease (*VHD*, $n = 10$), uncomplicated atrial septal defect (*ASD*, $n = 14$) or congestive heart failure (*CHF*, $n = 12$). The correlation is significant ($r = 0.53$, $P < 0.001$, $n = 55$). (After *Penttilä* et al.: Acta Med Scand 203: 161–166, 1978)

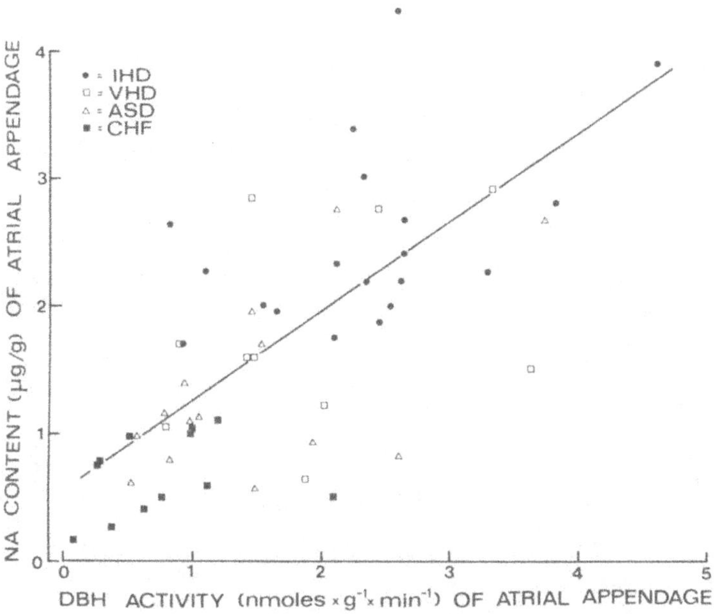

Fig. 3. Relationship between dopamine-β-hydroxylase (*DBH*) activity and noradrenaline (*NA*) content in the same samples as in Fig. 2. The correlation is significant ($r=0.66$, $P<0.001$, $n=55$). (After *Penttilä* et al.: Acta Med Scand *203:* 161–166, 1978)

Results

In the IHD group the NA content (μg/g wet tissue, mean ± SD) was 2.51 ± 0.71 and in the VHD group 1.78 ± 0.79 ($P<0.02$). The values in the ASD and CHF groups were 1.33 ± 0.70 and 0.64 ± 0.34, respectively, which were significantly lower than in the IHD group ($P<0.001$) (Table 1 and Fig. 1). The NA content in the CHF group was significantly lower than in the ASD and VHD group (ASD/CHF $P<0.01$, VHD/CHF $P<0.001$). There was no significant difference between the ASD and VHD group ($P<0.2$).

In the IHD group the TH activity (nmoles × g^{-1} × min^{-1} wet tissue, mean ± SD) was 0.48 ± 0.16 and in the VHD group 0.41 ± 0.21 ($P<0.3$). Compared with the IHD group, the TH activity was significantly lower both in the ASD and the CHF group, being 0.30 ± 0.15 ($P<0.01$) and 0.19 ± 0.11 ($P<0.001$), respectively (Table 1). The TH activity was significantly lower in the CHF than in the VHD ($P<0.01$) and the ASD group ($P<0.05$). The difference between the ASD and the VHD group was not significant ($P<0.2$).

In the IHD group the DBH activity (nmoles × g^{-1} × min^{-1} wet tissue, mean ± SD) was 2.33 ± 0.92 and in the VHD group 1.93 ± 0.96 ($P<0.3$). Compared with the IHD group, the DBH activity was significantly lower both in the ASD and the CHF group, being 1.46 ± 0.90 ($P<0.02$) and 0.90 ± 0.58 ($P<0.001$), respectively (Table 1). The DBH activity was significantly lower in the CHF than in the VHD group ($P<0.01$) but not significantly lower than in the ASD group ($P<0.1$). Neither was the difference between the ASD and the VHD group significant ($P<0.3$).

There was a significant positive correlation between NA content and TH activity in the whole material ($r=0.53$, $P<0.001$, $n=55$) (Fig. 2). A corresponding significant correlation was also found between NA content and DBH activity

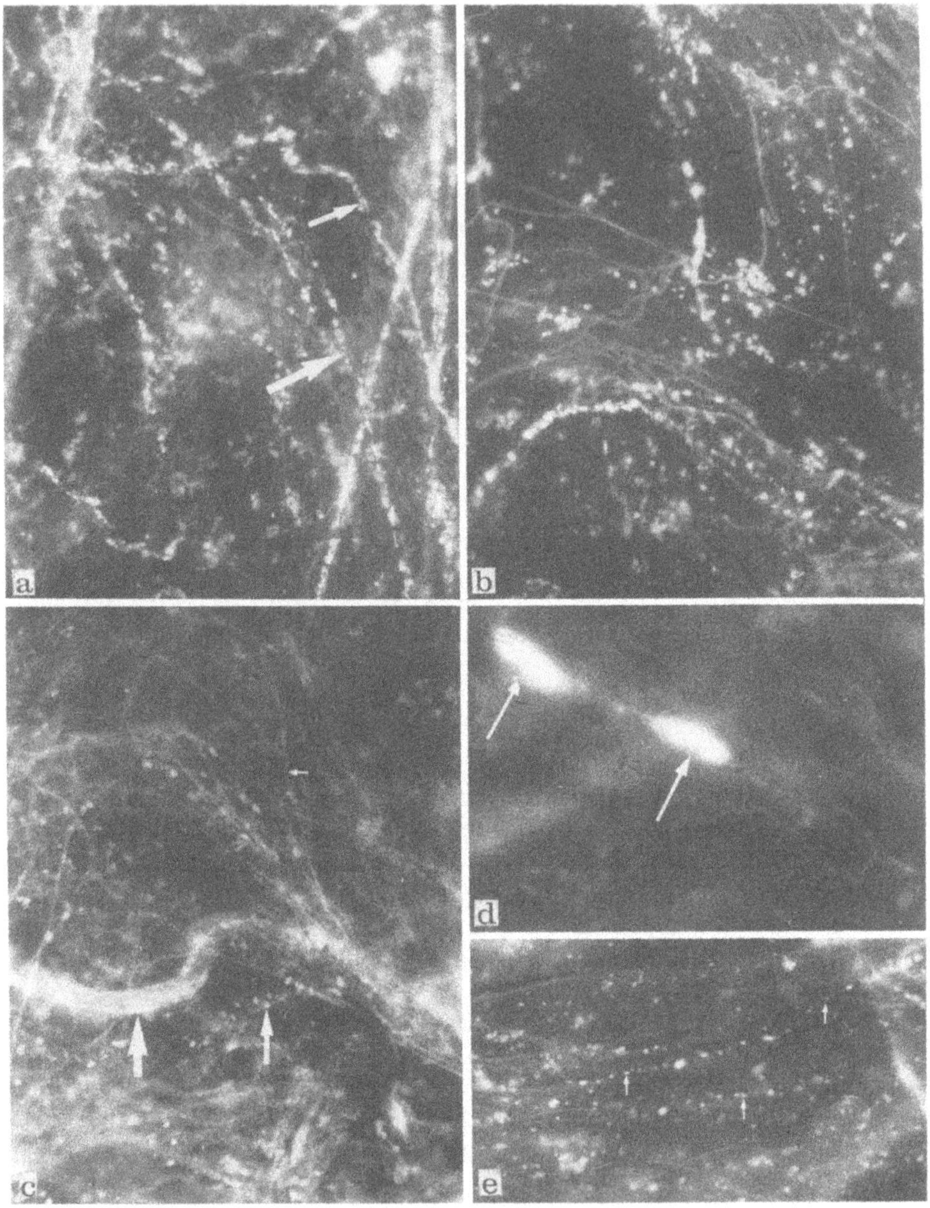

($r=0.66$, $P<0.001$, $n=55$) (Fig. 3). These correlations were insignificant within all the individual patient groups. On the other hand, in the total material there was a significant correlation between TH and DBH activities ($r=0.36$, $P<0.01$, $n=55$). The systemic diastolic blood pressure was not significantly correlated to the NA content or the enzyme activities. But the heart volume was in the whole material inversely correlated to the NA content ($r=-0.56$, $P<0.001$, $n=55$), and the TH ($r=-0.42$, $P<0.01$, $n=55$) and the DBH activity ($r=-0.42$, $P<0.01$, $n=55$).

Histochemically the intensity of the fluorescence of the adrenergic structures was generally higher in the IHD group than in the CHF group. In this respect no

◀ **Fig. 4a–e.** Glyoxylic acid-induced catecholamine fluorescence in stretch preparations taken from the right atrial appendage of adult human heart. In addition to the specific catecholamine fluorescence, lipofuscin granules and connective tissue fibres exhibit an unspecific autofluorescence. ×250. **a** Large numbers of strongly fluorescent adrenergic nerves in the heart of a patient suffering from IHD. Fluorescent varicose fibres run singly (*big arrow*) or course for long distances in close proximity to each other (*small arrow*). **b** Specifically fluorescent adrenergic nerves in the heart of a patient with uncomplicated ASD. The number of fluorescent nerves may be slightly smaller than in the heart of the patient with IHD (see **a**) but obviously more abundant than in an ordinary case of ASD (see **e**). **c** The picture demonstrates the absence of fluorescent nerves in a large area of myocardium from the heart of a patient suffering from CHF. In the *centre* of the picture there is a blood vessel without fluorescent nerves. Quite intense autofluorescence is seen in the connective tissure fibres in the myocardium (*small arrow*), in the wall of the blood vessel (*big arrow*) and lipofuscin granules (*large arrow*). **d** Large, strongly fluorescent "droplet fibres" (*arrows*), diameter of the varicosities about 10 μm, in the heart of a patient suffering from VHD. Compare the size of the droplets with the size of the varicosities of ordinary adrenergic axons in **e** (*small arrows*). **e** Single axons with many fluorescent varicosities of different size (*small arrows*) between the muscle fibres in the heart of a patient suffering from uncomplicated ASD. (After *Penttilä* et al.: Virchows Arch A Path Anat and Histol. 373:279–292, 1977)

certain difference could be detected between the IHD and the ASD group. In the IHD group the density of the perivascular as well as of the myocardial nerve net was greater than in the ASD and the CHF group. Further, the average size and the number of the varicosities per given length of axon and the proportional share of the largest varicosities were greater in the IHD than in the ASD and the CHF group. The difference between the IHD and the ASD group was not prominent. In some patients with CHF fluorescent axons were observed only occasionally, and the tiny varicosities exhibited a hardly detectable fluorescence. However, in some VHD patients without clinically overt failure the adrenergic network was relatively well visualized with axons and varicosities of ordinary fluorescence intensity. The impression was gained that the adrenergic nerve fibres supplying the myocardium disappeared earlier than the perivascular ones, i.e. they became depleted of NA to such an extent that the fluorescence could no longer be visualized. Summarizing all the histochemical results, the amount of NA in the adrenergic fibres seemed to be higher in the IHD than in the ASD and especially the CHF group (Fig. 4). In one patient suffering from IHD and in six patients with valvular heart disease axons with huge local accumulations of NA, i.e. so-called droplet fibres were observed.

Discussion

Considerable attention has been paid to the activity of the adrenergic nervous system in patients with heart failure. Several studies have revealed that the depression of the contractile state in the hypertrophied and failing ventricle represents an intrinsic property of the muscle [5]. The exact underlying mechanism of the depression of contractility is open but e.g. an abnormality in excitation-contraction coupling associated with a deficit in or a maldistribution of intracellular calcium has been suggested [5]. However, despite the depressed contractility the overall circulatory effectiveness is tried to be maintained by compensatory mechanisms as increased muscle mass and intensified adrenergic stimulation [5].

In patients with heart failure measurements of the level in arterial blood during exercise as well as of the urinary excretion of NA have shown marked elevations indicating an augmented adrenergic activity [8]. However, in spite of this the myocardial NA concentration is considerably reduced [3]. According to certain experimental evidence there even could exist a diminished release of NA from the myocardial adrenergic nerve endings [9], which could account for the declined adrenergic support. The severity of the failure may in this way be intensified and the myocardium may become more dependent upon circulating catecholamines in order to be able to maintain its appropriate performance [27].

There is biochemical, histochemical and other experimental evidence suggesting that the reduction of cardiac NA concentrations in failure is only in part explained in terms of simple dilution of adrenergic nerve endings in the hypertrophied muscle mass [2, 7] and that there is a true depletion of NA in the whole myocardium regardless of which heart chamber carries the primary haemodynamic burden [3]. The present results concerning the atrial myocardial NA content in the failing heart correspond to those presented in the literature [2, 6, 11, 24, 31]. Histochemically it was also possible to demonstrate a substantial depletion of NA from the adrenergic nerve network in human atrium, which is in agreement with earlier observations on failing bovine heart [33]. Our findings in man have been confirmed by *Borchard* [2].

It has been assumed that up to 90% of the NA present in the heart is synthetized locally [3]. TH, which converts tyrosine to dopa, normally is the rate limiting enzyme in the synthesis. Marked reduction in the activity of this enzyme is associated with myocardial NA depletion in experimental [26] as well as in human heart failure [11]. There is experimental evidence suggesting that in heart failure the rate limiting step in NA synthesis is shifted from hydroxylation of tyrosine to hydroxylation of dopamine [3, 30]. Already *DeQuattro* et al. [11] considered this possibility in their study on the determinants of cardiac NA depletion. The present results confirm earlier observations on low myocardial TH activity in human heart failure [11]. Also the activity of DBH was low in myocardial failure. A significant correlation existed between NA content and TH and DBH activity, as well as between the activities of both of the synthetizing enzymes. These results indicate that the reduction in the activities of both enzymes at least partly accounts for the depletion of the adrenergic neurotransmitter. The actual cause of the reduction in the activities of the enzymes remains to be solved. It appears that in addition to a defect in NA synthesis, also neuronal binding [26, 33] and uptake [25] of catecholamines may be impaired in failure. It is postulated that changes in MAO and COMT activities do not essentially contribute to the depletion of cardiac NA in hypertrophied and failing hearts [28].

One of our most prominent findings [21, 23] was that in IHD there is a relatively high myocardial NA level accompanied by high activities of TH and DBH. The high level of NA in IHD has been confirmed by *Petch* and *Nayler* [24]. But the exact relationship of the NA level and the activities of the biosynthetic enzymes in IHD to those in normal heart is unknown. The NA content seems also to be high in aortic valvular patients without failure [11, 21]. These patients often tend to have concomitant coronary heart disease. The lower NA content in patients with ASD compared to the IHD group could possibly be explained by the continuous increase in blood flow through the right heart consequently leading to sustained right atrial hypertrophy accompanied by progressive NA depletion. Contrary to this, the haemodynamic burden in patients suffering from aortic valvular lesion has not led to as low NA level as in patients with ASD in spite of very marked left ventricular and left atrial hypertrophy. Thus the exact phase of depletion of NA in man in relation to hypertrophy remains to be defined. There are experimental results, which suggest that the NA depletion is related to duration of hypertrophy [28].

By histochemistry there could be shown in atrial myocardium adrenergic fluorescent axons following the course of blood vessels as perivascular plexuses and others forming a three-dimensional nerve net with varicosities. All these structures were more numerous and more intensely fluorescent in IHD than in CHF. In some IHD and valvular patients huge local accumulations of NA were observed at random in atrial myocardium. These so-called droplet fibres [10] may be associated with axonal degeneration due to disease-induced disruption of axoplasmic flow. Electron microscopic studies [2, 13] have revealed, in addition to dense-cored vesicles, signs of axonal degeneration, which were considered a consequence of the underlying disease.

There are findings indicating that activation of the adrenergic nervous system may be responsible for the enhanced predisposition to ventricular fibrillation during the early period of experimental coronary artery occlusion, and that reflex diminution of the adrenergic tone may decrease myocardial vulnerability [16]. The electrical instability of the heart can also be induced by stimulation of the stellate ganglia, which convay the sympathetic traffic from the CNS to the heart [16]. It has been shown in dogs that central sympathetic stimulation results in coronary vasoconstriction mediated by alpha-adrenergic receptors [12]. Recently it has been suggested by *Mudge* et al. [19] that the adrenergically mediated coronary vasoconstriction may constitute a possible mechanism of myocardial ischaemia in coronary heart disease. They were able to show that the cold pressor test increased coronary vascular resistance in IHD and concluded that this was consistent with the hypothesis that the tone of coronary vasculature is regulated by alpha-adrenergic receptors. There is further a surge of adrenergic discharge during the onset of acute experimental myocardial ischaemia [16]. It is believed that in patients with ischaemic heart disease and electrical instability enhanced neuronal adrenergic tone may be sufficient to induce coronary spasm and provoke lethal ventricular arrhythmias [16].

Loss of stored myocardial NA takes place within a short time of the onset of ischaemia [29]. This has been suggested to be associated with electrophysiological and metabolic events leading to e.g. loss of intracellular potassium, decline in intracellular pH, and to changes in permeability and contractility [20]. The essential role of NA in the experimental myocardial ischaemic injury has been demonstrated [34]. Patients exhibit an elevated plasma level and increased urinary excretion of NA following myocardial infarction [32]. The concomitant decrease in myocardial NA concerns the non-infarcted areas as well as both atria but the NA content seems to be restored in a couple of weeks [17, 31].

It may be concluded that the human myocardial NA content varies considerably in different heart diseases. In CHF the NA stores are depleted and the activities of TH and DBH are low, whereas it is obvious that symptomatic IHD is associated with a high myocardial NA content as well as with high activities of the biosynthetic enzymes. The rather high auricular and probably also ventricular myocardial NA content as well as the dense adrenergic nerve network in IHD indicate that in this disease there are favourable local conditions for promoting adrenergic stimulation possibly resulting in or at least contributing to the electrophysiological, metabolic, vascular or other harmful events mentioned above.

References

1. Bertler Å, Carlsson A, Rosengren E (1958) A method for the fluorimetric determination of adrenaline and noradrenaline in tissues. Acta Physiol Scand 44:273–292
2. Borchard F (1978) The adrenergic nerves of the normal and the hypertrophied heart. In: Bargmann W, Doerr W (eds) Normal and pathological anatomy, vol 33. Thieme, Stuttgart
3. Braunwald E (1975) The autonomic nervous system in heart failure. In: Braunwald E (ed) The myocardium: failure and infarction. HP Publishing, New York, pp 59–69
4. Braunwald E (1980) Pathophysiology of heart failure. In: Braunwald E (ed) Heart disease, a textbook of cardiovascular medicine. Saunders, Philadelphia London Toronto, pp 453–471
5. Braunwald E, Sonnenblick EH, Ross J (1980) Contraction of the normal heart. In: Braunwald E (ed) Heart disease, a textbook of cardiovascular medicine. Saunders, Philadelphia London Toronto, pp 413–452
6. Chidsey CA, Braunwald E, Morrow AG, Mason DT (1963) Myocardial norepinephrine concentration in man. N Engl J Med 269:653–658
7. Chidsey CA, Kaiser GA, Sonnenblick EH, Spann JF, Braunwald E (1964) Cardiac norepinephrine stores in experimental heart failure in the dog. J Clin Invest 43:2386–2393
8. Chidsey CA, Braunwald E, Morrow AG (1965) Catecholamine excretion of cardiac stores of norepinephrine in congestive heart failure. Am Heart J 39:442–451
9. Covell JW, Chidsey CA, Braunwald E (1966) Reduction of the cardiac response to postganglionic sympathetic nerve stimulation in experimental heart failure. Circ Res 19: 51–56
10. Dahlström A (1970) Adrenergic neurons in mammals with special reference to fluorescence microscopical studies. In: Bargmann W, Scharrer B (eds) Aspects of neuroendocrinology. Springer, Berlin Heidelberg New York pp 55–78
11. DeQuattro V, Nagatsu T, Mendez A, Verska J (1973) Determinants of cardiac noradrenaline depletion in human congestive failure. Cardiovasc Res 7:344–350
12. Feigl EO (1967) Sympathetic control of coronary circulation. Circ Res 20:262–271
13. Kyösola K, Partanen S, Korkala O, Merikallio E, Penttilä O, Siltanen P (1976) Fluorescence histochemical and electron microscopic observations on the innervation of the atrial myocardium of the adult human heart. Virchows Arch [Pathol Anat] 371: 101–119
14. Levitt M, Spector S, Sjoerdsma A, Udenfriend S (1965) Elucidation of the rate-limiting step in norepinephrine biosynthesis in the perfused guinea-pig heart. J Pharmacol Exp Ther 148:1–8
15. Lindvall O, Björklund A (1974) The glyoxylic acid fluorescence histochemical method: a detailed account of the methodology for the visualization of central catecholamine neurons. Histochemistry 39:97–127
16. Lown B (1980) Cardiovascular collapse and sudden cardiac death. In: Braunwald E (ed) Heart disease, a textbook of cardiovascular medicine. Saunders, Philadelphia London Toronto, pp 778–817
17. Mathes P, Cowan C, Gudbjarnason S (1971) Storage and metabolism of norepinephrine after experimental myocardial infarction. Am J Physiol 220:27–32
18. Molinoff PB, Weinshilboum R, Axelrod J (1971) A sensitive assay for dopamine-β-hydroxylase. J Pharmacol Exp Ther 178:425–431
19. Mudge GH, Grossman W, Mills RM, Lesch M, Braunwald E (1976) Reflex increase in coronary vascular resistance in patients with ischaemic heart disease. N Engl J Med 295:1333–1337
20. Oliver MF (1975) The vulnerable ischaemic myocardium and its metabolism. In: Oliver MF (ed) Modern trends in cardiology – 3. Butterworths, London, Boston, pp 280–291
21. Penttilä O, Merikallio E, Siltanen P, Klinge E (1977) Auricular catecholamine content in ischaemic heart disease. Acta Med Scand 201:317–321
22. Penttilä O, Kyösola K, Partanen S, Merikallio E, Siltanen P (1977) Studies of auricular catecholamines by fluorescence histochemistry in various heart diseases of man. Virchows Arch [Pathol Anat] 373:279–292
23. Penttilä O, Merikallio E, Pispa J, Klinge E, Siltanen P, Kyösola K (1978) Auricular tyrosine hydroxylase and dopamine-β-hydroxylase activities and noradrenaline content in ischaemic heart disease. Acta Med Scand 203:161–166

24. Petch MC, Nayler WG (1979) Concentration of catecholamines in human cardiac muscle. Br Heart J 41:340–344
25. Petch MC, Nayler WG (1979) Uptake of catecholamines by human cardiac muscle in vitro. Br Heart J 41:336–339
26. Pool PE, Covell JW, Levitt M, Gibb J, Braunwald E (1967) Reduction of cardiac tyrosine hydroxylase activity in experimental congestive heart failure. Circ Res 20:349–353
27. Rutenberg HL, Spann JF (1973) Alterations of cardiac sympathetic neurotransmitter activity in congestive heart failure. Am J Cardiol 32:472–480
28. Sassa H (1971) Mechanism of myocardial catecholamine depletion in cardiac hypertrophy and failure in rabbits. Jpn Circ J 35:391–403
29. Shahab L, Wollenberger A, Krause E-G, Genz S (1972) Effect of acute ischaemia on catecholamines and cyclic AMP levels in normal and hypertrophied myocardium. In: Oliver MF, Julian DG, Donald KW (eds) Effect of acute ischaemia on myocardial function. Churchill – Livingstone, Edinburgh, London, pp 97–105
30. Sole MJ, Kamble AB, Hussain MN (1977) A possible change in the rate-limiting step for cardiac norepinephrine synthesis in the cardiomyopathic syrian hamster. Circ Res 41: 814–817
31. Tomomatsu T (1971) Study on myocardial content of catecholamine, with special reference to heart disease. Jpn Circ J 35:979–983
32. Valori C, Thomas M, Schillingford JP (1967) Urinary excretion of free noradrenaline and adrenaline following acute myocardial infarction. Lancet 1:127–130
33. Vogel JHK, Jacobowitz D, Chidsey CA (1969): Distribution of norepinephrine in the failing bovine heart. Correlation of chemical analysis and fluorescence microscopy. Circ Res 24:71–84
34. Waldenström AP, Hjalmarson ÅC, Thornell L (1978) A possible role of noradrenaline in the development of myocardial infarction. Am Heart J 95:43–51
35. Waymire JC, Bjur R, Weiner N (1971) Assay of tyrosine hydroxylase by coupled decarboxylation of dopa formed from 1-^{14}C-1-tyrosine. Anal Biochem 43:588–600

Discussion

Szekeres, Szeged: Is it possible that the different content of connective tissue in the eliminated appendages could be responsible for the different decrease of myocardial noradrenaline contents in ischemic heart disease and congestive heart failure?

Penttilä: The degree of myocardial hypertrophy and the relatively high connective tissue content found in human auricles should be taken into account in biochemical studies concerned with the noradrenaline (NA) content and the activities of the biosynthetic enzymes. However, there is evidence (e.g. *Chidsey* et al. 1964, *Borchard* 1978) that in cardiac hypertrophy and failure there is a true depletion of NA in the whole myocardium, which cannot only be explained by dilution of adrenergic nerve endings, but which is due to several factors. The supposedly different amount of connective and scar tissue hardly explains the variation of NA content and enzyme activities observed in the present study. There is no significant difference in the mean protein content of auricular samples from corresponding patient groups.

Kopin, Bethesda: The loss of norepinephrine, tyrosine hydroxylase, dopamine-β-hydroxylase and uptake of catecholamines suggests the disappearance of the nerve terminals, similar to the effects of chronic guanethidine treatment. Is there any evidence that the nerves are still present in congestive heart failure (CHF)?

Penttilä: To our knowledge no direct evidence exists that in CHF the nerves are still present in the myocardium. There are certain electron microscopical observations (*Kyösola* et al.: Virchows Arch A (1976) 371: 101–119) on auricular specimens from patients with CHF where the adrenergic axons are still present, although the axonal degeneration is evident. However, it must be taken into account that these patients underwent operation in a clinically stable state without obvious clinical signs of decompensation.

Starke, Freiburg: Your data impressively show the decrease in noradrenaline, tyrosine hydroxylase and DBH levels in congestive heart failure. You did not go into the problem of how these decreases are brought about. What is known today about the mechanism, what is your personal opinion? Do the neurones degenerate, or are existing neurones depleted? Does increased activity of the neurones play a role?

Penttilä: The exact underlying mechanism is unknown but several factors may be involved, such as defective synthesis, neuronal binding and uptake. It appears that the neurones in congestive heart failure (CHF) are really depleted of NA and that a diminished release of NA from the myocardial nerve endings could exist in this situation. On the other hand there is electron microscopic evidence for axonal degeneration in CHF as well as in certain other heart diseases. The proportional share of these different factors is unknown.

Dominiak, Regensburg: Did you try to correlate ejection fraction to noradrenaline content in patients with ischemic heart disease?

Penttilä: Such calculations were not made from the present material. However, in a corresponding study including patients with a variety of anatomic lesions, no significant correlations could be found between auricular NA content on the one side and dP/dt, left ventricular end diastolic pressure and ejection fraction on the other side. There was a significant inverse correlation between right auricular NA content and pulmonary artery diastolic and wedge pressure.

Delius, Kempfenhausen: What is your opinion concerning the influence of chronic beta blockade on myocardial catecholamine content?

Penttilä: According to experimental studies long-term treatment with different beta-adrenergic blocking drugs does not affect NA tissue content, perhaps with the exception of the central nervous system. Also the abrupt cessation does not have any major consequences on tissue NA content in rats.

In the present material these drugs were withdrawn 7 days before the operation, and in another similar study they were continued up to the operation. In both series the myocardial NA contents were of the same order.

Catecholamines in Patients with Coronary Heart Disease[1]

P. Dominiak, W. Schulz, W. Delius, G. Kober, H. Grobecker

Summary

Circulating catecholamines have been investigated in different vascular beds of patients with coronary heart disease (CHD), patients with congestive heart failure (CHF) and the respective control subjects (C). In 20 patients with CHD, 13 patients with CHF and 9 subjects without angiographic evidence of CHD or CHF noradrenaline (NA) concentrations were radioenzymatically assayed in samples obtained from a peripheral vein (PV), aorta (AO) and coronary sinus (CS) at rest and during physical exercise on an ergometric bicycle. At rest in patients with CHD and CHF significantly increased NA concentrations (pg/ml) could be observed in all vascular beds (C: PV = 216 ± 56, AO = 166 ± 37, CS = 217 ± 69; CHD: PV = 451 ± 48, AO = 449 ± 66, CS = 530 ± 102; CHF: PV = 761 ± 230, AO = 632 ± 102, CS = 750 ± 178). During exercise the increase in NA concentration in CS exceeded the increase in AO, indicating a release of NA by the myocardium. The release of NA by the myocardium was significantly increased in patients with CHD and CHF versus C (CS: Δ971 ± 151 pg/ml and CS: Δ2239 ± 447 pg/ml, respectively). A positive correlation between coronary score and circulating NA in CS during physical exercise could be observed ($r = 0.56$). There was an inverse relationship between ejection fraction and circulating NA in coronary sinus at rest and during exercise ($r = -0.62$ and $r = -0.70$). In another group of 11 patients with CHD changes in arterial and coronary sinus NA and adrenaline concentrations were measured before and after incremental pacing of the right atrium and graded physical exercise. Also in these patients with CHD, resting NA levels in brachial artery (318 ± 39 pg/ml) and CS (417 ± 72 pg/ml) were increased. In addition increased adrenaline levels in brachial artery (117 ± 28 pg/ml and CS 99 ± 19 pg/ml) were assayed. Whereas after pacing no further liberation of NA by the myocardium was observed, during physical exercise a significant release of NA by the myocardium occurred (100 Watt: ΔNA in CS: 1066 pg/ml). In contrast to a positive arterial coronary venous NA difference (AVD), a negative AVD for adrenaline at rest and during electric stimulation was measured, indicating an uptake of adrenaline by the heart. The uptake of adrenaline was blunted during physical exercise.

The results suggest that an increased myocardial and peripheral sympathetic tone in patients with CHD may contribute to the development of angina pectoris.

Introduction

In recent studies concerning the synthesis, storage, release and uptake of noradrenaline in cardiac sympathetic nerve endings of patients with congestive heart failure and coronary heart disease the following results have been obtained:

---

1 Supported by the Deutsche Forschungsgemeinschaft (Gr 274/10)

1. In patients with congestive heart failure activity of tyrosine hydroxylase and dopamine-β-hydroxylase was reduced [5, 12]. In addition the noradrenaline content of the atrium and myocardium was decreased [2, 5, 8, 11]. The reduction in activity of biosynthetic enzymes and the depletion of cardiac noradrenaline stores were associated with an increased urinary excretion of noradrenaline and increased circulating noradrenaline at rest and during exercise [2, 3, 6, 17]. Also an impaired neuronal uptake of noradrenaline can contribute to the depletion of cardiac noradrenaline stores in congestive heart failure [9].

2. In patients with coronary heart disease however, the activities of tyrosine hydroxylase and dopamine-β-hydroxylase, as well as the noradrenaline content of the atrium were increased, at least compared with a group of patients with congestive heart failure [11, 12]. These results suggest an increased cardiac sympathetic tone in patients with coronary heart disease based on a possibly enhanced noradrenaline synthesis and release. In contrast to this hypothesis in a recent paper it was claimed that during pacing-induced angina in patients with coronary heart disease, cardiac sympathetic tone was decreased [16]. Therefore the aim of our study was to determine the cardiac sympathetic tone in patients with coronary heart disease by measuring circulating catecholamines in different vascular beds at rest, during physical exercise and incremental atrial pacing.

Patients and Methods

Forty-two male patients with a mean age of 50.3±2.3 years (±S.E.M.) were studied during their admission to a cardiac unit. They had not been treated with drugs including β-adrenoceptor blocking agents before the study. The patients were informed of the nature and risks of the study and their consent was obtained. The haemodynamic investigation was performed in the morning after overnight fasting, with the patients in a supine position. Catheters were placed into the coronary sinus and the aorta through the right brachial artery and brachial vein. Another catheter was inserted into the left antecubital vein for blood sampling. Recordings of haemodynamics were performed at rest. After 30 min in supine position blood samples were obtained simultaneously from coronary sinus, aorta and peripheral vein for radioenzymatic assay of noradrenaline according to *Henry* et al. [7]. Exercise for 5 min was performed on a bicycle ergometer connected to the catheterisation table. The average work load was 65±7 Watt. At the end of the fifth minute again blood samples were obtained simultaneously from the different vascular beds. According to careful clinical examinations and to the results of haemodynamic and angiographic investigations the patients were separated into the following groups:

1. Controls: a group of 9 patients with normal coronary and left ventricular angiograms.
2. A group of 20 patients with coronary heart disease, who showed various grades of coronary artery stenosis (CHD).
3. A group of 13 subjects were classified as patients with congestive heart failure (CHF). They had impaired left ventricular function, but no signs of coronary artery disease.

A further group of 11 male patients with a mean age of 48.7±1.8 years (S.E.M.) with well documented coronary artery disease were investigated during their admission to a cardiac unit. Noradrenaline and adrenaline concentrations were assayed radioenzymatically (*Da Prada* and *Zürcher* [4]; *Peuler* and *Johnson* [13]) in blood samples obtained from the brachial artery and coronary sinus before and after icremental pacing of the right atrium and graded physical exercise. After a rest of 30 min heart rate was stepwise increased at 20 beats/min by pacing the right atrium for 2 min. Heart rate was raised from 90 beats/min up to 150 beats/min. At the interval of every 2 min blood samples from coronary sinus and the brachial artery were obtained. After a further rest of 15 min graded physical exercise in supine position on a bicycle ergometer was performed. Work load was increased stepwise by 25 Watts every 3 min from 25 Watts up to 100 Watts. At each interval blood samples were obtained from the vascular beds mentioned above.

Results, given as mean±SEM were statistically evaluated by means of the Student's *t*-test for paired and unpaired data, analysis of variance and linear regression analysis.

Results

Haemodynamics

Heart rate and blood pressure of the groups of patients investigated are summarized in Table 1. Patients with coronary heart disease and patients with congestive heart failure did not show significant differences in blood pressure at rest and during exercise, when compared with controls. Heart rates in patients with coronary heart disease and congestive heart failure at rest and during exercise were significantly raised compared to control subjects. Moreover, in patients with congestive heart failure, heart rates were significantly higher than in patients with coronary heart disease (Table 1). Calculated ejection fraction was significantly diminished in both groups of patients when compared with control subjects (Table 1).

Plasma Noradrenaline Concentrations

Circulating noradrenaline in the vascular beds investigated are depicted in Fig. 1 (cf. also Table 1). In the control group resting noradrenaline concentrations in the samples obtained from the anticubital vein, aorta and coronary sinus were not significantly different. During exercise the increase in noradrenaline concentration was more pronounced in coronary sinus. In patients with coronary heart disease at rest significantly increased noradrenaline concentrations were measured in all vessels. In addition the increase in circulating noradrenaline was significantly higher in these patients during physical exercise, when compared with the control group. Also in patients with congestive heart failure significantly increased circulating noradrenaline concentrations could be assayed at rest and during exercise in all vascular beds investigated. As observed in the control group also in patients with coronary heart disease and congestive heart failure the increase in circulating noradrenaline concentrations during physical exercise was more pronounced in the coronary sinus than in the aorta or peripheral vein (Fig. 1).

In agreement with this observation the aorto-coronary sinus noradrenaline difference during physical exercise was significantly higher in patients with

Table 1. Haemodynamic parameters and plasma noradrenaline concentrations in control subjects, patients with coronary heart disease and patients with congestive heart failure at rest and during physical exercise

	C		CHD		CHF			Analysis of variance
	R	PE	R	PE	R	PE		
RR	149/79 ± 7/3	179/90 ± 5/2	141/79 ± 5/3	167/94 ± 6/3	135/85 ± 5/6	168/93 ± 8/7	(mm Hg)	
HR	73 ± 2.9	102 ± 4.2	78 ± 2.2	116 ± 4.1	87 ± 5.5	139 ± 9.9	(b/min)	*/*****
EF	69.9 ± 3.4		56.8 ± 4.0		47.1 ± 5.5		(%)	**
CYSE			32.7 ± 3.0					
PV	216 ± 56	483 ± 150	451 ± 48	1201 ± 175	761 ± 230	1187 ± 140	(pg/ml)	*/**
AO	166 ± 37	529 ± 133	449 ± 66	1313 ± 237	632 ± 102	1935 ± 350	(pg/ml)	*****/**
CS	217 ± 69	891 ± 230	530 ± 102	1862 ± 381	750 ± 178	3130 ± 677	(pg/ml)	*/**
AVDNA	63 ± 39	315 ± 153	81 ± 56	564 ± 159	163 ± 129	1195 ± 415	(pg/ml)	n.s./*
PE		77.8 ± 6.6		64.0 ± 5.4		55.4 ± 8.6	(watt)	
Age	51.7 ± 2.2		53.6 ± 1.1		45.6 ± 3.5		(years)	*
(*n*)	(9)		(20)		(13)			

Results are expressed as $\bar{x} \pm$ SEM. Significance values: * $P < 0.05$, ** $P < 0.025$, ***** $P < 0.005$, R/PE

Abbreviations: BP, blood pressure; HR, heart rate; EF, ejection fraction; CYSE, coronary score; PV, peripheral vein; AO, aorta; CS coronary sinus; AVDNA, arterio-venous noradrenaline difference; C, controls; CHD, coronary heart disease; CHF, congestive heart failure; R, rest; PE, physical exercise

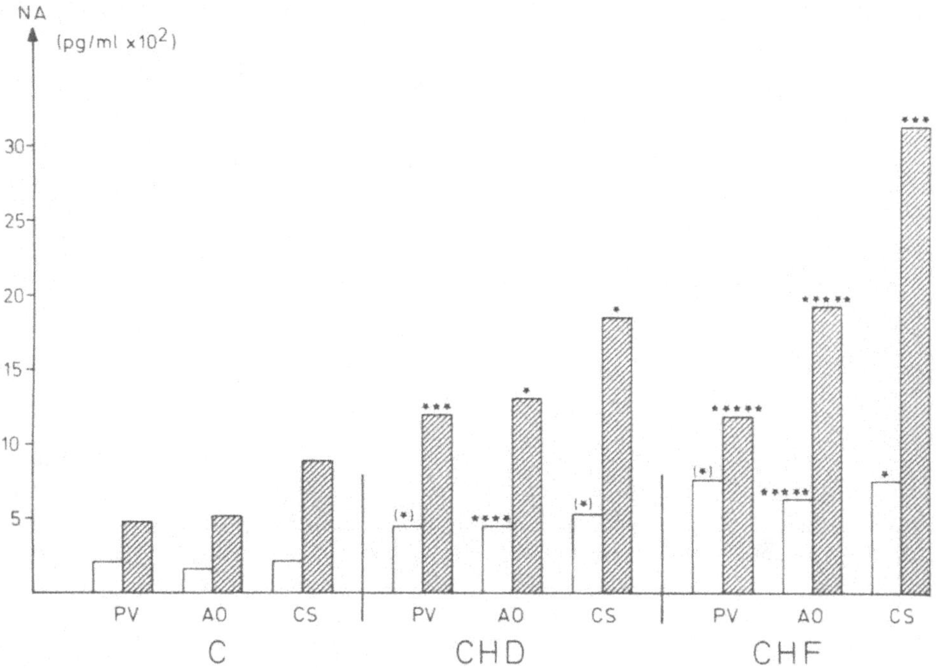

Fig. 1. Plasma noradrenaline concentrations in peripheral vein (*PV*), aorta (*AO*) and coronary sinus (*CS*) in control subjects (*C*), patients with coronary heart disease (*CHD*) and patients with congestive heart failure (*CHF*) at rest (*white columns*) and during physical exercise (*hatched columns*). Significance (t-test): (*) = $P<0.1$, * = $P<0.5$, *** = $P<0.02$, **** = $P<0.01$ and ***** = $P<0.005$

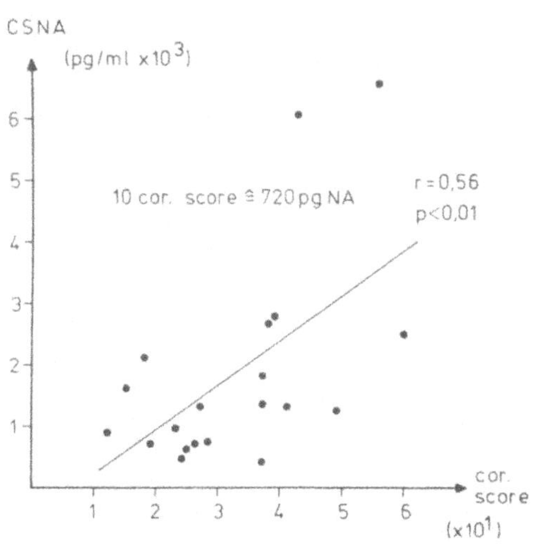

Fig. 2. Correlations between coronary score and coronary sinus noradrenaline concentrations (*CSNA*) during physical exercise in patients with coronary heart disease

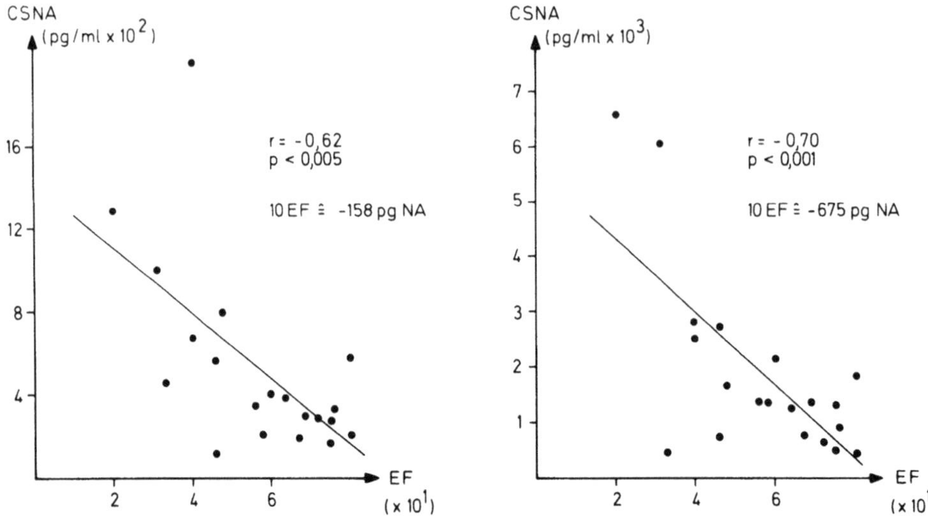

Fig. 3. Correlations between ejection fraction (*EF*) and noradrenaline concentrations in coronary sinus (*CS*) at rest (*left*) and during physical exercise (*right*) in patients with coronary heart disease

coronary heart disease and most pronounced in patients with congestive heart failure (Table 1).

The relationship between the degree of severity of the coronary artery disease (coronary score) and circulating noradrenaline concentrations in coronary sinus is depicted in Fig. 2. There was a positive significant correlation between these two paramters during physical exercise. A similar correlation was obtained between these parameters in patients with coronary heart disease at rest ($r=0.45$, $P<0.1$).

On the other hand we found a significant inverse correlation between circulating noradrenaline in coronary sinus of patients with coronary heart disease and the ejection fraction under resting and exercise conditions, indicating possibly a compensatory increase in noradrenaline release by the impaired myocardium (Fig. 3).

As shown in Fig. 4 also in the second group of patients with coronary heart disease, increased resting levels of noradrenaline in samples obtained from the brachial artery and coronary sinus were measured. Whereas during pacing of the right atrium no significant changes of noradrenaline concentrations in coronary sinus and brachial artery was seen, a steep increment in noradrenaline concentrations in both vessels occurred after graded physical exercise. The measured heart rate during pacing and physical exercise was comparable.

Increments in circulating adrenaline concentrations in the coronary sinus and brachial artery were significant only after a work load of 100 Watts (Table 2).

The release of noradrenaline by the myocardium during incremental pacing and graded physical exercise is shown in Fig. 5. There was a positive arterial coronary venous noradrenaline difference, but no further increase in the release of noradrenaline occurred during pacing. A negative arterial coronary venous adrenaline difference was calculated at rest and during the pacing period, indicating possibly an uptake of circulating adrenaline by the heart. During physical exercise the increments in the positive arterial coronary venous noradrenaline was more

Table 2. Plasma catecholamine concentrations in brachial artery and coronary sinus, atrest during incremental pacing of the right atrium (STIM/min) and during graded physical exercise (Watt), heart rate and arterio-venous catecholamine differences in patients with coronary heart disease.

STIM/min		0	90	110	130	150	0		Analysis of variance
	HR	76 ± 3.0	86 ± 1.8	108 ± 2.0	129 ± 2.1	145 ± 1.7	77 ± 2.0	(b/min)	******
NA	AR	318 ± 39	291 ± 35	321 ± 41	364 ± 35	478 ± 66	377 ± 50	(pg/ml)	(*)
	CS	417 ± 72	396 ± 49	442 ± 79	517 ± 67	622 ± 108	539 ± 69	(pg/ml)	
A	AR	177 ± 28	163 ± 31	180 ± 46	200 ± 57	243 ± 91	315 ± 136	(pg/ml)	
	CS	99 ± 19	115 ± 29	148 ± 49	154 ± 61	198 ± 106	258 ± 117	(pg/ml)	
	AVDNA	99 ± 50	105 ± 29	122 ± 42	153 ± 30	145 ± 61		(pg/ml)	
	AVDA	−77 ± 15	−48 ± 12	−32 ± 16	−45 ± 19	−33 ± 9		(pg/ml)	
	(*n*)	(11)	(8)	(9)	(8)	(8)	(9)		
Watt		**0**	**25**	**50**	**75**	**100**			
	HR	76 ± 3.0	107 ± 6.4	109 ± 4.3	126 ± 4.5	133 ± 6.1		(b/min)	******
NA	AR	318 ± 39	490 ± 67	532 ± 75	940 ± 227	1023 ± 133		(pg/ml)	******
	CS	417 ± 72	846 ± 206	882 ± 201	1082 ± 206	1483 ± 165		(pg/ml)	*****
A	AR	177 ± 28	309 ± 140	310 ± 97	407 ± 149	646 ± 217 xx		(pg/ml)	(*)
	CS	99 ± 19	315 ± 179	336 ± 133	425 ± 182	645 ± 240 xx		(pg/ml)	
	AVDNA	99 ± 50	357 ± 143	351 ± 135	142 ± 92	461 ± 133		(pg/ml)	
	AVDA	−77 ± 15	−40 ± 13	27 ± 41	18 ± 57	−1 ± 56		(pg/ml)	
	(*n*)	(11)	(5)	(11)	(7)	(7)			

Results are expressed as $\bar{x} \pm$ SEM. Significance values: (*) $P < 0.1$, ***** $P < 0.005$, ****** $P < 0.001$. xx $P < 0.025$ (*t*-test).
Abbreviations: NA, noradrenaline; A, adrenaline; AR, brachial artery; CS, coronary sinus; O, rest; HR, heart rate; AVD, arterio-venous catecholamine differences

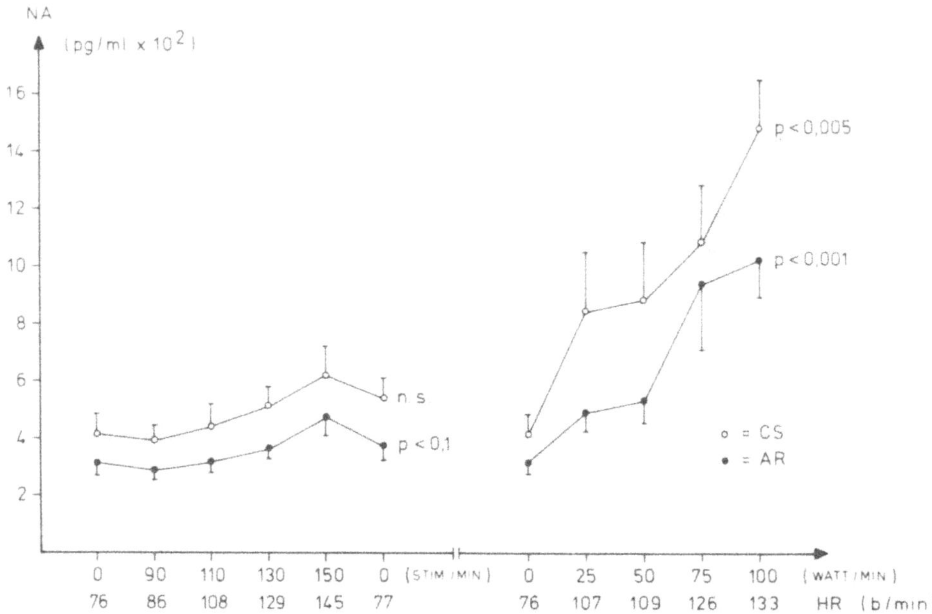

Fig. 4. Changes in arterial (*open circles*) and coronary sinus (*black circles*) plasma noradrenaline concentrations (*NA*) during incremental pacing of the right atrium (*left*) and during graded physical exercise (*right*) in patients with coronary heart disease. Significance was calculated by the analysis of variance

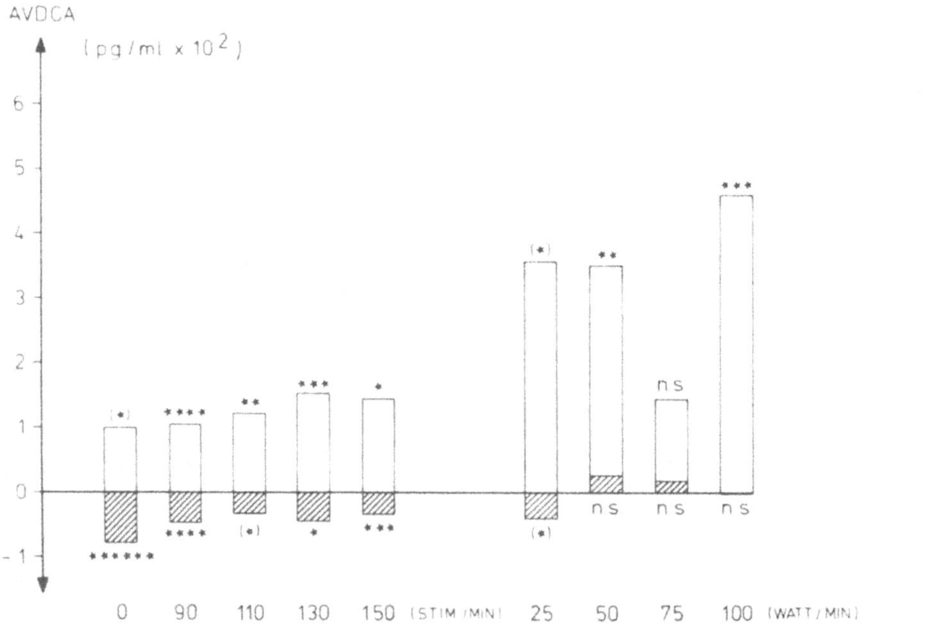

Fig. 5. Arterial coronary sinus catecholamine differences (*AVDCA*, noradrenaline = *white columns*, adrenaline = *hatched columns*) in patients with coronary heart disease during pacing (*left*) and during exercise (*right*). Significance values (*t*-test): (*) = $P<0.1$, * = $P<0.05$, ** = $P<0.025$, *** = $P<0.02$, **** = $P<0.01$ and ***** = $P<0.001$

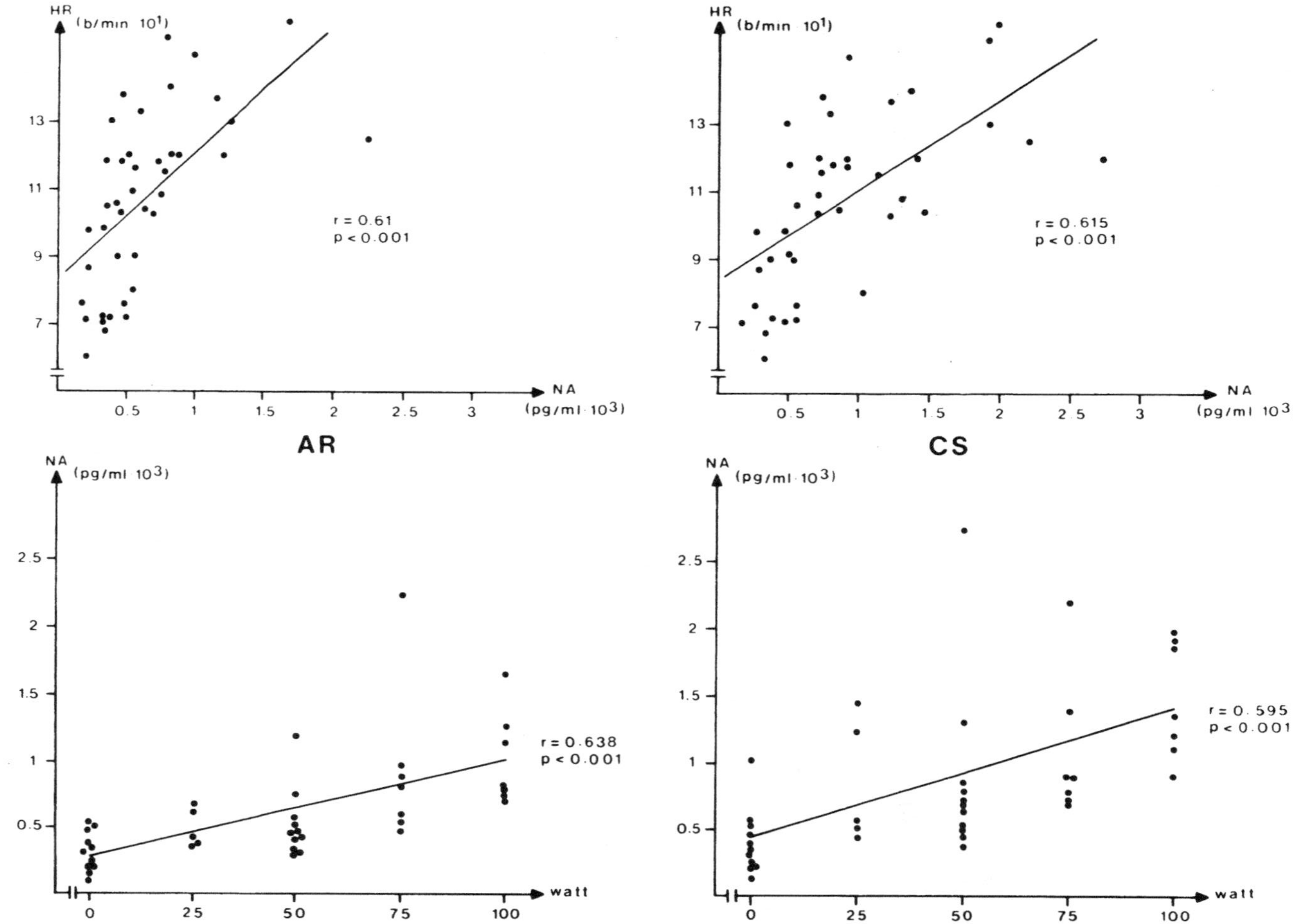

Fig. 6. Correlations between arterial (*AR*) and coronary sinus (*CS*) noradrenaline concentrations (*NA*) and heart rate (*HR*) during physical exercise (*top*) and between work load (watt) and noradrenaline concentrations in these vessels (*bottom*)

pronounced compared to the increase during electric stimulation. The negative arterial coronary venous adrenaline difference disappeared during exercise.

Whereas no correlations between heart rate and catecholamine concentrations during incremental pacing could be obtained, there was a positive significant correlation between heart rate and noradrenaline concentrations in the brachial artery and coronary sinus (Fig. 6). In addition a significant positive correlation between work load and circulating noradrenaline in both vessels was calculated (Fig. 6).

Discussion

In the present paper it could be demonstrated that in patients with coronary heart disease at rest circulating levels of noradrenaline in different vascular beds (coronary sinus, aorta , peripheral vein) are increased. During physical exercise the aorto-coronary sinus noradrenaline difference in patients with coronary heart disease is significantly higher than in control subjects, indicating a more pronounced release of noradrenaline from cardiac sympathetic nerve endings. Since approximately 90% of the noradrenaline in the heart is synthesized locally [1], and evidence for an increased synthesis and content of noradrenaline in cardiac tissue of patients with coronary heart disease was presented [11, 12], it is likely, that in these patients cardiac sympathetic activity is augmented. In addition, the elevated circulating noradrenaline concentrations in the peripheral vein in patients with coronary heart disease indicate an increased peripheral sympathetic activity. From our finding of a positive correlation between coronary score and high circulating levels of noradrenaline in coronary sinus during exercise it can be concluded that the degree of severity of coronary artery disease is possibly associated with the increased cardiac sympathetic tone. On the other hand the results of an inverse relationship between circulating noradrenaline in the coronary sinus at rest and during exercise and ejection fraction may reflect a compensatory augmented cardiac sympathetic activity due to intrinsic inotropic weakness of the failing myocardium. In patients with heart failure an increased basal excretion of noradrenaline [3] as well as an excessive elevation of circulating noradrenaline during exercise [2] could be demonstrated. In agreement with these results we measured high circulating noradrenaline concentrations in coronary sinus, aorta and peripheral vein in patients with congestive heart failure at rest and during exercise. Thus it appeared likely that the increased circulating levels in coronary sinus in patients with coronary heart disease are the result of an increased synthesis and release of the transmitter due to the coronary artery disease, and possibly also due to the beginning heart failure with compensatory enhanced sympathetic activity. Our results obtained with sensitive and specific radioenzymatic assays for circulating catecholamines do support the classic studies by *Raab* [14, 15] in which an association between an increase in circulating catecholamines and angina pectoris was postulated. In this connection the reflex increase in coronary vascular resistance in patients with ischemic heart disease after cold pressure test mediated by activation of coronary α-adrenergic receptors [10] may be also caused by an increased release of noradrenaline.

The results of our investigations in patients with coronary heart disease after incremental pacing and graded physical exercise do not support the hypothesis of a decreased cardiac and peripheral sympathetic tone in patients with cardiac ischemia [16]. Comparing the effects of incremental pacing and graded physical exercise in the same group of patients with coronary heart disease we observed a further release of noradrenaline from the heart only after physical exercise. Therefore in contrast to physical exercise we consider atrial pacing as an inadaequate stimulus to an overall increase of cardiac and peripheral sympathetic function.

In conclusion the results suggest that an increased myocardial und peripheral sympathetic tone in patients with coronary heart disease may contribute to the development of angina pectoris.

Acknowledgement. The skilful technical assistance of Miss Eva Stocker is gratefully appreciated.

References

1. Braunwald E (1975) The autonomic nervous system in heart failure. In: E Braunwald (ed) The myocardium: failure and infarction. HP Publishing, New York, pp 59–69
2. Chidsey ChA, Harrison DC, Braunwald E (1962) Augmentation of the plasma norepinephrine response to exercise in patients with congestive heart failure. N Engl J Med 267:650–654
3. Chidsey ChA, Braunwald E, Morrow AG (1965) Catecholamine excretion and cardiac stores of norepinephrine in congestive heart failure. Am J Med 39:442–451
4. Da Prada M, Zürcher G (1976) Simultaneous radioenzymatic determination of plasma and tissue adrenaline, noradrenaline and dopamine within the femtomole range. Life Sci 19:1161–1174
5. De Quattro V, Nagatsu T, Mendez A, Verska J (1973) Determinants of cardiac noradrenaline depletion in human congestive heart failure. Cardiovasc Res 7:344
6. Dominiak P, Schulz W, Kober G, Grobecker H (1978) Simultane Bestimmung von Plasmanoradrenalin in Aorta, peripherer Vene und Koronarvenensinus in Ruhe und unter Belastung. Z Kardiol [Suppl 5]:73
7. Henry DP, Starman BJ, Johnson DG, Williams RH (1975) A sensitive radioenzymatic assay for norepinephrine in tissue and plasma. Life Sci 16:375–384
8. Kramer RS, Mason DT, Braunwald E (1968) Augmented sympathetic neurotransmitter activity in the peripheral vascular bed of patients with congestive heart failure and cardiac norepinephrine depletion. Circulation 38:629–634
9. Mäurer W, Ablasser A, Opherk D et al. (1981) Verminderung der kardialen Noradrenalin (NA)-Aufnahme bei Patienten mit kongestiver Kardiomyopathie (CMP). Z Kardiol 70:276
10. Mudge GH, Grossman W, Mills RM, Lesch M, Braunwald E (1976) Reflex increase in coronary vascular resistance in patients with ischaemic heart disease. N Engl J Med 295:1333–1337
11. Penttilä O, Merikallio E, Siltanen P, Klinge E (1977) Auricular catecholamine content in ischaemic heart disease. Acta Med Scand 201:317–321
12. Penttilä O, Merikallio E, Pispa J, Klinge E, Siltanen P, Kyösola K (1978) Auricular tyrosine hydroxylase and dopamine-β-hydroxylase activities and noradrenaline content in ischaemic heart disease. Acta Med Scand 203:161–166
13. Peuler JD, Johnson GA (1977) Simultaneous single isotope radioenzymatic assay of plasma norepinephrine, epinephrine and dopamine. Life Sci 21:625–636
14. Raab W (1962) The sympathogenic biochemical trigger mechanism of angina pectoris. Am J Cardiol 9:576–590
15. Raab W (1963) The non-vascular metabolic myocardial vulnerability factor in 'coronary heart disease'. Am Heart J:66:685–706
16. Schwartz L, Sole MJ, Vaughan-Neil EF, Hussain NM (1979) Catecholamines in coronary sinus and peripheral plasma during pacing-induced angina in man. Circulation 59:37–43
17. Thomas JA, Marks BH (1978) Plasma norepinephrine in congestive heart failure. Am J Cardiol 41:233–242

Discussion

Rahn, Maastricht: Could the differences in noradrenaline levels you observed during physical exercise be at least partly due to different work capacity?

Dominiak: The work load was different in our 3 groups of patients (control group ≈ 75 Watt, coronary heart disease group (CHD) ≈ 68 Watt and congestive heart failure group (CHF) ≈ 53 Watt). Despite this, the patients with CHD and CHF revealed increased NA-concentrations during physical exercise when compared with control subjects.

Keul, Freiburg: If patients suffering from coronary heart disease are divided according to their left ventricular myocardial function, there is a clear differentiation in relation to their catecholamine levels. This is particularly noticeable under exercise conditions:

Group 1 (I): normal cardiac performance at rest, but an increased enddiastolic pressure during exercise.
Group 2 (II/III): an increased enddiastolic pressure both at rest and during exercise, but normal cardiac output.
Group 3 (II/III): an increased enddiastolic pressure both at rest and during exercise, and a limitation of cardiac output during exercise. Since group 2 and 3 are difficult to separate they were taken together.
Group 4: increased enddiastolic pressure and limited output at rest.

As a result the individual patient groups show a typical response in their plasma catecholamine levels, corresponding to their myocardial function. Thus, dependent on the plasma catecholamine levels, especially during exercise, additional information about the functional condition of the myocardium can be obtained.

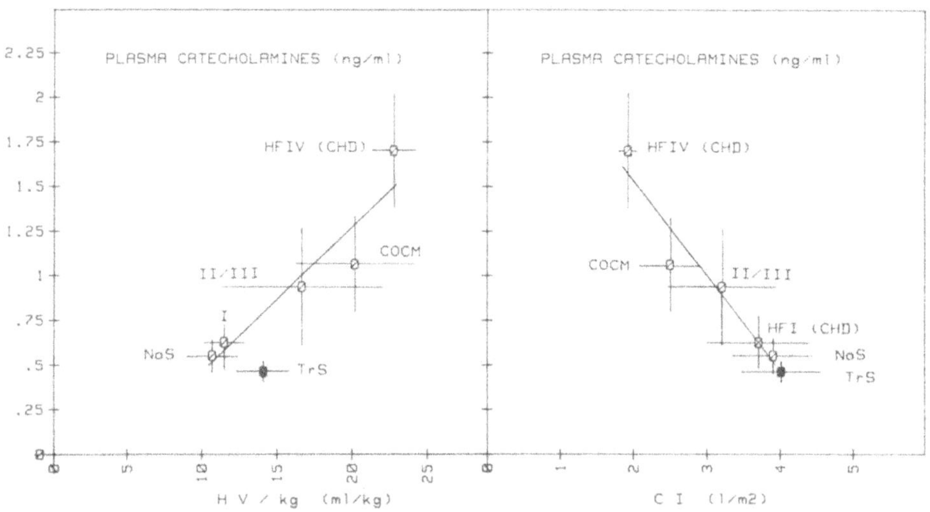

Fig. 1. Discussion. The venous catecholamine levels increase along with the enlargement in heart size as an indication of myocardial damage in coronary patients. At the same time the *PCPm* (mean pulmonary capillary pressure) increases and the cardiac output of the heart decreases. Trained athletes show the lowest catecholamine levels and the highest cardiac output along with an increase in heart size (*Lehmann* and *Keul* 1981). (*COCM*, congestive cardiomyopathia; *NoS*, normal subjects; *TrS*, trained subjects)

Dominiak: According to our studies there is an inverse correlation between the left ventricular function (ejection fraction) and the NA-concentrations at rest as well as during exercise.

Poole-Wilson, London: In your initial studies the groups of patients were exercised against a fixed work load. The A-V differences for noradrenaline were obtained at different heart rates. Heart rate influences coronary blood flow and possibly ejection fraction. Would your results and the correlations be affected if measurements had been made at the same heart rate?

Dominiak: The mean heart rate did not differ in patients with coronary heart disease and in control subjects, neither under resting conditions nor during physical exercise. Therefore, the conditions for measuring NA-concentrations were comparable in these two groups.

Feigl, Seattle: *Mudge* et al. [10] (1333–1337: Circulation (1979) *59:*544–550) have shown a hyperresponsive coronary vasoconstriction in patients with coronary heart disease during a cold pressure test. Have you examined patients with a cold pressure test to elicit sympathetic activation?

Dominiak: We did not perform the cold pressure test in patients with CHD. To provoke the sympathetic activity we studied physical exercise only. The results of *Mudge* et al. are in agreement with our conclusions.

Koch, Berlin: I think it is extremely difficult to compare sympathetic activity in terms of plasma noradrenaline (NA) levels during exercise in different groups of patients by putting them all on the same work load. In similar studies on different patient groups we found similar NA responses when relating them to the same *relative* work load (e.g. 80% of maximal work capacity). Conversely, different NA levels occurred, as in your study, at the same absolute work load. It thus appears mandatory to study adrenergic responses in different patients at similar *relative* work loads.

Dominiak: Our control subjects performed submaximal exercise whilst patients with CHD and CHF performed symptom limited exercise, which was lower than that of control subjects. Nevertheless, these patients showed two- to threefold greater increases of NA-concentration in all vascular beds investigated.

Starke, Freiburg: *Haneda* et al. (Am. Heart J. (1980) 100:465–472) found that in congestive heart failure (CHF) the aortocoronary sinus NA difference is decreased. Could this have been another state of CHF? Is the increase in overflow the result of a decrease in uptake?.

Dominiak: Our subgroup of patients with CHF was not a homogeneous group. As Dr. Mäurer has shown (Z. Kardiol.), the uptake of NA by the heart was diminished in patients with CHF.

Kopin, Bethesda: If in CHF there is a decrease in norepinephrine (NE) content and biosynthetic capacity of the nerves in the heart, the greater exercise-induced release of NE measured as differences in coronary sinus-aortic levels of norepinephrine, are difficult to explain. Where does the norepinephrine come from?

Dominiak: It is possible that the enhanced NA-concentration in coronary sinus blood in patients with CHF is the result of a decreased uptake of NA.

Catecholamines in Patients with Heart Failure [1]

W. Mäurer, R. Tschada, J. Manthey, A. Ablasser, W. Kübler

Summary

1. In a group of 57 patients with heart disease, 24-h urinary excretion of norepinephrine, epinephrine and their O-methylated metabolites, normetanephrine and metanephrine, was found significantly increased in class III an IV (NYHA). The urinary excretion of dopamine did not show any significant change.

2. In 52 patients with heart failure, plasma levels of norepinephrine and epinephrine at rest were increased mainly in those patients, in whom cardiac index was depressed below the critical value of 2 liters/min/m². There was no strong relationship between left ventricular ejection fraction and plasma catecholamines.

In 11 patients with congestive cardiomyopathy and severe heart failure, increase in cardiac index due to therapeutic intervention with dihydralazine or/and dopamine revealed significant decrease in plasma catecholamines.

During exercise, in 12 patients with impaired left ventricular function an exaggerated increase in plasma catecholamines could be observed, probably as consequence of peripheral underperfusion due to low cardiac output. There was no strong relationship between plasma catecholamines and plasma levels of lactate.

3. In 17 patients suffering from severe chronic aortic regurgitation, cardiac tissue stores of norepinephrine, normetanephrine and epinephrine were significantly decreased. Myocardial dopamine level did not change.

The transformation of dopamine to norepinephrine seems to be the rate limiting step of myocardial norepinephrine synthesis. A defect of norepinephrine uptake represents an important mechanism of cardiac catecholamine depletion.

Introduction

In patients with heart failure, activity of the sympathoadrenergic system is known to be increased [5, 6, 8, 9, 14, 15, 22]. In order to study cardiac catecholamine metabolism in humans, 3 different approaches have been chosen:

1. A crude index to the level of sympathetic activity can be provided by measurements of the 24-h urinary excretion of catecholamines [9, 14].
2. Information about instantaneous sympathetic activity and its rapid changes, however, can only be obtained by determination of the plasma concentration of catecholamines [5, 8].
3. Further information concerning cardiac catecholamine metabolism in humans may be obtained from patients undergoing cardiac surgery [9, 16]. Right atrial biopsies may be taken for measurement of tissue levels of catecholamines in the myocardium. The values obtained can be related to the extent of impairment of cardiac function.

---

1 These investigations were supported by the Deutsche Forschungsgemeinschaft (SFB 90, Kardiovaskuläres System, University of Heidelberg)

Materials and Methods

Ad 1: In 57 patients suffering from different heart diseases and in 20 healthy persons, 24-h urinary excretion of catecholamines were measured using a fluorometric method [1, 3, 11, 19, 23]. The severity of heart disease was classified according to the classification of the New York Heart Association [17].

Ad 2: In 52 patients with heart disease of different origine, plasma concentration of catecholamines under resting condition were determined, using a sensitive radioenzymatic method [10]. The obtained values were correlated to the corresponding values of left ventricular ejection fraction as determined by angiography, and to the cardiac index as determined by thermodilution.

In 11 patients with congestive cardiomyopathy and severe heart failure, effects of a therapeutic intervention on hemodynamic parameters and on the plasma concentration of catecholamines were investigated. Positive inotropic intervention was performed using dopamine [13] in the relatively low dose of 2.5–5 μg/kg/min. In the same patients, a second therapeutic trial was undertaken using the vasodilator dihydralzine [12] in a dose of 225 mg daily. Finally, combined therapy using dihydralazine and dopamine was applied.

In 12 patients with congestive cardiomyopathy and a mean left ventricular ejection fraction of 23% as determined by angiography, effects of bicycle exercise on the plasma concentration of norepinephrine, epinephrine, lactate and on oxygen saturation in the pulmonary artery, were assessed. Mean values obtained in the patients group with congestive cardiomyopathy were compared to corresponding results obtained from 14 patients with valvular heart disease and normal left ventricular function.

Ad 3: In 17 patients suffering from severe chronic aortic regurgitation, during aortic valve replacement right atrial biopsies were taken. In these samples myocardial contents of norepinephrine, normetanephrine, epinephrine and dopamine [1, 3, 11, 16] were determined. Fifteen patients with atrial septum defect of foramen secundum type served as controls [16].

In 6 patients with congestive cardiomyopathy and in 5 patients with normal left ventricular function, during routine heart catheterization myocardial noerpinephrine uptake was assessed from values of arteriocoronary venous norepinephrine difference and coronary blood flow. During infusion of norepinephrine in a dose of 0.1 μg/kg/min into the pulmonary artery, blood samples were taken in short intervals simultaneously from the descending aorta and the coronary sinus for measuring plasma norepinephrine concentration. Coronary blood flow was determined using the argon method of *Bretschneider* [7].

Results and Discussion

Ad 1: As can be seen in Fig. 1, patients of class III and IV, according to the New York Heart Assossation [17], exhibited a significantly higher excretion of norepinephrine than normals or patients of class I and class II. It could also be demonstrated that in patients of class IV the excretion of epinephrine was significantly increased. The O-methylated metabolites of norepinephrine and epinephrine – normetanephrine and metanephrine – did show a corresponding increase in

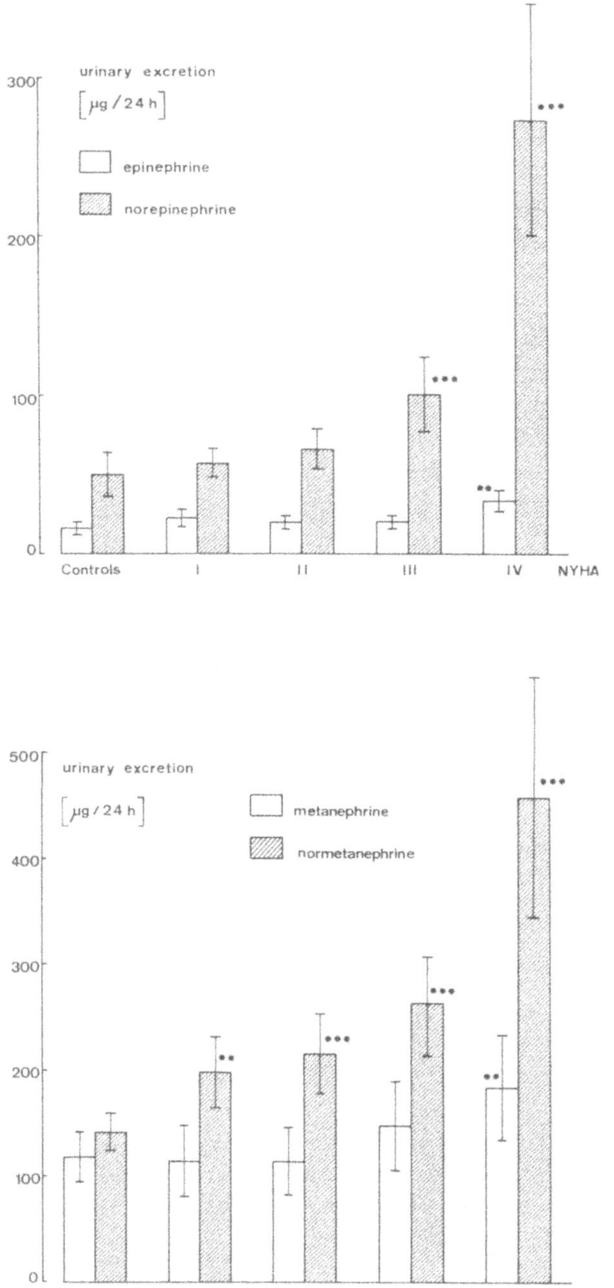

Fig. 1. Twenty-four hour urinary excretion of epinephrine (*empty columns*), norepinephrine (*hatched columns*), metanephrine (*empty columns*) and normetanephrine (*hatched columns*) in normals ($n=20$) and in 57 patients suffering from different heart disease. *I–IV* refers to the classification of the New York Heart Association. The *columns* represent mean values, the *vertical lines* with *cross bars* equal ±1 S.E.M.

urinary excretion. From these data it follows that the increased urinary excretion of norepinephrine and epinephrine in patients with severe heart failure is not due to a reduced catecholamine degradation but rather to increased sympathetic activity related to heart failure. In contrast to norepinephrine and epinephrine, the urinary excretion of dopamine was not related to heart failure, but to the daily produced urine volume [14].

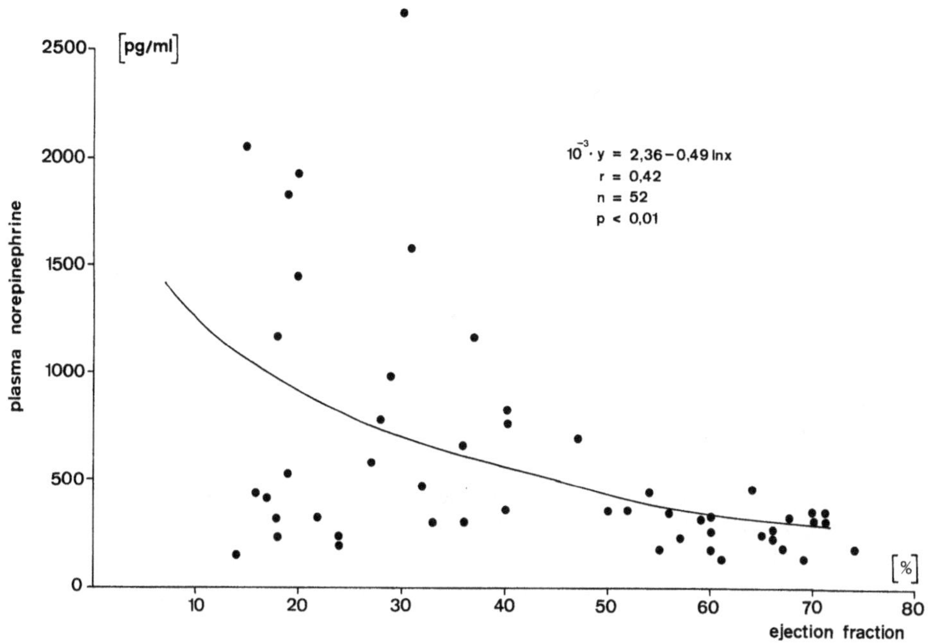

Fig. 2. Relationship between plasma norepinephrine levels at rest and left ventricular ejection fraction in 52 patients with left heart disease of different origin

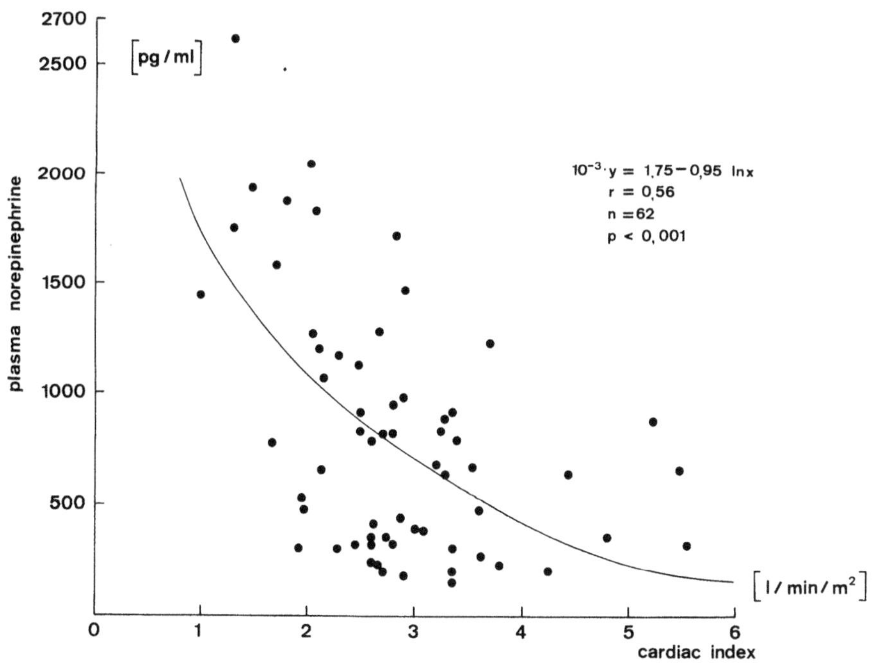

Fig. 3. Relationship between plasma norepinephrine levels at rest and cardiac index in 52 patients with left heart disease of different origin

From these data it may be inferred that a rather good correlation should exist between plasma norepinephrine levels and left ventricular function.

Ad 2: As can be seen in Fig. 2, there was no strong relationship between left ventricular ejection fraction and plasma norepinephrine. High plasma levels of norepinephrine were almost always found in patients with depressed ejection fraction below 30%; however, in some of these patients plasma norepinephrine was found to be even in the normal range.

In the same group of patients, no significant relationship could be found between epinephrine and left ventricular ejection fraction.

If, however, depressed left ventricular function was assessed not by diminished ejection fraction but by reduced cardiac index, a clearer relationship to resting levels of norepinephrine could be obtained (Fig. 3). In almost all cases, a reduction of cardiac index below the critical value of 2 liters/min/m² was associated with an increase in plasma norepinephrine above 800 pg/ml [15].

A rather good correlation was also evident if, in the same group of patients, values of cardiac index were related to the corresponding values of plasma epinephrine. High plasma levels of epinephrine were found in nearly all cases with reduction of cardiac index below 2 liters/min/m², indicating that in these critically ill patients not only the sympathoneuronal but also the sympathoadrenal system is stimulated.

These results suggest that increased plasma levels of catecholamines in heart failure may be related more to reduced peripheral perfusion [25] than to depressed left ventricular function itself.

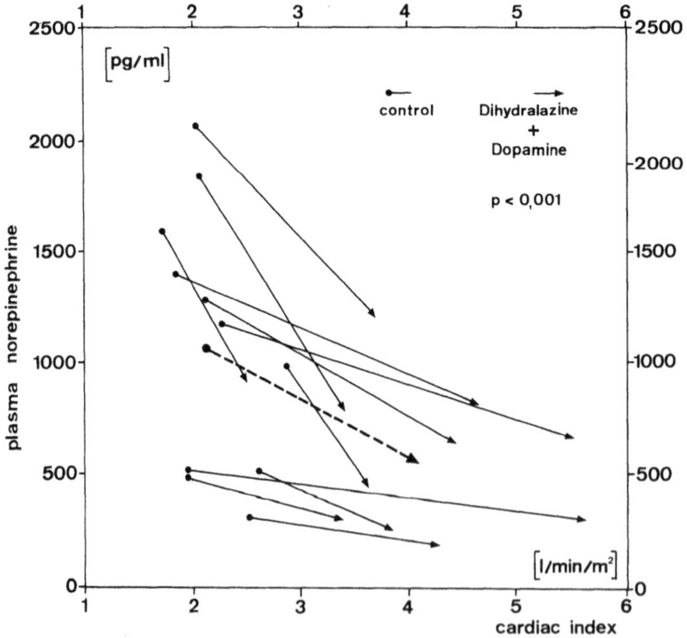

Fig. 4. Relationship between plasma norepinephrine and cardiac index in 11 patients with congestive cardiomyopathy during control and after combined therapeutical administration of dihydralazine (225 mg daily) and dopamine (2.5–5 µg/kg/min)

An inverse relationship between cardiac index and plasma catecholamine levels should be demonstrated preferentially in those cases with heart failure in whom cardiac index was altered by pharmacological intervention.

In the patients group with congestive cardiomyopathy and severe heart failure, the administration of dopamine or dihydralazine revealed in a significant improvement of cardiac index. This increase in cardiac index in all cases was associated with a reduction of the prior elevated plasma levels of norepinephrine. If both agents were administered in combination (Fig. 4), cardiac index rose to an even greater extent, and the reduction of norepinephrine plasma concentration was most pronounced. After combined administration of dihydralazine and dopamine, in addition, elevated plasma levels of epinephrine decreased significantly.

During physical exercise, patients with impaired left ventricular function exhibit an excessive augmentation of plasma catecholamine concentration [6, 8].

In Fig. 5, the effects of bicycle exercise on plasma concentration of norepinephrine, epinephrine, lactate and on oxygen saturation in the pulmonary artery, in patients with congestive cardiomyopathy are shown. Mean values obtained from 12 patients with congestive cardiomyopathy are compared to corresponding results obtained from 14 patients with normal left ventricular function.

Already during exercise with 25 or 50 Watts, a highly significant difference in the plasma catecholamine concentration and in central oxygen saturation was found between the two groups. In contrast to the findings obtained in patients with normal cardiac function, in patients with congestive cardiomyopathy the exercise induced exaggerated increase in plasma catecholamines always precede the increase in plasma lactate levels.

These results may indicate that in patients with severe left ventricular dysfunction, sympathoadrenergic stimulation during exercise may be rather related to underperfusion of the periphery [25] than to exhaustion of the skeletal muscle. Therefore, exercise-induced augmentation of plasma catecholamines in patients with heart failure primarily reflects peripheral underperfusion as consequence of inadequate augmentation of cardiac output during exercise.

Ad 3: In congestive heart failure increased concentration of catecholamines in plasma are accompanied by reduced myocardial catecholamine levels [5, 9, 16, 20].

The depletion of catecholamine stores in the myocardium in patients with impaired ventricular function is not yet fully understood. Several mechanisms are discussed:

- A reduced activity of the tyrosinehydroxylase [18] which is considered the rate limiting enzyme of catecholamine synthesis [2].
- A reduced activity of the dopamine-β-hydroxylase which enables transformation of dopamine to norepinephrine [16].
- A defect of uptake processes [20], mainly of the uptake 1.
- A defect of storage in the specific granulas [24].

As can be seen in Fig. 6, in patients with aortic regurgitation demonstrating no clinical signs of congestive heart failure, right atrial contents of norepinephrine, normetanephrine and epinephrine were significantly reduced compared to the myocardial tissue levels being found in patients with atrial septum defect. Right atrial content of dopamine did not reveal any difference between the two groups.

The lack of change in dopamine tissue levels, in spite of the significant decrease in myocardial norepinephrine content in patients with chronic aortic regurgitation,

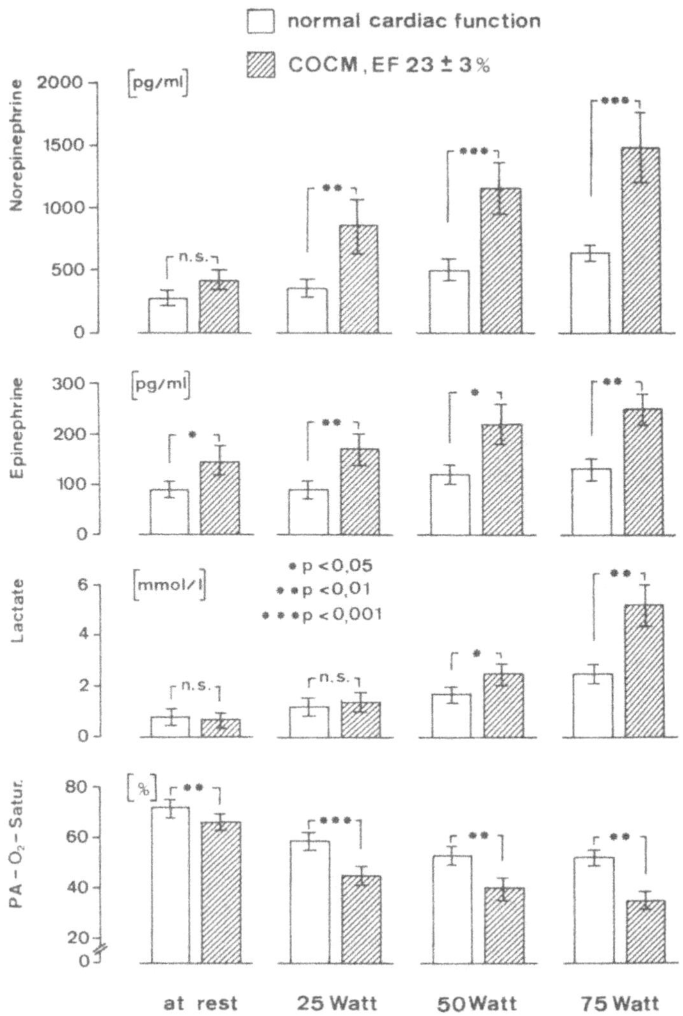

Fig. 5. Plasma levels of norepinephrine, epinephrine, lactate and oxygen saturation in the pulmonary artery (*PA-02-Satur.*) in 14 patients with normal cardiac function (*empty columns*) and in 12 patients with congestive cardiomyopathy (*COCM, hatched columns,* ejection fraction =23%) at rest and during ergometric exercise with 25, 50 and 75 Watts. The *columns* represent mean values, the *vertical lines* with *cross bars* equal ± 1 S.E.M.

may indicate that not the hydroxylation of tyrosine, but rather the transformation of dopamine to norepinephrine [4] does represent a rate limiting step for myocardial norepinephrine synthesis in patients with left heart disease. Recently, *Sole* et al. [21] demonstrating similar results in the experimental cardiomyopathy, postulated an inhibition of dopamine transport into the specific granulas.

As epinephrine cannot be synthetized in the myocardium due to lack of the phenylethanlamine-*N*-methyltransferase in cardiac tissue [2], reduction of right atrial epinephrine levels can only be explained by an impaired myocardial uptake.

A depressed catecholamine uptake in patients with heart failure can be directly demonstrated by simultaneous determination of myocardial blood flow and arterio-coronary venous catecholamine difference.

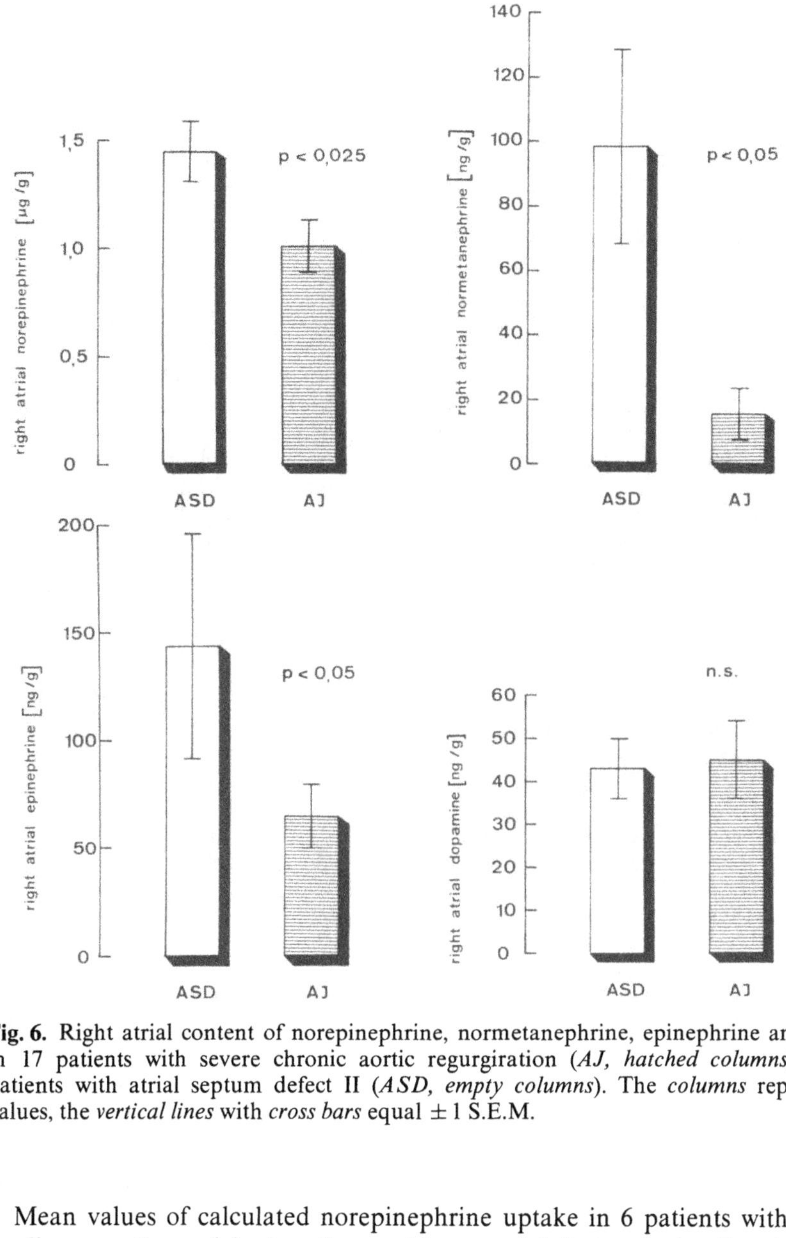

Fig. 6. Right atrial content of norepinephrine, normetanephrine, epinephrine and dopamine in 17 patients with severe chronic aortic regurgiration (*AJ, hatched columns*) and in 15 patients with atrial septum defect II (*ASD, empty columns*). The *columns* represent mean values, the *vertical lines* with *cross bars* equal ± 1 S.E.M.

Mean values of calculated norepinephrine uptake in 6 patients with congestive cardiomyopathy and in 5 patients with normal left ventricular function revealed significant reduction of norepinephrine uptake in the patients group with congestive cardiomyopathy. These data may indicate a defect of norepinephrine uptake as main mechanism of myocardial catecholamine depletion in patients with impaired ventricular function.

References

1. Anton AH, Sayre DF (1964) The distribution of dopamine and dopa in various animals and method for their determination in diverse biological material. J Pharmacol Exp Ther 145:326
2. Axelrod J (1962) Purification and properties of phenylethanolamin-*N*-methyltransferase. J Biol Chem 237:1657
3. Bertler A, Carlsson A, Rosengren E (1959) Fluorometric method for differential estimation of 3-O-methylated derivatives of adrenaline and noradrenaline (Metanephrine and Normetanephrine). Clin Chim Acta 4:456
4. Blaschko H (1973) Catecholamine biosynthesis. Br Med Cull 29:105
5. Braunwald E (1974) The autonomic nervous system in heart failure. In: Braunwald E (ed) The myocardium: failure and infarction. HP Publishing, New York, pp 59–69
6. Braunwald E, Chidsey CA (1965) The adrenergic nervous system in the control of the normal and failing heart. Proc R Soc Med 58:1062
7. Bretschneider HJ, Cott L, Hilgert G, Probst R, Rau G (1966) Gaschromatographische Trennung und Analyse von Argon als Basis einer neuen Fremdgasmethode zur Durchflußmessung von Organen. Verh Dtsch Ges Kreislaufforsch 32:267
8. Chidsey CA, Harrison DC, Braunwald E (1962) Augmentation of plasma norepinephrine response to exercise in patients with congestive heart failure. N Engl J Med 267:650
9. Chidsey CA, Braunwald E, Morrow AG (1969) Catecholamine excretion and cardiac stores of norepinephrine in congestive heart failure. Am J Med 39:442
10. Da Prada M, Zürcher G (1976) Simultaneous radioenzymatic determination of plasma and tissue adrenaline, noradrenaline and dopamine within the femtomole range. Life Sci 19:1161
11. v. Euler US, Lishajko F (1961) Improved technique for the fluorometric estimation of catecholamines. Acta Physiol Scand 51:348
12. Franciosa JA, Pierpont G, Cohn JN (1977) Hemodynamic improvement after oral hydralazine in left ventricular failure. A comparison with nitroprusside infusion in 16 patients. Ann Intern Med 86:388
13. Goldberg LI (1974) Dopamine – clinical use of an endogenous catecholamine. N Engl J Med 291:707
14. Mäurer W, Yoshida Y, Kübler W (1976) Die Urinausscheidung der Katecholamine Adrenalin, Noradrenalin und Dopamin sowie der Abbauprodukte Metanephrin und Normetanephrin bei Herzkranken. Z Kardiol 65:1124
15. Mäurer W, Hausen M, Schwarz F, Mehmel HC, Kübler W (1980) Veränderungen der Katecholaminkonzentration im Plasma in Abhängigkeit von der linksventrikulären Funktion in Ruhe und bei Belastung. Z Kardiol 69:242
16. Mäurer W, Ablasser A, Saggaau W, Storch H, Hausen M, Helmus G, Kübler W (1981) Veränderungen des myokardialen Katecholamin-Stoffwechsels und Verhalten der Plasma-Katecholamine bei Patienten mit chronischer Aorteninsuffizienz. Z Kardiol 70:540
17. Nomenclature and criteria for diagnosis of diseases of the heart and blood vessels (1963) New York Heart Association, New York

17. a Criteria Commitee of the New York Heart Association, Inc.: "Diseases of the Heart and Blood Vessels" (Nomenclature and Criteria for Diagnosis), 6th ed., Little Brown and Company, Boston, 1964

18. Pool PE, Covell JM, Levitt M, Gibb J, Braunwald E (1967) Reduction of cardic Tyrosinhydroxylase activity in experimental congestive heart failure. Circ Res 20:349
19. Sandhu RS, Freed RM (1968) An improved method for the determination of urinary catecholamines. Clin Chem 14:824
20. Spann JF, Chidsey CA, Pool PE, Braunwald E (1965) Mechanism of norepinephrine depletion in experimental heart failure produced by aortic constriction in the guinea pig. Circ Res 17:312
21. Sole JM (1980) Dopamine in the failing hamster heart: transvesicular transport limits the rate of norepinephrine synthesis. Circulation 62 [Suppl III]: 178
22. Starling EH (1897) Some points in the pathology of heart disease. Lancet I:569
23. Udenfriend S (1964) In: Fluorescence assay in biology and medicine. Academic Press, New York Vol I, 1962; Vol II, 1969

24. Vogel JHK, Jacobowitz D, Chidsey CA (1969) Distribution of norepinephrine in the failing bovine heart correlation of chemical analysis and fluorescence microscopy. Circ Res 24:71
25. Zelis R, Flaim SF, Nellis S, Longhurst J, Moskowitz R (1978) Autonomic adjustments to congestive heart failure and their consequences. In: Fishman AP (ed) Heart failure. Hemisphere Washington London p237

Discussion

Starke, Freiburg: Can a reduced uptake of noradrenaline be responsible for the increased overflow into the coronary sinus?
Diminished uptake can only be a minor component for a lowered NA content when tyrosine hydroxylase (TH) and dopamine β-hydroxylase (DBH) are proportionally reduced. Degeneration of nerves seems more likely.

Mäurer: Reduced uptake of catecholamines represents one important mechanism for producing catecholamine overflow into the coronary sinus. Enhanced release of catecholamines from the myocardium may play an additional role. Degeneration of adrenergic nerves seems also to be an important factor for the catecholamine overflow. However, it is still not possible to directly measure the myocardial density of adrenergic nerves in humans.

Sleight, Oxford: This morning Prof. Starke suggested that we do not know the afferent stimulus for the rise in circulating catecholamines in heart failure. Two candidates might be the arterial baroreceptors or the vagal mechanoreceptors in the left ventricle.
Was there a significantly reduced blood pressure rise in response to exercise in the patients with cardiac failure?

Mäurer: As a matter of fact, there was a reduced blood pressure rise in patients with congestive cardiomyopathy during exercise.

Kopin, Bethesda: Only 20% of the norepinephrine (NE) content of heart is derived from uptake. Complete inhibition by drugs of NE uptake does not deplete catecholamines. Increased release of catecholamines is usually attended by *increases* in tyrosine hydroxylase activity. The phenomenon of increased release, decreased reuptake, and *decreased* activities of biosynthetic enzymes appears unique for heart sympathetic fibres in congestive heart failure. I wonder what mechanisms may account for this unique situation.

Mäurer: In patients with congestive heart failure disturbance of myocardial catecholamine metabolism leading to a decrease in norepinephrine stores is surely a complex phenomenon, which cannot be compared to the situation in healthy rats. Therefore, I doubt whether a value of 20% which you mentioned, might also be valid for the situation of a patient with severe congestive heart failure.
I agree that probably several mechanisms are responsible for the myocardial catecholamine depletion: decreased activities of the biosynthetic enzymes, a storage defect and, in addition, a significant reduction of the norepinephrine uptake.

The demonstration of a marked overflow of norepinephrine in patients with heart failure, that means high values in the coronary sinus as compared to the corresponding values obtained from arterial blood, favours the assumption of a significantly reduced myocardial uptake of norepinephrine.

Schümann, Essen: Can you explain why dopamine content of the heart does not decrease?

Mäurer: An explanation for the unchanged level of dopamine in the myocardium might be that the transformation of dopamine to norepinephrine is inhibited due to depressed activity of the dopamine-β-hydroxylase. Therefore, this last step of the norepinephrine synthesis seems to be rate limiting.

Lymphocyte Beta-Adrenoreceptor Function in Congestive Heart Failure: Modulation by Dopamine-beta-Hydroxylase

J. A. Thomas, D. V. Unverferth, B. H. Marks

Summary

Congestive heart failure (CHF) is clinically characterized by a reduction in exercise capacity, which, in the presence of severe ventricular impairment, forces the patient to remain at complete bed rest. We have previously reported that biochemically, CHF is characterized by elevated levels of plasma norepinephrine (NE). The greater the degree of left ventricular dysfunction, the higher the plasma NE level. Peripheral blood lymphocytes from these patients with the highest NE levels and the greatest degree of cardiac decompensation failed to generate normal amounts of cyclic-3′,5′-adenosine monophosphate (cAMP) following stimulation with a beta adrenoreceptor (BAR) agonist.

Patients with the severest cardiac decompensation also demonstrated a reduction in plasma dopamine-beta-hydroxylase (DBH) enzymatic activity in conjunction with the high NE. When the ratios of NE/DBH in heart failure patients were compared to controls, they were significantly greater in the patients with the most severe heart failure. Pre-incubation of lymphocytes with purified human DBH prior to incubation with isoproterenol (ISO) produced a significant increase in the amount of cAMP generated. The ability of DBH to enhance ISO stimulated cAMP production was dependent upon the integrity of the amino acid sequence of DBH. DBH did not significantly increase cAMP in the absence of ISO.

We have concluded from our data that low DBH enzyme activity in patients with CHF may reflect the failure of the sympathetic nerve terminals to secrete physiologic amounts of DBH into the synaptic cleft, and that such a failure of secretion of this neuro-secretory protein may result from prolonged hyperfunction of the sympathetic system. We further propose that this "neuronal DBH deficiency response" may function to uncouple the catecholamine overload generated by the stress, from the intracellular structures which amplify and transmit this stress to the effector units within the cell. In this instance, the DBH deficiency may promote the downregulation of BAR.

Introduction

Patients with congestive heart failure (CHF), regardless of the etiology, clinically demonstrate a reduced exercise capacity, resting tachycardia, and a reduced cardiac output. These signs and symptoms of CHF are characteristically exacerbated by physical and emotional stress. The degree of such stress needed for their provocation depends on the extent to which cardiac performance is impaired [13]. With mild impairment of ventricular function, the signs and symptoms of heart failure are absent under ordinary conditions, appearing only during extreme stress. With further deterioration of ventricular performance, the degree of stress needed to provoke these signs and symptoms becomes less. In the case of severe ventricular impairment, this clinical picture becomes manifest even at rest.

We have previously assessed parameters of sympathetic nervous system function in persons with mild, moderate, and severe CHF [22]. Those patients whose failure was most severe, as determined by changes in systolic time intervals, had the highest plasma norepinephrine (NE) levels, and reduced plasma dopamine-beta-hydroxylase (DBH) enzyme activity. In addition, the heart rates and mean arterial pressures were inappropriately low considering the large circulating NE concentration. An analysis of the systolic time interval data in these patients demonstrated that the pre-ejection period was prolonged rather than shortened as would be expected if NE were exerting its inotropic effect, that the left ventricular ejection time was inappropriately shortened compared to its normal response to inotropic stimuli, and that the pre-ejection period/left ventricular ejection time ratio demonstrated a progressive increase with increasingly higher levels of plasma NE. These findings led us to study the ability of peripheral blood lymphocytes to generate cyclic-3′,5′-adenosine monophosphate (cAMP) in vitro. Peripheral blood lymphocytes from these patients with increased plasma NE, low plasma DBH, and severe left ventricular decompensation also failed to generate normal amounts of cAMP following beta-adrenoreceptor (BAR) stimulation by isoproterenol (ISO) in vitro.

We chose the parameters of plasma NE content and DBH enzyme activity, and lymphocyte BAR responsiveness as variables which we could measure since we believe that they reflect an adrenergic triad which collectively describe the dynamic character of the sympathetic nervous system. Therefore, in the severely decompensated individual with CHF, the least amount of stress worsens the clinical picture, and the relationships between the adrenergic triad are abnormally altered continuously. In contrast, patients with milder degrees of ventricular impairment have lower levels of plasma NE, higher levels of plasma DBH enzyme activity, and their lymphocytes generate more cAMP in response to BAR stimulation than do cells from patients with severe impairment.

In the person with severe CHF, NE apparently ceases to function primarily as a neurotransmitter, and appears to function as a hormone which is then transported via the blood from its various sites of origin to its various sites of action. The question posed by *Silverberg* et al. [20] concerning the role of circulating NE is quite relevant to the patient with CHF: "Is norepinephrine in plasma merely spilled neurotransmitter in route to metabolic degradation and excretion or is it a biologically active hormone in route to its target cells?" Do the clinical signs and symptoms of CHF reflect the direct effects of NE acting as a hormone, or do they reflect abnormalities in the function of NE as a neurotransmitter within the synaptic cleft?

Presumably, enhanced neurosecretory activity of the sympathetic nervous system contributes to the high plasma levels of NE, although delayed clearance from the plasma may also contribute to the elevation [20]. Low plasma DBH enzyme activity is more difficult to explain. *Weinshilboum* [26] has reported that genetic factors may be of importance in the determination of serum DBH activity. Their results from population and family studies are compatible with the existence of an allele for very low serum DBH activity, which appears to be inherited in an autosomal recessive manner. *Weinshilboum* concluded from his study that genetically mediated differences in enzymatic DBH activity in human serum are due to differences in the quantity of enzyme molecules released into the circulation.

Whether or not genetic factors contribute to the low DBH enzyme activity found in patients with CHF is not known. Likewise, we do not yet know the significance to the individual subject of having an intraneuronal biosynthetic enzyme like DBH circulating freely in the plasma. Understanding the significance of low DBH enzyme activity in persons with CHF therefore, remains enigmatic.

This study was designed to examine the significance of low DBH enzyme activity in patients with CHF. Peripheral blood lymphocytes were isolated from control subjects and from patients with varying degrees of heart failure (New York Heart Association Classification), and were incubated in vitro, with and without added purified human DBH, in the presence of high concentrations of the BAR agonist ISO. Lymphocyte BAR responsiveness to stimulation was determined by measuring the amount of cAMP generated during a 10 minute period at 37 °C. BAR responses were diminished in persons with more severe cardiac impairment in the absence of added purified human DBH. The addition of the purified enzyme to the incubation mixture prior to the addition of the BAR agonist was associated with a restoration of cAMP generation toward near normal levels. DBH did not, however, increase cAMP in the absence of ISO, and the ability of DBH to enhance ISO stimulated cAMP production was found to be dependent upon the integrity of the amino acid sequence of DBH.

Study Subjects and Methods

Seventeen subjects were included in this study, 5 with heart failure as a result of congestive cardiomyopathy, and 12 with failure as a result of atherosclerotic heart disease. All patients were receiving a digitalis preparation, a diuretic, and a potassium supplement during the course of the observation period. The criteria for patient inclusion into this study have been described in detail elsewhere [22]. The functional cardiac status (FCS) at the time of admission of each patient, as described by the New York Heart Association Criteria Committee, ranged from slightly compromised (class II) to severely compromised (class IV). Eight of the patients were women and 9 of the patients were men. Control subjects were hospital employees without any prior history of heart disease, and control subjects were age matched to eliminate the effect of increasing age on the plasma NE concentration, and to eliminate the effect of loss of lymphocyte BAR with increasing age. Blood specimens for lymphocyte purification were obtained on the day of admission to the study, with the patient at rest for at least five minutes. Plasma NE was measured by the method of *Henry* et al. [8] as previously described, and is reported as picograms of NE per milliliter of plasma. DBH enzyme activity was measured by the radioisotopic method of *Weinshilboum* and *Axelrod* [27]. DBH enzyme activity is reported as units of activity, with one unit representing 1 n*M* of product formed per milliliter per hour. Lymphocytes were purified from whole blood on a Ficoll-Hypaque gradient as described by *Boyum* [1], and were then incubated in the presence or absence of 0.05 m*M* ISO for varying time intervals at 37 °C as described by *Parker* et al. [17]. Cell viability, as determined by trypan blue dye exclusion, was greater than 95%. Preparations having greater than 10% polymorphonuclear leukocytes were excluded. In all experiments in which DBH was added, additions were made and pre-incubations carried out at 4 °C. ISO was then added at timed intervals, and cells were incubated at 37 °C for 10 min, in a final volume of 625 μliters. Reactions were stopped by the addition of trichloroacetic acid to a final concentration of 7%. Cells were then frozen in liquid nitrogen and stored at –70 °C until cyclic nucleotide determinations were performed. Cyclic nucleotides were measured by immunoassay according to the techniques of *Steiner* et al. [21], modified by acetylation.

Human DBH was purified from a pheochromocytoma (1920 units/mg protein), as described by *Goldstein* et al. [4]. Gel electrophoresis of the DBH enzyme in 0.1% sodium dodecyl sulfate revealed one band. When this purified enzyme was tested against specific antibody against DBH, only one precipitin band was observed.

The duration of the phases of the left ventricular systole were measured from simultaneous recordings of the electrocardiogram, phonocardiogram, and carotid arterial pulse tracing as previously described by *Weissler* et al. [28].

The level of statistical significance of data was evaluated with the use of Student's *t*-test for individual and group differences and paired data analysis, and by the methods of least squares for calculating regression coefficients as described by *Delaunois* [3].

Results

When the ratios of NE/DBH activity were analyzed in a series of patients with CHF, the ratios were significantly greater in patients than in a large group of control subjects without heart failure. The ratio of NE/DBH was significantly greater, furthermore, in those patients with severe decompensation as compared to patients with lesser degrees of heart failure (Table 1).

Figure 1 shows the regression of the pre-ejection period/left ventricular ejection time (PEP/LVET) plotted as a function of the NE/DBH ratio for controls and patients with heart failure. The significant positive correlation suggests that an increase in the NE/DBH ratio in plasma is associated with a decrease in left ventricular contractility, although this must be viewed cautiously because of the inherent limitation of systolic time intervals as predictors of ventricular performance at values greater than 0.60.

Table 2 shows the response of isolated lymphocytes to ISO stimulation in control subjects and in patients with (FCS III–IV) CHF. There is a significant difference in the basal cAMP levels in these two cell populations in the absence of ISO. The

Table 1. A comparison of admission to discharge plasma NE/DBH ratios in patients with congestive heart failure

Cardiac status	Plasma NE/DBH (mean ± SEM × 10^{-6})	
	Admission	Discharge
Control (*n* = 20)	2.68 ± 0.61	–
FCS-II (*n* = 4)	4.29 ± 1.8 [b]	3.41 ± 0.43 [a]
FCS-III (*n* = 6)	22.32 ± 8.6 [a]	12.25 ± 5.4 [a]
FCS-IV (*n* = 7)	14.34 ± 3.1 [a]	6.24 ± 1.78 [a]

[a] $P < 0.01$ compared to control
[b] $P < 0.02$ compared to control
Abbreviations: FCS, functional cardiac status (New York Heart Association Classification), NE, norepinephrine; DBH, dopamine-beta-hydroxylase; NE/DBH, n*M* NE-ml^{-1}/n*M* DBH-ml^{-1}

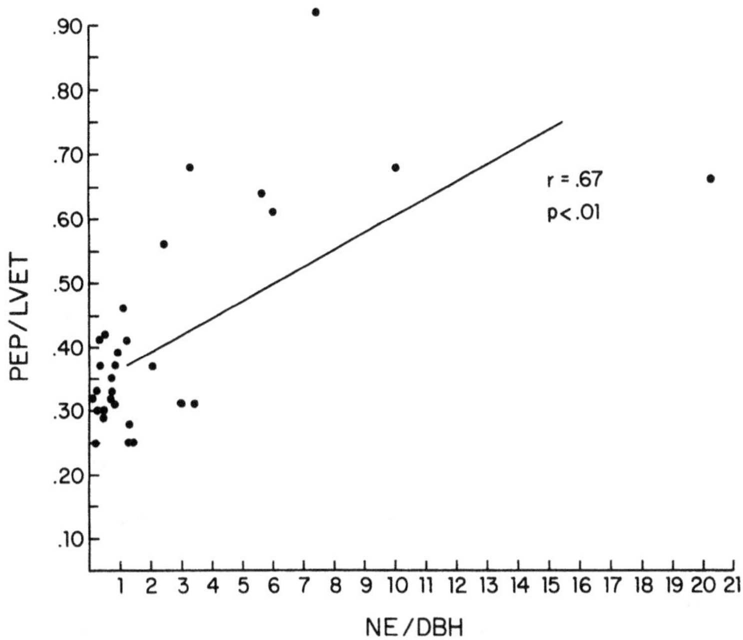

Fig. 1. Regression of the ratio of plasma norepinephrine concentration (*NE*) to plasma dopamine-beta-hydroxylase enzyme activity (*DBH*) on the ratio of pre-ejection period to left ventricular ejection time (*PEP/LVET*). The regression line represents the total spectrum (controls and patients with heart failure, $n = 37$). The regression line and correlation coefficient (r) is indicated ($P < .01$). This "cardiac-sympathetic index" described the normal response of the sympathetic nervous system to sustained alterations in cardiac ejection

basal level of cAMP in cells from CHF patients is significantly reduced compared to cells from controls ($P < 0.05$), and may be related to the prior in vivo exposure to high concentrations of NE. Following ISO treatment for 10 min, cells from patients with CHF generate significantly less cAMP than cells from normal subjects ($P < 0.005$).

When cells from both control subjects and from patients with heart failure were pre-incubated with human DBH, and then exposed to ISO, there was a significant

Table 2. Lymphocyte cAMP generation in control subjects and in patients with congestive heart failure[a]

Subjects	Lymphocyte cAMP (pM/10^6 cells ± SEM)		
	10 min + KH		10 min + 0.05 mM ISO
Control	9.37 ± 1.61 ($n = 27$)	$P < 0.005$	18.13 ± 1.36 ($n = 67$)
	$P < 0.05$		$P < 0.005$
CHF (FCS III, IV)	1.15 ± 0.59 ($n = 5$)	$P < 0.01$	7.85 ± 1.04 ($n = 24$)

[a] All incubations were carried out at 37° C

Abbreviations: cAMP, cyclic-3′,5′-adenosine monophosphate; KH, Krebs-Henseleit buffer, pH 7.4; pM, picomoles; SEM, standard error of the mean; CHF, congestive heart failure; ISO, isoproterenol; FCS, functional cardiac status, New York Heart Association Classification

Table 3. Isoproterenol-stimulated generation of cAMP in lymphocytes pre-incubated with human dopamine-beta-hydroxylase

Incubation conditions	Lymphocyte cAMP (pM/10^6 cells) (mean ± SEM)		
	Control		CHF
10 min + isoproterenol (0.05 mM)	27.44 ± 2.5 (n = 14)	$P < 0.001$	9.21 ± 1.49 (n = 8)
	N.S.		$P < 0.005$
10 min + isoproterenol + DBH	37.63 ± 3.42 (n = 14)		21.63 ± 3.08 (n = 8)
% increase cAMP	31.8		134.8

The concentration of human DBH used was 1 unit enzyme activity per incubation tube.
Abbreviations: cAMP, cyclic-3′,5′-adenosine monophosphate; DBH, dopamine-beta-hydroxylase; CHF, congestive heart failure; pM, picomoles; SEM, standard error of the mean

increase in the amount of cAMP generated in cells from failure patients as compared to pre-incubation with Krebs-Henseleit buffer alone (Table 3). The percent increase in cAMP generation in the presence of DBH was 134% for cells from patients with CHF, as compared to 31% for cells from controls. Control cells did not consistently demonstrate a significant increase in cAMP generation in the presence of DBH and ISO, and only minor increases in cAMP have been noted in cells incubated with DBH in the absence of ISO. In preliminary experiments, addition of approximately 1 unit of DBH enzyme activity per milliliter of incubation volume appeared to produce the maximum potentiation of cAMP accumula-

Table 4. The effect of tryptic digestion on the trophic action of HDBH

Incubation condition	Lymphocyte cAMP (pM/10^6 cells) (mean ± SEM)		
1. KH alone	9.21 ± 0.74 (n = 6)		
2. KH + ISO (0.05 mM)	23.93 ± 0.38 (n = 6)	} $P < 0.05$ (2 vs. 3)	} NS[a] (2–5)
3. KH + ISO + HDBH	34.10 ± 2.18 (n = 6)		
4. KH + ISO + HDBH + trypsin-chymotrypsin	27.63 ± 1.87 (n = 6)		
5. KH + ISO + trypsin-chymotrypsin	26.54 ± 2.86 (n = 6)		

[a] Not significantly different compared to conditions 4 or 5

The concentration of human DBH used in each incubation was 1 unit per incubation tube.
Abbreviations: KH, Krebs-Henseleit buffer, pH 7.4; ISO, isoproterenol; HDBH, human dopamine-beta-hydroxylase; pM, picomoles; SEM, standard error of the mean

tion. However, the DBH mediated increase was consistently greater in patients with CHF than in controls.

When DBH was incubated for 16 h at 4 °C with a mixture of trypsin-chymotrypsin, and then added to isolated lymphocytes in the presence of ISO, no significant increase in cAMP was noted (Table 4).

Discussion

Sympathetic nervous system activity and BAR function in CHF remain poorly understood. Patients with CHF, regardless of the etiology, clinically demonstrate a reduced exercise capacity, resting tachycardia, and reduced cardiac output [13]. Biochemically, these same patients manifest an adrenergic triad consisting of elevated plasma NE levels, reduced plasma DBH enzyme activity, and hyporesponsiveness of lymphocyte BAR. Patients whose myocardial contractility was most severely reduced, as measured by changes in systolic time intervals, had high plasma NE, low plasma DBH enzyme activity, and their lymphocytes failed to generate normal amounts of cAMP.

Horwitz and *Travis* [11] have reported low plasma DBH enzyme activity in patients with CHF, and they have suggested that low plasma DBH enzyme activity might serve as a marker for CHF. *Hortnagl* [10] has likewise reported that in patients with diseases resulting in high plasma levels of NE and epinephrine (E), that there is a progressive decrease in circulating DBH enzyme activity over time.

In contrast to the high plasma level of NE found in CHF, experimental and human models of CHF have demonstrated severely reduced myocardial levels of NE [17, 19], and there are reduced concentrations of NE found in coronary sinus effluents in several human models of CHF [5, 6, 7]. This implies a deficiency of NE at the adrenoreceptor level within the synaptic cleft where NE functions primarily as a neurotransmitter, and a marked excess of NE in the systemic circulation where NE functions biologically as a hormone. The functional significance of an apparent chemical signal deficit at the level of the myocardial pre-synaptic sympathetic nerve terminal, co-existing with a chemical signal overload manifested by high circulating levels of NE in persons with CHF is poorly understood.

Unverferth [25] has recently reported a significant reduction in left ventricular myocardial NE content and DBH enzyme activity in dogs with CHF due to cobalt induced myopathy. Urinary catecholamines were elevated in these animals, reflecting elevated plasma catecholamines. Peripheral lymphocytes isolated from these failed dogs exhibited a reduction in BAR density as measured by radioligand binding. Preliminary radioligand binding studies from myocardial biopsy specimens in these same animals also demonstrate significant reductions in BAR density (*D. Unverferth,* personal communication).

Penttilä [17] has likewise reported reduced myocardial content of NE, and reduced activity of the enzymes DBH and tyrosine hydroxylase (TH) in atrial biopsies from patients with CHF. BAR density was not measured in this study, however. In contrast, *Karliner* [12] has reported an increased BAR density in guinea pigs with heart failure induced by chronic pressure overload. Although myocardial NE content was reduced in these animals, no assessment was made of plasma catecholamine levels or of lymphocyte BAR density. *Tse* [24] has also reported that chronic administration of ISO to rats in vivo produces cardiomegaly, and a decreased magnitude and sensitivity of contractility in vitro to stimulation by ISO. These authors noted several biochemical alterations in this ISO induced model of cardiomegaly which included: (a) a decreased number of BAR; (b) a decreased sensitivity and magnitude of adenylate cyclase activity to stimulation by ISO or sodium fluoride (NaF); and (c) decreased cAMP formation associated with decreased cAMP dependent protein kinase activation in minced hearts incubated with ISO.

Both the available clinical evidence and the experimental data presented suggest that a diminution in the responsiveness of the heart to catecholamines may be of importance in the production of myocardial failure, although the mechanism of this postulated desensitization of the heart to catecholamines remains poorly understood. Radioligand binding studies of adrenergic receptors under both physiologic and pathophysiologic conditions, as recently reviewed by *Hoffman* and *Lefkowitz* [9] support the view that modification of receptor sensitivity is one of the methods by which hormone action and/or chemical transmission is regulated.

At synapses involving NE for example, long term stimulation in vivo desensitizes BAR, whereas transmitter depletion or long term receptor blockade supersensitizes BAR. The diminution of the biochemical effects of hormone stimulation which occurs with BAR desensitization has been accompanied by a decrease in the apparent number of BAR in some systems [14], or accompanied by "uncoupling" of the BAR adenylate cyclase complex in others.

Marsh [14] has observed that short term (i.e., 30 minutes) exposure of chick embryo ventricular tissue to 1 μM ISO resulted in a reduction in adenylate cyclase activity, but with no change in adrenoreceptor density. Since the adenylate cyclase response to a non-BAR stimulus (NaF) was unaltered, this indicated that the catalytic site of adenylate cyclase was intact. They concluded that the short term desensitization to the inotropic effect of ISO is accompanied by uncoupling of the receptor from adenylate cyclase activation, and that this may be the initial step in a sequence of events resulting in long term desensitization and actual loss of receptors from the cell surface.

The functional state of myocardial BAR in patients with CHF cannot be clearly determined from the available data. An assessment of BAR function in clinical or experimental CHF should include an analysis of factors which regulate the concentration of agonist available in the micro-environment surrounding the receptor, an analysis of those factors which influence the receptor affinity, and an assessment of those factors which affect the fluidity of the adjacent membrane structures within the cell membrane. Finally, those factors which influence the receptor's ability to couple to the putative membrane nucleotide regulatory protein and adenylate cyclase, and hence affect the transmission of the agonist signal to the intracellular effector units, must be considered.

Our present report that lymphocytes isolated from patients with CHF fail to generate normal amounts of cAMP in response to BAR stimulation with ISO in vitro, and that the addition of purified human DBH to these same cells restored the ISO mediated generation of cAMP toward normal, suggests that soluble DBH may perform a trophic function between the BAR and adenylate cyclase. The data reported in this study is also supported by data from similar work performed with isolated rat pineals [23].

In this model which we have proposed, sympathetic nerve terminals are viewed as exocytotically releasing catecholamines and a receptor-modulatory protein, DBH, which then acts to modulate the response of the catecholamine as it interacts with the target BAR. The failure of the sympathetic nerve terminal to secrete adequate amounts of DBH into the synaptic cleft as we have proposed, may represent a neuronal adaptive response to repeated or prolonged stress. In the presence of prolonged stress, a deficiency of DBH may serve to promote the down-regulation of BAR, particularly in the presence of high concentrations of catecholamines.

References

1. Boyum A (1968) Isolation of mononuclear cells and granulocytes from human blood. Scand J Clin Lab Invest 21:77–89
2. Covell JW, Chidsey CA, Braunwald E (1962) Reduction of the cardiac response to postganglionic sympathetic nerve stimulation in experimental heart failure. N Engl J Med 267:650–654
3. Delaunois AL (1973) Biostatistics in pharmacology. Pergamon, Oxford
4. Goldstein M, Lauber E, McKereghan M (1965) Studies on the purification and characterization of 3,4-dihydroxy-phenylethylamine-beta-hydroxylase. J Biol Chem 240:2077–2072
5. Haneda T, Miura Y, Arai T et al. (1980) Norepinephrine levels in the coronary sinus in patients with cardiovascular diseases at rest and during isometric handgrip exercise. Am Heart J 100:465–472
6. Haneda T, Miura Y, Miyazawa K et al. (1978) Plasma norepinephrine concentrations in the coronary sinus in cardiomyopathies. Cathet Cardiovasc Diagn 4:399–405
7. Hansen JF, Christensen NJ, Hesse B (1978) Determinants of coronary sinus noradrenaline in patients with ischaemic heart disease: coronary sinus catecholamine concentration in relation to arterial catecholamine concentration, pulmonary artery oxygen saturation and left ventricular end-diastolic pressure. Cardiovasc Res 12:415–421
8. Henry DP, Starman BJ, Johnson DG (1975) A sensitive radioenzymatic assay for norepinephrine in tissues and plasma. Life Sci 16:375–384
9. Hoffman BB, Lefkowitz RJ (1980) Radioligand binding studies of adrenergic receptors: New insights into molecular and physiological regulation. Ann Rev Pharmacol Toxicol 20:581–608
10. Hortnagl H, Stadler-Wolffersgrun R, Brucke T, Hammerle AF, Hackl JM (1978) Changes of dopamine-beta-hydroxylase activity in human plasma during prolonged overactivity of the sympathetic nervous system in various diseases. Naunyn-Schmiedebergs Arch Pharmacol 303:235–242
11. Horwitz LD, Travis VL (1978) Low serum dopamine-beta-hydroxylase activity: A marker of congestive heart failure. J Clin Invest 62:899–906
12. Karliner JS, Barnes P, Brown M, Dollery CT (1980) Chronic heart failure in the guinea pig increases cardiac alpha-1 and beta-adrenoreceptors. Eur J Pharmacol 67:115–118
13. Katz AM (1977) Physiology of the heart. Raven, New York
14. Mukherjee C, Caron MC, Lefkowitz RJ (1975) Catecholamine-induced subsensitivity of adenylate cyclase associated with loss of beta-adrenergic binding sites. Proc Natl Acad Sci USA 72:1945–1949
15. Marsh JD, Barry WH, Neer EJ, Alexander RW, Smith TW (1980) Desensitization of chick embryo ventricle to the physiological and biochemical effects of isoproterenol. Circ Res 47:493–501
16. Newman WH (1977) A depressed response of ventricular contractile force to isoproterenol and norepinephrine in dogs with congestive heart failure. Am Heart J 93:216–221
17. Penttila O, Merikallio E, Pispa J, Klinge E, Siltanen P, Kyosola K (1978) Auricular tyrosine hydroxylase and dopamine-beta-hydroxylase activities and noradrenaline content in ischaemic heart disease. Acta Med Scand 203:161–166
18. Parker CW, Smith JW (1973) Alterations in cyclic adenosine monophosphate metabolism in human bronchial asthma: leukocyte responsiveness to beta adrenergic agents. J Clin Invest 52:48–59
19. Rutenberg HL, Spann JF (1973) Alterations of cardiac sympathetic neurotransmitter activity in congestive heart failure. Am J Cardiol 32:472–480

20. Silverberg AB, Shah SD, Haymond MW, Cryer PE (1978) Norepinephrine: hormone and neurotransmitter in man. Am J Physiol 243: E252–E256
21. Steiner AL, Wehmann RE, Parker CW et al. Radioimmunoassay for the measurement of cyclic nucleotides. Adv Cycl Nucl Res (1972), Vol 2, pp 51–62
22. Thomas JA, Marks BH (1978) Plasma norepinephrine in congestive heart failure. Am J Cardiol 41: 233–243
23. Thomas JA, Sakai KK, Holck MI, Marks BH (1980) Dopamine-beta-hydroxylase: A modulator of beta adrenergic receptor activity. Res Comm Chem Pathol Pharmacol 29: 3–14
24. Tse J, Powell JR, Baste CA, Priest RE, Kuo JF (1979) Isoproterenol: induced cardiac hypertrophy: modifications in characteristics of beta-adrenergic receptor, adenylate cyclase, and ventricular contraction. Endocrinology 105: 246–255
25. Unverferth DV, Miller M, Thomas JA et al. (submitted) The evolution of beta adrenergic dysfunction during the induction of heart failure in dogs
26. Weinshilboum RM (1979) Serum dopamine-beta-hydroxylase. Pharm Rev 30: 133–166
27. Weinshilboum RM, Axelrod J (1971) Serum dopamine-beta-hydroxylase activity. Circ Res 28: 307–315
28. Weissler AM, Harris WS, Schoenfeld CD (1968) Systolic time intervals in heart failure in man. Circulation 37: 140–159

Discussion

Krawietz, Munich: You only showed a catecholamine-stimulated adenylate cyclase activity. If there is a coupling defect between β-receptor and adenylate cyclase, it would be of interest to know if the nucleotide or fluoride stimulatory effect is also decreased.

Thomas: Sodium fluoride stimulated activation of adenylate cyclase was not reduced in lymphocytes from patients with heart failure.

Starke, Freiburg: How did non-DBH-proteins influence adenylate cyclase?

Thomas: Non-DBH-proteins similar in molecular weight, did not potentiate, but instead further decreased agonist-mediated generation of cAMP.

Kopin, Bethesda: Is there a relationship between dopamine β-hydroxylase-stimulated ligand binding to lymphocyte β-adrenoceptors and the patient's endogenous DBH-levels?

Thomas: No direct relationship was found between the level of plasma DBH and the ability of DBH to potentiate agonist-mediated generation of cAMP, although the higher the plasma level of norepinephrine, the lower the amount of cAMP which could be generated in the presence of isoproterenol. The increase in cAMP generation in the presence of DBH was highest in this group.

Provoked Coronary Arterial Spasm

M. E. Bertrand, J. M. Lablanche, P. Y. Tilmant, F. P. Thieuleux

Summary

This study has established the incidence of coronary artery spasm provoked by ergonovine in 1089 consecutive patients undergoing coronary angiography. The test was performed following a routine coronary arteriogram and comprised intravenous injection of 0.4 mg ergometrine. The population studied included patients with angina, both typical and atypical, patients following recent myocardial infarction and patients with either valvular disease or congestive cardiomyopathy. Patients with spontaneous spasm, left mainstem narrowing or severe three vessel disease were excluded. 134 patients experienced focal spasm and this was observed in exceptionally few patients with atypical precordial pain (1.2%), exercise-induced angina of effort (4.3%), valvular disease (1.95%) or cardiomyopathy (0%). Spasm was observed mostly in patients with angina at rest and less often in association with angina both at rest and induced by exercise. Spasm was provoked in 20% of patients with recent transmural infarction, but in only 6.2% of patients studied at a longer interval following infarction. Spasm was superimposed on fixed atherosclerotic lesions in 60% of cases. No serious complications were encountered.

Although the patients who underwent provocation tests in this study are not representative of all patients with coronary artery disease, we estimae that spasm occurs in 20% of patients who have experienced a coronary event and in 15% of patients complaining of chest pain.

Introduction

In the last few years, it has been demonstrated that coronary artery spasm could play a significant role throughout the wide spectrum of ischemic heart disease [7, 10, 14, 17, 19]. However, the transience of coronary artery spasm makes it difficult to diagnose. Various interventions have been used to provoke coronary arterial spasm during coronary arteriography. These provocation tests include the administration of metacholine [11] epinephrine [21] in combination with propranolol [27], cold pressor test [21], alkalinisation [28] (infusion of Trisbuffer+ hyperventilation) but at present only ergonovine is widely used. Almost 30 years ago, ergonovine was introduced by *Stein* [23] as a provocative test in patients with suspected coronary artery disease. Renewed interest in this agent developed following the initial report of *Heupler* [12] suggesting that ergonovine may provoke attacks of variant angina. This study reports our experience with this provocation test and aims to establish the incidence of coronary artery spasm provoked pharmacologically in a group of 1089 patients who underwent coronary arteriography.

Methods

In May 1979, it was decided to systematically perform provocation tests in all patients undergoing coronary angiography. Subjects were excluded and the provocation test not performed if: (a) left main narrowing, (b) very severe three vessel disease was detected or (c) if spontaneous spasm was observed.

From June 1st, 1979 to October 31st 1980, provocation tests, using ergonovine, were performed in 1089 consecutive patients undergoing coronary angiography.

The procedure was explained to each patient and it was indicated that detection of coronary artery spasm could lead to the use of special drugs, i.e. calcium antagonists [1, 4, 5]. All gave their informed consent.

Patients

Patients were categorised according to their symptoms. Typical angina was defined as pain which was retrosternal and radiating down the inner aspect of the left arm, to the lower jaw, neck or teeth and described as strangling, constricting or crushing and quickly relieved by nitroglycerin. The ad hoc [24] committee recommended that exercise-induced angina should be characterised by transient episodes of angina precipitated by exercise; angina at rest, by episodes of pain occurring without apparent relation to increased myocardial oxygen demand. When these criteria were not met, the symptoms were classified as atypical chest pain.

Clinically, the patients were found to fall into six discernable groups:

Group A included 248 patients with atypical chest pain. They did not have the characteristics of typical chest pain and presented only one or two features suspicious of angina. Most of these had normal or near normal coronary arteries.

Group B included 255 patients with typical angina of effort. The majority of these had exercise tests in which there were ECG signs of exercise-induced ischaemia. Group B could be further subdivided into two subgroups:

Subgroup B1 included 117 patients with exertional angina alone;

Subgroup B2 included 138 patients in whom angina was mainly evoked by exercise, but occasionally occurred at rest. Long term ambulatory Holter monitoring was performed during admission in these cases. Transient ST segment changes were observed in only 14 of these (ST elevation in 8, ST depression in 5 and T wave changes in 1).

Group C included 203 patients who suffered one or more episodes of typical angina which were always unrelated to exercise and most often nocturnal. During admission, 24 h Holter monitoring was performed and complete 12 leads tracings were recorded when the patient experienced an attack of angina. Some did not suffer further attacks during admission. During attacks, transient ST segment changes were demonstrated in 78 patients. (ST elevation in 59, ST depression in 11 and T wave changes in 8). In the 125 remaining patients who were painfree throughout admission, no changes were observed despite continuous Holter monitoring.

Group D included 180 patients who had suffered myocardial infarction. All were asymptomatic, but in our institution, as previously published, coronary angiography was systematically performed following myocardial infarction in order to document the state of the coronary arteries and to establish the prognosis. Again there were two subgroups:

Subgroup D1: 116 patients in whom coronary angiography was performed within 6 weeks of the onset of the infarction.

Subgroup D2: 65 patients who were investigated at a later date.

Group E consisted of 154 patients with valvular disease. Some had pain, but all were at least 40 years old and undergoing catheterisation to evaluate their valve disease. This includes coronary angiography in patients over 40 years old, in our institution.

Seventy-one patients had mitral valve disease (stenosis, regurgitation, or both). Fifty had aortic valve disease (stenosis, regurgitation or both). Thirty-three had mitral disease with associated aortic disease.

Group F included 49 patients with congestive cardiomyopathy. The protocol included right and left heart catheterisation, left ventriculography and coronary angiography. These patients also underwent provocation testing.

Provocation Test

Provocation was performed following routine coronary angiography. This included measurement of the left ventricular pressure, aortic pressure and duplicate estimation of the cardiac output by a dye dilution technique. Left ventriculography was carried out in both a right anterior oblique projection of 30° and in the lateral position. Cineangiography was performed using 50 frames per second. The technique adopted for the coronary angiography was that of Judkin, using Bourassa's catheters and an introducer sheath incorporating haemostatic valves (Cordis 501–608). The left coronary artery was examined in five radiological projections including two with a cranial tilt. The right coronary artery was examined in four projections including two with a saggital tilt. Prior to the provocation test, a preliminary angiogram was performed in the lateral projection. An ECG was recorded from leads 1, 111 and V2.

A simultaneous recording was made of the aortic pressure. 0.4 mg ergometrine (Methergine®) were then given as a bolus via the femoral vein [2]. Following this, the coronary artery was opacified three and 5 min later. After opacification of the vessel (in general the right coronary artery) at the 5th minute, the alternative catheter for the left coronary artery was introduced in order to visualise this vessel. The rapid changeover was facilitated by the use of the sheath. If prior to the third minute, chest pain and electrocardiographic changes supervened (elevation or depression of the ST segment), the vessel corresponding to the electrocardiographic abnormality was immediately injected with dye. If the criteria of spasm, described below, were fulfilled, an immediate dose of isosorbide dinitrate (3–5 mg) was administered either intravenously or directly into the affected vessel. The opacification was then repeated every minute to monitor the disappearance of the localised spasm. During the entire procedure, the ECG and aortic pressure were closely monitored.

Angiographic Analysis

The coronary angiograms were analysed separately by two independant observers. The coronary artery spasm was assessed according to the criteria described by *Chahine* [8] in 1975:

1. Appearance of total occlusion of a segment of normal coronary artery or at the site of atherosclerotic narrowing.
2. Appearance of significant narrowing (more than 75%) of a segment of coronary artery which was initially or subsequently considered to be normal.
3. The disappearance, either spontaneous or facilitated by pharmacological agents (in particular nitroglycerin and its derivatives), of the narrowing or occlusion.

The catheter-induced spasm (localised to the segment of artery adjacent to the catheter-tip) and the diffuse insignificant narrowing seen throughout any vessel, were excluded from this study. Thus, focal spasm fulfilling the above criteria was the only spasm considered. The severity of the vessel disease was classified as follows: grade 0: lumen perfectly smooth and even, grade 1: irregularities, grade 2: reduction of the lumen diameter by 25%–50%, grade 3: lumen reduction of 50%–75%, grade 4: lumen reduction of 75%–90% and grade 5: total occlusion. Lesions greater than a 50% reduction in the lumen were considered to be significant.

Group Results

The results are summarized in Table 1.

Group A: this group included 248 patients with atypical chest pain. 221 had normal or near normal vessels. Twenty-two had significant disease of one vessel, 5 had significant disease of two vessels and none had three-vessel disease. In this group ergonovine provoked spasm in three patients (1.2%) all of whom were women. All had insignificant vessel disease, angiography showing only small irregularities. Spasm was located in the mid-portion of the RCA in 2 cases and on the left circumflex artery in the third.

Group B: ($n=255$)

Table 1. Vessel disease and frequency of coronary artery spasm in the study population

Group	n	Vessel disease				Spasm
		Non-signif.	1	2	3	
A	248	221	22	5		3 (1.2%)
B1	117	23	38	26	30	5 (4.3%)
B2	138	58	22	34	24	19 (13.8%)
C	203	93	55	29	26	77 (38%)
D1	116	7	54	31	24	23 (20%)
D2	64	3	26	20	15	4 (6.2%)
E	154	132	16	5	1	3 (2%)
F	49	44	4	1		
Total	1089	581	237	151	120	134

Group A, atypical chest pain; group B1, exertional angina pectoris; group B2, angina on effort and at rest; group C, angina at rest; group D1, recent transmural myocardial infarction; group D2, "old" myocardial infarction; group E, valvular disease; group F, cardiomyopathy

In *subgroup B1* ($n = 117$), the patients experienced angina during exercise without any rest pain. 38 (32.5%) had single vessel disease, 26 (22.2%) had two vessel disease and 30 (25.6%) three vessel disease. Twenty-three had insignificant disease. Only 5 patients were observed to have focal spasm. Hence the frequency of the occurrence of spasms appears to be very low in patients with exercise-induced angina (4.3%). Three of these five had insignificant vessel disease. One had single vessel disease and one had three vessel disease. Coronary spasm was in the RCA in 4 cases (mid-part in 3 and distally, proximal to the crux in 1). One patient had spasm totally occluding the first diagonal artery. It is interesting to observe that in the 23 patients with exertional angina and angiographically normal coronary arteries, only 3 had spasm, while 13 had no spasm, but a myocardial lactate production during the tachycardia induced by atrial pacing.

Subgroup B2 ($n = 138$): the patients experienced exertional angina with some episodes of angina at rest. ST segment changes during attacks at rest were recorded in 14 patients. One, two or three vessels were involved in 16%, 24.6% and 17.4% respectively. Fifty-eight (=42%) had insignificant vessel disease. Spasm was, however, more frequent in this subgroup than in the preceding subgroup. Nineteen developed angiographic changes satisfying the criteria above, an incidence of 13.8%. Of the 14 patients with documented ST segment changes during angina at rest, 8 had coronary artery spasm. If we consider the 58 patients of this subgroup with normal vessels or insignificant stenosis, only 4 had spasm. In the other patients a coronary sinus sampling catheter with pacing electrodes was positioned in the mid part of coronary sinus. The heart rate was increased by coronary sinus pacing and simultaneous arterial and coronary sinus blood samples for determination of lactate were obtained: these 30 patients had myocardial lactate production indicating anaerobic metabolism. This association of angina pain with normal or near normal angiograms and myocardial lactate production, sometimes called syndrom X, seems very frequent in this group.

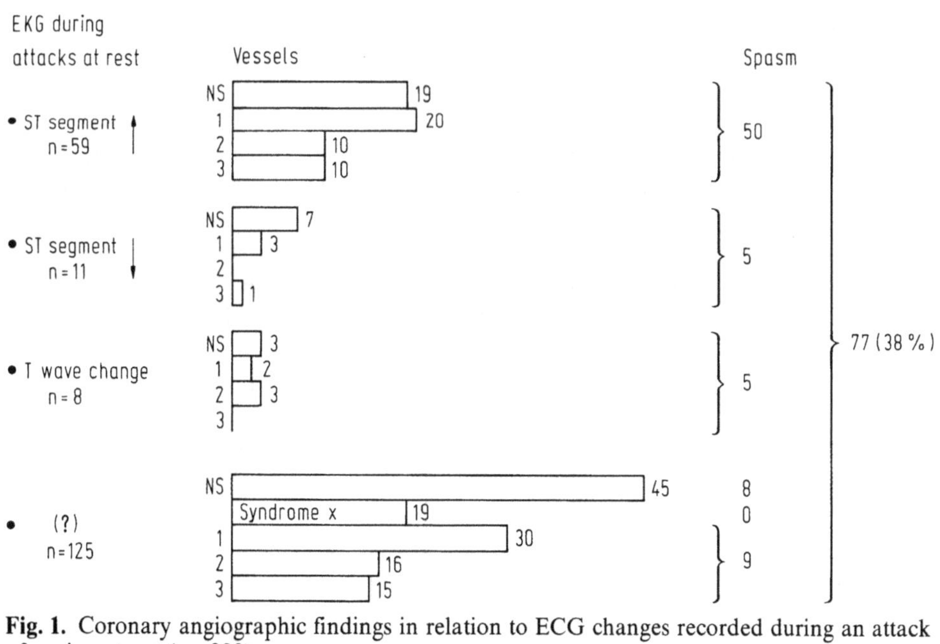

Fig. 1. Coronary angiographic findings in relation to ECG changes recorded during an attack of angina at rest ($n = 203$)

Finally it was noted that when episodes of angina at rest were associated with exercise-induced angina (subgroup B2) the frequency of coronary artery spasm was greater than in subgroup B1.

Group C: among 203 patients with angina at rest, 110 (54%) had significant coronary artery lesions. These involved one, two or three vessel disease in 27%, 14.2% and 13% of cases respectively. Ninety-three patients or 45.8% had only insignificant lesions. It was in this group that the provocation test was most successful in inducing coronary artery spasm, this being observed in 77 patients (38%). When ST segment elevation was recorded during angina at rest (59 cases), provocation testing induced spasm in 50 cases (85%) (Fig. 1). In contrast, spasm was provoked less frequently with ST segment depression or T wave changes, 45.4% and 62.5% respectively. In the 125 patients in whom it was impossible to record the ECG during angina at rest (Fig. 1), we observed that 64 (51%) had normal or near normal coronary angiograms. Syndrome X was demonstrated in 19 patients in this group and coronary artery spasm was documented in 8 (12.5%). 61 (=49%) had significant coronary artery lesions: one, two or three vessel disease in 49%, 26% and 25% of cases, respectively. Provoked coronary artery spasm was documented in 9 of these patients.

Group D included 180 patients with transmural myocardial infarction. It was subdivided into two subgroups:

Subgroup D1 included 116 patients with recent infarction. All had a typical clinical history, development of new Q or QS waves and unequivocal serum enzyme changes. The coronary angiogram and provocation test were performed within 6 weeks of these symptoms. Seventy-one patients had an anterior infarct, 43 inferior and 2 a lateral infarct. Fifty-four had single vessel, 31 had two vessel and 24 had three vessel disease. Seven had minimal vessel disease. In 23 (20%), ergonovine testing provoked focal spasm. The spasm was located on the vessel responsible for the infarction in 14 cases (12%). Spasm was located on a vessel unrelated to the area of the infarct in 9 cases (8%). None of the 7 patients with insignificant disease developed focal spasm, but it should be noted that in 4 patients ergonovine induced very severe diffuse narrowing in the vessel responsible for the infarction.

In *subgroup D2* ($n=64$) the patients were investigated at a longer interval. One, two or three vessels were involved in 41%, 31% and 23% of cases respectively. Only three patients had insignificant disease. In subgroup D2, coronary artery spasm was observed in only 4 patients (6.2%) and was, in all these, located on vessels unrelated to the area of infarct.

Group E consisted of 154 patients with valvular disease. Most had normal coronary angiograms (85.7%). Sixteen (= 10%) had single vessel disease, 5 had two and one had three vessel disease. In these 154, ergonovine induced spasm in 3 cases (1.5%). Two of these were women (52 and 59 years old) with mitral valve disease. They were painfree and in class II of NYHA. One had a normal coronary angiogram and the other had single vessel disease. The third was a man of 60 years with aortic stenosis (peak systolic gradient of 76 mm Hg) associated with regurgitation. He experienced angina at rest. During an attack, transient ST segment depression in leads III and aVF were recorded. Coronary angiograms showed some irregularities of the right coronary artery. Coronary spasm with ST elevation and pain were induced by ergonovine, spasm being located on the mid-portion of the RCA.

Group F included 49 patients with congestive cardiomyopathy. 44 had normal coronary angiograms and none were found to have spasm during the provocation tests.

Overall Results

Figure 2 summarises the results from the study of all 1089 patients. The provocation test succeeded in causing coronary artery spasm in 134 patients (12.3%).

Considering only the 886 patients with chest pain or a history of myocardial infarction, the frequency of coronary artery spasm was 15%. The majority of cases with spasm (96/134) occurred among patients complaining of angina at rest, whether or not associated with angina during exercise. In those cases where angina at rest was associated with exercise-induced angina, spasm was less frequent (13.8%) than when angina occurred only at rest (38%). In this context it should be noted that in 67 patients the diagnosis of Prinzmetal's angina was made prior to coronary angiography and on the basis of pain at rest with associated ST segment elevation. In this group, the provocation test achieved a particularly high incidence of coronary spasm (85.1%). It should be noted that the frequency of spasm in patients with recent transmural infarction was 20%, in contrast to the low incidence (5.2%) among patients investigated at a longer interval.

Characteristics of the Observed Spasm

In 79 cases (59%) the spasm was superimposed on fixed organic atherosclerotic narrowings. On the other hand, 55 patients (41%) in whom provoked spasm was observed were found to have insignificant lesions or even normal coronary angiograms.

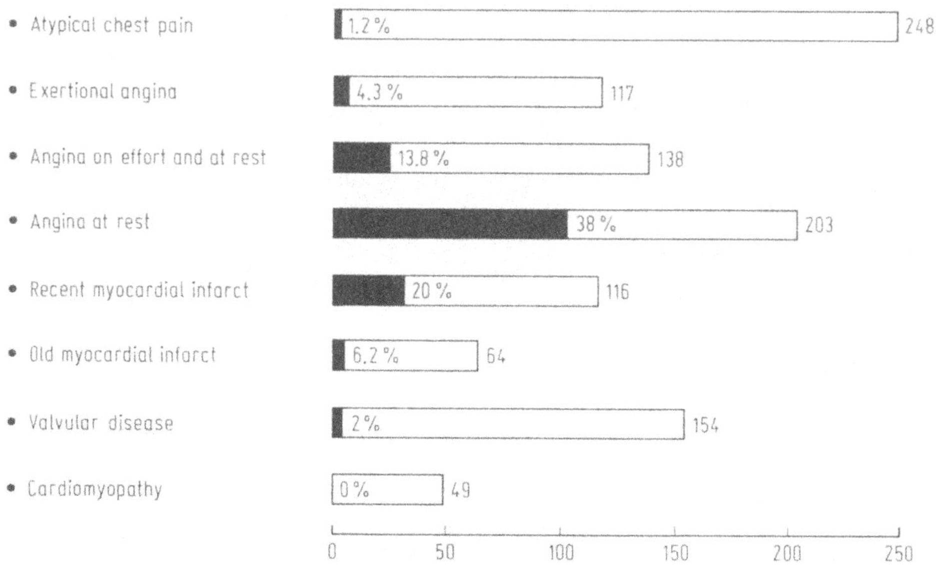

Fig. 2. Frequency of provoked spasm in 1089 patients undergoing coronary arteriography

The spasm occurred most often in the RCA (68 cases or 51%) and in the LAD in 41 cases (30%). The circumflex artery was the site of spasm in only 15 (11%) cases. Three patients (2.16%) were found to have spasm in the RCA and LAD simultaneously. Three others suffered spasm in both LAD and circumflex, and two patients had spasm in both LAD and diagonal arteries. One patient had spasm in the circumflex and RCA, while one had spasm in all three vessels.

Complications

No serious irreversible complications were observed in any of the 1089 patients. Four patients developed ventricular fibrillation which was immediately reverted by cardioversion. One patient developed a short run of ventricular tachycardia which resolved immediately with the treatment of the spasm. Another patient was observed to have third degree atrioventricular block with a heart rate of 40 beats per minute, reversed by atropine. Another developed a sinus bradycardia, easily reversed by 0.75 mg of atropine. Altogether, seven complications were noted (0.64%) but none were irreversible or lead to serious consequences. It should, however, be noted that all the complications were seen in patients with coronary artery spasm; in this subgroup the incidence was 5.2%.

Many authors [12, 22, 26] use increasingly large doses of ergonovine, sometimes reaching doses higher than the dose we used in a single bolus. In our experience, a single bolus of 0.4 mg was without substantial risk: the overall incidence of arrhythmias was 0.64% and these were easily reversed. Although these complications were minor and reversible, they were noteworthy. While some authors insist that the provocation test should be carried out in the intensive care unit, we strongly advise that the following requirements are met. In agreement with previous recommendations [20] the test should always be carried out in a room with angiographic facilities where it is possible to recognise severe atherosclerotic lesions, such as extreme narrowing of the left mainstem coronary artery, and thus to avoid performing the test in these cases. Angiography is also the only means of obtaining direct proof of spasm and it is essential to terminate the spasm as quickly as possible once it has been angiographically documented. When the test is performed without angiographic facilities, reversal of spasm by nitrates is only done, when ECG changes have been observed, and this is always later than anatomical changes in the coronary vessel. Moreover, when the test is performed during coronary angiography, relief of spasm is easily achieved by direct injection of nitroglycerin into the affected vessel. The absence of refractory spasm [6] and the low incidence of complications are evidence that our procedure is a safe one.

Discussion

Coronary artery spasm was proposed as a mechanism for angina more than a century ago. It was neglected in the light of pathological studies, and more recently, of coronary angiographic demonstration of fixed atherosclerotic lesions. In recent years it became increasingly apparent that myocardial ischaemia was not necessarily preceded by increased oxygen demand [13]. Moreover, it was demonstrated that in certain clinical circumstances a decreased oxygen supply resulted from a reduction in coronary blood flow. A revival of interest in spasm has occured during the last four years and it is now suggested that coronary artery spasm is the major factor in the pathophysiology of Prinzmetal's angina. The incidence of spasm in other manifestations of coronary insufficiency is, however, less well documented. The purpose of our study was to establish the frequency of spasm in a population of patients undergoing coronary angiography. The first important question was whether spasm provoked by ergonovine was similar to that occurring spontaneously. Previous studies of ours and of *Curry* [9] have confirmed that the provoked and spontaneous spasm are similar. The second important question concerns the specificity of the phenomenon. For this reason we systematically performed the test on patients with various cardiac diseases including valvular disease and cardiomyopathy. It should be noted that the incidence of provoked spasm was very low in these patients. Among 203 patients, 175 of whom had normal coronary angiograms, only 3 (1.5%) developed focal coronary spasm. Furthermore, one of them had angina at rest with ST segment depression. In the population of patients who experienced coronary events, the overall incidence of spasm was 15.4%. This only represents the incidence in a population of patients undergoing coronary angiography, i.e. in a group of patients with severe clinical manifestations and not the true incidence. A proof of this bias is supported by the large number of patients with severe pain at rest, because patients with severe pain at rest are admitted to hospital, whereas those with exertional angina are treated on an outpatient basis and only admitted to hospital when pain is refractory to medical treatment. Moreover, 248 patients had atypical chest pain and 221 had normal coronary arteries. If we consider the 665 patients with evidence of coronary disease, 131 (20%) had provoked spasm. This percentage can only be considered as a guide to the frequency of the phenomenon, the problems of the selection of the study population being born in mind.

During this study it was found that the incidence of spasm was very low (4.3%) in patients with exercise-induced angina. Recently, some authors [29–30] reported coronary artery spasm during episodes of angina during exercise in small groups of patients. Some had ST segment depression, but others had variant angina with exercise-induced chest pain and ST segment elevation. The mechanism whereby spasm is involved in response to exercise is uncertain and subject to speculation at present. In our study, it was found that most patients in whom spasm was induced were those suffering attacks of angina at rest. For example the incidence of spasm was considerably higher (13.7%) when angina at rest was associated with angina of effort, than when angina was only induced by effort, the incidence then being 4.3%. However, when patients experienced angina at rest, only 38% were found to have coronary artery spasm. Spasm was more frequently demonstrated (85%) in cases where rest angina was associated with ST segment elevation.

Our data can also provide answers to the pathophysiological problems of patients with ST segment depression during angina at rest. Some authors have observed alternating ST segment changes (ST segment elevation and ST segment depression) in patients with angina at rest. However, the extrapolation of results in these patients to patients with ST segment depression alone is unclear. In our study, 45% of these patients had coronary artery spasm.

The most difficult problem concerns patients admitted in the coronary care unit with angina at rest, but where it was not possible to record ECG changes during episodes of angina following admission. In this group 49% had significant vessel disease and spasm was superimposed in only 15%. Sixty-four (51%) had normal, or near normal coronary angiograms. The so-called syndrome X was demonstrated in 19 and in the 45 others, spasm was induced in only 8 (18%). The mechanism of angina at rest in this group remains unclear. Was the pain really related to the coronary event? Finally, spasm could not be completely excluded, since there is great variability of symptoms, even a circadian variability, and one can speculate whether a positive provocation test would have been obtained on another occasion.

Another interesting finding was the frequency of coronary artery spasm provoked in patients who had recently suffered transmural myocardial infarction. Our results have demonstrated the reactivity of vessels just before an infarct, supporting the previous studies [19, 15] which have suggested that spasm could be one of the several pathogenetic mechanisms leading to acute myocardial infarction.

The last question concerns the indications for performing the provocation test. We have found that of the patients subjected to the test, those complaining of angina at rest were those in whom spasm was most frequently induced. Therefore it seems reasonable that the provocation test should be performed in these patients particularly if they have normal coronary angiograms or minimal lesions.

Although spasm was frequently observed superimposed on organic lesions, is it justified to subject patients with fixed lesions to the provocation test? Taking into account the incidence of coronary artery spasm and the relative safety of our procedure, this trial is not only of pathophysiological interest, but has also therapeutic implications. It has been suggested that beta-blockers might aggravate coronary artery spasm, thereby increasing its frequency. On the other hand, calcium antagonists are known to effectively relieve spasms. Thus recognition of spasm may be important in the management of patients with angina.

Acknowledgment. We thank Marie Madeleine *Cordier* and Catherine *Feucherolles* for their secretarial assistance.

References

1. Antmann E, Muller J, Golberg S et al. (1980) Nifedipine therapy for coronary artery spasm. Experience in 127 patients. N Engl J Med 302:1269–73
2. Bertrand ME, Rousseau MF, Lablanche JM, Warembourg H Jr, Carre AG, Lekieffre JP (1979) La détection du spasme des artères coronaires par le test à la méthylergométrine. Technique. Résultats. Indications. Arch Mal Coeur 72:123
3. Bertrand ME, Lablanche JM, Tilmant PY (1980) Frequency of provoked coronary arterial spasm in 273 patients with chest pain (Abstr). Am J Cardiol 45:390
4. Bertrand ME, Lablanche JM, Tilmant PY (1980) Treatment of spasm of the coronary artery with nifedipine. Europ Heart J 1 (Suppl B): 65
5. Bertrand ME, Lablanche JM, Tilmant PY (1981) Treatment of Prinzmetal variant angina. Role of medical treatment with nifedipine and surgical coronary revascularization combined with plexectomy. Am J Cardiol 47:174
6. Buxton A, Goldberg S, Hirshfeld J et al. (1980) Refractory induced coronary vasospasm: importance of intracoronary nitroglycerin. Am J Cardiol 46:329
7. Chahine RA, Luchi RJ (1976) Coronary arterial spasm: culprit or bystander. Am J Cardiol 37:936–937
8. Chahine RA, Raizner AE, Ishimori T, Luchi RJ, McIntosh HD (1976) The incidence and clinical implications of coronary spasm. Circulation 52:972
9. Curry RC, Pepine CJ, Sabom MB, Conti R (1979) Similarities of ergonovine induced and spontaneous attacks of variant angina pectoris. Circulation 59:307

10. De Servi S, Specchia G, Angoli L et al. (1978) Etude clinique, angiographique et mécanisme pathogénique de l'angine de Prinzmetal. A propos de 31 observations. Arch Mal Coeur 72:32
11. Endo M, Hirosawa K, Kaneko N, Hase K, Inove Y, Konno S (1976) Prinzmetal's variant angina. Coronary arteriogram and left ventriculogram during angina attack induced by metacholine. N Engl J Med 294:252
12. Heupler F, Proudfit WL, Razavi M, Shirey EK, Greenstreet R, Saeldon WC (1978) Ergonovine maleate provocative test for coronary arterial spasm. Am J Cardiol 41:631
13. Maseri A, Severi S, Chierchia S, Parodi O, Biagini A (1978) Characteristics and pathogenetic mechanism of "primary" angina at rest. In: Maseri A, Klassen GA, Lesch M (eds) Primary and secondary angina pectoris. Grune & Stratton, New York, p 265
14. Maseri A, Severi S, Denes M et al. (1978) "Variant" angina: one aspect of a continuous spectrum of vasospastic myocardial ischemia. Pathogenetic mechanisms, estimated incidence and clinical and coronary arteriographic findings in 138 patients. Am J Cardiol 42:1019–35
15. Maseri A, L'Abbate A, Baroldi G et al. (1978) Coronary vasospasm as a possible cause of myocardial infarction: a conclusion derived from the study of preinfarction angina. N Engl J Med 299:1271
16. Maseri A, L'Abbate A, Chierchia S et al. (1979) Significance of spasm in the pathogenesis of ischemic heart disease. Am J Cardiol 44:788
17. Meller J, Pichard A, Dack S (1976) Coronary arterial spasm in Prinzmetal's angina: a proved hypothesis. Am J Cardiol 37:938
18. Oliva PB, Breckinridge JC (1977) Arteriographic evidence of coronary arterial spasm in acute myocardial infarction. Circulation 56:366
19. Oliva PB, Potts DE, Pluss RG (1973) Coronary arterial spasm in Prinzmetal angina: documentation by coronary arteriography. N Engl J Med 288:745–751
20. Pepine CP, Feldman RL, Conti CR (1980) Recommendations for use of ergonovine to provoke coronary artery spasm. Catheterization and cardiovascular diagnosis 6:423
21. Raizner AE, Chahine RA, Ishimori T, Verani M, Zacca N, Jamal N, Miller R, Luchi RJ (1980) Provocation of coronary artery spasm by the cold pressor test. Hemodynamic, arteriographic and quantitative angiographic observations. Circulation 62:925
22. Schroeder JS, Bolen JL, Quint RA, Clark DA, Hayden WG, Higgins CB, Wexler L (1977) Provocation of coronary spasm with ergonovine maleate new test with results in 57 patients undergoing coronary arteriography Am J Cardiol 40:487
23. Stein I (1979) Observations on the action of ergonovine on the coronary circulation and its use in the diagnosis of coronary artery insufficiency. Am Heart J 66:36
24. Surawicz B, Uhley H, Boron R et al. (1978) Task force I: standardization of terminology and interpretation 41:130
25. Waters DD, Chaitman BR, Dupras G, Theroux P, Mizgala HF (1979) Coronary artery spasm during exercise in patients with variant angina. Circulation 59:580
26. Waters DD, Theroux P, Szlachcic J, Dauwe F, Crittin J, Bonan R, Mizgala HF (1980) Ergonovine testing in a coronary care unit. Am J Cardiol 46:922
27. Yasue H, Touyama M, Shimamoto, Kato H, Tanaka S, Akiyama F (1974) Role of autonomic nervous system in the pathogenesis of Prinzmetal's variant form of angina. Circulation 50:534
28. Yasue H, Nagao M, Omote S, Takizawa A, Miwa K, Tanaka S (1978) Coronary artery spasm and Prinzmetal's variant form of angina induced by hyperventilation and Tris-buffer infusion. Circulation 58:56
29. Yasue H, Omote S, Takizawa A, Nagao M, Miwa K, Tanaka S (1979) Exertional angina pectoris caused by coronary arterial spasm: effects of various drugs. Am J Cardiol 43:647
30. Yasue H, Omote S, Takizawa A, Nagao M, Miwa K, Tanaka S (1979) Circadian variation of exercise capacity in patients with Prinzmetal's variant angina: role of exercise induced coronary arterial spasm. Circulation 59:938

Discussion

Wezler, Frankfurt/M.: Do you know anything about differences in the mode of initiation, i.e., the special mechanisms underlying coronary spasms. For instance, one could assume that in the case of cold pressure provocation the spasm is caused by autoregulatory activation of the coronary arterial smooth muscles. In other cases of provocation the arterial vasculature could be activated directly or by heteroregulation. This difference is perhaps important for an effective therapy.

Bertrand: Until now, the mechanisms leading to coronary spasm are not well known. There is a lot of evidence that local factors play an important role, but various influences (neurogenic, metabolic, humoral factors) can be superimposed.

Meyer, Aachen: How often did you find spasms in other vessels of the circulatory system, especially in the group of patients with angina at rest?

Bertrand: In our experiments coronary arterial spasm is only associated with Raynaud's phenomenon in 5.5% of cases.

Bassenge, Freiburg:
1. In your introduction you briefly mentioned α-receptors or serotonin receptors as mediators for ergonovine-induced coronary spasm. Do you have any data or information on the provocative test performed under α-blockade or under antiserotoninergic agents?
2. We were able to induce constrictions of large coronary arteries in dogs with ergonovine. These constrictions were completely blocked by prior administration of methysergide, an antiserotoninergic agent. Have you tried this approach?
3. Were you ever able to induce spasm with metacholin? Most centers were not successful in using this test.

Bertrand: 1. There are some reports in the literature indicating that α-blockade could prevent changes obtained with the ergonovine test. We have observed that phentolamine (5 mg) directly injected into the affected vessel could relieve spasm, but this may be by virtue of an own vasodilatory effect.
2. We have not tried methysergide.
3. We have not used metacholin to induce spasm.

Hjalmarson, Göteborg: I would like to know of your experience with propranolol for provocation of coronary artery spasm. We have given i.v. injections of the β_1-blocking agents metoprolol and practolol to about 1000 patients with acute myocardial infarction without inducing or aggravating chest pain. This would indicate that either vasospasm does not play a role in the infarct development or that β_1-blockade can not cause vasospasm, in contrast to propranolol. β_2-blockade might be required for this effect.

Bertrand: Whilst I have no experience with β_1-blocking agents such as metoprolol or practolol, I have a lot of evidence that propranolol can impair patients with coronary spasm. In the particular case of your patients with recent myocardial infarction, I would like to add that in this group of patients spasm (angiographically documented) occurring in the responsible vessel is usually not accompanied by chest pain or ECG changes. Thus, it may be that some of your patients had spasm without clinical or ECG expression.

Catecholamines and Early Post-Infarction Arrhythmias: The Effects of α- and β-Adrenoceptor Blockade

J. R. Parratt, C. Campbell, O. Fagbemi

Summary

The effects of stimulation and blockade of α and β-adrenoceptors on early post-infarction ventricular arrhythmias have been examined in a standardised and reliable animal model (coronary artery ligation in male rats under pentobarbitone anaesthesia). Protection (decreases in number of ventricular ectopic beats, in the incidence and duration of ventricular tachycardia (VT) and of ventricular fibrillation, and increased survival) was observed following acute intravenous administration of β-adrenoceptor blocking drugs (e.g. oxprenolol) 15 min prior to coronary artery ligation and following chronic (6 week) oral treatment (oxprenolol). This protection depended solely on β-adrenoceptor blockade. Marked protection was also observed following acute pretreatment with prazosin (in doses as low as 200 µg/kg) and, to a less marked extent, with yohimbine and phentolamine. Marked suppression of ventricular ectopic activity also resulted from the continuous intravenous administration of either adrenaline or noradrenaline (0.1 µg/kg/minute). The possible mechanisms of protective effects are discussed.

Introduction

There is convincing evidence that increased sympathetic activity is involved in the life-threatening ventricular arrhythmias that arise soon after the onset of acute myocardial infarction both in experimental animals and in man. Thus experimentally, there is evidence for increased activity in both afferent and efferent (cardiac bound) cardiac sympathetic nerves and this activity is associated with ventricular ectopic activity and fibrillation [21]. Although there is as yet no conclusive evidence for noradrenaline release early in myocardial ischaemia, even from the developing infarct [15] the fact that both surgical and "pharmacological" cardiac denervation have been found in several studies [21] to reduce early post-infarction ventricular fibrillation and to dramatically increase survival, argues for important and immediate sympathetic involvement. The pharmacological evidence of protection in this situation with both β-adrenoceptor blockade (Table 4) and α-adrenoceptor blockade [26] also supports such an involvement.

The clinical evidence for early sympathetic involvement in acute myocardial infarction and sudden cardiac death has been recently reviewed by Pantridge et al. [20]. Of 89 patients seen within 30 min of the onset of acute myocardial infarction over one-third showed evidence of sympathetic overactivity (as indicated by sinus tachycardia and transient hypertension); it was particularly evident in patients with an anterior wall infarction. Additional support for early sympathetic involvement comes from secondary reinfarction trials with β-adrenoceptor blocking drugs. Although most long-term studies with these drugs have shown, at best, only a small decrease in mortality and in the rate of reinfarction [3, 8] the recent Norwegian multicenter trial with timolol [19] showed a significant reduction both in the rate of reinfarction and in cardiac deaths. This applied especially to sudden cardiac death (defined in this particular study as "death within 24 h of the onset of new symptoms of heart disease") and implies that timolol has an antiarrhythmic effect in this situation. Since timolol is devoid of appreciable "membrane stabilising properties", is not cardioselective and has no intrinsic sympathomimetic activity, this again argues for the involvement of released catecholamines in the ventricular fibrillation which is responsible for a substantial proportion of sudden cardiac deaths.

One of the problems with the extensive previous literature dealing with experimental studies of adrenoceptor blocking drugs early in myocardial infarction is the extensive use of canine models. However, unless a standardised type and size of dog is used, there is considerable variation in the incidence of serious ventricular arrhythmias following acute coronary artery occlusion. Some dogs (e.g. beagles) have very few arrhythmias after coronary artery ligation and much depends upon whether or not an anaesthetic is used, the site of the coronary artery occlusion and the extent of pre-existing coronary collaterals. The latter has been particularly well demonstrated by Meesmann's group in Essen [16]. One other practical problem involved in comparing the effect of many adrenoceptor blocking agents on early post-infarction arrhythmias is the time and cost involved. Although our previous studies have utilised extensively an anaesthetised dog (greyhound) model [15] the present experiments have been performed using an anaesthetised rat model [4] which has a high degree of reliability, enables several experiments to be performed on the same day and more readily enables comparisons to be made of several adrenoceptor blocking agents. The purpose of the studies reported here was to determine the involvement of cardiac α and β-adrenoceptors in the genesis of these early, and life-threatening, ventricular arrhythmias which result from acute myocardial ischaemia.

Materials and Methods

Experiments were performed using male rats of the Sprague-Dawley strain, weighin 250–450 g. Anaesthesia was induced with pentobarbitone sodium, 6 mg/100 g intraperitoneally, with small additional amounts administered intravenously as required.

Systemic arterial blood pressure was recorded from the left common carotid artery using a capacitance transducer (Elema-Schönander, Type EMT35). A catheter was also placed in a femoral vein for administration of drugs, and the trachea was cannulated to allow artificial ventilation. The electrocardiogram (ECG) was recorded using standard limb leads. Arterial blood pressure and the electrocardiogram were monitored continuously on an oscilloscope and recorded using a Mingograph 81 ink-jet recorder (Elema-Schönander, Stockholm). Rectal temperature was maintained at approximately 38 °C. The chest was opened by a left thoracotomy at the fifth intercostal space and the fifth and fourth ribs were sectioned approximately 2 mm from the left margin of the sternum. Immediately after opening the chest, the animals were ventilated with room air using a stroke volume of approximately 2 ml/100 g and a rate of 54 strokes/min. These ventilation parameters maintained arterial PO_2, PCO_2 and pH within the normal range. After opening the pericardium the heart was exteriorised by gentle pressure on the chest wall, and a 6/0 braided silk suture attached to a 10 mm micro-point reverse cutting needle (Mersilk W812, Ethicon) was placed under the left coronary artery as described in detail by *Clark* et al. [4]. The heart was replaced in the chest cavity and any animal in which this procedure itself produced arrhythmias or a sustained fall in mean arterial blood pressure (MABP) to less than 70 mmHg was discarded from the study at this point. After an equilibration period of 15 min the ligature was tied and the blood pressure and electrocardiogram recorded for at least 30 min. After this period there are very few ventricular arrhythmias until 1–2 h after ligation [4].

Evaluation of Post-ligation Arrhythmias

In this particular study only early (0–30 min) post-ligation arrhythmias were assessed. Following coronary artery ligation there are multiple ventricular ectopic beats commencing between 3 and 7 min and continuing for about 20 min; thereafter the majority of the animals are in normal sinus rhythm until a later and more prolonged series of ventricular arrhythmias begin at about 1–2 h post ligation (see Fig. 1). The total number of ventricular ectopic beats over the initial 30 min post-ligation period is counted together with the duration of ventricular tachycardia (VT; defined as 7 or more consecutive ventricular extrasystoles) and of ventricular fibrillation (VF) which, in this species, usually spontaneously reverts to sinus rhythm. The tables in the *Results* section give the total number of ventricular ectopic beats and their distribution during consecutive 5 min intervals (Tables 2 and 3 only), the percentage incidence and total duration (in seconds) of both ventricular tachycardia and ventricular fibrillation and the number of animals dying (almost always in VF) within the initial 30 min post-ligation period. In Tables 1–3 the figures given refer only to those animals that survived.

Control (drug solvent) studies were performed alongside each drug study, the usual procedure being to do (for example) three or four drug experiments and two solvent (control) experiments on the same day. This was to eliminate any possible variation in response due to the time of day, temperature, laboratory conditions or to seasonal and hormonal variations. In the intravenous studies the drug was administered, as a single bolus injection, 15 min before coronary artery ligation or by continuous infusion throughout the immediate 30 min post-ligation period. In the chronic β-adrenoceptor blockade studies, oxprenolol (50 mg/kg) or placebo, was given by mouth twice daily for 6 weeks; the animals were then anaesthetised and prepared as above. Coronary artery ligation was carried out either 1 or 16 h after the last oral dose.

Results

Effects of Coronary Artery Ligation in Control (Untreated) Anaesthetised Rats

Ventricular ectopic activity commenced 3–5 min after coronary artery ligation and there were about 1000 ventricular ectopic beats during the first 30 min post-ligation period ("early arrhythmias" Fig. 1). There was evidence, in some of the rats, for a biphasic distribution of ventricular ectopic activity during this early post-ligation

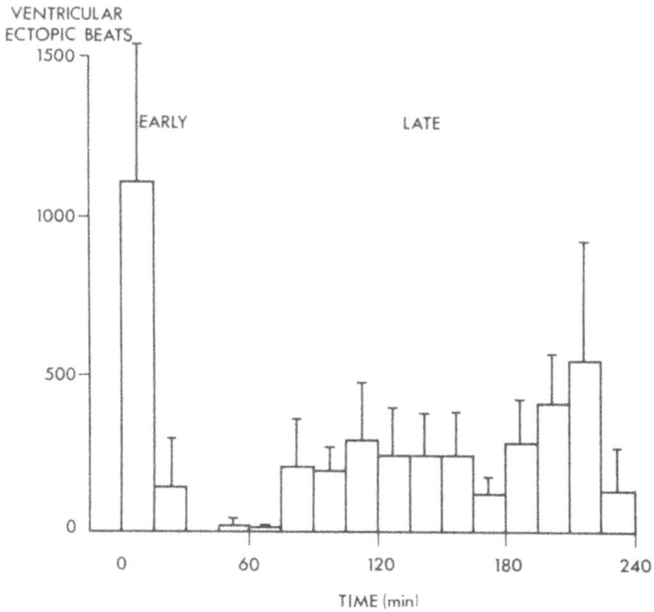

Fig. 1. The distribution with time of the number of ventricular ectopic beats following acute coronary artery ligation in pentobarbitone-anaesthetised male Sprague-Dawley rats. Notice the marked early ventricular ectopic activity (within 30 min of ligation). These arrhythmias have a different sensitivity to drugs, and probably arise from different mechanisms, compared with those occurring later after ligation (i.e. between 1 and 4 h). Only the early arrhythmias have been investigated in the present studies

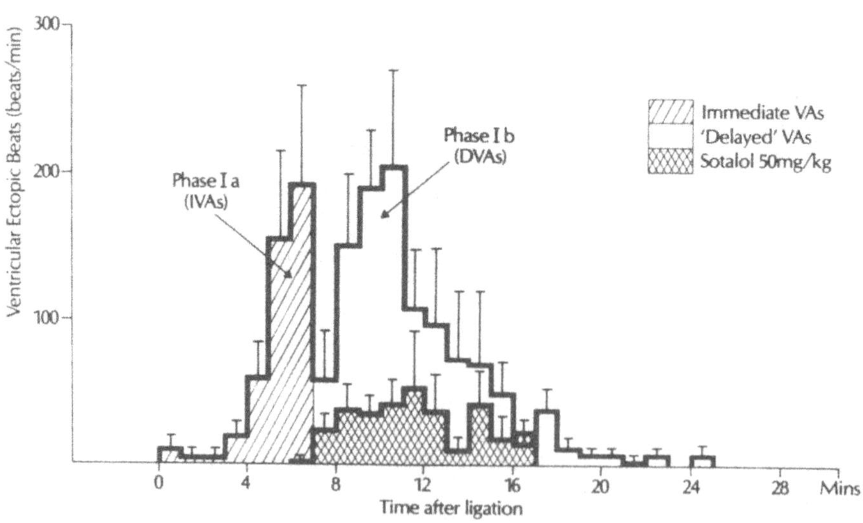

Fig. 2. Evidence for two distinct populations (phase Ia and phase Ib) of ventricular arrhythmias (*VA*) during the early (0–30 min) post-infarction period. The immediate ectopic activity appear to be preferentially suppressed by a large dose of the β-adrenoceptor blocking drug sotalol. From unpublished studies by Fiona M. *McDonald* (1979)

Table 1. The effect of chronic oxprenolol (50 mg/kg orally b.d. for 6 weeks) on arrhythmias[a] produced by coronary artery ligation in pentobarbitone-anaesthetised male rats either 1 or 16 h after the last oral oxprenolol dose

	n	VEBs	VT(s)	VF(s)	VF(total)	Mortality
i) Coronary artery ligation 1 h after last oral dose						
Control	30	927±211	67±19 (100%)	59±18 (36%)	56%	35%
Oxprenolol	36	624±106	48±12 (83%)*	11±3 (9%)**	11%***	5%**
ii) Coronary artery ligation 16 h after last oral dose						
Control	11	1144±314	78±27 (100%)	36±20 (63%)	73%	27%
Oxprenolol	11	1298±292	94±24 (100%)	64±13 (67%)	73%	18%

* $P<0.05$; ** $P<0.01$; *** $P<0.001$

[a] Total number of ventricular ectopic beats (VEBs) occurring in the first 30 min post-ligation period together with the duration (in seconds) and incidence (%) of ventricular tachycardia (VT) and ventricular fibrillation (VF), the total number of animals that fibrillated [including those that died; VF (total)] and the mortality (%).

period. This is illustrated, from some of Dr. Fiona McDonald's experiments in this laboratory, in Fig. 2. Although this evidence for two periods of early ventricular ectopic activity could be demonstrated in some of our control experiments it was not as clearly marked as in the dog studies reported by *Meesmann* et al. [17] and by *Kaplinsky* et al. [12]. It appeared (Fig. 2), however, that the β-adrenoceptor blocking drug sotalol suppressed the immediate (phase I a) arrhythmias rather more than the "delayed" (phase I b) arrhythmias. This is in contrast to the effects of nifedipine and structurally related calcium slow-channel blocking drugs which have a more marked effect on the "delayed" (phase I b) ventricular arrhythmias [6]. It should be noted however that although described as "delayed" [12] these ventricular arrhythmias still occur within minutes, rather than hours, of the onset of myocardial ischaemia.

In several separate groups of control experiments (e.g. Tables 1–3) the incidence and severity of the ventricular arhythmias resulting from coronary artery ligation was remarkably similar. The total number of ectopic beats during the early (30 min) post-ligation phase was about 1000, there was about 1 min of VT and between 25% and 40% of the animals fibrillated, with a mortality, in those different control groups (a total of about 60 rats) of between 12% and 35%.

Effects of Acute Treatment with β-Adrenoceptor Blocking Agents

These results will be presented in detail elsewhere but several important points emerge from just one of the studies (with the β-adrenoceptor blocking drug oxprenolol). These results (for the *dl* form of oxprenolol) are summarised in Figs. 3 and 4.

1. Pretreatment with oxprenolol (intravenously, 15 min prior to ligation) clearly suppressed ventricular ectopic activity, whether this was assessed from the number

Table 2. The effect of α-adrenoceptor blockade with yohimbine[a] on early post-infarction arrhythmias in anaesthetized rats

	Ventricular ectopic beats						Total	VT(s)	VF(s)	Mortality
	0–5	6–10	11–16	16–20	21–25	26–30 (min)				
Control	30±12	220±26	364±48	207±49	52±19	17±8	890±169	78±14	44±13 (25%)	12.5%
Yohimbine										
100 µg/kg	16±4	136±46	280±11	118±21	63±5	12±4	625±54	44±5	36±6 (25%)	12.5%
200 µg/kg	8±2*	31±6*	93±11*	42±16*	6±2*	0	279±28*	10±2*	23±4* (12.5%)	0
1 mg/kg	6±3*	97±16	159±34	125±48	27±16*	0	402±112*	27±7*	18±12* (25%)	0

* $P < 0.05$ ($n = 8$ in each group)
[a] Yohimbine given 15 min prior to coronary artery ligation

Table 3. The effect of adrenaline (0.1 µg/kg/min), phentolamine (90 µg/kg/min) and a combination of adrenaline and phentolamine on early post-infarction arrhythmias in anaesthetized rats

	Ventricular ectopic beats						Total	VT(s)	VF(s)	Mortality
	0–5	6–10	12–15	16–20	21–25	26–30 (min)				
Controls	11±3	190±31	348±52	229±31	115±26	42±13	935±109	67.6±13.6	18.6±8.4 (40%)	30%
Adrenaline	–	43±9*	112±32	28±9*	11±6*	4±1*	197±54*	15±3.1*	0*	0%*
Phentolamine	26±13	259±26	452±56	237±48	136±29	48±15	1144±159	66±18	6±3*	10%
Adrenaline + phentolamine	3±1	62±7*	103±17	16±3*	8±2*	2±1*	194±49*	3±1*	0*	0%*

* $P < 0.05$ ($n = 10$)

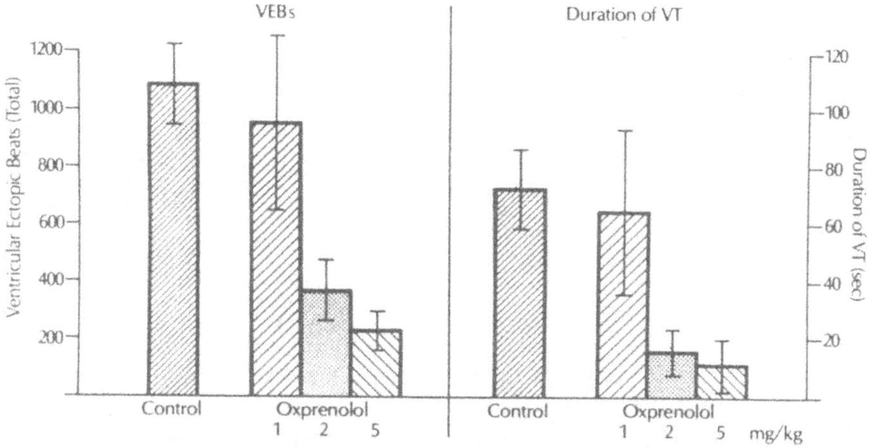

Fig. 3. The effect of oxprenolol (1, 2 or 5 mg/kg given intravenously 15 min prior to coronary artery ligation) on the number of ventricular ectopic beats (VEBs) and on the duration of ventricular tachycardia (VT) occurring in the first 30 min of acute myocardial infarction in anaesthetised rats

of VEBs or from the incidence and duration of either VF of VT. Fewer animals treated with oxprenolol died when compared to control (Fig. 4).

2. This protective effect is dose-related and is particularly evident in the reduction in the duration of VF (Fig. 4).

3. It is the serious, life-threatening ventricular arrhythmias (VT and especially VF) that are most markedly suppressed; ventricular ectopic activity (VEBs) is still present (a 60% suppression with the 2 mg/kg dose) when VF is very markedly reduced.

4. The protective effect of the oxprenolol resides almost entirely in the laevo isomer. For example 2 mg/kg of the dl form reduced the incidence of VF to 27%

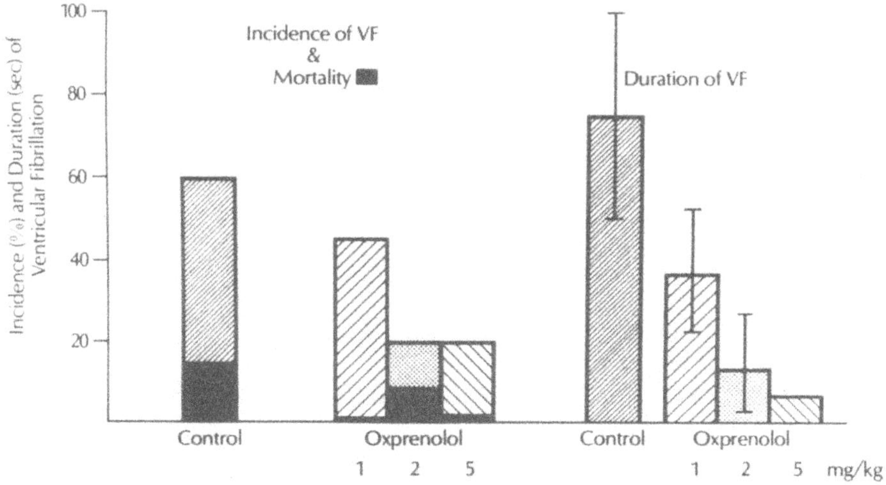

Fig. 4. The effect of d*l* oxprenolol on the incidence and duration of ventricular fibrillation (*VF*) and on mortality occurring in the first 30 min of acute myocardial infarction in anaesthetised rats (same experiments as Fig. 3)

and the duration to 14±13 s (75±26 s in the controls); 1.0 mg/kg of the laevo isomer reduced the incidence of total VF to 28% and *none* of the animals that survived the initial 30 min post-ligation period had any VF.

Other β-adrenoceptor blocking drugs (e.g. atenolol) also protect in this model [21], although propranolol had no effect on the total number of ventricular ectopic beats (1154±543 cp 1165±135 in the controls; [11]) on the incidence of VF or on mortality except in a dose (5 mg/kg) in which marked membrane stabilisation became evident.

Effects of Chronic (Oral) Treatment with Oxprenolol

These experiments were set up in order to determine whether additional benefit accrues from chronic, as compared to acute, treatment with β-adrenoceptor blocking drugs [31]. The results are given in Table 1 and demonstrate that, if the coronary artery is ligated at a time (1 h) after the last oral dose when considerable β-adrenoceptor blockade is present, there is a marked suppression of serious early post-infarction ventricular arrhythmias. Thus the incidence of VF in all the treated animals was reduced from 56% in the controls to only 11% and the duration of VF in the survivors was only 11±3 s compared with 59±18 s in the controls. Although mortality was markedly reduced there was no significant effect on the number of early ventricular ectopic beats. As with the acute studies, both with oxprenolol and propranolol, fibrillation was reduced more than ectopic activity. This is indeed a general finding in this model [6, 11]. At a time (16 h) after the last oral oxprenolol dose when there is no evidence of significant cardiac β-adrenoceptor blockade there is no protection against early post-infarction arrhythmias (Table 1). As is discussed below, this finding has considerable clinical repercussions.

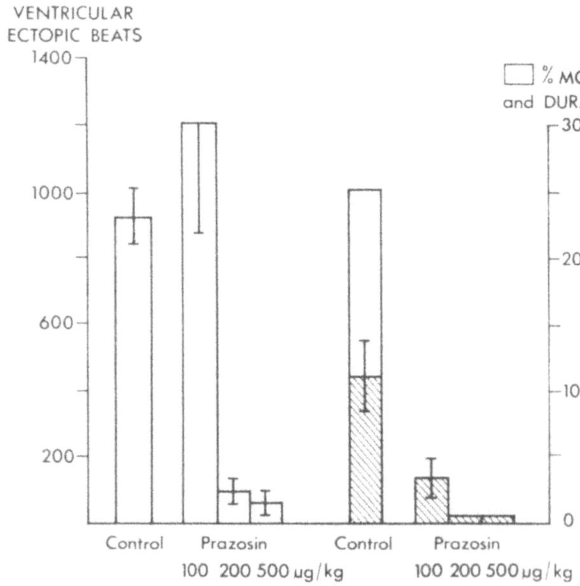

Fig. 5. The effect of prazosin (100, 200 or 500 μg/kg), given intravenously 15 min prior to coronary artery ligation, on early ventricular arrhythmias resulting from acute myocardial infarction in anaesthetised rats. Abbreviations as in Table 1

Effects of α-Adrenoceptor Blocking Agents

Three compounds have been examined, prazosin (more active on postsynaptic α_1-adrenoceptors), yohimbine (more active on presynaptic or α_2-adrenoceptors) and phentolamine (which appears to have little α-adrenoceptor selectivity). Prazosin and yohimbine were given as single intravenous injections and phentolamine as a continuous infusion throughout the early post-occlusion period. Prazosin, in doses above 100 μg/kg was especially effective (Fig. 5) and no VT or VF was observed at all with doses of 200, 500 or 2000 μg/kg. No animals pretreated with prazosin died, in contrast to the 25% mortality in the controls. Yohimbine was less active (Table 2) and 200 μg/kg appeared to be the optimum dose. Phentolamine, even in doses as high as 90 μg/kg/min was even less effective; it had no effect at all on ventricular ectopic activity or on VT although VF was reduced significantly (Table 3).

Effects of α and β- or α-Adrenoceptor Agonists

Perhaps surprisingly, in view of the effectiveness of either α- or β-adrenoceptor blockade, both adrenaline (Table 3) and noradrenaline (Fig. 6) were effective in this model in doses that increased arterial pressure by a mean of 20–25 mm Hg. When phentolamine was given together with noradrenaline, in a dose that reduced the hypertensive effect to only 10 mm Hg, this inhibition of post-ligation arrhythmias was still observed.

More selective α-adrenoceptor agonists (clonidine and cirazoline), both given by continuous intravenous infusion, also modified early post-ligation ventricular ectopic activity. Clonidine (0.6 μg/kg/min, a dose that decreased mean systemic arterial pressure by 28 mmHg) prevented VF and reduced the duration of VT from 63±7 s to 13±29 s without modifying the number of VEBs (1039±133 in the controls and 1647±329 in the clonidine-treated group). Cirazoline (0.3 μg/kg/min) significantly reduced the total number of VEBs (from 1039±133 to 417±166; $P<0.05$) and the duration of VT (from 63±7 s to 23±11 s; $P<0.05$), but did not significantly reduce the duration of VF (23±3 s in the controls to 17±13 s). A lower dose of cirazoline (0.15 μg/kg/min) had similar effects.

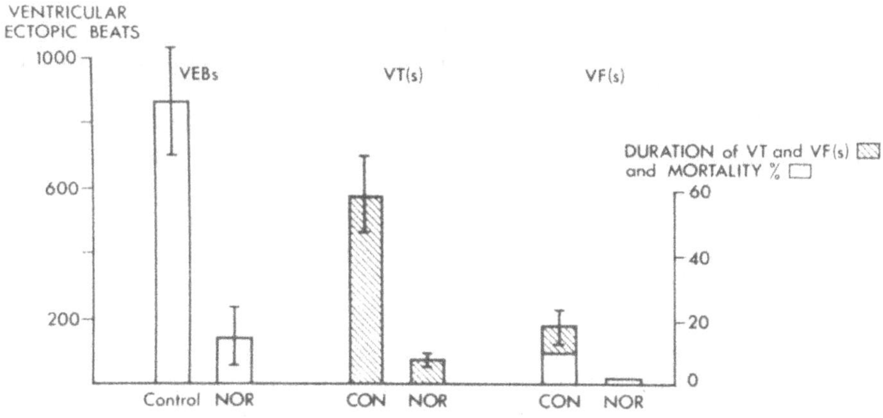

Fig. 6. The effect of infusing noradrenaline on the early ventricular arrhythmias resulting from coronary artery ligation in anaesthetised rats. Abbreviations as in Table 1

Effect of Reserpine Pretreatment

Reserpine was given in a dose of 2 mg/kg (intraperitoneal injection) for 2 days and the animals anaesthetised, and subjected to acute coronary artery ligation, 24 h after the last dose. This treatment *increased* the severity of early post-infarction arrhythmias (duration of VT 84±12 s compared to 63±7 s; duration of VF 70±8 s compared to 23±3 s in the controls, ($P < 0.05$) and increased mortality (from 20% in the controls to 50%). This pretreatment with reserpine had a profound hypotensive effect [(pre-ligation pressure of 49±6 mmHg (systolic) and 40±4 mmHg (diastolic) and reduced heart rate to 421±8 beats/min (476±8 beats/min in the controls)].

Discussion

The present studies in a highly reliable and reproducible model of early post-infarction ventricular arrhythmias, indicate that a variety of drugs interacting with cardiac α and/or β-adrenoceptors have the ability to suppress these life-threatening arrhythmias. The results with β-adrenoceptor blocking agents in general support previous studies, mainly in dogs, indicating that these drugs reduce ventricular fibrillation and subsequent mortality. Results, obtained from the literature, for propranolol and practolol (the two most commonly used β-adrenoceptor blocking agents in such studies) are summarised in Table 4. In only one study [22] was no protection obtained with propranolol but in this (and in the study of *Khan* et al. 1972 in conscious dogs) [13] the dose used was fairly high (0.5 mg/kg). Practolol reduced or prevented ventricular fibrillation in all reported studies and, in a direct comparative study, was more effective than propranolol [22]. Other β-adrenoceptor blocking agents (e.g. sotalol, tolamolol and oxprenolol) are also effective following single-dose administration in the dog model [21]. It is clear from the results presented in this paper with oxprenolol that this protection is dose-related (Fig. 3, 4) and is primarily due to β-adrenoceptor blockade since the *l*-isomer was much more active than the *d*-isomer. Further, it appears easier to suppress the more serious ventricular arrhythmias (VT and especially VF) with this β-adrenoceptor blocking agent than total ventricular ectopic activity (compare Figs. 3 and 4).

The long-term studies with oral oxprenolol are much more relevant to the problem, outlined in the *Introduction,* of the secondary prevention of reinfarction and/or reduction of sudden cardiac death in patients. The question as to whether chronic treatment with β-adrenoceptor blocking agents offers additional protection apart from β-adrenoceptor blockade has been suggested by *Vaughan-Williams* et al. [31]. These authors showed that, in rabbits, propranolol treatment for 6 weeks resulted in a significant prolongation of the cardiac muscle action potential assessed from intracellular recordings. This additional effect might itself be expected to result in antiarrhythmic activity. Certainly long-term treatment with either atenolol (*Menken* et al. [18] in dogs; 5 days treatment) or pindolol (*Szekeres* [29] in conscious rats; 14 days treatment) markedly reduced early ventricular fibrillation resulting from acute coronary artery ligation and increased survival (e.g. from 26% to 60% in the pindolol study). This protection can be adequately accounted for by β-adrenoceptor blockade alone. For example, in the atenolol dog study [18] no protection was observed if ligation was performed 18 h after the last oral dose and,

Table 4. Effect of intravenous propranolol or practolol on the incidence of ventricular fibrillation and ventricular ectopic beats and on mortality in dogs subjected to coronary artery ligation

Anaesthetic	Dose	Ventricular fibrillation	Ventricular ectopic beats	Mortality (controls in brackets)	Reference
i) Propranolol					
Thiopentone	0.08	Markedly reduced	?	14% (86%)	*Pentecost* and
Thiopentone	0.2	Reduced	?	50% (86%)	*Austen,* 1966
Diallylbarbituric acid	0.25	Prevented	Markedly reduced	0% (32%)	*Fearon,* 1967
None	0.1	Markedly reduced	?	24% (72%)	*Khan, Hamilton*
None	1.0	Reduced	?	70% (72%)	and *Manning,* 1972
Pentobarbitone	0.01	Markedly reduced	?	20% (78%)	*Evans* et al. 1976
Pentobarbitone	0.1	Markedly reduced	?	0% (78%)	
Pentobarbitone	0.2	Markedly reduced	?	0% (79%)	*Reynolds, Calzadilla* and *Lee,* 1978 [b]
Pentobarbitone	0.5	No effect	No effect	86% (86%)	*Pearle, Williford* and *Gillis,* 1978
ii) Practolol					
None	1.4	Reduced	?	31% (72%)	*Khan, Hamilton* and *Manning,* 1972
Pentobarbitone	3.0	Prevented	Prevented	0% (25%)	*Bocage, Otero* and *Harris,* 1971
Pentobarbitone	4.0	Prevented	Reduced	0% (45%)	*Sethi* et al. 1973
Morphine, urethane, chloralose	0.5	Reduced	Reduced	62% (100%)	*Stephan* and *Meesmann,* 1974
Pentobarbitone	1.5 to 2.5	Markedly reduced	Unaffected	14% (86%)	*Pearle, Williford* and *Gillis,* 1978
?	4.0	Reduced	?	73% (100%)	*Szekeres* (1977) [a]

[a] Occlusion and subsequent release; complete protection (mortality 0%) was afforded by practolol plus atropine (0.4 mg/kg)

[b] Dependent upon resting heart rate.

in the present oxprenolol rat series, which is one of the longest experimental studies so far performed (6 weeks), there was marked protection (Table 1) only when ligation was carried out at a time when there was significant β-adrenoceptor blockade. No protection was observed 16 h after the last oral dose, when tissue (myocardial) drug levels were not elevated and when there was no evidence, from isoprenaline challenge, of myocardial β-adrenoceptor blockade. Prolonged treatment with oxprenolol had no significant effect on the cardiac muscle action potential either at 1 or 16 h after the last oral dose (*Bullock, Campbell, Kane* and *Parratt,* unpublished). This implies that the protection observed 1 hour after the last oral dose is due solely to β-adrenoceptor blockade and not to effects on either Na^+ or Ca^{2+} inward currents.

The results with prazosin confirm, albeit in a different experimental model, the results of *Sheridan* et al. [26] in cats. In fact prazosin is the most active drug so far examined in the rat coronary artery ligation model with the exception on nifedipine and structurally related calcium slow-channel blocking drugs [6]. It completely prevented both VF and VT and markedly reduced VEBs (from 921 ± 90 to 99 ± 30) in a dose (200 μg/kg) which reduced systemic arterial pressure (by 18 mmHg) and decreased heart rate (by 40 beats/min). On theoretical grounds (decreased myocardial tissue perfusion) one might expect that this degree of systemic hypotension would increase, rather than decrease, the severity of post-infarction arrhythmias. In their cat study, *Sheridan* et al. [26] found prazosin (500 μg/kg) to be active both against arrhythmias resulting from reperfusion and against arrhythmias, especially VF, occurring whilst the occlusion was present. Like us they found that prazosin completely prevented early VF (0% compared with 28% in the controls cp. 0% compared with 25% in the present rat studies). We compared prazosin with yohimbine and phentolamine in order to determine whether α_1 or α_2-adrenoceptors were primarily involved; the results (Tables 2 and 3; Fig. 5) suggest that it is post-synaptic (α_1) adrenoceptor blockade that is especially important for this protection. This protection appears to be at the level of the myocardial cells rather than in the coronary vessels since, in Sheridan's study, α-adrenoceptor blockade did not markedly affect myocardial blood flow either in the normal or ischaemic zones. Our studies certainly support the view [26] that the deleterious electrophysiological alterations induced by ischaemia (and presumably mediated in part by the release of myocardial catecholamines) are dependent to a considerable extent on α-adrenoceptor mechanisms. The significance of myocardial α-adrenoceptors has been reviewed by *Benfey* [1].

The demonstrated protective effect of both α- and β-adrenoceptors can be explained by the known electrophysiological effects of catecholamines e.g. accelerated repolarization with subsequent shortening of the action potential and decrease in the refractory period; generation, in depolarised cells, of slow response action potentials; after depolarizations; increased automaticity [21]. What is more difficult to explain is the protective effect observed when either adrenaline or noradrenaline are infused immediately before and during the crucial, early post-ligation period (Fig. 6 and Table 3). Why should blockade of endogenous (cardiac?) catecholamines *and* infusion of exogenous catecholamines *both* be protective? There appear to be three possible explanations:

1. *Altered haemodynamic conditions.* This protection was seen with doses (0.1 μg/kg/min) that resulted in a 20–25 mmHg increase in systemic arterial pressure at the time the coronary artery was ligated. This probably resulted in an increased myocardial tissue perfusion; the dependence of the degree of conduction delay on local blood flow has been discussed by *Fagbemi* and *Parratt* (1981).

2. *Increase in the fibrillation threshold.* Although initially both adrenaline and noradrenaline reduce the fibrillation threshold in normal (e.g. non-ischaemic) cardiac muscle, the continued intravenous infusion of both amines causes a pronounced secondary increase in the fibrillation threshold both in atrial and ventricular muscle [30]. The mechanism is unclear since it is neither affected by β- nor by α-adrenoceptor blocking agents. Whether catecholamines also increase the ventricular fibrillation threshold in the acutely ischaemic myocardium is not known.

3. *Prolongation of action potential duration.* Detailed analysis of the effects of catecholamines on the cardiac muscle action potential show that, whereas action potential shortening is mediated through myocardial β-adrenoceptor stimulation, an increase in duration (and in refractory period) results from stimulation of myocardial α-adrenoceptors both in atrial and ventricular muscle [14, 30]. It could

perhaps be that these α-adrenoceptor mediated effects are more evident in ischaemic tissue; certainly there is evidence from our own laboratory that there is an increased preponderance of α-adrenoceptor activity early in myocardial ischaemia.

Although it is thus possible to reconcile the clear antiarrhythmic effects of β-adrenoceptor blocking drugs, both following single intravenous administration and chronic oral pretreatment, with those of infused adrenaline and noradrenaline (presumably mediated through α-adrenoceptor stimulation) it is more difficult at the moment to explain the marked protective effect of prazosin in this early post-infarction arrhythmia model. It suggests that a more detailed electrophysiological analysis of the effects of α- and β-adrenoceptor stimulation and blockade in *ischaemic* ventricular muscle is warranted.

References

1. Benfey BG (1980) Cardiac α adrenoceptors. Can J Physiol Parmacol 58:1145–1157
2. Bocage AJ, Otero H, Harris AS (1971) Preventive effect of beta-adrenergic blockade with practolol on early ventricular arrhythmias after coronary occlusion. Circulation 43/44 (Suppl 2): 142
3. Chamberlain DA (1978) Beta-adrenergic blocking agents in prevention of sudden death. Adv Cardiol 25:196–205
4. Clark C, Foreman MI, Kane KA, McDonald FM, Parratt JR (1980) Coronary artery ligation in anaesthetized rats as a method for the production of experimental dysrhythmias and for the determination of infarct size. J Pharmacol Methods 3:357–368
5. Evans DB, Peschka MT, Lee RJ, Laffan RJ (1976) Antiarrhythmic action of nadolol, a beta-adrenergic receptor blocking agent. Eur J Pharmacol 35:17–27
6. Fagbemi O, Parratt JR (in press) Calcium antagonists prevent early post-infarction ventricular fibrillation. Eur J Pharmacol
7. Fearon RE (1967) Propranolol in the prevention of ventricular fibrillation due to experimental coronary artery occlusion. Observations on its mode of action. Am J Cardiol 20:222–228
8. Fitzgerald JD (1976) The effect of β-adrenoreceptive antagonists on the morbidity and mortality in cardiovascular disease. Postgrad Med J 52:770–781
9. Kane KA, McDonald FM, Parratt JR (1979) Coronary artery ligation in anaesthetized rats as a model for the assessment of antidysrhythmic activity. Br J Pharmacol 66:436–464P
10. Kane KA, Lepran I, McDonald FM, Parratt JR, Szekeres L (1980) The effects of prolonged oral administration of a new antidysrhythmic drug (Org 6001) on coronary artery ligation dysrhythmias in conscious and anaesthetized rats. J Cardiovasc Pharmacol 2:411–423
11. Kane KA, McDonald FM, Parratt JR (1981) What pharmacological properties are necessary for the prevention of early post-infarction ventricular arrhythmias? Br J Pharmacol 72:512–513P
12. Kaplinsky E, Ogawa S, Balke CW, Dreifus LS (1979) Two periods of early ventricular arrhythmia in the canine acute myocardial infarction model. Circulation 60:397–403
13. Khan MI, Hamilton JT, Manning GW (1972) Protective effect of beta-adrenoceptor blockade in experimental coronary occlusion in conscious dogs. Am J Cardiol 30:832–837
14. Ledda F, Mugell A, Marshall L (1978) Myocardial adrenoceptors and cardiac arrhythmias. Pharmacol Res Commun 10:573–588
15. Marshall RJ, Parratt JR (1980) The early consequences of myocardial ischaemia and their modification. J Physiol (Paris) 76:699–715
16. Meesmann W, Schulz F-W, Schley G, Adolphsen P (1970) Überlebensquote nach akutem experimentellem Coronarverschluß in Abhängigkeit von Spontankollateralen des Herzens. Z Exp Med 153:246–264

17. Meesmann W, Stephan K, Abendroth RR, Menken U, Wiegand V (1978) Frühe Arrhythmien, insbesondere Kammerflimmern, nach akutem Koronarverschluß und Beta-Rezeptorenblockern. In: Mäurer W, Schömig A, Dietz R, Lichtlen PR (Hrsg) Beta-Blockade An international symposium 1977. Thieme, Stuttgart, S 244–258
18. Menken U, Wiegand V, Bucher P, Meesmann W (1979) Prophylaxis of ventricular fibrillation after acute experimental coronary occlusion by chronic beta-adrenoceptor blockade with atenolol. Cardiovasc Res 13:588–594
19. Norwegian Multicenter Study Group (1981) Timolol-induced reduction in mortality and reinfarction in patients surviving acute myocardial infarction. N Engl J Med 304:801–807
20. Pantridge JF, Webb SW, Adgey AA (1981) Arrhythmias in the first hours of acute myocardial infarction. Prog Cardiovasc Dis 23:265–278
21. Parratt JR (1980) Beta-adrenoceptor blockade and early post-infarction dysrhythmias. In: Burley DM, Birdwood GFB (eds) The clinical impact of beta-adrenoceptor blockade. Ciba, Horsham, pp 21–50
22. Pearle DL, Williford D, Gillis RA (1978) Superiority of practolol versus propranolol in protection against ventricular fibrillation induced by coronary occlusion. Am J Cardiol 42:960–964
23. Pentecost BL, Austen WG (1966) Beta-adrenergic blockade in experimental myocardial infarction. Am Heart J 72:790–796
24. Reynolds RD, Calzadilla SV, Lee RJ (1978) Spontaneous heart rate, propranolol, and ischaemia-induced ventricular fibrillation in the dog. Cardiovasc Res 12:653–658
25. Sethi V, Haider B, Ahmed SS, Oldewurtel HA, Regan TJ (1973) Influence of beta-blockade and chemical sympathectomy on myocardial function and arrhythmias in acute ischaemia. Cardiovasc Res 7:740–747
26. Sheridan DJ, Penkoske PA, Sobel BE, Corr PB (1980) Alpha adrenergic contributions to dysrhythmia during myocardial ischemia and reperfusion in cats. J Clin Invest 65: 161–171
27. Stephan K, Meesmann W (1974) Beeinflussung der frühen Arrhythmien nach akutem Koronarverschluß durch Beta-Blockade mit Practolol. Z Kardiol 7:603–611
28. Szekeres L (1977) Pharmacological principles of the therapy of early post-infarction arrhythmias. In: Schettler G, Horsch A, Morl H, Orth H, Weizel A (Hrsg) Der Herzinfarkt, p 216–229. Schattauer, Stuttgart, S 216–229
29. Szekeres L (1978) Theoretical considerations concerning drug treatment of dysrhythmias due to coronary insufficiency. Adv Pharmacol Ther 6:257–275
30. Szekeres L, Papp GyJ (1971) Experimental cardiac arrhythmias and antiarrhythmic drugs. Akademiai Kiado, Budapest
31. Vaughan Williams EM, Raine AEG, Cabrera AA, Whyte JM (1975) The effects of prolonged beta-adrenoceptor blockade on heart weight and cardiac intracellular potentials in rabbits. Cardiovasc Res 9:579–592

Discussion

Berne, Charlottesville: I believe you should add a third mechanism for an explanation of the arrhythmia following onset of myocardial ischemia. In addition to a role for increased vagal activity and catecholamine related arrhythmia one can observe 1st, 2nd and 3rd degree A-V blocks with hypoxia in isolated guinea pig hearts and intact dog hearts. These blocks are actually reversed by administration of theophylline (a known competitive inhibitor of adenosine). Also 2 patients with myocardial infarct and A-V nodal block had their A-V block restored to normal rhythm with theophylline. Hence, adenosine which can produce A-V block may well be responsible for this form of block seen in ischemic cardiac states and the treatment of choice would be theophylline or aminophylline. These studies have been carried out by Dr. L. Belardinelli in our laboratory.

Parratt: Thank you for that suggestion. We will certainly bear it in mind as we design future experiments.

Lichtlen: Hannover: If I understood your slide correctly you gave 100 mg oxprenolol per kg/rat/day. This would be equivalent to approx. 7 g (!) in a 70-kg patient. Why were such high doses applied? What were the blood levels and were there any side effects?

Parratt: The total dose of daily 100 mg/kg was given after discussions with Ciba-Geigy in Basle and after pilot studies in rats with oral oxprenolol. The dose was chosen to yield comparable plasma levels to those observed in the long-term oxprenolol trial in patients following infarction. Plasma and myocardial (normal versus ischaemic zone) levels have been determined at different times after the last daily oral dose; the results will be published elsewhere. There were no observable side effects and growth/body weight curves were similar to the controls.

Meyer, Aachen: Since in experimental animals arrhythmias occur immediately after opening coronary ligatures, we were afraid that the same could also happen in our patients after selective intracoronary streptolysis within the first 8 h following an acute myocardial infarction. Indeed, we observe ventricular arrhythmias and ventricular fibrillation when the vessel is reopened. The incidence, however, of these malignant disturbances is far less than in the animal experiments, at least in our present series of 68 patients.

Parratt: We were only dealing in these studies with arrhythmias occurring whilst the occlusion was present. As I pointed out in my second slide, reperfusion arrhythmias have a different genesis and a different sensitivity to drugs compared with early arrhythmias occurring whilst the occlusion is present. For example, they are resistant to β-blockade, but sensitive to calcium slow-channel blocking drugs.

Hjalmarson, Göteborg: Are the collateral vessels well developed in rats? If so, the effects of noradrenaline (NA), adrenaline (A) and prazosin (P) could be explained as due to improved collateral blood flow. Furthermore, did you look at the infarct size? This should be reduced by NA, A and P if the effects were due to better flow.

Parratt: Rats certainly have a high resting myocardial blood flow and you may well be right in suggesting these beneficial effects may be secondary to changes in blood flow within the ischaemic region. This is certainly markedly increased by noradrenaline, at least in dogs (*Marshall* and *Parratt* Am Heart J 1974). However, in Sheridans original study in cats [26] prazosin, in antiarrhythmic doses, did not alter flow within the ischaemic zone. Infarct size (assessed histologically) in our studies was reduced by long-term oxprenolol; we have no comparable studies with prazosin or with the catecholamines.

Different Types of Beta-Blocking Agents: Therapeutic Implications

K. H. Rahn

Summary

Several studies demonstrated that the direct cardiodepressant effect is without significance for the therapeutic actions and the side effects of beta receptor blocking agents. In hypertension and angina pectoris, the disease states where beta blockers are given most frequently, sympathomimetic activity and cardioselectivity don't play a role as far as the therapeutical efficacy is concerned. In hyperthyroidism, migraine and in cardiac arrythmias, where slowing of A-V conduction is desired, blockers without sympathomimetic effect are preferable. In patients with obstructive airways disease, a cardioselective antagonist should be used if treatment with a beta blocker is considered at all. In diabetic patients treated with insulin one would prefer a cardioselective beta-receptor antagonist.

Beta receptor blocking agents are drugs which antagonize the effects of catecholamines by blocking specific receptors in tissues. There is now considerable evidence that, as far as the therapeutic use is concerned, beta-receptor blockade is the most important property of these agents. Besides beta-adrenolytic activity some beta blockers possess a sympathomimetic effect. A number of beta receptor blocking drugs exert a direct cardiodepressant action. This means that – without catecholamines being present – the drugs have a negative inotropic, a negative chronotropic, a negative dromotropic and a negative bathmotropic action. Some beta blockers have a much higher affinity to the beta-1-receptors, which are for instance present in the heart, than for beta-2-receptors as present in vascular smooth muscle and in bronchial smooth muscle. This property is often classified as cardioselectivity.

In the following, the therapeutical implications of these different pharmacological properties will be discussed. Attention will be focussed on the most frequent indications for beta receptor blocking agents, that is therapy of hypertension and of angina pectoris. For these therapeutical indications the hemodynamic effects of beta blockers are of utmost importance.

Most beta receptor blocking agents, for instance propranolol, have typical effects on the circulation. After intravenous injection and after oral application of propranolol in men, heart rate and cardiac output decrease. Stroke volume does not change. Total peripheral resistance increases [6].

A few beta blockers such as oxprenolol and pindolol do not change significantly the just mentioned paramters in normal subjects under resting conditions. There are, however, discrepancies in the literature as far as hypertensive patients are concerned. During physical exercise all beta adrenolytic agents cause heart rate and cardiac output to decrease. These hemodynamic effects are due to inhibition of sympathetic drive of the heart, that is due to beta receptor blockade, and not due to a direct cardiodepressant action. This can be shown by comparing the hemodynamic effects of the optical isomers of beta blockers [8].

Clinical studies have demonstrated that the antihypertensive action of beta adrenergic antagonists, too, is independant of the cardiodepressant effect [7]. Taken together, on the basis of the experimental evidence available at present one must conclude that the direct cardiodepressive activity of beta blockers plays no role in situations where these agents are used therapeutically.

A number of beta adrenergic blocking agents are now being used in patients. Some of them, such as atenolol and metoprolol, are cardioselective. Several possess sympathomimetic activity. This is for instance true for acebutolol, alprenolol, bunitrolol, oxprenolol and pindolol.

In hypertension and in angina pectoris, the diseases where beta blockers are given most frequently, all beta blocking agents are similarly effective irrespective of whether they possess cardioselectivity or sympathomimetic activity [3, 9]. In other disease states, however, there are clinical observations which make a more selective use advisable. This is true for hyperthyroidism where apparently beta blockers with sympathomimetic activity are less effective than agents without this additional property. Hence, in patients with hyperthyroidism one would prefer a beta blocker without sympathomimetic effect. The same holds true for migraine. Beta adrenergic blocking agents with sympathomimetic activity cause less slowing of the A-V conduction than blockers without this property. Therefore, one would prefer beta adrenolytic agents without sympathomimetic activity if one aims to inhibit A-V conduction in patients with cardiac arrhythmias.

The most relevant side effects of beta receptor blocking agents are precipitation of congestive heart failure, abnormalities of cardiac rate and rhythm such as sinus bradycardia and A-V conduction disturbances as well as bronchial constriction. Not infrequently patients treated with beta blockers complain of cold extremities.

Different patterns of side effects could influence the choice of a beta adrenergic blocking agent. However, there are no differences betwen the different types of beta blockers in the ability to precipitate heart failure. There has been some arguing that as far as this side effect is concerned, blockers with sympathomimetic activity might be preferable. However, a comparative study of side effects performed by *Greenblatt* and *Koch-Weser* [4] revealed that the frequency of congestive heart failure is similar in patients treated with oxprenolol, a beta blocker with sympathomimetic activity, and in those given propranolol.

There is no doubt that cardioselective beta blockers cause less airways' obstruction than blockers without this property. Furthermore, cardioselective beta adrenolytics barely antagonize the effect of beta-2-sympathomimetic agents used in patients with obstructive airways disease. Therefore, in this condition a cardioselective antagonist should be used if treatment with a beta blocker is considered at all.

Very rarely, beta adrenergic blocking agents enhance and prolong hypoglycemia precipitated by insulin. This is due to blockade of beta-2-receptors in heart and skeletal muscle which mediate glycogenolysis. A controlled study by *Deacon* et al. [1] has shown that this unwanted effect is somewhat less when a cardioselective antagonist is given. Therefore, it is advisable to use a cardioselective beta blocker in insulin dependant diabetic patients, particularly in those tending to hypoglycemic episodes.

There is no unanimity as far as the use of beta receptor blocking agents in patients with sinus bradycardia is concerned. Particularly in young subjects with excellent physical fitness, sinus bradycardia reflects a low sympathetic influence on the heart. As could be expected, beta blockers do barely decrease heart rate in these subjects. However, in elderly patients beta receptor blocking drugs can substantially lower heart rate to a degree that symptoms of inadequate cerebral perfusion occur. In this situation, a blocker with sympathomimetic activity would be preferable.

Beta adrenergic blocking agents are contraindicated in patients with sick sinus syndrome as well as in patients with second and third degree A-V block. Patients with a first degree A-V block up to a PR interval of approximately 0.23 s may be given beta blockers if there is a therapeutical need. Drugs with intrinsic sympathomimetic activity are preferable in this situation since they have less influence on A-V conduction than agents without this property.

Besides cold extremities aggravation of peripheral vascular disease by different beta blockers has been reported. In a study by *Marshall* et al. [5], patients treated with propranolol complained more often of cold extremities than those treated with oxprenolol or atenolol. This study, however, is not convincing since the authors did not take into account differences in the severity and in the variable progress of the disease in the groups of patients compared.

So far, there are only few studies on the influence of beta receptor blocking agents on metabolic parameters during physical exercise. In patients with essential hypertension, *Franz* and *Lohmann* [2] found that blood glucose levels measured during physical exercise were lower during treatment with pindolol than at a time where no beta blocker was given. In contrast, the cardioselective antagonist metoprolol did not alter blood glucose concentrations. Apparently, the differences between the non-selective blocker pindolol and metroprolol are due to blockade of beta-2-receptors in heart and skeletal muscle by the former one. Although there is considerable overlapping of the blood glucose data it seems possible that patients treated with a cardioselective beta blocker have a higher working capacity during prolonged physical exercise than those given a non-selective antagonist. However, more studies appear warranted to support this idea.

References

1. Deacon SP, Karunanayake A, Barnett D (1977) Acebutolol, atenolol, and propranolol and metabolic responses to acute hypoglycemia in diabetics. Br Med J 2: 1255–1257
2. Franz I-W, Lohmann FW (1979) Der Einfluß einer chronischen sog. kardioselektiven und nichtkardioselektiven β-Rezeptoren-Blockade auf den Blutdruck, die O_2-Aufnahme und den Kohlenhydratstoffwechsel. Z Kardiol 68: 503–509
3. Frisk-Holmberg M (1979) Beta-adrenoceptor blocking drugs – comparative evaluation of their present clinical effectiveness and perspectives. Curr Ther Res 26: 1027–1045
4. Greenblatt DJ, Koch-Weser J (1974) Adverse reactions to beta-adrenergic receptor blocking drugs: a report from the Boston Collaborative Drug Surveillance Program. Drugs 7: 118–129
5. Marshall AJ, Roberts CJC, Barritt DW (1976) Raynaud's phenomenon as side effect of beta-blockers in hypertension. Br Med J 1: 1498–1499
6. Rahn KH (1978) Eigenschaften der verschiedenen Betarezeptoren-Blocker. Welche sind relevant? Cardiology 63 (Suppl 1): 1–8
7. Rahn KH, Hawlina A, Kersting F, Planz G (1974a) Studies on the antihypertensive action of the optical isomers of propranolol in man. Naunyn-Schmiedebergs Arch Pharmacol 286: 319–323

8. Rahn KH, Kersting F, Lamberts B, Seyberth HJ (1974b) Hemodynamic effects of the optical isomers of propranolol in man. In: Distler A, Wolff HP (Hrsg) Hypertension. Thieme, Stuttgart, S 248–250
9. Waal-Manning HJ (1976) Hypertension: which beta-blocker? Drugs 12:412–441

Discussion

Berne, Charlottesville: I wonder if you have considered the possibility of propranolol acting on the central nervous system? The group in the Department of Pharmacology at the Medical University of South Carolina has made some interesting observations on propranolol receptors in the brain (e.g. 4th ventricle) that exert a neurally mediated effect independent of the β-receptor blocking properties of propranolol.

Rahn: It has been mentioned sometimes that side effects on the central nervous system appear to occur more often during treatment with a lipophilic (e.g. propranolol) than with a hydrophilic blocker (e.g. atenolol). Our experience from the treatment of hypertensive patients seems to support this concept. However, I do not know of any well documented study concerning this matter.

Mattila, Helsinki: I slightly disagree with your view of β-blockers being equally effective in the treatment of hypertension. We found in a cross-over outpatient trial that atenolol (100 mg daily) alone and particularly in combination with prazosin (6 mg daily) was more effective than sotalol (150 mg daily) alone or together with prazosin. This was observed both at rest and during exercise (isometric and dynamic). I think there are differences provided high doses are not used.

Rahn: There are numerous studies demonstrating that β-blockers with sympathomimetic activity and/or cardioselectivity have an antihypertensive effect comparable with that of propranolol (e.g. papers by *Somer* et al., *Hansson* et al., Prichard's group and Bühler's group). Our own experience is in accordance with these observations. It may well be that your findings indicate differences in β-blocking activity at the respective doses. In addition, the concomitant use of prazosin may also be relevant.

Myocardial Infarction and Beta-Adrenoceptor Blockade

Å. Hjalmarson

Summary

The value of beta-blockade in patients with myocardial infarction is discussed in the present review. Animal experimental studies have shown that administration of beta-blockade can limit infarct size and protect from development of ventricular fibrillation. Both these effects seem to be mediated via $beta_1$-adrenoceptors. Various effects of beta-blockade in patients with acute myocardial infarction are demonstrated from studies during the last 10 years in Göteborg on the $beta_1$-adrenoceptor blocking agents practolol and metoprolol.

Using these $beta_1$-blockers, it is possible to reduce heart work and thereby myocardial ischemia, as indicated by decrease in the severity of chest pain and in ST-segment elevation. Furthermore, the treatment causes a reduction in anerobic myocardial metabolism and a beneficial effect on supraventricular tachyarrhythmias and possibly also on ventricular arrhythmias. It seems also possible to limit infarct size in man, as judged from changes in serum enzyme curves and Q- and R-wave changes from precordial mapping. From five studies using alprenolol, practolol and timolol it seems that it is possible to reduce sudden death and/or total mortality during 1–2 years in patients surviving their acute myocardial infarction. Whether it is possible to reduce also the early mortality (within 3 months), remains to be shown. A double-blind study in which metoprolol or placebo has been given to 1395 patients from time of hospital admission and during the first three months is aimed at infarct size limitation and early mortality. Preliminary analyses have shown that metoprolol is well tolerated and that there is a limitation in enzyme-estimated infarct size when the treatment is given within 12 h after onset of pain. Data on mortality will be published and available during the early fall this year.

The reason why our studies are performed with $beta_1$-blocking agents is that all positive effects in acute myocardial infarction seem to be due to $beta_1$-receptor blockade and furthermore that non-selective blockers have been reported to cause an increase in peripheral vascular resistance in contrast to the $beta_1$-blocking agents.

Introduction

Beta-adrenoceptor blocking agents have been widely used in various conditions for more than 15 years. Beta-blockade is today well established therapy in the treatment of patients with hypertension, angina pectoris and tachyarrhythmias. Although used in the treatment of acute myocardial infarction for as long as in other conditions, conflicting results have been obtained and the clinical value of this therapy is under debate. During the last few years it has been suggested that long-term treatment of patients surviving their acute myocardial infarction might reduce mortality as well as morbidity. Since that there are a great number of smaller studies showing negative results in post-infarction survival this indication is not yet generally accepted for beta-blockade. The aim of this presentation is to discuss the value of beta-blockade in patients with myocardial infarction and to discuss which property of the beta-adrenoceptor blocker seems to be involved in cardioprotection.

Animal Experimental Studies

During the last 10 years there has been a growing interest in prevention and limitation of the myocardial infarct development. Experimental studies have shown that the evolution of a myocardial infarct is a dynamic process which passes a stage of reversible ischemia before development of irreversible necrosis [25]. From experimental and clinical observations it has been suggested that sympathetic overactivity and local release of noradrenaline in the ischemic myocardium may play an important role in the infarction process [34]. Studies using animal experimental models demonstrated that early administration of beta-blocking agents could protect the ischemic myocardium and limit infarct size [17, 33]. From a number of experimental studies it is well clarified that sympathetic activity and the release of noradrenaline in the ischemic myocardium play a very important role in the development of serious ventricular arrhythmias and ventricular fibrillation [9]. Prophylactic administration of beta-blocking agents before induction of experimental ischemia has a marked protective effect against development of ventricular fibrillation. In Table 1 is summarized the possible role of catecholamines in experimental myocardial infarction and it seems that all these effects are mediated via $beta_1$-adrenoceptors and that protection can be obtained by $beta_1$-adrenoceptor blockade.

Hemodynamic Effects

In the early studies of propranolol in patients with acute myocardial infarction, it was found a significant fall in cardiac output and also induction of acute congestive heart failure [4]. Similar observations were also made after acute administration of alprenolol. Administration of nonselective beta-blockers, such as propranolol, has been reported to cause a fall in cardiac output and an increase in systemic vascular resistance [18, 19]. It was found later that a cardioprotective $beta_1$-blocker, such as

Table 1. Role of catecholamines (CA) in experimental myocardial infarction

- CA causes myocardial cell damage and ventricular fibrillation
- CA increases heart work and reduces diastolic perfusion period
- Direct myocardial cell damage by noradrenaline (NA) in isolated hearts
- Tyramine-released NA causes damage
- Ischemia releases NA from sympathetic nerve endings
- NA and high extracellular K^+ give coronary vasoconstriction
- NA causes thrombocyte aggregation
- All effects mediated via β_1-receptors and protection of β_1-blockade

practolol, in moderate doses could reduce heart rate without change in cardiac output or vascular resistance in normal subjects and in patients with heart failure [7, 8].

Our present treatment schedule for patients with acute myocardial infarction is to give 15 mg i.v. of the beta$_1$-blocking agent metoprolol immediately after arrival in hospital and 15 min later 50 mg metoprolol orally, which is repeated every 6 h for the first 48 h. As can be seen from Fig. 1, in patients with uncomplicated acute myocardial infarction this metoprolol treatment caused a significant fall in heart rate, systolic blood pressure with unchanged diastolic blood pressure and pulmonary artery end diastolic pressure [16]. The i.v. injection caused a rate-dependent drop in cardiac output with unchanged stroke volume and a very small increase in systemic vascular resistance, but these changes were restored to initial values within 6 to 24 h.

Despite the differences between nonselective and selective beta-blockers, both types seem to be well tolerated in the majority of patients with acute myocardial infarction. Even in patients with congestive heart failure and elevated filling pressures, both propranolol and metoprolol are well tolerated and there is a fall in filling pressure in patients with high pressures after i.v. administration [16, 19]. Even in patients with acute myocardial infarction and markedly elevated filling pressures and tachycardia there is a fall in this pressure after i.v. administration of metoprolol [26]. It seems likely that a beta$_1$-adrenoceptor drug without any peripheral vascular effect will be better tolerated in patients with acute myocardial infarction and latent heart failure. Furthermore, it is known that beta$_1$-blockade is the property of the drug causing the positive effects in tachyarrhythmias, angina pectoris and in experimental infarction and ventricular fibrillation. We have therefore chosen beta$_1$-blocking agents for our studies in patients with acute myocardial infarction, earlier practolol, and – from 1974 – metoprolol. Besides the advantage of having no or little effect on vascular resistance, a beta$_1$-blocker can also be given to patients with slight to moderate bronchial obstruction.

Effects on Myocardial Ischemia, Pain and ST-Segments

It has been reported that in patients with acute myocardial infarction, plasma catecholamine concentrations are elevated [28, 29]. The elevated plasma concentrations reflect an increased sympathetic nervous activity causing an increase in heart rate and contractility and thereby in metabolic demand. In a group of patients with uncomplicated acute myocardial infarction, heart rate and systolic blood pressure were significantly higher and contractility indices, such as preejection period, were markedly shorter during the first day in the coronary care unit in comparison with the situation after one week [30]. This was thought to be due to an inappropriate sympathetic overactivity and practolol was therefore given i.v. causing the same changes in heart rate, systolic blood pressure and preejection period within 30 min in 10 patients with uncomplicated acute myocardial infarction to the same extent as was found spontaneously during 1 week in 20 untreated patients [31]. After injection of propranolol in a dose of 0.1 mg/kg body weight to patients with acute myocardial infarction, *Mueller* et al. [19] noted a reduction in heart rate and mean arterial pressure, a modest decrease in cardiac output and a more marked reduction in coronary blood flow. However, there was a more marked depression in

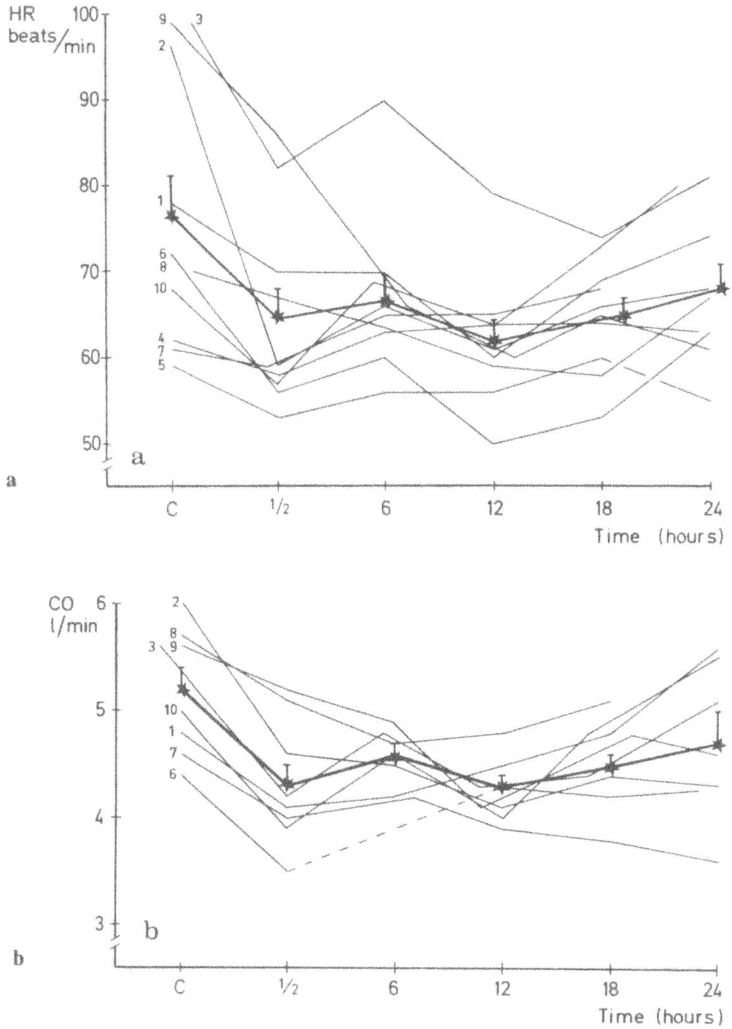

a

b

Fig. 1 a–d. Hemodynamic changes in 10 patients with acute myocardial infarction during 24 h before and after 15 mg metoprolol i.v. and 50 mg metoprolol orally every 6 h. *HR,* heart rate; *BP,* blood pressure; *CO,* cardiac output; PAEDP, pulmonary artery end diastolic pressure [32]

myocardial oxygen consumption causing, in almost all patients, an improved myocardial lactate utilization, indicating a beneficial effect on myocardial ischemia. Besides the changes in cardiac work and myocardial metabolic demand, there was a reduction in plasma free fatty acid levels in the majority of patients, also reducing myocardial oxygen consumption [18, 19].

In 1970 beta-blockade was well established therapy in the treatment of angina pectoris. We therefore asked ourselves whether it was possible to reduce ischemic chest pain with beta-blockade also in patients with acute myocardial infarction. It was found that many of the patients with uncomplicated myocardial infarction had a markedly elevated heart rate, systolic blood pressure and contractility. Intravenous injections of beta-blocking agents have been found to reduce external

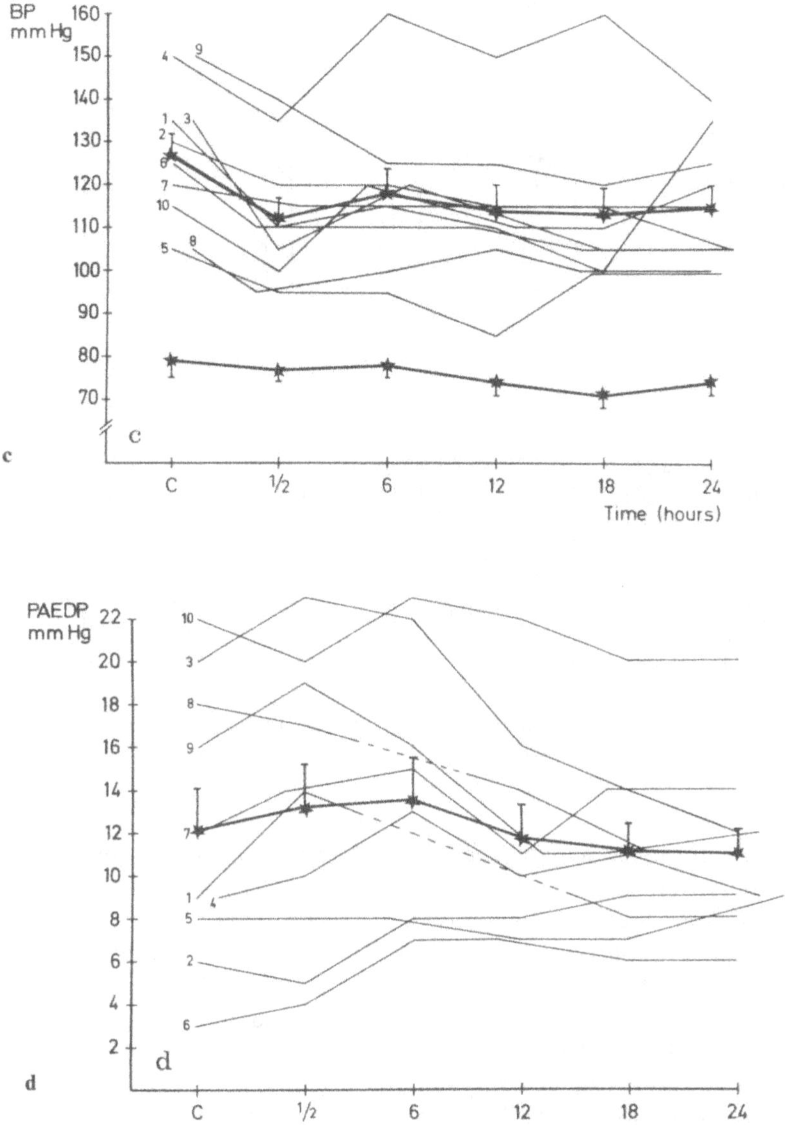

Fig. 1c–d.

cardiac work and metabolic demand. Practolol was given i.v. in an average dose of 18 mg (range 5–30 mg) to 17 patients with severe chest pain during the first day in the coronary care unit [30]. Chest pain was measured on a 5-point scale and the scores were calculated before and after beta-blockade. Practolol caused a reduction of pain from average score 3.9 to 0.9 ($P<0.001$). This was, however, an open study with psychological side-effects that might influence the chest pain of the patients. A double-blind study was therefore performed in 54 patients with acute myocardial infarction and severe chest pain [31]. All patients were treated immediately after arrival in hospital and all within 24 h of onset of severe pain. Eighteen patients were given saline and the remaining patients received either practolol with moderate intrinsic stimulatory activity (30 mg), H 87/07 – a $beta_1$-selective blocker

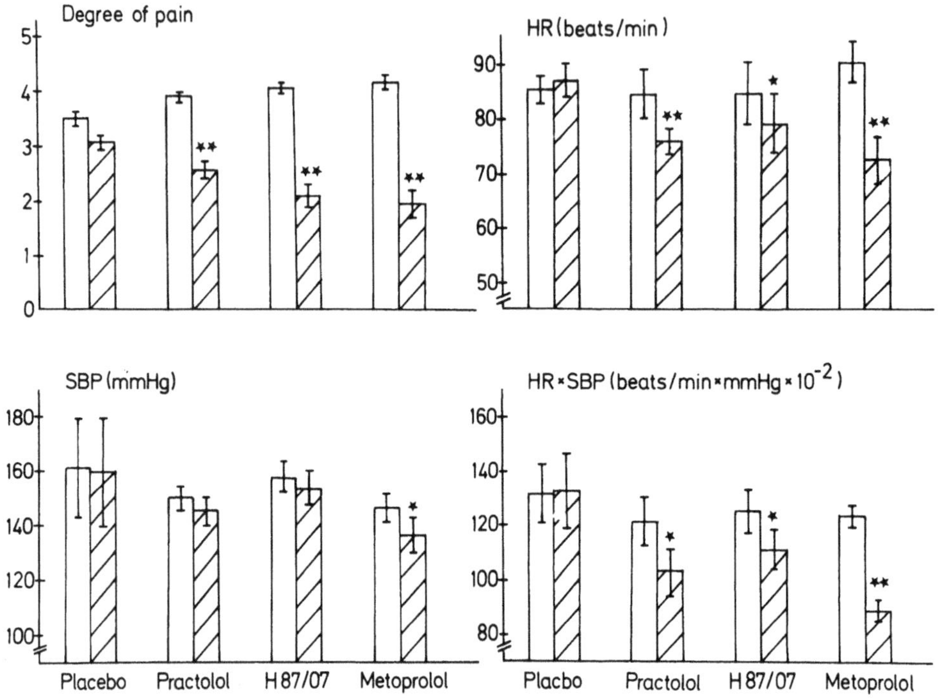

Fig. 2. Scores for chest pain, heart rate, blood pressure and rate-pressure product before and after injection of saline and three different beta-blockers in patients with acute myocardial infarction. Values are means ± SE. * $P<0.02$; ** $P<0.01$ [32]

with high intrinsic stimulatory activity (comparable to pindolol, 30 mg) – or metoprolol, which has no intrinsic stimulatory activity (15 mg). As can be seen from Fig. 2, all beta-blockers caused a significant fall in the degree of chest pain. There was also a fall in rate-pressure product for all beta-blockers, mostly marked for metoprolol without intrinsic activity, and the least marked for H 87/07, the beta-blocking agent with the highest intrinsic activity. The administration of beta-blocking agents was well tolerated and a reduction of heart work and chest pain lasted for more than 60 min in the majority of patients [32]. It was thought that the reduction of chest pain was concomitant with the improvement of myocardial anaerobic metabolism, as reported by *Mueller* et al. [19].

Changes in ST-segment elevation after pharmacological intervention have been shown to correlate well with enzymatic histological signs of myocardial necrosis under experimental conditions [14] and have been used in man with the so-called precordial surface mapping method. This method seems to correlate well with epicardial mapping. *Pelides* et al. [23] have shown, very impressively, in a 72-electrode surface mapping study that practolol can reduce the electrocardiographic changes of ischemia in the center and the borderline zones, both as regards extent and intensity.

In the 54 patients discussed above with acute myocardial infarction and chest pain given the three beta-blockers metoprolol, H 87/07 and practolol for pain relief, ST-segment analyses were also done. As can be seen from Table 2 there was a significant reduction in ST-segment elevation after injections of all three beta-blockers. The effect seemed to be somewhat more pronounced with metoprolol

Table 2. ST-segment changes in patients with acute myocardial infarction after administration of three beta-blockers i.v. (see Fig. 2) [31]

Injection	%	*n*
Saline	– 1.9 ±1.7	10
Analgesics after saline	– 0.7 ±2.3	10
Practolol	–29.0 ±9.2	6
H 87/07	–31.6 ±6.4	8
Metoprolol	–40.0 ±6.2	7
Analgesics after beta-blockade	+17.53±6.3	13

Mean±SE

which also showed the most marked reduction in external heart work, expressed from the rate-pressure product (Fig. 2). There seems to be good correlation in other studies between reduction in heart work and ST-segment elevation [31].

Effects on Infarct Size

A number of studies from various animals have demonstrated that beta-blockade administered prophylactically or within the first few hours can limit the size of infarction [5, 33]. The value of such therapy for limitation of infarct size in man is not yet clear. The problem in this field of research is that there is no accurate method for infarct size measurements in man. The most widely used method is the evaluation of serum enzyme curves. Based on analyses of enzyme curves there are four reports on beta-blockers aimed at infarct size limitation in man. It was first reported in 1978 [24] that propranolol given within 4 h of onset of chest pain could reduce a release of CK compared to 19 control patients. Preliminary reports from a double-blind study on metoprolol in Göteborg showed that early start of treatment lowered the serum curves of LD I+II in 183 patients compared to 177 placebo-treated patients [11]. In a double-blind study from Copenhagen [13], it was reported that alprenolol given within 12 h after onset of clinical symptoms caused lower CK levels compared to controls. More recently it was reported that alprenolol given to patients within 12 h after onset of chest pain reduced infarct size, as estimated by cumulative CK-enzyme release and also by Q- and R-wave changes [36]. Thus, it seems that the institution of beta-blockade early after onset of pain can preserve the ischemic myocardium and limit infarct size also in man.

Effects on Arrhythmias and Mortality

The use of beta-blocking agents in the treatment of cardiac arrhythmias is well established and has been extensively reviewed [27]. The treatment with beta-blocking agents is also effective on arrhythmias induced by increased sympathetic activity due to, for example, psychic stress. It is therefore not surprising that this treatment is very effective on supraventricular arrhythmias in patients with acute myocardial infarction, especially in the early stage [12, 15]. In randomized

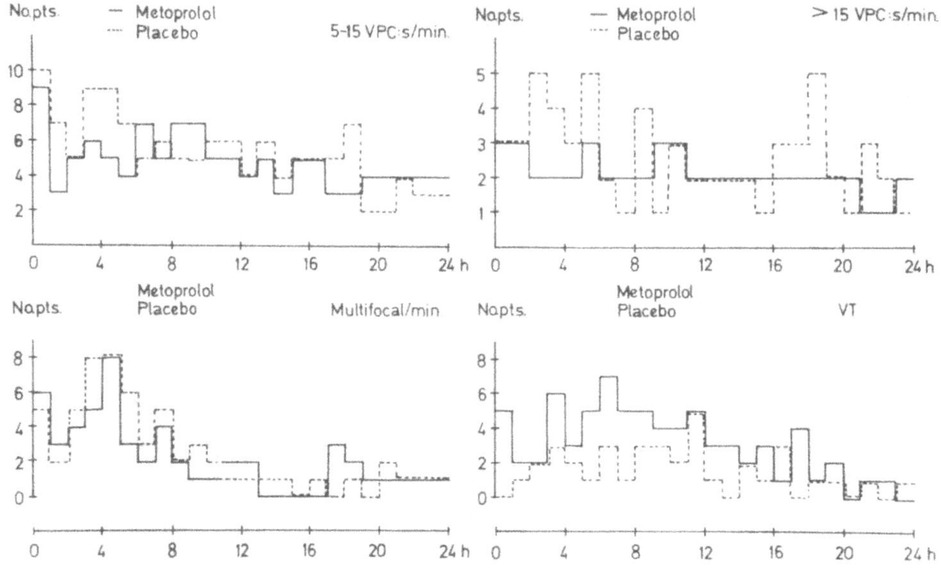

Fig. 3. Number of patients with any of the ventricular arrhythmias indicated in the figure during at least 1 min for each hour. VT means three VPC:s in a row with a rate of >100 beats/min. There are 53 and 50 patients with acute myocardial infarction in the metoprolol and the placebo groups respectively

prospective studies, beta-blocking agents given prophylactically where reported to have little or no effect on ventricular arrhythmias [3, 6, 21]. In the double-blind study on metoprolol in Göteborg, an analysis has been done on the first 103 consecutive patients with definite acute myocardial infarction in which ECG was continuously recorded on paper for 24 h. Metoprolol was given 15 mg i.v. immediately after arrival in the unit, and 15 min later 50 mg metoprolol was given orally, which was repeated every 6 h. As can be seen from Fig. 3, metoprolol had no significant effect on the various ventricular arrhythmias. All the patients were treated within 24 h after onset of pain and 75% of the patients within 8 h. Figure 4 shows that this treatment with metoprolol caused a 15% reduction in heart rate during the whole period of 24 h. This is in agreement with the experience that beta-blockade exerts a good control of heart rate in patients with sinus rhythm or supraventricular ectopic rhythms. In view of present clinical reports in patients with acute myocardial infarction beta-receptor drugs can be useful for the treatment of supraventricular

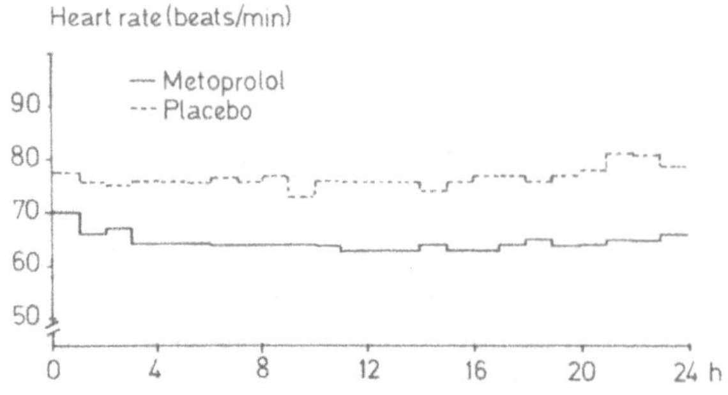

Fig. 4. Changes in heart rate in the same patients as in Fig. 3

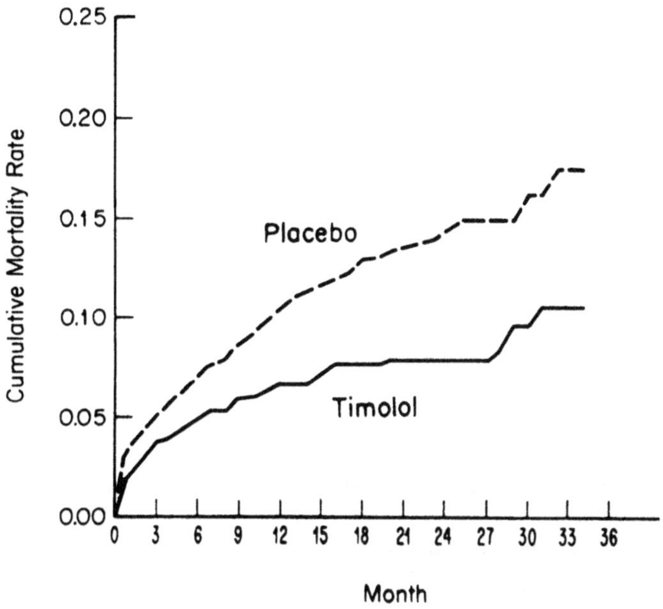

Fig. 5. Life table. Accumulated rates of deaths from all causes. [22]

arrhythmias while the effect on ventricular arrhythmias is less marked. This is in agreement with observations in animal experimental studies.

A great number of studies have been performed on beta-blockade in patients after myocardial infarction to study effects on survival [10]. Only five of the studies utilizing alprenolol, practolol and timolol have demonstrated a reduction in sudden death and/or total mortality [1, 2, 20, 22, 35]. The studies showing positive results of beta-blockade on survival are the two with the largest patient material (multicenter studies with practolol and timolol) and the studies in which a prognostic stratification of the patients into various risk groups was done (the three studies with alprenolol). The most convincing of these studies is the Norwegian study where there was a significant reduction on total mortality (Fig. 5) as well as on sudden cardiac death in all patients randomized, including patients with blind tablet withdrawals. In this study there was also a significant reduction in number of reinfarctions, supporting the idea that beta-blockade prevents and limits myocardial infarct development.

The Metoprolol Study in Göteborg

In 1975, experimental and clinical studies supported the idea that there were a number of possible beneficial effects of beta-blockade on acute myocardial infarction. The possible clinical effects are listed in Table 3. A larger study on metoprolol was therefore planned to test the hypothesis that early instituted beta-blockade also in man could reduce the incidence of infarctions, reduce infarct size and early mortality. Based on statistical estimations, it was judged that at least 800 patients with definite acute myocardial infarction had to be included in order to detect a possible effect on three-month survival. The design of the study and the main objectives are given in Table 4. The study started in August 1976 and lasted

Table 3. Possible clinical effects of β-blockade in AMI

Reduction in HR × SBP product
Reduction of ST-segment elevation
Reduction of ischemic chest pain
Reduction of ventricular arrhythmias
Reduction of infarct size
Reduction of mortality, early and late

Table 4. Metoprolol Study in Göteborg

Main objectives:	Mortality Infarct size Arrhythmias Tolerance
Design:	Double-blind placebo-metoprolol from arrival in hospital and 3 months.
Patient material:	All patients 40–74 years old with AMI or suspected AMI without contraindications.
Number of patients:	800 AMI (WHO criteria).
Metoprolol dose:	15 mg i.v. (5 mg every 2 min) and 200 mg daily for 3 months.

Table 5. Metoprolol Study in Göteborg (March 12 1981)

Preliminary results on max LD I + II.			
1359 patients studied (all with accept. curves; 97% of all)			
Placebo	678 patients	12.3 + 0.49 ukat/l	↓ 8.9%
Metoprolol	681 patients	11.2 ± 0.43 ukat/l	
923 patients treated within 12 h after onset of pain; 66%			
Placebo	469 patients	13.3 ± 0.62 ukat/l	↓ 15.0%
Metoprolol	454 patients	11.3 ± 0.54 ukat/l	

until January 1981. It included 1395 patients out of whom 809 patients developed a definite acute myocardial infarction and 162 a probable myocardial infarction whereas 424 patients did not develop a myocardial infarction. The compliance and tolerance for the treatment was found to be very good and less than 19% of the patients had to stop their blind tablet treatment during the three-month follow-up. The analysis of data are now in progress and the results on overall mortality will be reported in the early fall 1981. A preliminary analysis of the enzyme curves has shown a significant reduction of enzyme-estimated infarct size when the treatment was given within 12 h after onset of pain. As can be seen from Table 5, there were acceptable enzyme curves for LD I + II in 97% of all patients and for the whole material there was an 8.9% reduction in maximal enzyme value with the treatment of metoprolol, a reduction that did not reach statistical significance. However, in the patients given metoprolol within 12 h after onset of pain there was a 15% reduction in the maximal LD I + II ($P < 0.05$).

Possible Mechanisms for the Beneficial Effects of Beta-Blockade in Acute Myocardial Infarction

From a large number of studies, both in experimental animals and patients, it is known that beta-blocking agents can reduce the severity of myocardial ischemia. As discussed, all the beneficial effects seem to be induced by $beta_1$-blocking agents (e.g. metoprolol) as well as by other beta-blockers. Also a cardioprotective effect against death in post-infarction patients seems to be due to $beta_1$-blockade. The three beta-blocking agents with positive effect on survival, alprenolol, practolol and timolol, have only one property in common – the $beta_1$-blockade. It seems that $beta_2$-blockade, intrinsic stimulatory activity and membrane stabilizing property are of no importance for this cardioprotective effect. Possible mechanisms involved in cardioprotection are listed in Table 6.

Table 6. Possible mechanisms for beneficial effect of beta-blockade in acute myocardial infarction

1. All effects are due to $beta_1$-blockade
2. Reduced myocardial ischemia
3. Reduced heart work
4. Blockade of locally released noradrenaline
5. Direct antifibrillatory effects
6. Reduced thrombocyte aggregation
7. Unknown mechanisms

Conclusion

It can be summarized that there are a number of possible beneficial clinical effects of beta-blockade in acute myocardial infarction. All these effects seem to be due to $beta_1$-adrenoceptor blockade. It is possible to reduce heart work and thereby myocardial metabolic demand, causing an improvement of ischemic metabolism. The reduction of myocardial ischemia will cause positive electrophysiological changes, reduction of chest pain and improvement of left ventricular pump function. It seems possible to prevent and limit infarct development also in man. The results of available studies on survival in post-infarction patients suggest that beta-blockade can reduce mortality when given for 2 years after acute myocardial infarction. The results from several on-going studies will be available this year and it will be clear whether or not the suggestive beneficial effect on survival will hold true. The metoprolol study in Göteborg will give answer to the question whether it is possible to reduce also early mortality after acute myocardial infarction.

References

1. Ahlmark G, Saetre H, Korsgren M (1974) Letter: Reduction of sudden deaths after myocardial infarction. Lancet 2:1563
2. Andersen MP, Bechsgaard P, Frederiksen J et al. (1979) Effect of alprenolol on mortality among patients with definite or suspect acute myocardial infarction. Lancet 2:865–868
3. Barber JM, Boyle D McC, Chaturvedi NC, Singh N, Walsh MJ (1975) Practolol in acute myocardial infarction. Acta Med Scand [Suppl] 587:213–216
4. Bay G, Lund-Larsen P, Lorentsen E, Sivertssen E (1967) Haemodynamic effects of propranolol (Inderal) in acute myocardial infarction. Br Med J 1:141–143
5. Braunwald E, Maroko PR (1974) The reduction of infarct size – an idea whose time (for testing) has come. Circulation 50:206–209
6. Briant RB, Norris RM (1970) Alprenolol in acute myocardial infarction: double-blind trial. NZ Med J 71:135–138
7. Gibson DG (1971) Haemodynamic effects of practolol. Postgr Med J [Suppl] 47:16
8. Gibson DG, Coltart J (1972) Haemodynamic effects of practolol. Br Heart J 34:95–99
9. Hjalmarson Å (1980) Myocardial metabolic changes related to ventricular fibrillation. Cardiology 65:226–247
10. Hjalmarson Å (In press) β-blockers – current status in the prevention of sudden coronary death. In: Greenberg H, Dwyer EM Jr (eds) Conference on sudden coronary death, The New York Academy of Sciences, New York 1981. Ann NY Acad Sci
11. Hjalmarson Å, Ariniego R, Herlitz J et al. (1979) Limitation of infarct size in man by the $beta_1$-blocker metoprolol (Abstr). Circulation 60 [Suppl II]: 164
12. Jewitt D, Croxson R (1971) Practolol in the management of cardiac dysrhythmias following myocardial infarction and cardiac surgery. Postgr Med J 47 [Suppl Jan]:25–29
13. Jürgensen HJ, Frederiksen J, Hansen DA, Pedersen-Bjergaard O (1979) Effekten af alprenolol på myokardeinfarktstørrelsen bedømt ud fra seriebestemmelser af serumkreatinkinase. Ugeskr Laeg 141:3089–3091
14. Kjekshus JK, Maroko PR, Sobel BE (1972) Distribution of myocardial injury and its relation to epicardial ST-segment changes after coronary artery occlusion in the dog. Cardiovasc Res 6:490–499
15. Lemberg L, Arcebal AG, Castellanos JRA, Slavin D (1972) Use of alprenolol in acute cardiac arrhythmias. Am J Cardiol 30:77–81
16. Málek I, Waagstein F, Hjalmarson Å, Holmberg S, Swedberg K (1978) Hemodynamic effects of the cardioselective beta-blocking agent metoprolol in acute myocardial infarction. A 24-hour catheterization study. Acta Med Scand 204:195–201
17. Maroko PR, Kjekshus JK, Sobel BE, Watanabe T, Covell JW, Ross J Jr, Braunwald E (1971) Factors influencing infarct size following experimental coronary artery occlusion. Circulation 43:67
18. Mueller HS, Ayres SM (1980) Propranolol decreases sympathetic nervous activity reflected by plasma catecholamines during evolution of myocardial infarction in man. J Clin Invest 65:338–346
19. Mueller HS, Ayres SM, Religa A, Evans RG (1974) Propranolol in the treatment of acute myocardial infarction. Effect on myocardial oxygenation and hemodynamics. Circulation 49:1078–1087
20. Multicentre International Study (1975) Improvement in prognosis of myocardial infarction by long-term beta-adrenoceptor blockade using practolol. Br Med J 3:735–740
21. Norris RM, Caughey DE, Scott PJ (1968) Trial of propranolol in acute myocardial infarction. Br Med J 2:398–400
22. Norwegian Multicenter Study Group (1981) Timolol-induced reduction in mortality and reinfarction in patients surviving acute myocardial infarction. N Engl J Med 304:801–807
23. Pelides LJ, Reid DS, Thomas M, Shillingford JP (1972) Inhibition by beta-blockade of the ST segment elevation after acute myocardial infarction in man. Cardiovasc Res 6: 295–301
24. Peter T, Norris RM, Clarke ED et al. (1978) Reduction of enzyme levels by propranolol after acute myocardial infarction. Circulation 57:1091–1095
25. Reimer KA, Lowe JE, Rasmussen MM, Jennings RB (1977) The wavefront phenomenon of ischemic cell death. 1. Myocardial infarct size vs. duration of coronary occlusion in dogs. Circulation 56:786–794

26. Reiz S, Hjalmarson Å (in press) Hemodynamic effects of metoprolol and nitroprusside in patients with acute myocardial infarction and heart failure. Acta Med Scand
27. Singh BN, Jewitt DE (1974) Beta-adrenergic receptor blocking drugs in cardiac arrhythmias. Drugs 7:426–461
28. Valori C, Thomas M, Shillingford J (1967) Free noradrenaline and adrenaline excretion in relation to clinical syndromes following myocardial infarction. Am J Cardiol 20:605–617
29. Videbaek J, Christensen NJ, Sterndorff B (1972) Serial determination of plasma catecholamines in myocardial infarction. Circulation 46:846–855
30. Waagstein F, Hjalmarson ÅC (1975a) Effect of cardioselective beta-blockade on heart function and chest pain in acute myocardial infarction. Acta Med Scand [Suppl] 587: 193–200
31. Waagstein F, Hjalmarson ÅC (1975b) Double-blind study of the effect of cardioselective beta-blockade on chest pain in acute myocardial infarction. Acta Med Scand [Suppl] 587: 201–206
32. Waagstein F, Hjalmarson Å, Málek I (1978) Effect of β-blockers in reducing infarct size. In: Hjalmarson Å, Wilhelmsen L (eds) Acute and long-term management of myocardial ischaemia. AB Hässle, Mölndal, Sweden, pp 346–357
33. Waldenström AP, Hjalmarson ÅC (1977) Factors modifying ischemic injury in the isolated rat heart. Acta Med Scand 201:533–537
34. Waldenström A, Hjalmarson Å, Thornell L (1978) Factors of importance for infarct size in animals – Findings of the Göteborg Group. In: Hjalmarson Å, Wilhelmsen L (eds) Acute and long-term management of myocardial ischaemia. AB Hässle, Mölndal, Sweden, pp 221–228
35. Wilhelmsson C, Vedin JA, Wilhelmsen L, Tibblin G, Werkö L (1974) Reduction of sudden deaths after myocardial infarction by treatment with alprenolol. Lancet 2: 1157–1159
36. Yusuf S, Ramsdale D, Peto R et al. (1980) Early intravenous treatment in suspected acute myocardial infarction: preliminary report of a randomized trial. Lancet 2:273–276

Discussion

Mattila, Helsinki: The changes of enzyme rises associated with infarction were not much lower after metoprolol than after placebo. What is the correlation between infarct size and enzyme rise? It it linear or logarithmic? If it is logarithmic, your finding is perhaps not significant, and you may rather explain the protective drug effect with an antiarrhythmic action.

Hjalmarson: The problem is that we do not have good clinical methods for measuring infarct size. We use release of various enzymes and Q-R-changes recorded from precordial electrodes [24]. A 15% reduction in peak enzyme value was found with metoprolol compared to placebo. Our findings are in agreement with the observation in the Norwegian timolol study, which showed a markedly lower number of reinfarctions in the timolol group during 12–33 months of treatment. Dose-response studies need to be done before the question concerning the type of correlation can be answered.

Sleight, Oxford: In relation to the last question I would like to support your suggestion that β-blockade reduces infarct size. Dr. Yusuf in Oxford has shown a reduction in infarct size (either by ECG or enzyme methods) by about 30%. In this trial of 400 patients randomised to atenolol or not, 75% were treated within 6 h of the onset of pain.

However, what I really wished to comment upon was the incidence of arrhythmias. Dr. Rossi in my department found no reduction of total ectopics but further analysis showed a highly significant reduction in R on T ectopics, and a balancing increase in late ("escape") ectopic peaks in the atenolol group. This reduction in R on T ectopics might be more relevant to ventricular fibrillation.

Hjalmarson: I agree with you. The possibility of an effect of metoprolol on the ventricular ectopic beats occurring early in the cycle has not been evaluated. The fact that metoprolol reduced heart rate by 15% will unmask some ventricular ectopic activity that will occur late in the cycle. We will study this more in detail in our complete analysis of 200 patients with acute myocardial infarction.

Kübler, Heidelberg: In discussing this excellent study we probably have to distinguish between 3 different mechanisms of action.
1. Reduction in sudden death in the postinfarction period, which is evidenced in all studies and most likely due to an antiarrhythmic effect.
2. Prevention of new infarcts or reinfarctions. From invasive studies in patients with acute myocardial infarction we learnt that about 70%–75% of the acutely occluded vessels can be reopened by local application of streptokinase. Hence, thrombotic occlusion of a stenotic vessel seems to be the most likely cause.
3. Reduction in infarct size. If this reduction is calculated to be 15%–30% from plasma enzyme levels, a border zone of jeopardized myocardium of at least the same size has to be discussed. Since a border zone of this size has not yet been proven by direct biochemical techniques, impaired wash-out of enzymes due to reduced perfusion after β-blockade may be considered as an alternative explanation.

Hjalmarson: It is possible that prevention of development of severe myocardial ischemia and blockade of locally released noradrenaline can explain the effects of β-blockade on infarct size, reinfarction and ventricular fibrillation by a common mechanism. The "border zone" can be looked upon in different ways. We must assume that there is a dynamic infarction starting from no infarction and ending up in a very large infarction. Optimal preventive treatment could cause a salvage of, e.g., the whole anterior wall (= "border zone").

Sleight, Oxford: Prof. Kübler states that a 30% reduction of infarct size seems very unlikely (i.e. too large) in view of all the work on coronary ligation and infarct size with and without drug intervention in the experimental animal. One possible difference between man and experimental animals is the presence or absence of collateral channels. If these were substantial, we might have not a "border zone" at the edge of an infarct, but rather a reduced blood supply from collateral channels. These channels may be either constricted by catecholamines, or may be insufficiently large to supply myocardium stimulated by catecholamines. Under these circumstances β-blockade in man may have a much greater effect than in animal models. A better model might be Prof. Kübler's work with hypoxic perfusion: under these conditions cell death is a much slower process, and would allow a few hours when early post-occlusion interventions might be feasible.

Trap-Jensen, Copenhagen: Could you please comment on the possible effects of β-blockers with intrinsic stimulatory activity (ISA) in preventing sudden death after the first acute myocardial infarction. Until the recent study with timolol in Oslo only β-blockers-β_1-selective (with ISA) or non-selective – have been shown to reduce sudden death (practolol and alprenolol).

Hjalmarson: I am not convinced that ISA is of importance in clinical practice. Anyhow, it is not essential for the prevention of cardiac death, since timolol was found to do so and this β-blocker lacks ISA. I have not found any clear difference between practolol (with ISA) and metoprolol (without ISA) in the treatment of 1000 patients with acute myocardial infarction.

The Role of Beta-Adrenoceptor Blockade in Congestive Cardiomyopathy

F. Waagstein, K. Swedberg, I. Wallentin, U. Baandrup, Å. Hjalmarson

Summary

Prognosis in congestive cardiomyopathy (COCM) with marked dilatation and heart failure is poor. A low rate of spontaneous recovery was observed. A high level of circulating catecholamines is often found, which may explain the common finding of resting tachycardia. An early series of COCM patients with tachycardia responded surprisingly well when chronic beta-blockade was added to the conventional treatment of failure. Later, even patients without tachycardia were included with a period of placebo preceding beta-blocker treatment in order to exclude patients with spontaneous recovery. Acute beta-blockade was surprisingly well tolerated, possibly because the reduction in contractility was compensated for by an increase in myocardial compliance. After long-term beta-blockade, improvement was observed after 6 months and further improvement was seen after 24 months. In patients improving functionally, there was reduction in heart size and left ventricular end-diastolic diameter and increase in working performance. For the whole group, there was a decrease in rapid filling wave in apex cardiogram, reduction in third heart sound and increase in ejection fraction. Changes in the reverse way were seen in these three variables after beta-blocker withdrawal and relapse into heart failure occurred in 6/15 patients. Changes in diastolic function may be an early important effect of beta-blockade in COCM. Increase in survival was seen when beta-blocker-treated patients with COCM were compared to well-matched COCM controls.

It is suggested that increased sensitivity to sympathetic stimulation contributes to the pathogenesis in some patients with COCM and that improvement in myocardial function explains prolonged survival.

Congestive cardiomyopathy (COCM) with dilatation and clinical heart failure has an extremely poor prognosis with one-year mortality exceeding 50% [5]. Increased heart size on X-ray, low cardiac index, high filling pressure and old age are strong indicators of a poor prognosis [2]. Digitalis, diuretics and especially vasodilators may sometimes cause a marked clinical improvement. However, these drugs have not yet been proven to prolong life in patients with COCM.

It has been shown that there is a high level of circulating catecholamines in COCM patients in functional group[1] (FG) IV which is increased further during exercise [1, 6]. This may cause the often observed finding of resting tachycardia in these patients. In 1975 we reported [10] that chronic beta-blockade in COCM patients produced a dramatic clinical improvement within 2–6 months confirmed by increased working performance and ejection fraction, absence of third heart sound and pathological pulse curves, indicating improved systolic and diastolic function.

1 Functional classification was based on N.Y.H.A. criteria

Patients, Materials and Methods

In a prospective study, 28 patients in FG II, 19 in FG III and 8 in FG IV were investigated before and after beta-blocker treatment. Basically, all patients were on digitalis and varying doses of diuretics. In 20 of the total 28 patients, a period with placebo treatment was planned before beta-blocker administration and this was done in 16 of the 20 patients. In 4 patients in FG IV, the clinical condition was so poor that it was considered unethical not to give beta-blockade immediately.

Three different beta-blockers were used, alprenolol in 9 patients, practolol in 7 and metoprolol in 12 patients. Some patients were switched over from practolol to alprenolol or metoprolol after withdrawal of practolol from the market. In a few patients, alprenolol was substituted by metoprolol and vice versa due to drug-specific side-effects. Titration of the drug was performed by peroral administration of beta-blockers in doses equivalent to metoprolol 12.5 mg twice a day. The final

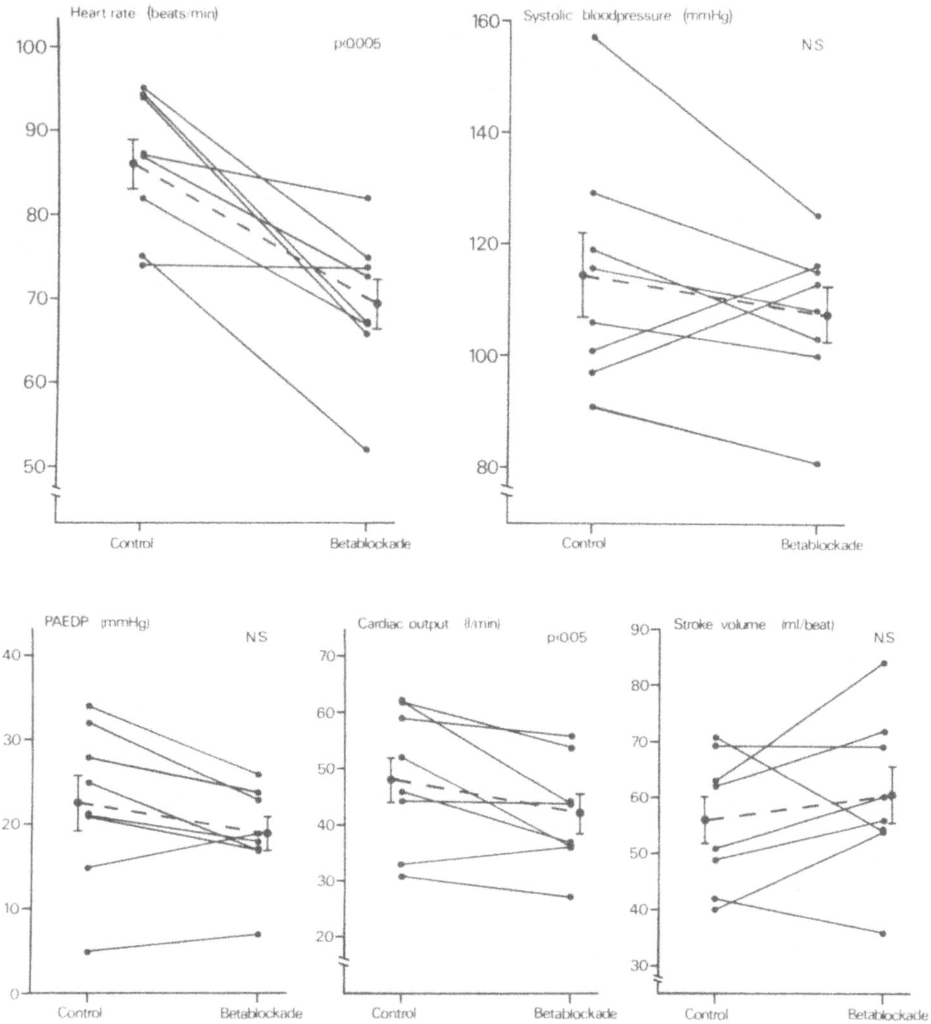

Fig. 1 a, b. Hemodynamic effects of metoprolol in congestive cardiomyopathy at rest ($n=8$)

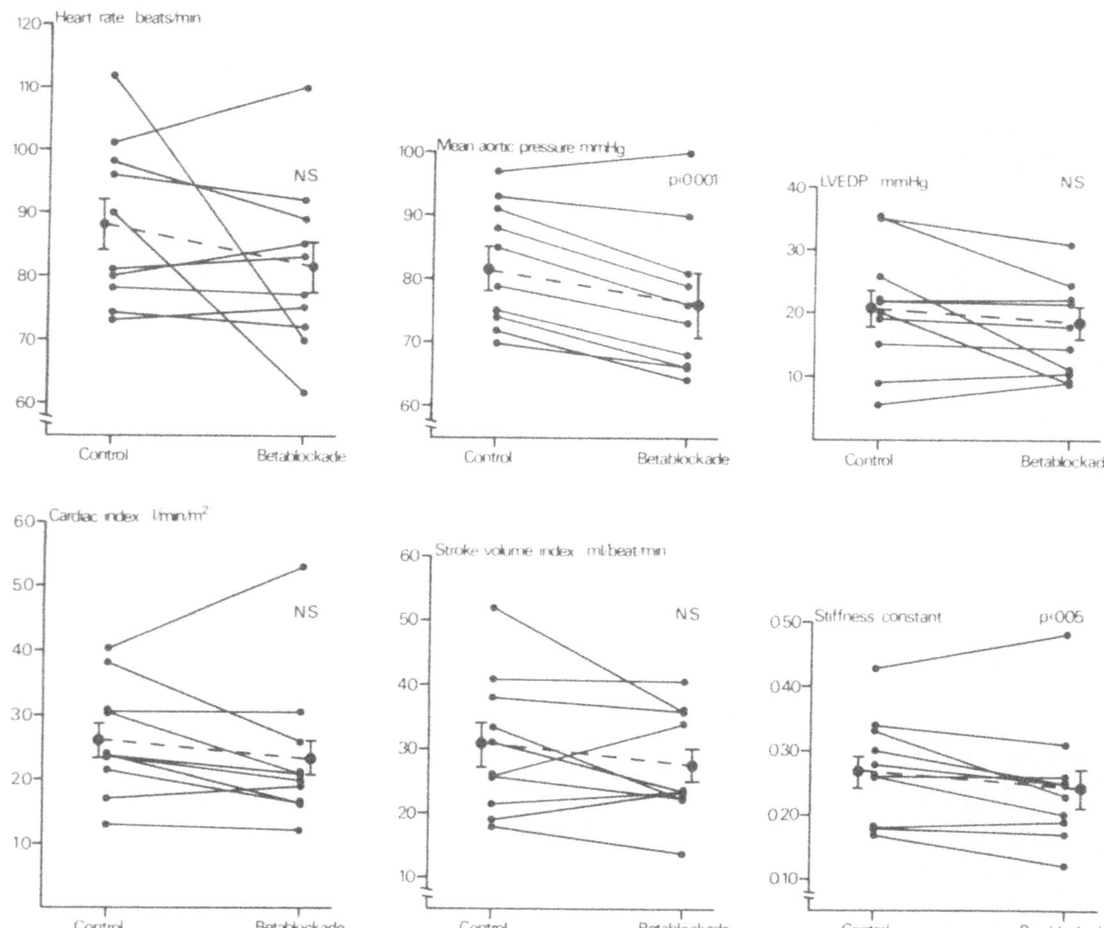

Fig. 2. Changes in heart rate (*HR*), mean aortic pressure (*MAP*), left ventricular end diastolic pressure (*LVEDP*), cardiac index (*CI*), stroke volume index (*SVI*) and stiffness constant after acebutulol 25 mg i.v. in 10 patients with congestive cardiomyopathy. (Based on data from *Ikram* et al. 1979)

dose was titrated in order to normalize heart rate and varied with a beta-blocker equivalent to metoprolol from 25 mg×2 to 100 mg×2 in different patients. In patients belonging to FG IV, the dosage was increased with caution and in some cases the final dose was achieved after as long as one month, whereas patients in FG III reached the final dose after 1–2 weeks only. In 3 patients beta-blockade was not tolerated and had to be stopped after a short trial period of up to 2 weeks.

Acute Hemodynamic Effects of Beta-Blockade in COCM

In a subgroup of patients, titration was performed under strict hemodynamic control and heart rate was found to decrease significantly. There was a slight fall in systolic blood pressure but no change in stroke volume or filling pressure after a dose of 15 mg of metoprolol given intravenously [6]. A marked rise in the level of circulating noradrenaline was observed after acute beta-blockade. This was considered to be a competitive release due to the cardiodepressive effect of beta-blockade [6]. In a later series of investigations these results were confirmed (Fig. 1 a, b). The explanation of the unexpected good tolerance to beta-blockade in these seriously ill patients might be an acute improvement of the diastolic properties of the myocardium (decreased stiffness) (Fig. 2) [3]. The acute effect of beta-blockade was found to decrease the myocardial oxygen consumption (MV_{O_2}) (Fig. 3) in proportion to the reduction in rate-ressure product. There was no change in lactate extraction ratio, indicating increased anerobic glycolysis. The oxygen-sparing effect of beta-blockade is mainly due to a reduction in heart rate since an increase in MV_{O_2} was noted also in patients paced to the same heart rate after beta-blockade.

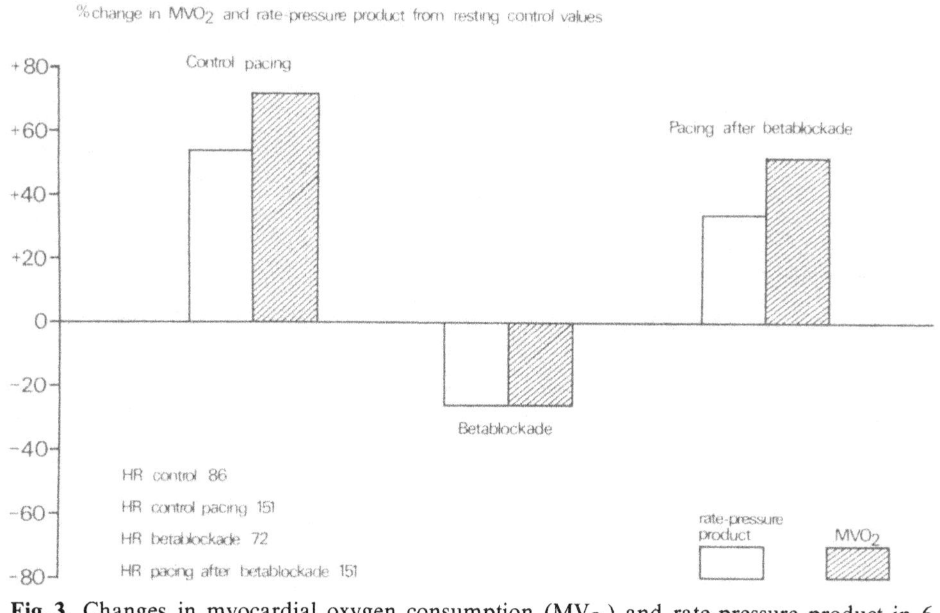

Fig. 3. Changes in myocardial oxygen consumption (MV_{O_2}) and rate-pressure product in 6 patients with congestive cardiomyopathy during pacing at rest after beta-blockade and during pacing after beta-blockade. HR = heart rate

Noradrenaline was measured in arterial and sinus coronary blood at rest and during cardiac pacing before and after 15 mg metoprolol. Increased production of noradrenaline was found in all patients except after beta-blockade plus pacing. Arterial as well as sinus coronary concentrations were then markedly increased parallel to an increase in adrenaline production, speaking in favor of increased release from the suprarenals.

Effect of Chronic Treatment with Beta-Blockade in COCM

Follow-up of the 28 patients with COCM on chronic beta-blockade over a period of 24 months revealed that functional state was improved after 6 months and continued to improve further during the next 18 months (Figs. 4, 5). There were three early deaths in progressive heart failure. The progression of heart failure continued despite withdrawal of beta-blockade. In the remaining patients, death occurred suddenly in 3 patients during a 2-year follow-up and another 4 patients died suddenly during a follow-up lasting up to 5½ years. All patients who died suddenly, except one, were in FG III and none of those who had improved to FG II (except one) or FG I died suddenly. This may indicate that beta-blockade, indirectly, by improving myocardial function protects patients against sudden death. However, in those with no or minimal improvement additional treatment with a potent antiarrhythmic drug may be necessary.

The whole group showed improvement of ejection fraction, as demonstrated in Table 1, reflecting an improved systolic function. Reduction and disappearance of a calibrated – third heart sound as well as reduction in the ratio of rapid filling wave to the total reflection of the apex curve indicated improvement in the early diastolic function which might correspond to a fall in left ventricular filling pressure observed in a few patients undergoing recatheterization [8]. For the whole group,

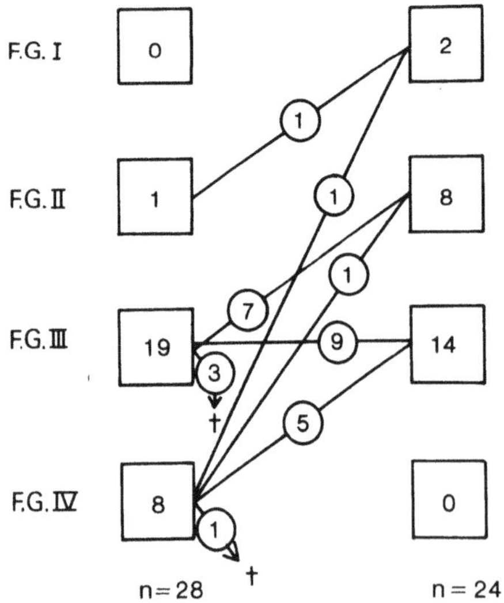

Fig. 4. Effect on functional group in congestive cardiomyopathy after 6 months on beta-blockade

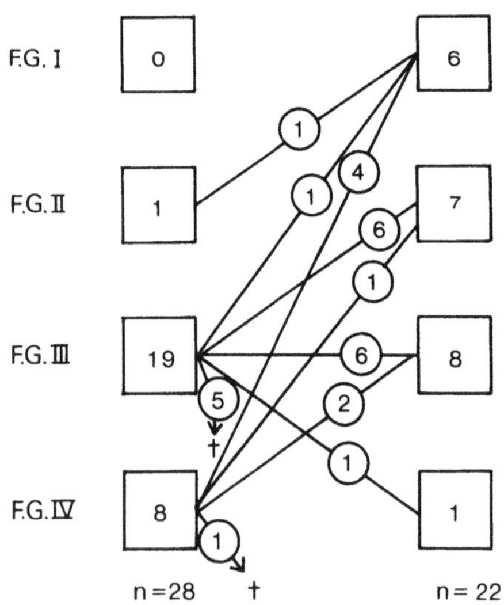

Fig. 5. Effect on functional group in congestive cardiomyopathy after 24 months on beta-blockade

Table 1. Effects of chronic beta-blockade in COCM

	Number of patients	Control	6 months after beta-blockade	P value
Ejection fraction	18	0.32	0.43	$P < 0.001$
Third heart sound (mm)	22	10	2.2	$P < 0.001$
Rapid filling wave (%)	23	20.9	10.5	$P < 0.001$
Left ventricular ejection time (%)	23	80	83	$P < 0.05$
Left ventricular enddiastolic diameter (mm)	22	73.1	73.6	NS
Heart size (ml/m²/BSA)	21	797	770	NS
Working performance (Watt)	19	92.4	100.8	NS

there was also improvement in left ventricular ejection time, probably reflecting an increase in stroke volume. There was no difference in left ventricular end-diastolic diameter and no significant difference in heart volume or exercise performance, which may reflect the fact that the treated population consisted of patients who responded to beta-blockade as well as those who did not respond to chronic beta-blockade. By subgrouping the material into patients who responded and those who did not, the identification of a group who decreased their left ventricular end diastolic diameter and heart size and increased working capacity were identified [8].

When 24 patients from the beta-blocker group were compared to 13 retrospective controls [7], these two groups were found to match well with regard to age, functional classification and noninvasive variables describing the myocardial function. Significant differences in survival rate after 1, 2 and 3 years were observed

Table 2. Survival in COCM patients with and without chronic beta-blockade

	No. of patients	Survival (%)		
		1-year	2-year	3-year
Chronic beta-blockade treated patients	24	83	66	$P < 0.01$
Retrospective controls	13	46	19	

Table 3. Effect of chronic beta-blockade in COCM. Male 58 years. Anamnesis of COCM: 3 years

	Before beta-blockade	8 months after beta-blockade
Functional group (N.Y.H.A)	III B	II
Left ventricular end diastolic diameter (cm)	9.3	9.1
Left atrium (cm)	6.4	5.9
Left ventricular end diastolic pressure (mmHg)	20	6
Ejection fraction	0.38	0.45
Fiber diameter (μ)	18.3	14.8
Nucleus diameter (μ)	5.5	6.2
Ratio fiber diam./nucleus diam.	3.3	2.5
Collagen volume fraction (%)	2.8	2.9
Smooth muscle in endocardium	+	–

(Table 2). Analyzing the noninvasive data, it was found that the one-year survival was significantly increased in the beta-blocker treated patients with a left ventricular end-diastolic diameter of more than 7 cm [7]. Even in patients with consistently increased left ventricular end-diastolic diameter significant improvement may be obtained after beta-blockade. Table 3 illustrates a case with a moderate increase in ejection fraction and improvement in diastolic function, as reflected by a fall in filling pressure, disappearance of third heart sound and a marked reduction in rapid filling wave in the apex curve. This may reflect an adaptation of the heart to marked dilatation, less hypertrophy of the fibers, as well as smooth muscle in the endocardium and might be the consequence of chronic beta-blockade [4]. Therefore, lack of decrease of heart size, as measured by echocardiogram or X-ray, does not necessarily mean lack of functional myocardial improvement since left ventricular systolic diameter may have shortened. In summary, early effects of beta-blockade seem mainly to affect the diastolic function which may explain the early improvement after start of beta-blockade with less dyspnea observed in some patients. In many patients even systolic function improved after a longer period of treatment with beta-blockers (Table 4).

Table 4. Effects of beta-blockade in COCM

	Acute effects	Long-term effects
Ejection fraction	↓	↑
Filling pressure	–	↓
Third heart sound	–	↓
Heart size	– (↑)	↓
Working capacity	– (↓) (↑)	↑
Survival	–	↑ ?

Withdrawal of Beta-Blockade

When beta-blocker treatment was withdrawn in 15 patients on chronic beta-blockade, clinical deterioration was observed after 7–59 days (mean 23 days), whereas the clinical condition remained unchanged in 9 patients. Among these, however, increase in third heart sound and left filling wave ratio was seen in three patients and a fall in ejection fraction in 6 of 8 investigated patients [9].

Conclusion

Our data strongly support the hypothesis that sympathetic activity in some way plays a role in the development of congestive cardiomyopathy and that beta-blockade may be beneficial in the management of selected cases of congestive cardiomyopathy.

Furthermore, these data indicate that the observed improvement after beta-blockade in the majority of patients did not occur by chance but as a consequence of chronic beta-blockade.

References

1. Chidsey CA, Braunwald E, Morrow AG (1965) Catecholamine excretion and cardiac stores of norepinephrine in congestive heart failure. Am J Med 39:442–451
2. Fuster V, Gersh BJ, Giuliani ER, Tajik AJ, Brandenburg RO, Frye RL (1981) The natural history of idiopathic dilated cardiomyopathy. Am J Cardiol 47:525–531
3. Ikram H, Chan W, Bennett SI, Bones PJ (1979) Haemodynamic effects of acute beta-adrenergic receptor blockade in congestive cardiomyopathy. Br Heart J 42:311–315
4. Källfelt BJ, Waldenström AP, Hjalmarson ÅC (1977) Effects of adrenaline *in vivo* on protein synthesis, mechanical performance and sensitivity to ischemia in the perfused rat heart. J Mol Cell Cardiol 9:383–398
5. Kuhn H, Breithardt G, Knieriem H-J, Köhler E, Lösse B, Seipel L, Loogen F (1978) Prognosis and possible presymptomatic manifestations of congestive cardiomyopathy (COCM). Postgrad Med J 54:451–459
6. Swedberg K, Hjalmarson Å, Holmberg S (1979a) Effects of work and acute beta-receptor blockade on myocardial noradrenaline release in congestive cardiomyopathy. Clin Cardiol 2:424–430
7. Swedberg K, Hjalmarson Å, Waagstein F, Wallentin I (1979b) Preliminary communication: Prolongation of survival in congestcve cardiomyopathy by beta-receptor blockade. Lancet 1:1374–1376
8. Swedberg K, Hjalmarson Å, Waagstein F, Wallentin I (1980a) Beneficial effects of long-term beta-blockade in congestive cardiomyopathy. Br Heart J 44:117–133
9. Swedberg K, Hjalmarson Å, Waagstein F, Wallentin I (1980b) Adverse effects of beta-blockade withdrawal in patients with congestive cardiomyopathy. Br Heart J 44:134–142
10. Waagstein F, Hjalmarson Å, Varnauskas E, Wallentin I (1975) Effect of chronic beta-adrenergic receptor blockade in congestive cardiomyopathy. Br Heart J 37:1022–1036

Discussion

Henderson, Cardiff: I was interested in your comment that the improvement may be attributable at least in part to an increase in diastolic compliance. Clinical measurements of so-called complicance are, however, based on different points in time during diastole as well as different points of the pressure-volume relationship. We must bear in mind then that these changes might reflect earlier systolic relaxation rather than a true increase in resting diastolic compliance.

Waagstein: I agree with your comment. The consequence may be, however, that an increase in filling pressure and wall tension is avoided despite the negative inotropic effect of β-blockers.

Fox, London: Why were the patients in your β-blocker group who died, not on beta blockers?

Waagstein: Those who died of progressive heart failure early after the onset of β-blockade had been withdrawn from β-blocker treatment and continued to deteriorate. Those who died later were all on chronic treatment and all succumbed to sudden death.

Opie, Cape Town: Could you explain in greater detail the apparently improved survival in β-blocked patients? Is it a real phenomenon?

Waagstein: I must emphasize that we have compared one group studied retrospectively, a so-called historical control, with a group studied prospectively. This is always a problem. The two groups were, however, well matched with regard to all the studied variables.
The difference in survival observed between the two groups may therefore be a true phenomenon, but has of course to be confirmed in a controlled retrospective study.

Kuhn, Düsseldorf: The mortality of your patients in the β-blocker treated group was 50% 3 years after the onset of β-blocker therapy. That is the same mortality rate observed in our patients, who were not treated with β-blockers. Our study is based on 258 patients.

Waagstein: I am familiar with your results. However, in your paper in the Postgraduate Medical Journal you have a graph showing survival data in a subgroup with an ejection fraction of about 0.3. In this group you have a similar survival time as in our control group. The mean ejection fraction in both our control and treatment groups was also on the average 0.3. Therefore, our data may well be in accordance with yours.

Simon, Bonn: Was β-blockade in patients with congestive heart failure and a high heart rate the only form of treatment?

Waagstein: No. All patients had conventional treatment with digitalis and diuretics before entering the study.

Comparison of Withdrawal Phenomena After Propranolol, Metoprolol and Pindolol

Robert E. Rangno, Serge Langlois [1]

Summary

Three groups of hypertensive patients were studied after they had received one of three pharmacologically different beta blockers for at least one month: PIND 10 mg bid ($n=10$), PROP 80 mg tid ($n=9$), MET 150 mg bid ($n=8$). After abrupt withdrawal of drug and replacement with placebo we measured the following on days zero (0) and about every 2nd day for up to 3 weeks: beta adrenergic sensitivity (βAS) by the chronotropic dose of isoproterenol required to increase heart rate (HR) by 25 beats/min (CD_{25}); resting HR, exercise HR (PIND only), resting blood pressure (BP), and symptoms. Values are medians. On day 0, beta blockade was evident from increased CD_{25} values of 181 µg for PIND, 57 µg for PROP and 8.7 µg for MET, compared to 3,0 µg, 2,4 µg and 2,6 µg, respectively, at days 14–20. After PIND day 0, the CD_{25} slowly decreased to baseline by day 10 to 20. In contrast, after PROP day 0, the CD_{25} decreased significantly 50% below baseline from day 4 to 14 (i.e. ↑ βAS) and after MET 52% below baseline from day 2 to 8. After PIND all patients had a gradual increase of BP to baseline. A significant rebound of BP occurred in 6/9 patients after PROP and 1/8 after MET all during the period of ↑ βAS (initial = 135/85, peak = 154/98, baseline = 139/86). After PIND a significant rebound of resting HR occurred in 9/10 patients (initial = 79, peak = 88, baseline = 75) and in exercise HR in 10/10 patients. After MET a rebound of resting HR occurred in 8/8 patients during the period of ↑ βAS (initial = 61, peak = 86, baseline = 75). After PROP the resting HR rapidly returned to baseline with a rebound of 10 beats/min in 4/9 patients. We also developed a gradual withdrawal schedule for PROP and MET aimed at preventing the changes described. Studies as described were performed on the last day of full dose, serially during gradual withdrawal and serially during placebo. Gradual withdrawal of PROP and MET either prevented or markedly diminished the changes in CD_{25}, HR, BP and symptoms observed after abrupt withdrawal. Conclusions: Withdrawal of PIND, PROP and MET produced differing evidence of increased cardiac responsiveness which may reflect their different effects on cardiac receptors influenced primarily by neuronal versus circulating catecholamine. The events after abrupt withdrawal of PROP and MET were prevented by gradual dose reduction.

Introduction

A withdrawal syndrome after abrupt cessation of the beta adrenergic blocker, propranolol, is gaining progressive recognition [9, 24]. It has been observed mainly in patients with coronary artery disease, manifesting as worsening of angina, myocardial infarction, atrioventricular arrhythmias and sudden death [15, 16]. In hypertensive patients, abrupt withdrawal of beta blocking drugs has resulted in

1 Dr. Langlois is a recipient of a Canadian Heart Foundation Fellowship

symptoms of malaise, headache, tremor, sweating, anxiety and palpitations [13, 20]. These initial reports led to several prospective studies which attempted to determine the incidence, time course and mechanism of the phenomena. The most viable theory, tested by our group and others [4, 17] is an adaptive increase in cardiac beta adrenergic sensitivity perhaps as a result of increased receptor number [1, 10]. The majority of case reports and formal studies are relative to withdrawal from propranolol, but other beta blockers may also be culpable. Indeed a few reports suggest withdrawal features after metoprolol [13], practolol [19], oxprenolol [19], and atenolol [12]. It is unknown whether any difference in frequency or severity of the withdrawal phenomena is due to limited experience with other beta blockers or to real differences in their pharmacology. We have completed studies in three groups of hypertensive patients and have shown differences after withdrawal from propranolol (non-selective), metoprolol (selective) and pindolol (non-selective with intrinsic sympathomimetic activity). This presentation is limited to a comparison of four measurements made in those three studies, namely cardiac beta adrenergic sensitivity to isoproterenol, resting blood pressure, resting heart rate and symptoms. Gradual dose reduction schedules for propranolol and metoprolol were developed to prevent withdrawal manifestations.

Methods

Subjects

Ethical considerations precluded studies in patients with angina, therefore the study population consisted of patients treated for mild to moderate essential hypertension, who were free of historical or ECG evidence of coronary artery disease. Table 1 shows the patient characteristics and drug doses before abrupt withdrawal. Some patients participated in more than one study. All patients were on therapy for at least one month before withdrawal. Seven patients receiving propranolol continued to take a thiazide diuretic throughout the study, while others received no other antihypertensive medication.

Design – Abrupt Withdrawal

Patients were studied initially, on day 0, 12 h after their last full dose of beta-blocker, and serially thereafter at the same time of the day at approximately two

Table 1. Patient characteristics and dose of beta-blocker

	Number of Patients	Age		Dose [a] (mg/day)	Post-withdrawal Placebo
		Median	Range		
Propranolol	9	48	(26–79)	240	No
Metoprolol	8	52	(42–82)	300	Yes
Pindolol	10	51	(28–64)	20	Yes

[a] Daily doses were divided into TID for propranolol and BID for metoprolol and pindolol.

day intervals for 2–3 weeks. Patients refrained from tobacco and caffeine beverages for at least 12 h before each study.

Design – Gradual Withdrawal

After completion of the abrupt withdrawal studies some patients again received chronic full dose propranolol or metoprolol as described before abrupt withdrawal. After 1 month of full dose propranolol 3 patients had a stepwise 50% dose reduction at 3 day intervals and 6 patients received a prolonged low dose of 10 mg T.I.D. for 14 days. After one month of full dose metoprolol 7 patients received a prolonged low dose of 25 mg b.i.d. for 10 days. Placebo was substituted after propranolol and metoprolol were completely discontinued. Studies were performed on the last day of full dose and serially for 10–14 days during dose reduction and an additional 10–14 days during placebo.

Measurements

On each study day, the same procedure was performed. A brief history was taken using a checklist to elicit complaints possibly related to withdrawal. A 19-guage butterfly needle was inserted into an antecubital vein. After 30 min of bed rest in a quiet room, blood pressure was measured three times by sphygmomanometer. The heart rate changes produced by small incremental doses of isoproterenol were measured to obtain an index of cardiac beta-adrenergic sensitivity, according to the method of *Cleaveland, Rangno,* and *Shand* [6]. Before each dose of isoproterenol, baseline heart rate was obtained by measuring the two shortest RR interval on a resting ECG strip. Isoproterenol 200 μg/ml was used, with 20 μg/ml and 2 μg/ml solutions obtained by serial 1:10 dilutions. Each bolus dose of isoproterenol was administered by a rapid intravenous push and followed by a rapid infusion of 20–30 ml of 5% dextrose solution. The logarithm of each isoproterenol dose was plotted aginst the resulting change in heart rate, to obtain a log dose-response curve. A minimum of 4 points was obtained and plotted by linear leastsquares regression. The isoproterenol dose required to produce an increase in heart rate of 25 beats/min was obtained from this line of best fit. This dose will be referred to as the isoproterenol CD_{25} (chronotropic dose 25). The median CD_{25} on days 14 to 20 after withdrawal were considered the "baseline CD_{25}" reflecting beta-adrenergic sensitivity of beta-blocker. A decrease of CD_{25} below baseline reflects a withdrawal induced increase in cardiac beta adrenergic sensitivity. The medians of heart rate and blood pressure on days 14–20 were termed "baseline heart rate" and "baseline blood pressure".

Analysis of Data

All data for the groups is expressed as median and range. Changes in measurements over time were analyzed by the Friedman non-parametric analysis of variance with multiple comparison and one tailed probability [11]. Statistical analysis was performed within each of the three drug groups but not between the groups.

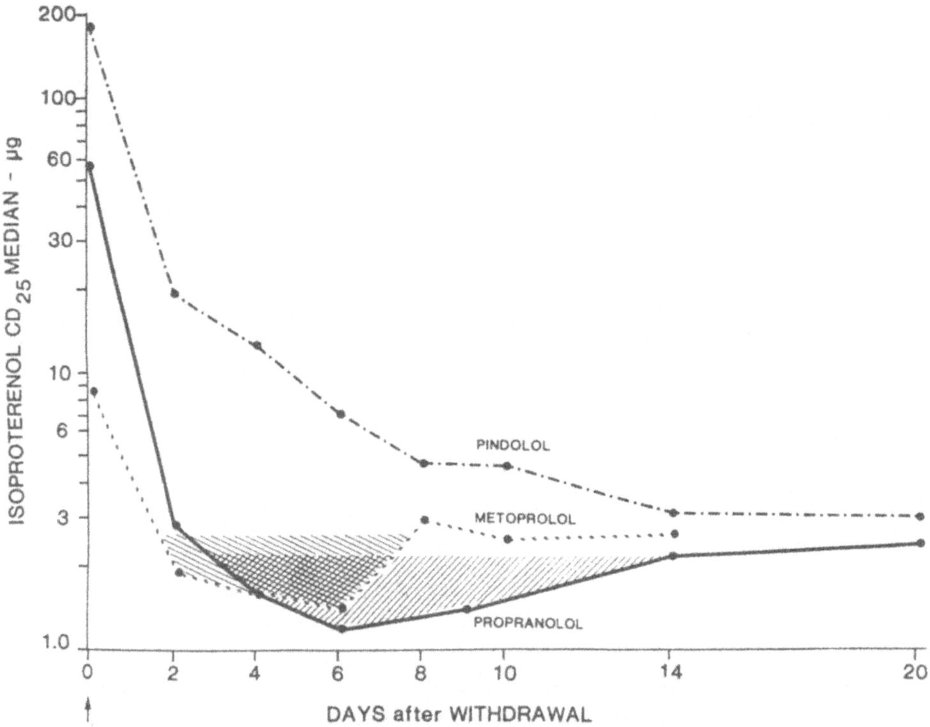

Fig. 1. Median isoproterenol CD_{25} values (µg) after abrupt withdrawal of propranolol, metoprolol and pindolol. The dashed areas represent the period of increased beta-adrenergic sensitivity after propranolol and metoprolol. After propranolol and metoprolol, day 2–6 significantly less than day 20 ($P<0.05$)

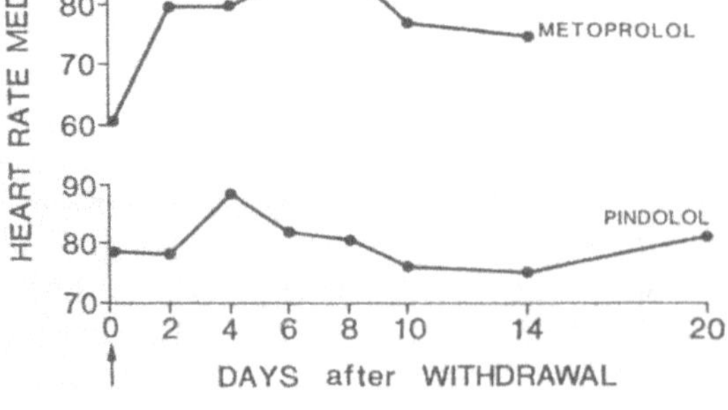

Fig. 2. Median heart rates after abrupt withdrawal of propranolol, metoprolol and pindolol. After metoprolol, days 2–8 significantly greater than day 14 ($P<0.05$). After pindolol days 2–6 significantly greater than days 14–20 ($P<0.05$)

Results

Isoproterenol CD_{25} values – Abrupt Withdrawal

The comparative changes in CD_{25} values after withdrawal from propranolol, metoprolol and pindolol are shown in Fig. 1. Comparable and reproducible stable baseline CD_{25} values were acheived by days 14–20. The values were 2.4 μg (range 1.4 to 7.6) after propranolol, 2.6 μg (range 0.9 to 5.1) after metoprolol and 3.0 (range 1.6 to 16.7) after pindolol. This contrasts with initial CD_{25} of 57 μg (range 4.6 to 355), 8.7 μg (range 3.1 to 24) and 181 μg (range 16.0 to 800) measured 12 h after the last dose of propranolol, metoprolol and pindolol respectively. Not only these initial values but also the subsequent course of CD_{25} values differed notably between drugs. After propranolol, a period of increased beta-adrenergic sensitivity extended from day 3 to 13. This was manifested by a median CD_{25} value below baseline during this interval, with all nine subjects having a value 2 to 5 fold below baseline (i.e. increased sensitivity) somewhere between day 5 and 9 ($P<0.05$). A similar pattern in CD_{25} values was observed in each patient after withdrawal of metoprolol, but was different from propranolol by an earlier onset and shorter duration of increased beta-adrenergic sensitivity. Median CD_{25} values were significantly below baseline between day 2 and 8, with individual values 2 to 3 fold below baseline for the eight subjects ($P<0.05$). After pindolol, the median CD_{25} fell gradually to the ultimate baseline, and in none of the ten subjects was the CD_{25} on days 0 to 10 lower than or day 14 or 20.

Heart Rate – Abrupt Withdrawal

Comparative changes in resting heart rate after drug withdrawal are shown in Fig. 2. Two days after propranolol, heart rate had returned to baseline of 78 beats/min (range 60 to 105), from an initial value of 55 beats/min (range 48 to 70) on day 0. Four patients showed a rebound in heart rate of about 6 beats/min between days 2 to 8. After metoprolol, patients had a comparable initial heart rate of 61 (range 54 to 79) on day 0, but 8 of 8 patients had a significant rebound between days 2 to 8 during the period of increased beta adrenergic sensitivity (median 6 beats/min above ultimate baseline range 3 to 18, $P<0.025$). Pindolol did not produce the same bradycardia as propranolol and metoprolol, the median initial heart rate being 79 beats/min (range 74 to 86). After pindolol withdrawal, a significant rebound of heart rate occurred in 9 of 10 patients between day 2 and day 6 varying from 6 to 22 beats/min (median 10) ($P<0.05$).

Blood pressure – Abrupt Withdrawal

Comparative changes in resting blood pressure after drug withdrawal are shown in Fig. 3. After propranolol, 6 of the 9 patients experienced a significant rebound of both systolic and diastolic pressures ($P<0.05$); the median of these six subjects were 135/85 initially, 154/98 at peak and 139/86 at ultimate baseline. Median pressures for all 9 patients were 140/87 $\left(\text{range } \frac{97}{74} \text{ to } \frac{159}{100}\right)$ initially, 149/95 $\left(\text{range } \frac{128}{84} \text{ to } \frac{180}{107}\right)$ on day 6 and 140/89 $\left(\text{range } \frac{112}{79} \text{ to } \frac{180}{120}\right)$ at baseline. After metoprolol,

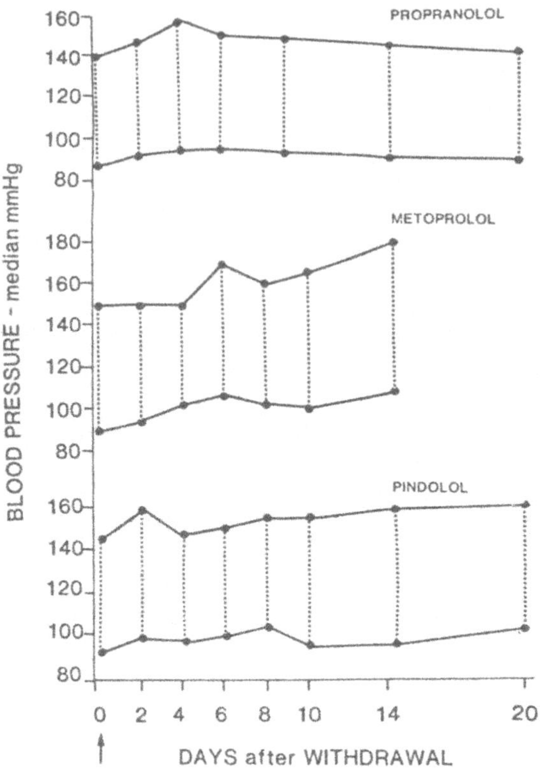

Fig. 3. Median systolic and diastolic blood pressures after abrupt withdrawal of propranolol, metoprolol and pindolol. After propranolol, systolic and diastolic pressure on day 6 significantly greater than on day 20 ($P < 0.05$)

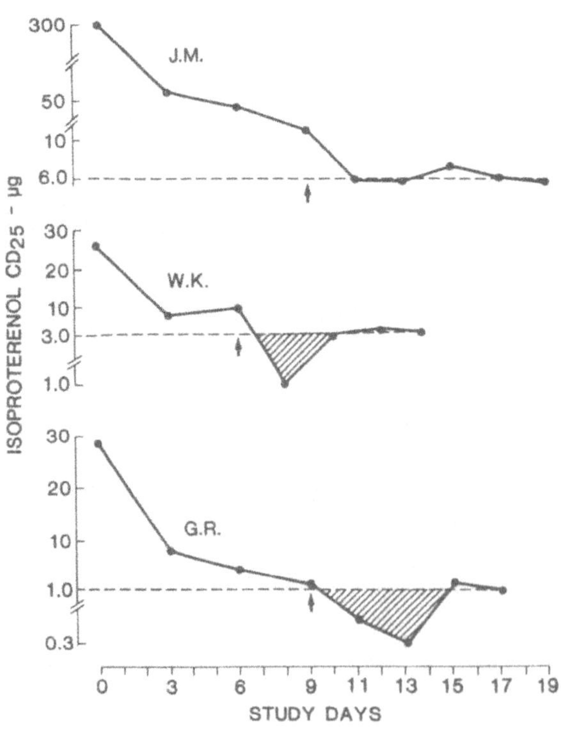

Fig. 4. Serial changes in beta adrenergic sensitivity, as measured by isoproterenol CD_{25} (μg) in patients withdrawn from propranolol by serial dose reduction (group A). The dashed horizontal line indicates the ultimate stable CD_{25}. Values below the stable CD_{25} (*hatched area*) indicate increasing beta adrenergic sensitivity seen in patients WK and GR. *Arrows* indicate the end of propranolol dose tapering schedule

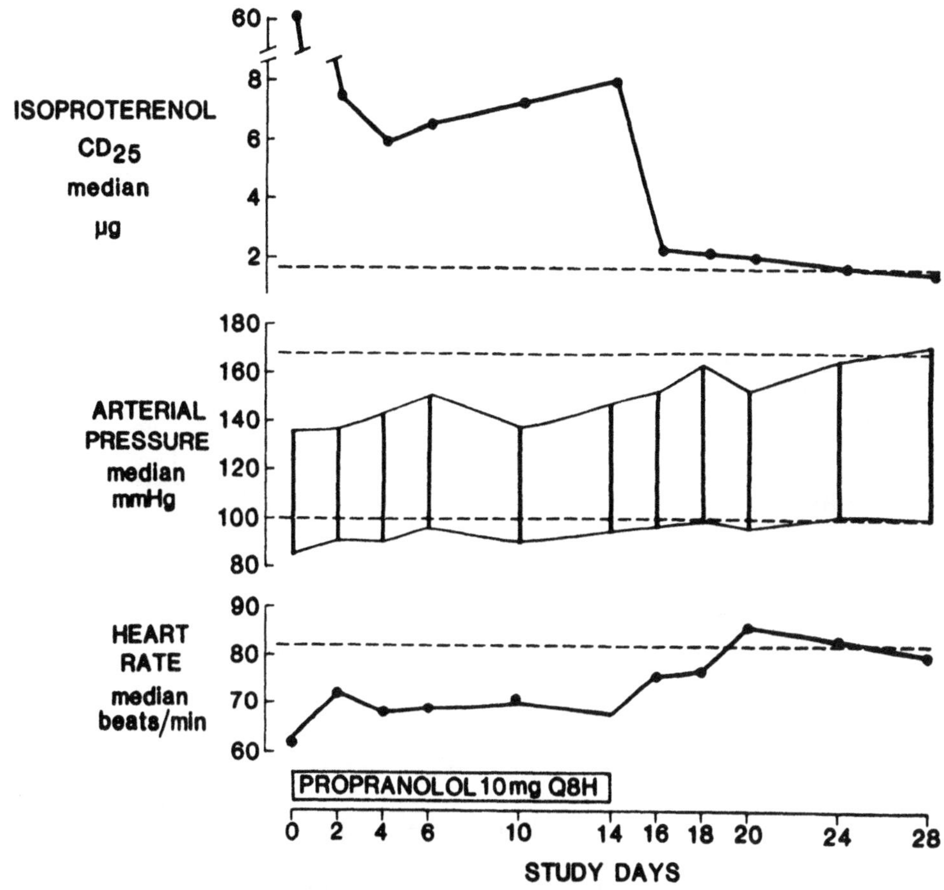

Fig. 5. Median isoproterenol CD_{25} (μg) median arterial blood pressure (mmHg) and median heart rate (beats/min) during and after prolonged low dose propranolol withdrawal in group B patients ($n = 6$). *Dashed horizontal lines* represent the ultimate stable values

systolic and diastolic pressure rose progressively from 149/90 $\left(\text{range } \frac{131}{74} \text{ to } \frac{187}{108}\right)$ on day 0 to 178/107 $\left(\text{range } \frac{147}{91} \text{ to } \frac{206}{125}\right)$ on day 14. There was no rebound except for one subject, on day 8 (systolic and diastolic). The same progressive rise was observed after pindolol from 144/90 $\left(\text{range } \frac{130}{70} \text{ to } \frac{184}{116}\right)$ on day 0 to 161/102 $\left(\text{range } \frac{140}{80} \text{ to } \frac{190}{122}\right)$ on day 20.

Table 2. Symptoms after withdrawal of Beta-Blockers

	Propranolol ($n=9$)	Metoprolol ($n=8$)	Pindolol ($n=10$)
Headache	3	1	1
Tremulousness	1	1	1
Anxiety	1	1	0
Chest pain	1	0	0
Total	6	3	2

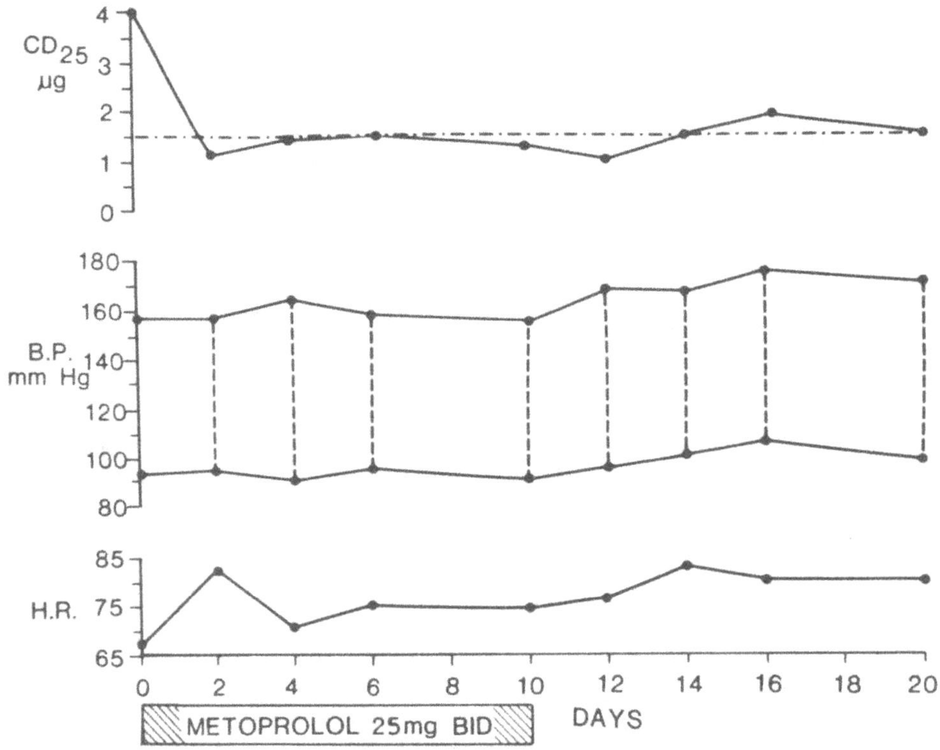

Fig. 6. Effect of gradual metoprolol withdrawal on the dose of isoproterenol required to increase chronotropy 25 beats/min (CD_{25}, top), blood pressure (BP, *middle*) and resting heart rate (HR, *bottom*). Full dose metoprolol (150 mg BID) was gradually reduced to 25 mg BID for 10 days before being completely withdrawn and replaced with placebo. All the data points are median values for the group ($n=7$)

Symptoms – Abrupt Withdrawal

Clinical symptoms suggestive of withdrawal were described by six patients after propranolol, by three after metoprolol and by two after pindolol (Table 2). These were of variable duration, but did not last for more than four days except in two patients who had progressive headache concomitant with a rise in their blood pressure after withdrawal of metoprolol or pindolol.

CD_{25}, Heart Rate, Blood Pressure and Symptoms – Gradual Withdrawal

Figure 4 shows that stepwise dose reduction of propranolol prevented hypersensitivity to isoproterenol in only 1 of 3 patients. Figure 5 shows that prolonged low dose largely prevented all the changes in isoproterenol hypersensitivity and rebound in heart rate and blood pressure.

Figure 6 shows that prolonged low dose metoprolol largely prevented all of the changes in isoproterenol hypersensitivity and rebound in heart rate.

Discussion

There were substantial differences in cardiac responsiveness as measured by isoproterenol and resting heart rate between withdrawal of propranolol, metoprolol and pindolol. The explanation for these divergent observations may lie in the hypothesis of Ariens and *Simmons* [2] and the recent observations of *Bryan* et al. [5] and *Bake* et al. [3]. These investigators propose that cardiac beta-1 receptors are primarily stimulated by local adrenergic neurone release of neurotransmitter norephinephrine (hence "beta-T" receptors) while cardiac beta-2 receptors are primarily stimulated by the circulating hormone epinephrine (hence "beta-H" receptors). Resting heart rate would be a measure of beta-1 receptor responsiveness and the isoproterenol test a measure of beta-2 receptor responsiveness. Differing adaptive changes may occur in these receptors after chronic exposure to beta blockers with different pharmacologic properties. After abrupt withdrawal and rapid elimination of beta blocker these changes in receptor responsiveness may be unmasked.

After propranolol withdrawal there was a common, marked and prolonged rebound sensitivity to isoproterenol but a less common, mild and short lived rebound in resting heart rate. This would be consistent with an increase in beta receptor density as shown by others [1, 10], but with a greater change of beta-2 versus beta-1 receptors. This suggests that propranolol's greater antagonist action is at the beta-2 receptor and that it is non-selective.

After metoprolol withdrawal there was a common, marked and prolonged rebound sensitivity to both isoproterenol and resting heart rate. This would be consistent with a more equal increased density of both beta-1 and beta-2 receptors. This is in keeping with the known beta-1 selectivity of metoprolol and the fact that progressive beta-2 blockade occurs at greater doses such as used in this study.

After pindolol withdrawal there was a common and marked rebound in resting heart rate at the same time as directly opposite common, marked and persistent decrease in isoproterenol responsiveness. This would be consistent with enhanced beta-1 receptor density and suppressed beta-2 receptor density. This latter phenomenon has been shown after chronic administration of the beta-2 agonist terbutaline [8, 25]. The known partial agonist activity of pindolol at higher concentrations could occur predominantly at beta-2 receptors and result in down regulation. There could be a much lower effective concentration of pindolol at the beta-1 receptor since there is a relatively much greater concentration of agonist (norepinephrine) at this site [7]. Thus a predominant antagonist action of pindolol at low concentration could result in up regulation of the beta-1 receptor.

The rebound in heart rate or blood pressure and the symptoms as part of the withdrawal features after propranolol, metoprolol and pindolol were statistically significant in some instances but we believe were of potential clinical significance even if they occurred in only one patient. Because the patients were resting during our observations, both the frequency and magnitude of the changes could have been underestimated. Some authors have suggested that increased adrenergic sensitivity will only become clinically manifest when the sympathetic drive is elevated, as for example by assuming a standing position, exercising or using a vasodilator [14, 23]. In addition, our relatively healthy patients without measurable cardiac disease, probably had a higher threshold to possible life-threatening consequences of these changes. The clinical implications of these withdrawal features are quite different in patients with ischemic heart disease or arrhythmia. It has been shown that in a given patient, the onset of angina tends to occur at the same degree of myocardial oxygen demand, of which systolic pressure and heart rate are important determinants [22]. For ethical reasons, we limited our study to a hypertensive population, but the results can be extrapolated to all patients with overt cardiac disease and to hypertensive patients who may have ischemic heart disease masked by their beta-blocker therapy.

The duration of increased cardiac sensitivity, 2–8 days after metoprolol, 3–13 days after propranolol and 2–6 days after pindolol suggests that none of these drugs should be discontinued rapidly. We have shown that tapering of propranolol in less than 9 days does not prevent withdrawal phenomena [18, 21], while a prolonged low dose over two weeks prevents the increase in cardiac sensitivity and most of the other withdrawal features [7, 21]. Similarly, prolonged low dose metoprolol prevented most of the changes observed after abrupt withdrawal. We did not study a gradual withdrawal schedule for pindolol but based on the rebound in resting heart rate we would suggest a dose of 2–3 mg daily for 2 weeks before complete discontinuation.

The controversy over the prevalence, or even the existence, of a beta-blocker withdrawal phenomenon is partly explained by our data. The magnitude of increased cardiac responsiveness after propranolol, metoprolol or pindolol would probably go unnoticed in an individual without heart disease. So also the mild, infrequent and almost nonspecific symptoms would be unlikely to raise suspicion. The same may be so for a patient with mild ischemic heart disease. However, the patient with more severe ischemic or arrhythmic heart disease, through a further decrease in cardiac threshold, could experience an adverse cardiac event either at rest or after a previously tolerated physical-emotional stress. Such an event would be typical of, and therefore attributed to, the underlying disease without question of medication withdrawal. Patients should be informed of these risks to enhance medication compliance.

References

1. Aarons RD, Nies AS, Gal J, Hegstrand LR, Molinoff PB (1980) Elevated of β-adrenergic receptor density in human lymphocytes after propranolol administration. J Clin Invest 65:949
2. Ariens EJ, Simonis AM (1976) Receptors and receptor mechanisms. In: Saxena PR, Forsyth RP (eds) Beta adrenoreceptor blocking agents. Elsevier/North Holland, Amsterdam, pp 4–27
3. Bake SP, Boyd HM, Potter LT (1980) Distribution and function of beta adrenoreceptors in different chambers of the canine heart. Br J Pharmacol 68:57–63
4. Boudoulas H, Lewis RP, Kates RE, Dalamangas G (1977) Hypersensitivity to adrenergic stimulation after propranolol withdrawal in normal subjects. Ann Intern Med 87:433
5. Bryan LJ, Cole JJ, O'Donnel SR, Wanstall JC (1981) A study designed to explore the hypothesis that beta-1 adrenoreceptors are "innervated" receptors and beta-2 adrenoreceptors are "hormonal" receptors. J Pharmacol Exp Ther 216:395–400
6. Cleaveland CR, Rangno RE, Shand DG (1972) A standardized isoproterenol sensitivity test: the effects of sinus arrhythmias atropine and propranolol. Arch Intern Med 130:47
7. Frishman W, Kostis J, Strom J et al. (1979) Clinical Pharmacology of the new beta-adrenergic blocking drugs. Part 6. A comparison of pindolol and propranolol in treatment of patients with angina pectoris. The role of sympathomimetic activity. Am Heart J 98:526

8. Galant SP, Duriseti L, Underwood S, Insel PA (1978) Decreased beta-adrenergic receptors on human polymorphonuclear leukocytes after adrenergic therapy. N Engl J Med 299:933
9. Gerber JG, Nies AS (1979) Abrupt withdrawal of cardiovascular drugs. N Engl J Med 301:1234
10. Glaubiger G, Lefkowitz RJ (1977) Elevated beta-adrenergic receptor number after chronic propranolol treatment. Biochem Biophys Res Commun 78:720
11. Hollander M, Wolfe DA (1973) Non-parametric statistical methods. Wiley & Sons, New York, pp 138–155
12. Jackson G, Schwartz J, Kates RE, Harrison DC (1979) Physiologic basis for beta blockade rebound: atenolol vs propranolol. Clin Res 27:444
13. Lederballe Pedersen O (1976) Comparison of metoprolol and hydrochlorothiazide as antihypertensive agents. Eur. J Clin Pharmacol 1:381
14. Lewis MJ, Ross PJ, Henderson AH (1979) Rebound effect after stopping beta-blockers. Br Med J 2:606
15. Miller RR, Olson HG, Amsterdam EA, Mason DT (1975) Propranolol-withdrawal rebound phenomenon. Exacerbation of coronary events after abrupt cessation of antianginal therapy. N Engl J Med 293:416
16. Mizgala HF, Counsell J (1976) Acute coronary syndromes following abrupt cessation of oral propranolol therapy. Can Med Assoc J 114:1123
17. Nattel S, Rangno RE, Van Loon G (1979) Mechanism of propranolol withdrawal phenomena. Circulation 59:1158
18. Nattel S, Shanks J, Rangno RE (1978) Propranolol withdrawal. Ann Intern Med 89:288
19. Nellen MS (1969) Trial of two beta-blockade drugs, Ciba 39089 and ICI 50172 in angina pectoris. S Afr Med J 2:155
20. O'Brien ET, MacKinnon J (1972) Propranolol and polythiazide in treatment of hypertension. Br Heart J 34:1042
21. Rangno RE, Nattel S, Lutterodt A (in press) Prevention of propranolol withdrawal phenomena by gradual dose reduction. Am J Cardiol
22. Robinson BF (1967) Relation of heart rate and systolic blood pressure to the onset of pain in angina pectoris. Circulation 35:1073
23. Ross PJ, Lewis MJ, Henderson AH (1979) Beta-blockers withdrawal syndrome. Lancet 1:875
24. Shand DG, Wood AJJ (1978) Propranolol withdrawal syndrome – Why? Circulation 58:202
25. Wolfe BB, Harden TK, Molinoff PB (1977) In vitro study of β-adrenergic receptors. Ann Rev Pharmacol Toxicol 17:575

Discussion

Rahn, Maastricht: Your observation of a blood pressure overshoot after cessation of therapy with propranolol is at variance with a study by *Dollery* et al. Do you have an explanation?
What is your definition of hypertension, since in your first propranolol study the median ultimate stable blood pressure is about 140/90 mmHg?

Rangno: The rebound of blood pressure only occurred in 6 out of 9 patients and peaked between the 4th and 9th day. Our patients may have been different from those of Dollery or our observations were extended further after cessation of propranolol treatment.
The rebound of blood pressure above ultimate baseline may have appeared more significant, since the ultimate stable values were lower than expected. This failure

of all patients to return to higher pressure values is consistent with data on the slow return of pressure to previous hypertensive levels after chronic control medication is stopped.

Henderson, Cardiff: I was interested to note the later occurrence in time of the rebound peak in the case of resting responses than of stressed responses, which presumably relates to the degree of stimulation of the β-receptors. It may be of interest to add that in our study of stressed heart rate responses (British Heart Journal [in press]) we found rebound increases following withdrawal in the treatment with two other β-blockers with partial agonist activity (oxprenolol and acebutolol), similar to your findings with pindolol.

Rangno: The time course for hypersensitivity to isoproterenol does not have to be the same as for the rebound in resting heart rate. Different β-blockers probably have varying degrees of action on transmitter (β_1) versus hormonal (β_2) receptors; therefore, the degree and duration of hyperresponsiveness after withdrawal will probably be different. We believe that resting (or exercise) heart rate is a measure of β_1 response and the isoproterenol test is a measure of β_2 response.

Sympathomimetic Drugs in the Treatment of Congestive Heart Failure

G. Klein, A. Wirtzfeld

Summary

In 18 patients presenting with chronic intractable low-output cardiac failure secondary to congestive cardiomyopathy (12 pat.) or ischemic heart disease (6 pat.) the hemodynamic effects of the adrenoceptor agonists dopamine, dobutamine and prenalterol were compared (Swan-Ganz-thermodilution catheter).

In a first study performed on 10 patients dopamine at a dosage of 4 µg/kg/min and dobutamine at 7.5 µg/kg/min were administered. There was a comparable rise in cardiac output (CO) by about 60% with a fall in peripheral resistance (TPR) by about 33% on both drugs. Heart rate rose by 12% on dopamine but remained unchanged on dobutamine. The filling pressures of the right and left ventricles fell by 50% and 25%, respectively, on dobutamine, whereas dopamine did not alter the average pressures but even caused a further increase in some patients.

In a second study 8 patients received, in addition to dopamine and dobutamine the new beta-1-adrenoceptor agonist prenalterol. This compound was administered in doses of 20 and 40 µg/kg given over a period of 5 min and hemodynamic measurements were performed 15, 30 and 60 min after the end of each infusion. The effects of dopamine and dobutamine were similar as in the first study. Prenalterol caused an increase of CO by 32% with a fall in TPR by 18%. Right atrial mean pressure diminished by 20% and pulmonary wedge pressure by 15%.

In summary, our studies have shown that in the treatment of severe congestive heart failure with low-output syndrome demanding inotropic support, a beta-1-selective adrenoceptor agonist as dobutamine or prenalterol is preferable compared with the less-selective dopamine, as any alphamimetic effects on the venous circulation is undesirable. Prenalterol resembles dobutamine in its hemodynamic effects demonstrating a selective beta-1-stimulating action on the heart. Although the effectiveness of prenalterol in the dose range given was smaller compared with the other compounds, the pharmacokinetic properties of prenalterol, i.e. the oral bioavailability and the longer half-live are desirable features and warrant further studies of the drug in the management of severe cardiac failure.

Introduction

Of the many agents capable of improving cardiac function the catecholamines are among the most effective ones and the inotropic potency of these sympathomimetic amines far exceeds that of digitalis glycosides [5]. For the treatment of severe low-output cardiac failure dopamine has become the drug of choice as it exerts less undesirable chronotropic and peripheral vascular effects than the older catecholamines and, in addition, specifically stimulates renovascular receptors (dopaminergic receptors) thus improving renal perfusion [11]. Dobutamine is a synthetic catecholamine recently developed in an effort to more selectively stimulate the cardiac β_1-receptors and to reduce the deleterious effects associated with other sympathomimetic drugs [25].

Prenalterol was synthesized in 1970 by *Haessle* AB (Sweden). Animal pharmacological studies have demonstrated selective β_1-adrenoceptor stimulating properties of the racemic form H 80/62 and its levo-isomer H 133/22. The inotropic response was more pronounced than the chronotropic effects in comparison to terbutaline and isoprenaline [8, 10].

Recently, in some human pharmacological studies a positive inotropic effect of prenalterol using non invasive methods was also demonstrated in man [13, 14, 15, 17, 20, 21, 23]. After i.v.-administration as well as after oral application total electromechanical systole was shortened as evidenced by a shorter pre-ejection period and left ventricular ejection time. Heart rate and systolic blood pressure increased significantly, diastolic blood pressure remaining unchanged. Furthermore it was possible to completely reverse the cardiodepressive effects of metoprolol in healthy volunteers [20] and in patients with acute myocardial infarction [3]. In patients with coronary heart disease Prenalterol produced considerable improvement of cardiocirculatory performance [12], even in severe congestive heart failure [4].

Our investigations were designed as comparative cross-over studies to assess the quantitative and qualitative effectiveness of dopamine, dobutamine and prenalterol in patients with severe chronic decompensated heart failure.

Dopamine/Dobutamine Study

Methods and Materials

The first study was performed in 10 patients who were admitted to our intensive care unit for chronic refractory heart failure (NYHA III–IV) secondary to congestive cardiomyopathy (6 cases) or coronary heart disease (4 cases) (Table 1). Besides the symptoms of chronic pulmonary and systemic venous congestion all

Table 1. Clinical and hemodynamic characteristics of patients in the dopamine/dobutamine study

Case	Sex	Age (yr)	Cardiac diagnosis	NYHA class	PCPm (mm Hg)	CI (l/min · m²)	SWI (g · m/m²)	ECG
1	m	58	CHD	IV	26	1.45	15.3	AF
2	m	64	COCM	IV	25	1.40	17.1	AF
3	m	54	CHD	IV	30	1.30	18.1	SR
4	m	51	CHD	III	30	2.20	18.3	SR
5	m	38	COCM	III	32	2.65	25.5	SR
6	m	24	COCM	III	31	2.25	20.4	SR
7	m	25	COCM	III	30	2.05	14.8	SR
8	m	54	COCM	III	25	2.35	20.5	SR
9	m	36	COCM	IV	35	1.25	11.0	SR
10	m	65	CHD	III	20	1.90	24.2	PM

Abreviations: CHD, coronary heart disease; COCM, congestive cardiomyopathy; PCPm, pulmonary capillary wedge pressure; SWI, stroke work index; AF, atrial fibrillation; SR, sinus rhythm; PM, pacemaker patients exhibited signs of low-output state with more or less marked renal and hepatic insufficiency and general cachexia. All patients were on chronic digitalis and diuretic treatment which was maintained during the study. No patient was in state of acute pulmonary edema or cardiogenic shock.

For determination of the hemodynamic parameters a Swan-Ganz-thermodilution catheter was placed via subclavian puncture into the pulmonary artery under X-ray-control. Arterial pressure was recorded either directly by radial cannula or non-invasively by blood-pressure cuff. Heart rate was taken from ECG. An Edwards-Cardiac-Output Computer-Model 9510 was used.

The statistical evaluations were performed using the Wilcoxon-test.

Protocol

After control measurements had been obtained and hemodynamic stability over a period of at least 30 min had been established an infusion of dopamine at a constant rate with 4.0 μg/kg/min was started. Hemodynamic parameters were measured after 15, 30 and 60 min. After 60 min the infusion was stopped and the patients were allowed to return to control status. This always occurred within 30 min, but a period of at least 60 min free of medication was maintained before the dobutamine infusion was started at a dosage of 7.5 μg/kg/min. Hemodynamic measurements were again performed at 15, 30 and 60 min. To rule out mutual influence the sequence of the two substances was randomized. The dosages of dopamine and dobutamine were chosen because preliminary studies had demonstrated a good comparable increase of cardiac output with the two drugs at the dosages applied.

Results

The effects of dopamine and dobutamine on hemodynamics are presented in Tables 2 and 3. The cardiac index increased from a control value of 1.9 l/min/m² to 3.0 (+59%) on dopamine and 3.1 (+64%) on dobutamine. This augmentation of cardiac output was largely due to an increase in stroke volume which increased from a control of 21 ml/m² to 30 (+41%) on dopamine and 33 (+59%) on dobutamine. There was no significant change in heart rate on dobutamine but on

Table 2. Hemodynamic effects of dopamine (4 μg/kg/min) in 10 patients with severe congestive heart failure

Parameter	Control (mean ± SD)	Dopamine (mean ± SD)	Percent change	*P*
Heart rate (min^{-1})	87.9 ± 12.6	101.2 ± 13.6	+12	< 0.01
Stroke volume (ml/m^2)	21.2 ± 4.4	29.9 ± 8.7	+41	< 0.01
Cardiac output ($l/min \cdot m^2$)	1.87 ± 0.47	2.97 ± 1.0	+59	< 0.001
PC wedge pressure (mm Hg)	28.4 ± 4.4	29.7 ± 6.8	–	n.s.
RA mean pressure (mm Hg)	9.6 ± 5.5	10.5 ± 6.0	–	n.s.
Stroke work ($g \cdot m/m^2$)	18.7 ± 4.4	27.1 ± 8.7	+45	< 0.01
Arterial mean pressure (mmHg)	90.7 ± 7.4	94.2 ± 13.7	–	n.s.
Peripheral resistance ($dyn \cdot sec \cdot cm^{-5}$)	2014 ± 653	1386 ± 648	–31	< 0.001

Table 3. Hemodynamic effects of dobutamine (7.5 μg/kg/min) in 10 patients with severe congestive heart failure

Parameter	Control (mean ± SD)	Dobutamine (mean ± SD)	Percent change	P
Heart rate (min^{-1})	92.6 ± 14.2	92.3 ± 15.2	–	n.s.
Stroke volume (ml/m^2)	20.6 ± 5.0	32.8 ± 8.3	+59	< 0.001
Cardiac output ($l/min \cdot m^2$)	1.89 ± 0.53	3.11 ± 1.22	+64	< 0.001
PC wedge pressure (mmHg)	28.5 ± 5.0	21.5 ± 7.3	–25	< 0.05
RA mean pressure (mmHg)	10.0 ± 6.0	5.0 ± 5.0	–50	< 0.001
Stroke work ($g \cdot m/m^2$)	18.4 ± 4.7	31.1 ± 10.1	+69	< 0.001
Arterial mean pressure (mmHg)	90.9 ± 11.3	87.2 ± 10.4	–	n.s.
Peripheral resistance ($dyn \cdot sec \cdot cm^{-5}$)	2047 ± 827	1326 ± 545	–35	< 0.001

dopamine heart rate rose by a mean of 13 beats/min ($P < 0.01$) as compared to pretreatment levels.

Systemic arteriolar resistance which was markedly elevated in most patients fell during infusion of both agents by 31% on dopamine and 35% on dobutamine. Pulmonary arteriolar resistance fell by 29% on dopamine and 35% on dobutamine.

LV stroke work was markedly diminished in every patient prior to therapy, the mean value being $18.4 \pm 5\ g \cdot m/m^2$. This parameter increased to 27.1 ± 9 during dopamine and 31.1 ± 10 during dobutamine infusion. This is an increase in LV stroke work by 45% and 69%, respectively. The difference between the two drugs was statistically significant ($P < 0.05$).

A significant and clinically quite relevant differential effect of the two catecholamines was seen on the right and left ventricular filling pressures. The mean pulmonary capillary wedge pressure on dopamine rose by more than 3 mmHg in six patients, the maximum increment being 10 mmHg, it remained unchanged in

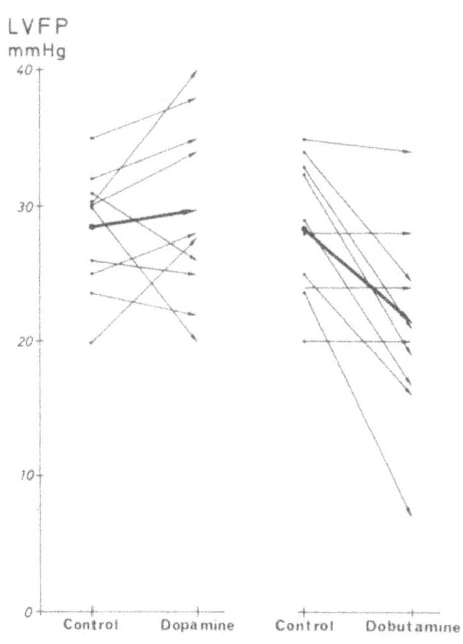

Fig. 1. Left ventricular filling pressure (pulmonary capillary wedge pressure) on dopamine and dobutamine

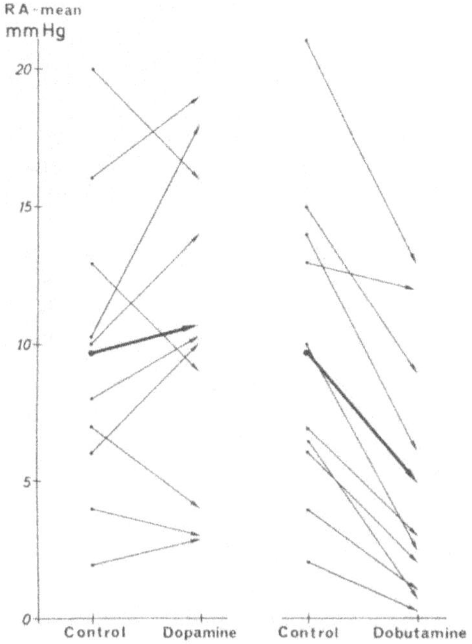

Fig. 2. Right atrial mean pressure on dopamine and dobutamine

one and diminished slightly in three cases (Fig. 1). On the average the pulmonary capillary wedge pressure increased from 28.4±4 to 29.7±7 mmHg on dopamine. On dobutamine the wedge pressure fell in six patients and remained unchanged in four; a rise in wedge pressure was never observed. There was an average fall in pressure by 7 mmHg (–25%). Thus the average pulmonary capillary wedge pressure on dopamine was 9 mmHg higher as compared to dobutamine ($P < 0.001$).

The same differential effects of the two drugs were observed on the right ventricular filling pressures. Right atrial mean pressure fell significantly on dobutamine in every patient, from 10.0±6 to 5.0±5 mmHg (–50%). On dopamine there was a rise in pressure in six and only an insignificant fall in the remaining four, on the average there was an increase in right atrial mean pressure from 9.6±5 to 10.5±6 mmHg (Fig. 2).

Dopamine/Dobutamine/Prenalterol Study

Methods and Materials

In the second study 8 patients with severe congestive heart failure (NYHA III–IV) in a functional and clinical state comparable to the patients discussed in the first study were treated (Table 4). Also the methods of hemodynamic measurements were identical.

Protocol

In addition to dopamine and dobutamine the patients received prenalterol in an intraindividual cross-over study. Dopamine and dobutamine were given in the

Table 4. Clinical and hemodynamic characteristics of patients in the dopamine/dobutamine/prenalterol study

Case	Sex	Age (yr)	Cardiac diagnosis	NYHA class	PCPm (mm Hg)	CI (l/min · m²)	SWI (g · m/m²)	ECG
1	m	36	COCM	IV	20	1.25	11.9	SR
2	m	64	COCM	IV	25	1.40	17.8	AF
3	m	51	CHD	III	30	2.20	18.3	SR
4	m	38	COCM	III	32	2.65	25.5	SR
5	m	24	COCM	III	31	2.25	20.4	SR
6	m	54	CHD	III	17	2.64	27.3	SR
7	m	25	COCM	III	30	2.05	14.8	SR
8	m	62	COCM	IV	32	2.02	17.9	SR

Abreviations: COCM, congestive cardiomyopathy; CHD, coronary heart disease; PCPm, pulmonary capillary wedge pressure; CI, cardiac index; SWI, stroke work index; SR, sinus rhythm; AF, atrial fibrillation

same dosages as in the first study, i.e. 4.0µg/kg/min of dopamine and 7.5µg/kg/min of dobutamine as constant infusions. Thereafter prenalterol was administered in doses of 20 and 40 µg/kg given as infusions within a period of 5 min each. Hemodynamic measurements were performed 15, 30 and 60 min after the end of each prenalterol infusion.

As the elimination half-life of prenalterol is relatively long, an interval of 4 h free of therapy was maintained between the two prenalterol doses.

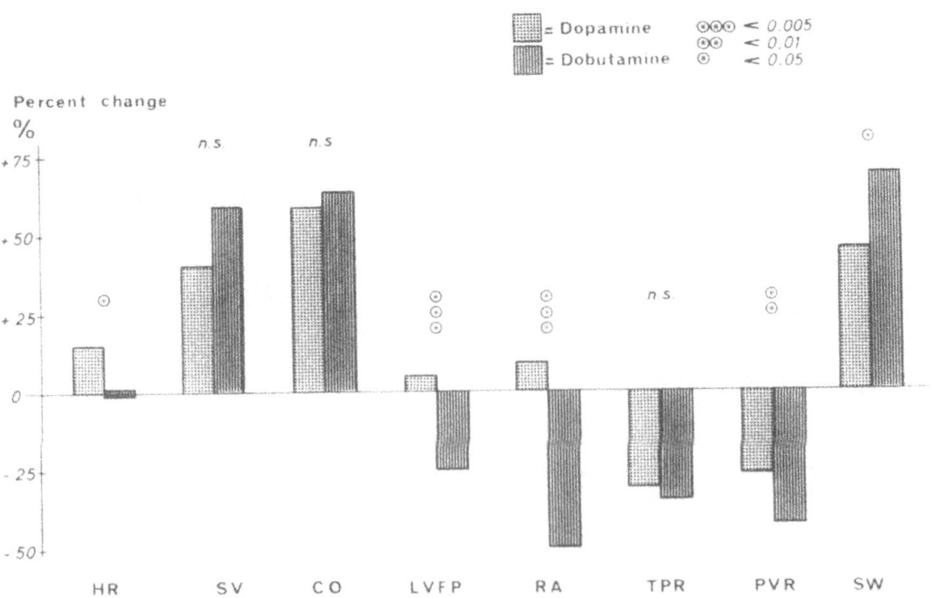

Fig. 3. Percent change of hemodynamic parameters during dopamine and dobutamine infusion. *HR*, heart rate; *SV*, stroke volume; *CO*, cardiac output; *LVFP*, left ventricular filling pressure; *RA*, right atrial mean pressure; *TPR*, total peripheral resistance; *PVR*, pulmonary vascular resistance; *SW*, stroke work of the left ventricle

Results

Similar as in the first study heart rate rose significantly on dopamine (+14%, $P < 0.05$) whereas there were no significant changes on dobutamine and prenalterol.

Mean right and left ventricular filling pressures were not significantly altered by dopamine while on dobutamine and prenalterol there was a significant fall in pressure as compared to control values.

The effect of prenalterol on stroke volume with a mean increase of 20% after 20 µg/kg and 23% after 40 µg/kg was statistically significant ($P < 0.05$), but compared with dopamine (+62%) and dobutamine (+70%) the magnitude of improvement was much smaller (Fig. 4).

Following prenalterol, there was a significant increase in cardiac output 15 min after each dose which, however, was less pronounced than on the other compounds tested: cardiac output increased by 26% after 20 µg/kg and 32% after 40 µg/kg of prenalterol (Fig. 5).

Only on dopamine systolic arterial pressure rose significantly (+14%, $P < 0.01$), on dobutamine and after prenalterol no significant changes in blood pressure were observed.

Systemic arteriolar resistance decreased by 28% on dopamine and 36% on dobutamine, while after prenalterol the fall was only about 18% after 20 µg/kg and 17% after 40 µg/kg ($P < 0.05$). Pulmonary arteriolar resistance fell during infusion of dopamine by 28% and dobutamine by 38%. The decrease after prenalterol at both dosages applied was about 20% ($P < 0.05$).

Discussion

Summarizing the hemodynamic effects of dopamine and dobutamine in patients with severe low-output failure, it can be stated that at a dosage of 4.0 µg/kg/min for dopamine and of 7.5 µg/kg/min for dobutamine

- there is a comparable augmentation of cardiac output by roughly 60% accompanied by a fall in peripheral resistance by about 30% with both drugs,
- but there is a clinically relevant and statistically significant difference in the effects of the two agents on heart rate and on right and left ventricular filling pressures.

Similar hemodynamic effects were also presented by others [1, 2, 6, 9, 18, 19, 22, 24].

Prenalterol resembles dobutamine in the profile of its hemodynamic effects demonstrating a beta-1-stimulating action on the heart. Although the effectiveness of prenalterol in the dose range given was smaller compared with the other compounds, improvement of cardiac output with a relative increase of stroke volume by about 30% seems clinically quite relevant. Our results are in good accordance to the findings of *Awan* et al. [4].

As could be expected from the pharmacokinetic properties of prenalterol with mean half-lives of the distribution phases of 1.7 and 11.2 min and a mean elimination half-life of 2.0 h [16] we found a maximal hemodynamic effect at the time of our first measurement, i.e. 15 min after the end of each infusion; 30 and 60 min after prenalterol no significant changes compared to control values could be

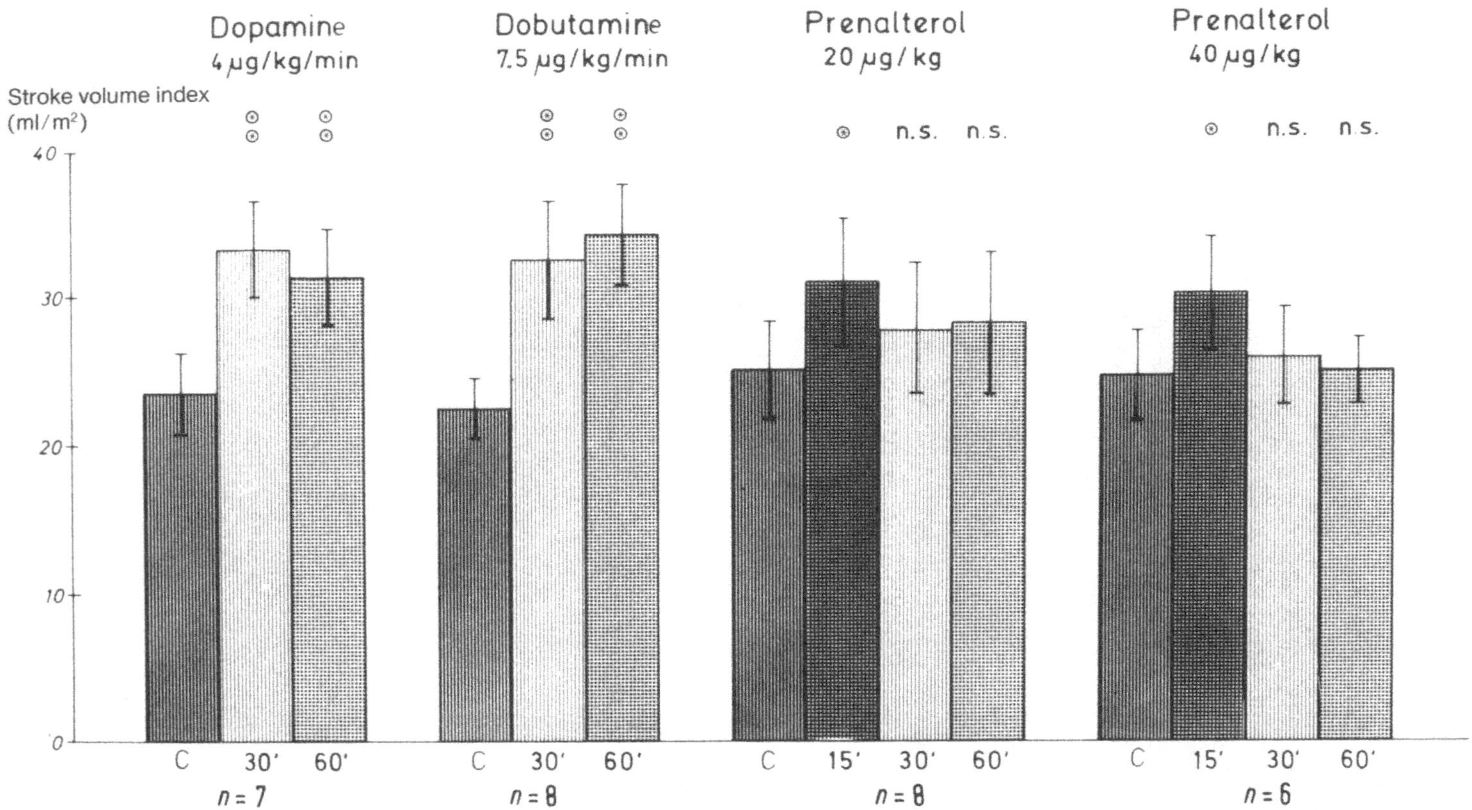

Fig. 4. Stroke volume index on dopamine, dobutamine and after prenalterol. C = control; mean ± SEM; n.s. = non-significant. * = $P < 0.05$, ** = $P < 0.01$

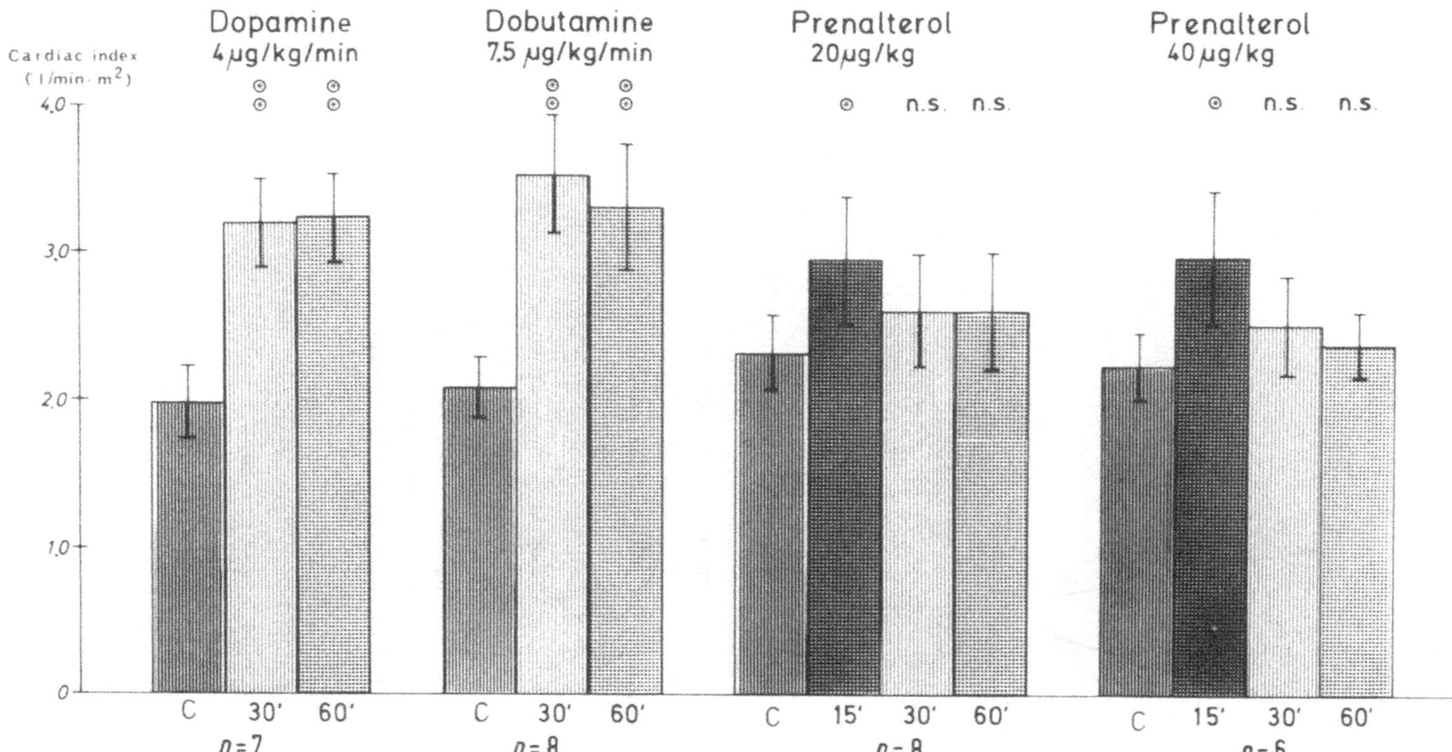

Fig. 5. Cardiac index on dopamine, dobutamine and after prenalterol. C = control, mean ± SEM; n.s. = no significant, * = $P < 0.05$; ** = $P < 0.01$

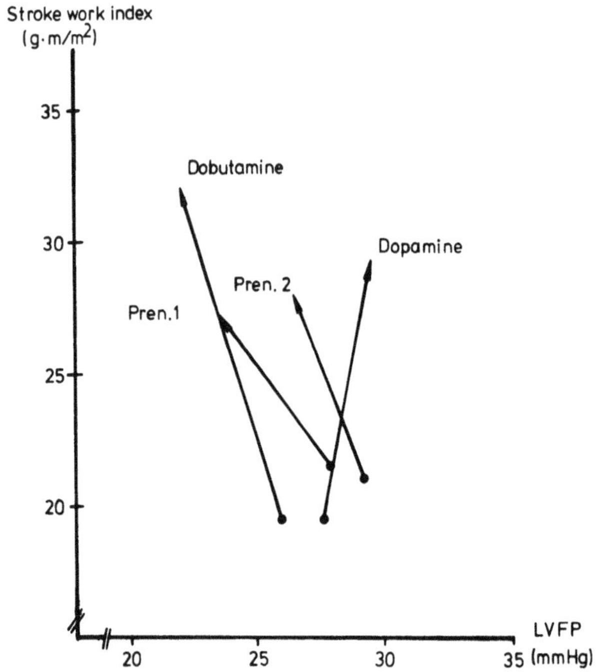

Fig. 6. Effect of dopamine, dobutamine and prenalterol on LV function

detected, which again is well in accordance with the low plasma levels to be expected at this time [16].

The observed difference in the effect on ventricular filling pressures of the three drugs studied deserves special attention and has to be taken into consideration when comparing the improvement of ventricular function.

As shown in Fig. 6, dobutamine shifted the relationship of stroke work to filling pressure upward and to the left whereas on dopamine the improvement of left ventricular function is partly paid for by a further increase in preload as the left ventricular function curve is shifted upward and to the right. Prenalterol in this respect resembles dobutamine shifting the ventricular function curve upward and to the left but the improvement in cardiac performance is less pronounced. The mechanism for the rise in filling pressures on dopamine may be explained by the alpha-mimetic action [7] of the drug which by causing constriction of systemic veins augments the return of venous blood to the heart. The failing ventricles despite their improved contractility are often unable to cope with this additional amount of blood and pulmonary and venous congestion will become worse.

In summary our studies have shown that the positive inotropic agents dopamine, dobutamine and prenalterol can considerably improve cardiac output in patients with so called intractable, low-output cardiac failure. In addition dobutamine and prenalterol may also diminish right and left ventricular preload, whereas on dopamine aggravation of pulmonary and systemic congestion may occur. Furthermore, in patients with diminished coronary flow dopamine may aggravate myocardial ischemia by exerting an unfavorable effect on the relationship between myocardial oxygen delivery and demand: myocardial oxygen delivery may be diminished because of a further rise in enddiastolic ventricular pressure and an aggravation of hypoxemia secondary to pulmonary congestion: oxygen demand increases with the rise in heart rate and stimulation of contractility. In contrast myocardial oxygen demand following dobutamine and prenalterol is only increased by the enhancing effect on contractility which may even be compensated for by a favorable effect on myocardial oxygen delivery secondary to a diminished LV enddiastolic pressure.

In the treatment of severe congestive heart failure, therefore, a beta-1-selective adrenoceptor agonist as dobutamine or prenalterol is to be preferred to the less selective dopamine, as any alpha-mimetic effect is undesirable. Although prenalterol in the dose range given was less effective than dobutamine the pharmacokinetic properties of prenalterol, i.e. oral bioavailability and the longer elimination half-life are desirable features and warrant further studies of prenalterol in the management of severe cardiac failure.

References

1. Akhtar N, Mikulic E, Cohn JN, Chaudhry MH (1975) Hemodynamic effect of dobutamine in patients with severe heart failure. Am J Cardiol 36:202
2. Andy JJ, Curry CL, Ali N, Mehrotra PP (1977) Cardiovascular effects of dobutamine in severe congestive heart failure. Am Heart J 94:175
3. Ariniego R, Waagstein F, Mombay B, Hjalmarson Å (1979) Hemodynamic effects of a new β_1-receptor agonist in acute myocardial infarction. A useful antidote to unwanted cardiac effects of β-blocking agents. Br Heart J 42:139
4. Awan NA, Needham KE, Evenson MK, Win A, Mason DT (1981) Hemodynamic actions of prenalterol in severe congestive heart failure due to chronic coronary disease. Am Heart J 101:158
5. Beiser GD, Epstein SE, Goldstein RE, Stampfer M, Braunwald E (1970) Comparison of the peak inotropic effects of a catecholamine and a digitalis glycoside in the intact canine heart. Circulation 42:805
6. Beregovich J, Bianchi Ch, D'Angelo R, Diaz R, Rubler S (1975) Haemodynamic effects of a new inotropic agent (dobutamine) in chronic cardiac failure. Br Heart J 37:629
7. Braunwald E, Ross J Jr, Kahler RL, Gaffney TE, Goldblatt A, Mason DT (1963) Reflex control of the systemic venous bed. Effects on venous tone of vasoactive drugs and of baroreceptor and chemoreceptor stimulation. Circ Res 12:539
8. Carlsson E, Dahlöf C-G, Hedberg A, Persson H, Tangstrand B (1977) Differentiation of cardiac chronotropic and inotropic effects of β-adrenoceptor agonists. Naunyn-Schmiedebergs Arch Pharmacol 300:101
9. Francis G, Sharma B, Hodges M (1979) Comparison of dobutamine and dopamine in acute low-output cardiac failure. Circulation 60 [Suppl II]:183
10. Hedberg A, Carlsson E, Tangstrand B (1977) H 80/62, a new selective β_1-adrenoceptor agonist. Acta Pharmacol Toxicol [Suppl] 41:51
11. Herbert P, Tinker J (1980) Inotropic drugs in acute circulatory failure. Intensive Care Med 6:101
12. Hutton I, Murray RG, Boyes RN, Rae AP, Hillis WS (1980) Haemodynamic effects of prenalterol in patients with coronary heart disease. Br Heart J 43:134
13. Johnsson G, Jordö L, Lundborg P, Rönn O, Welin-Fogelberg I, Wikstrand J (1978) Haemodynamic and tolerance studies in man of a new, orally active, selective β_1-adrenoceptor agonist H 80/62. Eur J Clin Pharmacol 13:163
14. Klein G, Schnelle K, Wirtzfeld A, Schinz A, Himmler FCh, Holzmüller W (1979) Pharmakodynamik des neuen β_1-spezifischen Agonisten Prenalterol (H 133/22) verglichen mit Dopamin und Dobutamin. Z Kardiol 68:282
15. Klein G, Schnelle K, Holzmüller W (1979) Pharmakodynamik des β_1-Adrenoceptor Agonisten Prenalterol verglichen mit Dobutamin. Naunyn-Schmiedebergs Arch Phrmacol [Suppl] 308:R20
16. Klein G, Wirtzfeld A, Himmler FCh (1980) Pharmakokinetik des β_1-Adrenoceptoragonisten Prenalterol (H 133/22). (Abstr 25) Z Kardiol 69:183
17. Knaus M, Pfister B, Dubach UC, Imhof PR (1978) Human pharmacology studies with a new orally active stimulant of cardiac adrenergic beta-receptors. Am Heart J 95:602

18. Leier CV, Webel J, Bush ChA (1977) The cardiovascular effects of the continous infusion of dobutamine in patients with severe cardiac failure. Circulation 56:468
19. Loeb HS, Bredakis J, Gunnar RM (1977) Superiority of dobutamine over dopamine for augmentation of cardiac output in patients with chronic low output cardiac failure. Circulation 55:375
20. Rönn O, Graffner Ch, Johnsson G, Jordö L, Lundborg P, Wikstrand J (1979) Hemodynamic effects and pharmacokinetics of a new selective $beta_1$-adrenoceptor agonist, Prenalterol, and its interaction with Metoprolol in man. Eur J Clin Pharmacol 15:9
21. Rönn O, Fellenius E, Graffner Ch, Johnsson G, Lundborg P, Svensson L (1980) Metabolic and haemodynamic effects and pharmacokinetics of a new selective $beta_1$-adrenoceptor agonist, Prenalterol in man. Eur J Clin Pharmacol 17:81
22. Sakamoto T, Yamada T (1977) Hemodynamic effects of dobutamine in patients following open-heart surgery. Circulation 55:525
23. Scott HH, Arthur GR, Boyes RN, Scott DB (1979) Cardiovascular effects of Prenalterol (H 133/22) in normal man. Br J Clin Pharmacol 7:365
24. Stoner JD, Bolen JL, Harrison DC (1977) Comparison of dobutamine and dopamine in treatment of severe heart failure. Br Heart J 39:536
25. Tuttle RR, Mills J (1975) Dobutamine: Development of a new catecholamine to selectively increase cardiac contractility. Circ Res 36:185

Discussion

Trap-Jensen, Copenhagen: We have determined dose-response curves with respect to the effects of i.v. prenalterol on heart rate (HR), cardiac output (CO), mean blood pressure (BP), forearm blood flow (FBF) and forearm vascular-resistance (FVR) in healthy young subjects and in patients with coronary artery disease. In both groups of subjects FBF was *increased* and FVR *decreased* after prenalterol. This strongly suggests that prenalterol acts as a peripheral vasodilator at the same time as it stimulates the myocardium.

Klein: Total peripheral resistance significantly decreased in our patients, but we interpreted this effect to be secondary to the improved cardiac output and not to be a direct vasodilating action. In a study on the effects of prenalterol in healthy subjects there was an increase in peripheral blood flow due to increased cardiac output, the diastolic blood pressure being only insignificantly lowered.

Bödigheimer, Munich: In your study there was no increase in heart rate following dobutamine. We found heart rate to be highly elevated after application of dobutamine and half of the increase in cardiac output was caused by the rise in heart rate.

Klein: The patients in our study had an initial heart rate of about 90 min^{-1}. When dobutamine was given in a dosage of 7.5 μg/kg/min there was no further increase in this elevated heart rate and the improvement of cardiac output was mainly due to an increase in stroke volume. However, in another study we infused dobutamine at higher doses. At 15 μg/kg/min there was a positive chronotropic effect of dobutamine indicating a dose dependency of this effect.

Zimmer, Munich: You have presented data concerning the effects of acute infusions of dopamine and dobutamine. Did you also apply the catecholamines for longer periods of time? I am asking because in our experience continuous infusion of dopamine or dobutamine for 48 h in rats leads to a consistant elevation of heart rate and dP/dt_{max}. Furthermore, under these conditions cardiac adenine nucleotide synthesis is stimulated. This effect may also be of importance in prolonging the hemodynamic actions.

Klein: We did not study systematically the hemodynamic effects of longer lasting infusions of dopamine or dobutamine. From clinical experience I can say, however, that these adrenergic drugs remain effective when infused for several days or discontinuously infused for weeks.

Effect of Vasodilator Therapy on Sympathetic Activity in Patients with Impaired Left Ventricular Function *

J. Manthey, R. Dietz, E. Hackenthal, Y. N. Ke, H. Leinberger, A. Schömig, F. Schwarz, W. Kübler

Summary

To determine whether changes in sympathetic nervous activity may counteract or modify the effects of vasodilator therapy in congestive heart failure, 31 patients with impaired left ventricular function were studied. The patients received either oral hydralazine (1 mg/kg) in combination with isosorbide dinitrate (40 mg) (15 patients, group I) or intravenous hydralazine (0.4 mg/kg) without isosorbide dinitrate (16 patients, group II). Circulating catecholamines were measured before and after drug administration and compared with hemodynamic data.

Eight patients of group I (=responders) showed a decrease in systemic vascular resistance $\geqq 10\%$ of pretreatment values and an increase in cardiac index from 2.6 ± 1 to 3.6 ± 0.3 ml/min/m^2 after therapy whereas the remaining 7 patients of group I (=nonresponders) responded poorly to vasodilator administration (decrease in systemic vascular resistance $< 10\%$, no significant increase in cardiac index). After treatment, plasma norepinephrine increased in nonresponders from 274 ± 40 to 396 ± 54 pg/ml ($P < 0.001$). Epinephrine levels were not significantly affected by the drugs. In group II, all patients showed a marked decrease in systemic vascular resistance and an increase in cardiac index.

Norepinephrine levels increased in 11 and decreased in 5 patients. When patients of group II were subdivided according to whether plasma norepinephrine increased (group IIA) or decreased (group IIB), a significant difference was found in pretreatment norepinephrine concentrations (group IIA: 130 ± 5, group IIB: 440 ± 46 pg/ml, $P < 0.001$). Patients of group IIA had less depressed stroke volume index (40 ± 2 ml/m^2) and less elevated diastolic pulmonary artery pressure (15 ± 1 mmHg) than patients of group IIB (stroke volume index: 23 ± 2 ml/m^2 ($P < 0.001$); diastolic pulmonary artery pressure: 28 ± 4 mmHg ($P < 0.001$)). These findings suggest that in a significant proportion of patients vasodilator administration is accompanied by an increase in sympathetic neuronal activity which may counteract and limit the beneficial effects of vasodilator therapy in these patients.

Introduction

Augmented activity of the sympathetic nervous system as assessed by elevated levels of circulating catecholamines has been demonstrated in patients with heart failure [1]. This compensatory sympathetic stimulation plays an important role in maintaining contractility when myocardial function is depressed. In addition, this increase in sympathetic drive has been considered to be responsible in part for the

* Supported by the Deutsche Forschungsgemeinschaft (SFB 90, Kardiovaskuläres System, University of Heidelberg)

elevated systemic vascular resistance observed in cardiac failure. Because increased vascular resistance may further depress left ventricular pump function, vasodilator therapy has been used to counteract this vasoconstriction and to improve cardiac performance [2, 4]. Little attention has been focused, however, on the sympathetic response to such vasodilator therapy. Administration of vasodilators may cause a further stimulation of the sympathetic nervous system which may limit the beneficial effects of these drugs [7]. Therefore, in the present study we evaluated the effects of vasodilator therapy on the sympathetic activity in patients with chronic congestive heart failure.

Methods

Thirty-one patients (27 men, 4 women) ranging in age from 21 to 58 (mean 47) years with functional class III or IV congestive heart failure (CHF) were studied. All patients had undergone right and left heart catheterization and coronary angiography. The cause of heart failure was congestive cardiomyopathy in 28 and coronary heart disease in 3 patients. Left ventricular ejection fraction as determined by contrast angiography ranged from 0.08 to 0.46 (mean 0.27).

Measurements of pulmonary artery pressure and cardiac output were performed using a Swan-Ganz catheter and a bedside computer (Edwards Laboratory). Arterial pressure was measured by the cuff technique. Systemic vascular resistance was calculated as the mean arterial pressure divided by cardiac output. Blood specimens were obtained for measurement of plasma catecholamines [3], plasma renin activity [5] and plasma vasopressin [10]. Measurements were made at supine rest before and 60–90 min after vasodilator administration. After control measurements, the patients were given either hydralazine (1 mg/kg) orally and isosorbide dinitrate (40 mg) orally (15 patients, group I) or hydralazine (0.4 mg/kg) intravenously (16 patients, group II).

Statistical analysis was performed using Student's t-test for paired or unpaired data. All results are given as mean $\pm$SEM.

Results

Oral Vasodilator Administration

After hydralazine-nitrate administration, mean cardiac index increased from 2.6 ± 0.1 to 3.1 ± 0.2 l/min/m² ($P < 0.01$) as a consequence of a reduction in mean systemic vascular resistance from 21 ± 1 to 17 ± 1 units ($P < 0.01$). Comparison of individual values, however, demonstrated that several patients did not show a marked hemodynamic improvement after vasodilator therapy. Therefore, the patients were classified into two groups (IA and IB) according to whether the vasodilator regimen produced hemodynamic improvement or not. The responders (8 patients, group IA) were defined by a decrease in sympathetic vascular resistance of 10% or more. The nonrespondes (7 patients, group IB) had a decrease in systemic vascular resistance of less than 10% of pretreatment values. The hydralazine-nitrate combination produced a significant increase in cardiac index from 2.6 ± 0.1 to 3.6 ± 0.3 ml/min/m² ($P < 0.01$) and in stroke volume index from 32 ± 2 to 41 ± 4 ml/m² ($P < 0.01$) in group IA. No increase occurred in group IB patients [car-

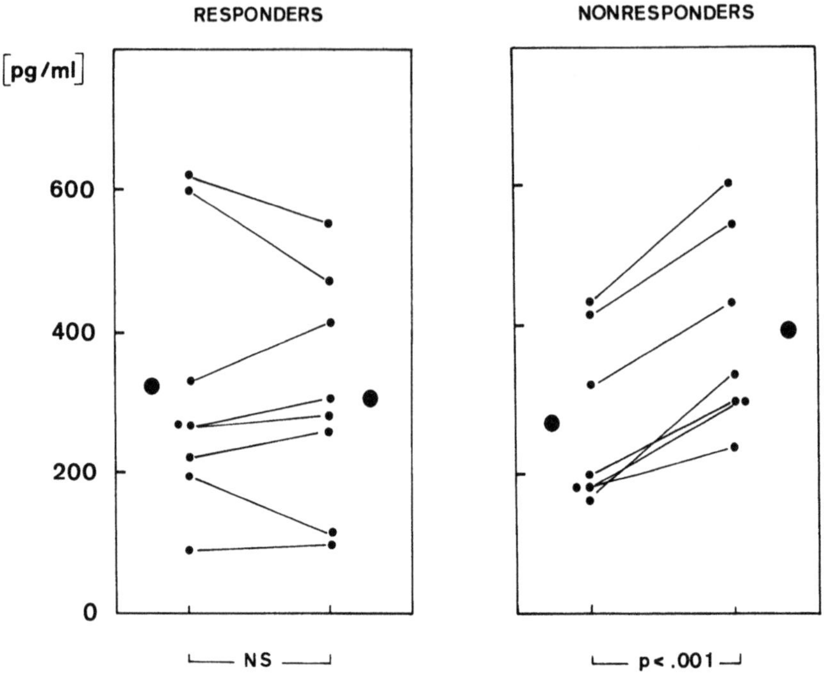

Fig. 1. Plasma norepinephrine concentrations before and after administration of a hydralazine-nitrate combination in 15 patients (group I). *Large points* = mean values. NS = not significant

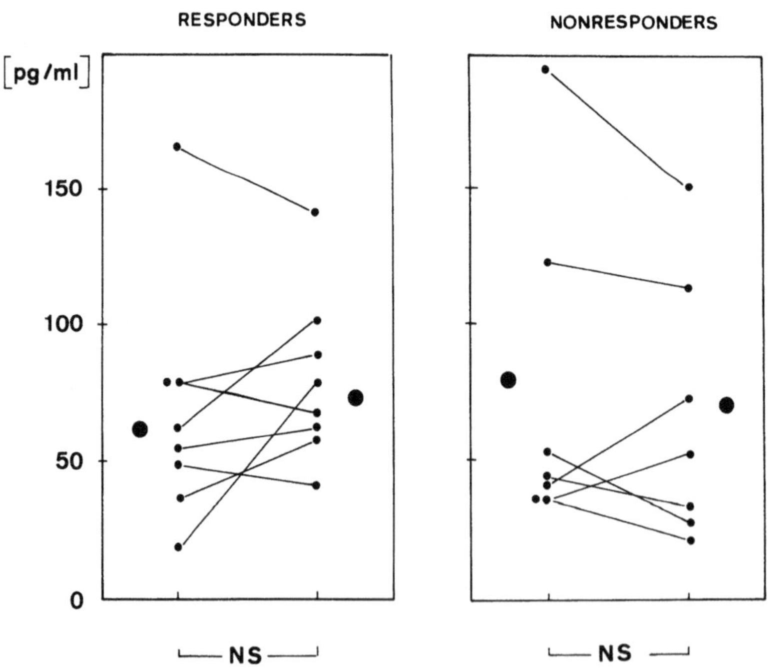

Fig. 2. Plasma epinephrine concentration before and after administration of a hydralazine-nitrate combination in 15 patients (group I). *Large points* = mean values. NS = not significant

diac index: 2.8±0.1 (control) vs 2.6±0.2 ml/min/m² (vasodilators), stroke volume index: 41±4 (controls) vs 37±4 ml/m² (vasodilators)]. Diastolic pulmonary artery pressure fell slightly from 24±3 to 19±3 mmHg ($P<0.05$) in group IA and from 25±3 to 21±2 mmHg ($P<0.05$) in group IB patients.

The individual values for plasma catecholamine concentrations are given in Figs. 1 and 2. Mean plasma norepinephrine levels did not change after vasodilator therapy in group IA but increased significantly in group IB patients ($P<0.001$). In contrast, mean plasma epinephrine concentrations were unchanged in both groups after the hydralazine-nitrate combination. Plasma renin activity and plasma vasopressin levels increased in all patients but no significant differences between responders and nonresponders were observed.

Intravenous Vasodilator Administration

Intravenous hydralazine administration resulted in hemodynamic improvement in all patients. Systemic vascular resistance decreased from 19.5±1.4 to 9.7±0.6 units ($P<0.001$). Cadiac index rose from 2.8±0.2 to 4.6±0.2 l/min/m² ($P<0.001$) and stroke volume index from 34±3 to 49±2 ml/m² ($P<0.001$). Diastolic pulmonary artery pressure decreased from 19±2 to 15±1 mmHg ($P<0.001$). No significant changes in mean plasma catecholamine levels after hydralazine could be observed when all patients were considered together (Fig. 3). Patients of group II were subdivided into those showing an increase in plasma norepinephrine levels (11 patients, group IIA) or a decrease (5 patients, group IIB). A significant difference was found in pretreatment norepinephrine concentrations in the 2 groups (group IIA: 130±5, group IIB: 440±46 pg/ml ($P<0.001$). The patients of group IIA had less depressed stroke volume index (40±2 ml/m²) and less elevated diastolic pulmonary

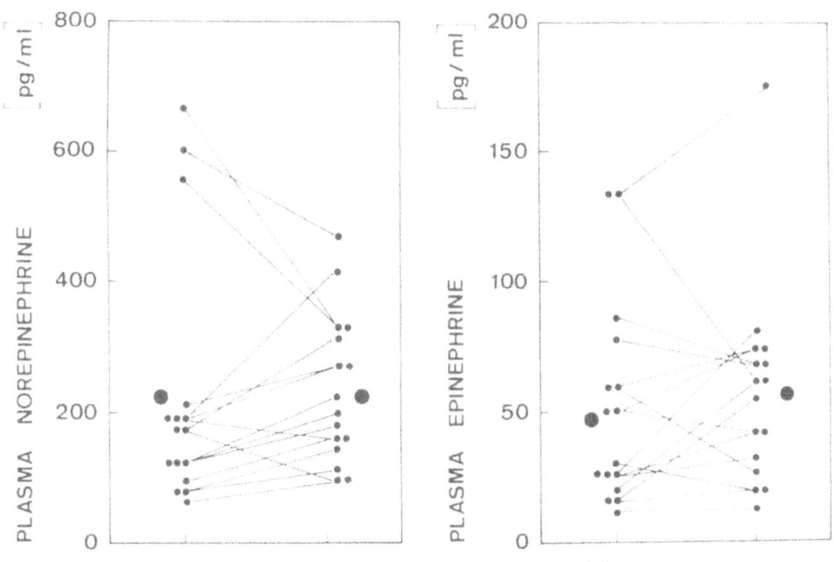

Fig. 3. Plasma norepinephrine and epinephrine concentrations in 16 patients (group II) before and after intravenous hydralazine. *Large points* = mean values. NS = not significant

artery pressure (15±1 mmHg) than patients of group II B [stroke volume index: 23±2 ml/m² ($P<0.001$), diastolic pulmonary artery pressure: 28±4 mmHg ($P<0.001$)]. The fall in systemic vascular resistance was less pronounced in group II A (–43% of pretreatment values) than in group II B [–59% of pretreatment values ($P<0.001$)].

Discussion

Several vasodilators improve hemodynamics in patients with CHF by decreasing pulmonary artery pressure and by increasing cardiac output [2]. More recent studies, however, have indicated that the response to vasodilator therapy in heart failure is variable and that adverse cardiovascular reactions may occur [6, 7, 8]. This study shows that vasodilator administration to patients with CHF may be accompanied by a marked increase in the activity of vasoconstrictor systems.

Sympathetic Activity After Oral Vasodilator Administration

Our finding of a high proportion of nonresponders among the patients treated with an oral hydralazine-nitrate combination is in accordance with previous studies [4, 7, 9]. It should be stressed that patients of group I A (responders) and I B (nonresponders) has markedly different changes in systemic vascular resistance after therapy but had a similar fall in left ventricular filling pressure. The reasons for the

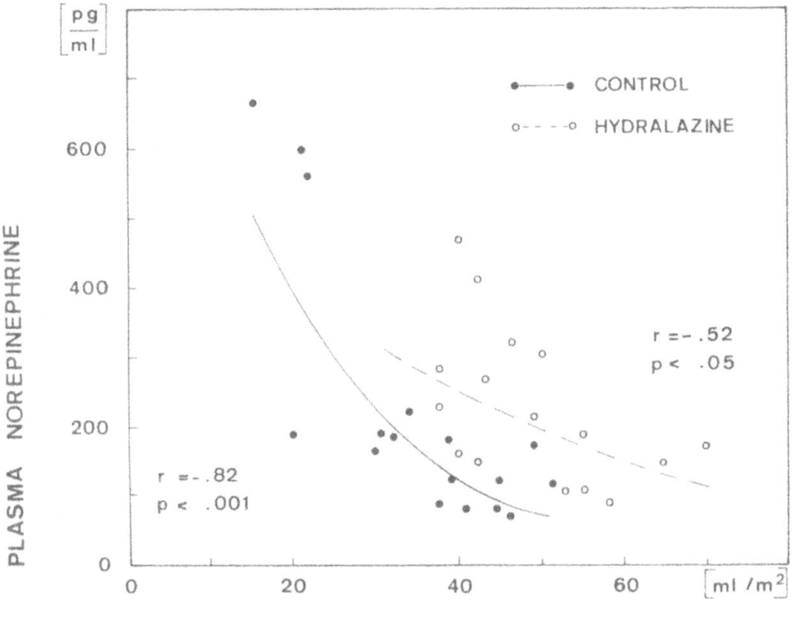

Fig. 4. Relationship between stroke volume index and plasma norepinephrine concentrations in group II patients before and after intravenous hydralazine. The regression equations are significantly different in slope ($P<0.05$)

poor hemodynamic improvement in the responder group after oral vasodilator administration are not fully understood. Malabsorption as well as enhanced metabolic inactivation of hydralazine may be responsible in part for the failure to respond adequately in group I B patients. The significant increase in plasma norepinephrine observed in nonresponders but not in responders suggests that the increase in sympathetic neuronal activity in responders may contribute to the failure of oral vasodilator therapy to produce a fall in systemic vascular resistance in these patients.

Sympathetic Activity After Intravenous Hydralazine

After intravenous hydralazine therapy, all patients showed a marked hemodynamic improvement. However, our data suggest, that the change in plasma norepinephrine levels observed after hydralazine may modulate the drug mediated effects. In patients with severely depressed stroke volume and elevated pulmonary artery pressure the hemodynamic improvement is accompanied by a fall in plasma norepinephrine levels (Fig. 4). This depression of sympathoneuronal activity further increases the vasodilator effect of hydralazine by additional reduction of catecholamine induced peripheral vasoconstriction. In this way, the beneficial effect of hydralazine is increased in the subgroup of severely ill patients. In contrast, the increase in plasma norepinephrine levels observed in patients with mildly depressed stroke volume and slightly elevated pulmonary artery pressure may counteract and limit the drug mediated effects.

References

1. Chidsey CA, Braunwald E (1966) Sympathetic activity and neurotransmitter depletion in congestive heart failure. Pharmacol Rev 18:685
2. Cohn JN, Franciosa JA (1977) Vasodilator therapy of cardiac failure. N Engl J Med 297:27
3. Da Praea M, Zürcher G (1976) Simultaneous radioenzymatic determination of plasma and tissue adrenaline, noradrenaline, and dopamine within femtomol range. Life Sci 19:1161
4. Franciosa JA, Cohn JN (1979) Immediate effects of hydralazine-isosorbide dinitrate combination on exercise hemodynamics in patients with left ventricular failure. Circulation 59:1085
5. Haber E, Koerner T, Page LB, Kliman B, Purnode A (1969) Application of a radioimmunoassay for angiotensin I to the physiologic measurements of plasma renin activity in normal human subjects. J Clin Endocrinol 29:1349
6. Laslett LJ, DeMaria AN, Amsterdam EA, Mason DT (1978) Hydralazine-induced tachycardia and sodium retention in heart failure: hemodynamic and symptomatic correction by prazosin therapy. Arch Intern Med 138:819
7. Manthey J, Dietz R, Leinberger H, Schmidt-Gayk H, Schömig A, Schwarz F, Kübler W (1980) Vasodilator therapy in heart failure: limited by activation of vasoconstrictor mechanisms? (Abstr) Circulation [Suppl III] 62:259
8. Packer M, Meller J, Medina N, Gorlin R, Herman MV (1980) Importance of left ventricular chamber size in determining the response to hydralazine in severe chronic heart failure. N Engl J Med 303:250
9. Packer M, Meller J, Medina N, Gorlin R, Herman MV (1980) Dose requirements of hydralazine in patients with severe chronic congestive heart failure. Am J Cardiol 45:655
10. Uhlich E, Weber P, Gröschel-Stewart U, Röschlau T (1975) Radioimmunoassay of argininvasopressin in human plasma. Horm Metab Res 7:501

Discussion

Liebau, Würzburg: Your results are specific for hydralazine therapy in patients with congestive heart failure because baroreceptor reflex mechanisms are left intact. When captopril or prazosin are employed these increases in vasopressor hormone activity are not observed.

Manthey: An increase in vasopressin levels is only observed after oral administration of a hydralazine-nitrate combination and it is most likely mediated by the nitrate-induced preload reduction and low pressure baroreceptor stimulation. Whether changes in sympathetic activity after prazosin or captopril are similar to those observed after hydralazine administration remains to be clarified.

Kopin, Bethesda: It appears that vasodilators will only reduce plasma catecholamine levels in patients in whom high initial plasma norepinephrine (NE) levels are found. This is due to a reduction in the degree of cardiac failure which then reduces plasma NE levels and masks the increased levels which would result if blood pressure fell. Was there a fall of blood pressure in your patients which was of the same magnitude as in patients in whom vasodilation evoked an increase in plasma NE?

Manthey: The fall in blood pressure observed in the patients who showed a decrease in norepinephrine after treatment with hydrazaline was less pronounced than in patients with an increase in NE after therapy. I agree that in the first group (with high NE) the decrease in blood pressure does not cause an increase in NE levels because the effect of the baroreceptor reflex is counterbalanced by a marked improvement in cardiac function.

Betablockers and Calcium Antagonists: Mode of Action in Angina pectoris

P. R. Lichtlen, H. J. Engel, I. Amende

Summary

The different effects of betablockers and calciumantagonists on LV hemodynamics and coronary blood flow (CBF) are discussed based on studies performed with atenolol, propranolol and nifedipine. LV hemodynamics were recorded in coronary patients at rest, measuring LV volumes and pressures simoultaneously during angiography with the use of Millar tip manometers. CBF was assessed at rest and during atrial pacing up to ischemia (ST depression ± angina) without and with the drug using the precordial xenon clearance technique (gamma camera, CDC 1700 computer). Basic differences were present both for hemodynamics as for coronary dynamics: under betablockade LV contractility, cardiac output and ejection fraction decreased, enddiastolic volume increased and the pressure-volume relation was shifted upwards and flattened, indicating both an increase in diastolic wall stress and a decrease in passive elastic stiffness. During atrial pacing CBF decreased both in the normal as well as in the ischemic (poststenotic) zone, when compared to control, a result of the decrease in MVO_2 due to the reduced contractility. After Nifedipine LV function improved slightly, contractility and cardiac output increasing, and peripheral resistance decreasing; enddiastolic volume and pressure remained unchanged. After the drug, CBF during pacing showed a further (not sign.) increase in the ischemic zone and remained unchanged in the normal area, inspite of a sign. decrease in MVO_2 i.e. in the rate pressure product. Hence, calcium antagonists improve stable angina, i.e. ischemia due to increased O_2 demand (exercise, pacing) by increasing O_2 supply in the ischemic zone and decreasing demand through afterload reduction, whereas betablockers decrease both O_2 demand and supply, the latter by autoregulatory changes of CBF.

Today, the treatment of angina pectoris is based on nitrates, betablockers and most recently, on calcium antagonists, three different therapeutic principles with different indications. Their mode of action varies both with regard to their influence on myocardial oxygen consumption as well as on coronary blood flow. The following presentation, which concentrates mainly on the difference between betablockers and so-called calcium antagonists, describes the hemodynamic effects and the action on the coronary circulation of both groups of drugs in patients with coronary artery disease.

Hence, in studying these drugs, four systems have to be taken into consideration:

1. *The peripheral circulation,* that is, the effect on venous tone (preload) as well as on arteriolar tone (afterload), the former being best studied by measuring left ventricular enddiastolic volume and pressure, the latter by recording changes in systemic blood pressure.

2. *The myocardium,* especially contractility (dp/dt max and dp/dt min), wall motion (V_{CF}), and left ventricular geometry.
3. *The coronary circulation,* i.e., the action on coronary flow in the normal, yet even more importantly, in the poststenotic, eventually ischemic zone, and of special interest since several years, the behavior of smooth muscle tone in extramural coronary arteries, especially in high-grade obstructions.
4. *The conduction system,* i.e., sinus node- and av-node-function.

During the last few years, we have studied both central and peripheral hemodynamics as well as coronary dynamics of several betablockers and calcium antagonists. In the following presentation, as typical examples, the effects of two betablockers, i.e., atenolol (Tenormin), a beta-1-blocker, and propranolol (Dociton), a beta-1- and beta-2-blocker, both without intrinsic activity, and of one calcium antagonist, i.e., nifedipine (Adalat) will be demonstrated more extensively.

Hemodynamic Changes

After Betablockade with Atenolol (5 mg intravenously)

Methodology. The effect both of atenolol and nifedipine on left ventricular function was studied using Millar PC 481 tip manometers, allowing simultaneous pressure measurements and contrast injections into the left ventricle. High fidelity LV

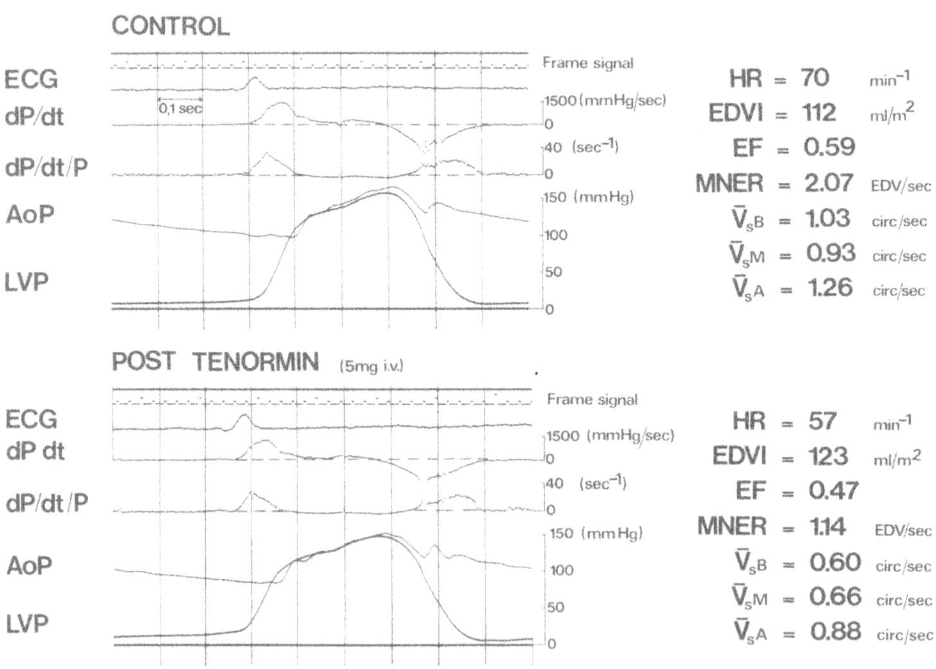

Fig. 1. Effect of atenolol (Tenormin) 5 mg i.v. on hemodynamics and left ventricular wall motion in a 41-year-old patient (♂) without coronary artery disease. *Left:* the original tracing; *from top to bottom:* ECG, dp/dt max and dp/dt min, dp/dt/P, aortic and left ventricular pressure. *Right:* specific data before and after the betablockade. (For details, see text.) (From: Postgraduate Medical Journal (1977), Suppl 3, Vol 53, p 86)

pressures were sampled at cine frame rate, digitized and stored on the same cine frame. LV angiograms were obtained in 40° right anterior oblique projection at 80 frames/second.

LV volumes were calculated by the use of a computer-assisted video system (Grafomed, Philips) at 25 msec interval, applying Simpson's rule for single plane projection (*Simon* 1979). The computer provided printouts of the pressure-time-, volume-time relation and of the pressure-volume curve.

Diastolic compliance, that is, pressure-volume relation was calculated according to *Mirsky* et al. [20] and *Gaasch* et al. [8].

Results. Figure 1 shows a typical example of the changes in left ventricular and aortic pressure before and 15 min after administration of 5 mg atenolol i.v. The betablocker led to a marked decrease in heart rate, systolic LV and aortic pressure, ejection fraction and mean velocity of fiber shortening in the basal, mid- and apical segment; in contrast, there was a marked increase in enddiastolic pressure as well as in enddiastolic volume. Table 1 summarizes the results found in 15 patients with severe coronary artery disease, mainly triple vessel disease, yet normal or only mildly hypokinetic left ventricles. On average, a significant decrease in heart rate, dp/dt max, ejection fraction and mean normalized ejection rate as well as in circumferential fiber shortening of the basal, middle and apical segment was observed, associated with an increase in systolic and enddiastolic ventricular pressure as well as volumes.

A typical example of the diastolic logarithmic pressure-volume relation before and after atenolol obtained by computer analysis is shown in Fig. 2.

In this typical and representative example, two changes can be observed after administration of atenolol:

1. A marked increase in the intercept of the regression line of the log diastolic pressure-volume relation on the log pressure scale (ordinate) suggesting a marked increase in diastolic wall stress.

Table 1. Average changes of the most important parameters before and after atenolol (5 mg i.v.) in 15 patients with coronary artery disease, mostly triple vessel disease, and normal or hypokinetic left ventricular wall motion. (Patients with large akinetic areas were excluded)

	Control	Atenolol	*P*
HR (min^{-1})	68±10	60±7	<0.001
LVSP (mmHg)	131±24	137±22	n.s.
LVEDP (mmHg)	15±6	19±6	<0.001
dp/dt max (mmHg/sec)	1710±328	1505±282	<0.025
EDVI (ml/m^2)	88±18	95±19	<0.001
ESVI (ml/m^2)	35±11	42±15	<0.001
EF (%)	61±8	56±8	<0.001
MNER (sec^{-1})	2.05±0.41	1.70±0.32	<0.001
V_sB (circ/sec)	1.17±0.25	0.99±0.23	<0.001
V_sM (circ/sec)	1.03±0.43	0.85±0.3	<0.001
V_sA (circ/sec)	1.21±0.47	1.01±0.37	<0.001

Abbreviations: HR, heart rate; LVSP, left ventricular systolic pressure; LVEDP, left ventricular enddiastolic pressure; dp/dt max, maximum rise of systolic pressure; EDVI, enddiastolic volume index; ESVI, endsystolic volume index; EF, ejection fraction; MNER, mean normalized ejection rate; V_S, velocity of circumferential fiber shortening (B, basal; M, middle; A, apical segment)

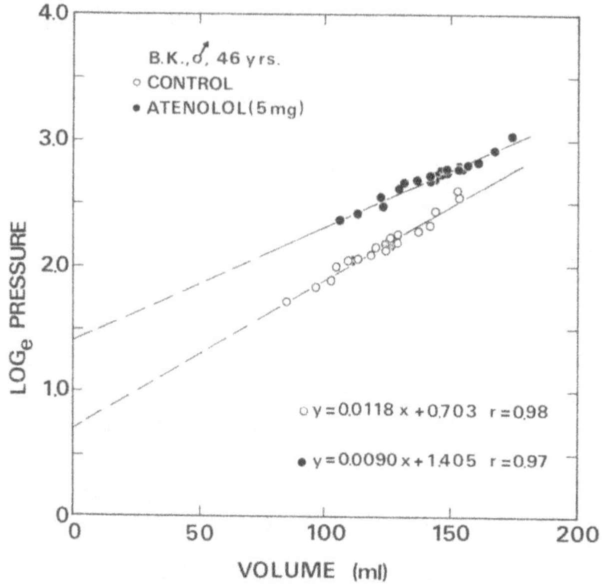

Fig. 2. Effect of atenolol on the pressure-volume relation in a 46-year-old patient (♂) with normal coronary arteries. *Ordinate:* diastolic pressure in logarithmic scale. *Abscissa:* diastolic volume. Note the upward-shift of the pressure-volume relation and the decrease in the slope of the regression line after the betablocker (changes from at the lowest diastolic pressure until the beginning of the a-wave). (For details, see text.) (From: *Amende I, Simon R, Hood WP, Lichtlen PR:* The effects of the betablocker Atenolol and Nitroglycerin on left ventricular function and geometry in man. Circulation 60:836, 1979)

Table 2. Hemodynamic changes in 7 patients following Nifedipine. (From: *Lichtlen P, Engel HJ, Amende I, Rafflenbeul W, Simon R:* Mechanisms of various antianginal drugs. Relationship between regional flow behaviour and contractility. Third International Adalat Symposium. New therapy of ischemic heart disease. Ed. by *AD Jatene* and *PR Lichtlen,* Excerpta Medica, Amsterdam 1976, p. 14)

	HR (min^{-1})	CO (l/min)	Mean AO (mm Hg)	PVR (dyn sec cm^{-5})	dp/dt_{max} (mm Hg/sec)
Before	62.6	5.87	87.8	1217	1421
After	75.8	7.33	89.1	1011	1606
Diff.	+13.2	+1.46	+2.3	−206	+185
P<	0.001	0.01	n.s.	0.005	0.05

	V_{pm} (ML/sec)	EDVI (ml/m²)	EF (%)	MNER EF/ET (sec^{-1})	$\overline{V_{CF}}$ (circ/sec)	Segmental changes %B	%M	%A
Before	36.0	89.6	58	1.65	0.958	29.8	29.6	33.2
After	35.0	89.6	62	1.83	0.986	30.6	30.4	37.2
Diff.	−1.0	0	+4	+0.18	+0.028	+0.8	+0.8	+4.0
P<	n.s.	n.s.	n.s.	n.s.	n.s.	n.s.	n.s.	n.s.

Abbreviations as in Table 1; CO, cardiac output; PVR, peripheral vascular resistance; V_{pm}, velocity of fiber shortening expressed in muscle length (ML); V_{CF}, mean velocity of fiber shortening

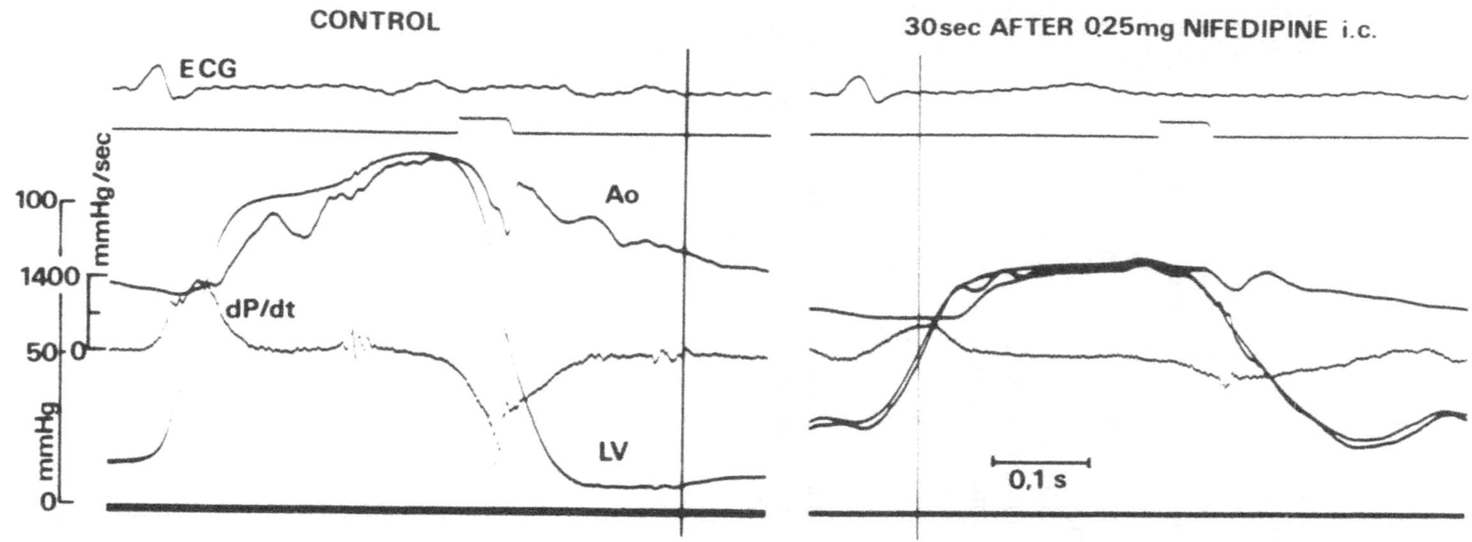

HR	71	min^{-1}
LVP	117/16	mmHg
LV pos. dP/dt	1382	mmHg/sec
LV neg. dP/dt	-1927	mmHg/sec
CS-FLOW	109	ml/min

HR	88	min^{-1}
LVP	82/29	mmHg
LV pos. dP/dt	618	mmHg/sec
LV neg. dP/dt	-691	mmHg/sec
CS-FLOW	163	ml/min

Fig. 3. Effect of intracoronary administration of nifedipine on left ventricular dynamics in a 49-year-old patient (♂) with coronary artery disease. LV pressure was measured with a Millar-tip Manometer; coronary sinus flow (*CS*) was assessed by means of the thermodilution technique. (For details, see text)

2. A marked decrease in the slope of the regression line indicating a decrease in diastolic passive elastic stiffness (volume stiffness). Such a behavior was observed in six patients after atenolol [16].

After Administration of 20 mg Nifedipine Sublingually

Table 2 summarizes the hemodynamic changes after administration of nifedipine observed in seven patients with coronary disease, yet again normal or only mildly hypokinetic left ventricles, ten minutes after administration of 20 mg nifedipine sublingually. On the average, a significant increase in heart rate and cardiac output as well as dp/dt max, and a significant decrease in peripheral vascular resistance were seen, whereas enddiastolic volume, ejection fraction, and mean normalized ejection rate remained unchanged. There were also no significant alterations in mean velocity of circumferential fiber shortening as well as in percent shortening of the basal, middle and apical segment.

A completely different behavior, however, was observed when nifedipine was administered directly into the left coronary artery, a typical example being shown in Fig. 3. Thirty seconds after intracoronary administration of 0.25 mg nifedipine – a rather large dose – a massive decrease in systolic left ventricular pressure as well as max positive and negative dp/dt could be observed; on the other hand, heart rate increased markedly and so did coronary sinus blood flow measured by the thermodilution technique.

Figure 4 demonstrates a typical example of the behavior of diastolic functions after administration of nifedipine: the pressure-volume relation changed only minimally, the intercept of the regression line of the log diastolic pressure-volume

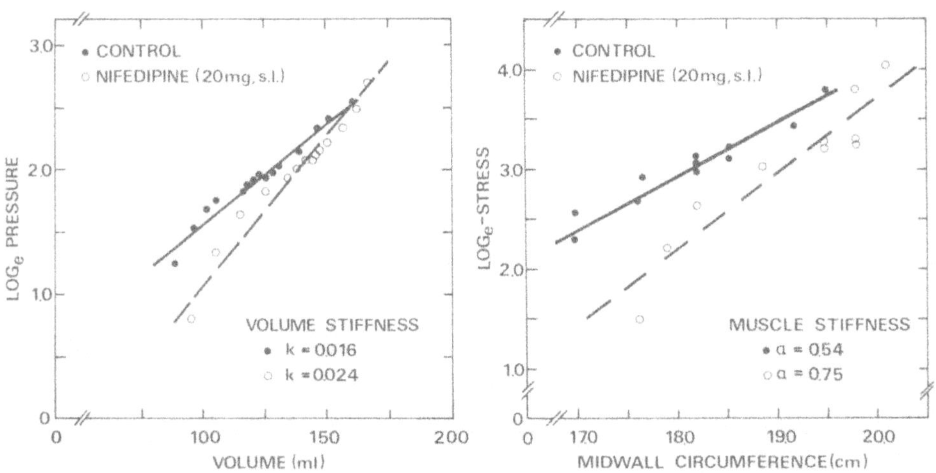

Fig. 4. Effect of nifedipine on the pressure-volume and the stress-circumference relationship in a 40-year-old patient (♂) with triple vessel coronary disease and normal left ventricle. *Left:* pressure-volume relation. Abbreviations as in Fig. 3. Note the mild, not significant increase in early diastolic pressure; at enddiastole, the diastolic pressures before and after the drug are again identical. *Right:* stress-midwall-circumference relation. Note the increase in wall stress for a given midwall circumference associated with a decrease in the regression line after nifedipine. (For details, see text)

relation on the log pressure scale (ordinate) increasing mildly, enddiastolic volumes, however, being almost identical due to a mild decrease of the slope of the regression line throughout diastole. The study of six patients, all with normal or only mildly hypokinetic left ventricles, revealed no significant changes in preload, that is, enddiastolic volume (*Amende,* 1980).

Comparison Between the Hemodynamic Changes Observed During Betablockade and Treatment with Calcium Antagonists

The following hemodynamic differences between calcium antagonists and betablockers could be observed: When administered systemically, i.e., sublingually or intravenously, the calcium antagonist shows a mild, yet significant increase in dp/dt max, probably due to the increase in heart rate, caused by the raised baroreceptor activity following the decrease in blood pressure, i.e., in afterload. In contrast, the betablocker leads to a significant decrease in dp/dt max, that is, contractility associated with a profound decrease in heart rate, whereas afterload in the acute investigation more or less remains unchanged.

On the other hand, the calcium antagonist produces a decrease in contractility only when administered directly into the coronary system, where the drop in blood pressure is secondary. On the diastolic performance, the two drugs also act in an opposite way; nifedipine mildly increases passive elastic volume- and muscle-stiffness, whereas atenolol tends to decrease both of them. Finally, with the calcium antagonists, peripheral resistance, i.e., afterload decreases markedly, whereas cardiac output increases; in contrast, the betablocker – at least in the acute study – tends to increase peripheral resistance while it clearly reduces cardiac output due to the marked decrease in contractile force. Finally, it is important to note that nifedipine does not influence preload, i.e., venous return, at least not in situations of normal enddiastolic volumes, whereas the betablocker is followed by an increase both in enddiastolic volume and pressure, this, however, being a result of decreased myocardial contractility and not of primary changes in venous tone.

Coronary Dynamics, Regional Myocardial Blood Flow

Methods of Flow Recording

Measurements of regional myocardial blood flow were performed by the use of the precordial xenon residue detection technique in left anterior oblique projection of 40° following intracoronary injections of 10 to 20 mCi of xenon 133. Recordings of the precordial isotope activity were done by a gamma camera with 4096 matrix points (approximately 400–600 matrix points/heart) and low energy collimation, connected to a CDC 1700 computer system (32 K, 16 bits, deadtime of the whole system 4 μs). Flow calculations were performed in a monoexponential fashion following the Kety-Schmidt formula (for details, see: [13, 18, 19]). Only the first 30 s after peak activity has been reached were calculated; matrix points with less than 15% of maximum peak activity and reaching the peak later than 15 s after maximum peak activity were excluded from the study; furthermore, the standard deviation of the washout for each matrix point had to be less than 30%. The

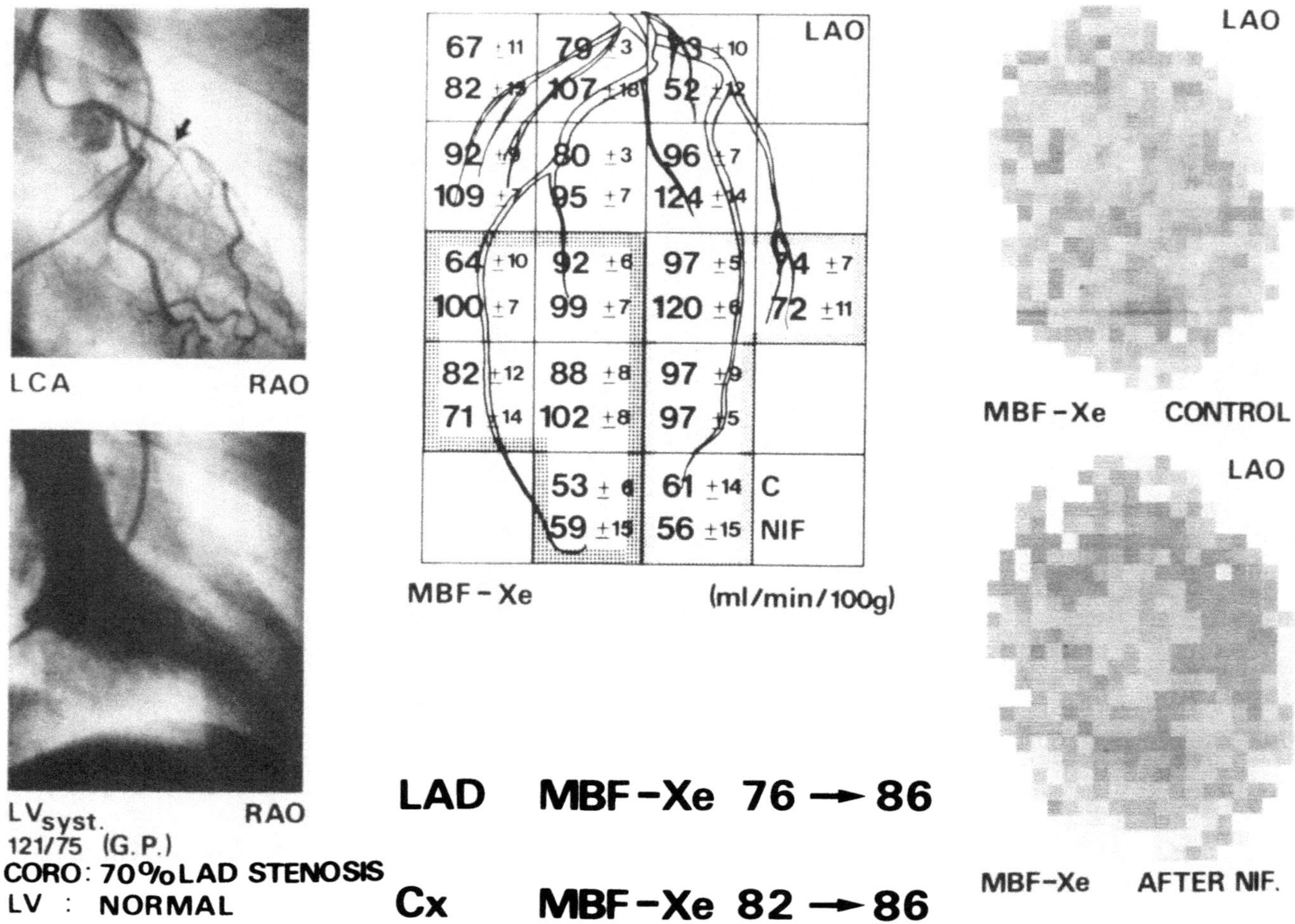
LCA
RAO
LV syst.
RAO
121/75 (G.P.)
CORO: 70% LAD STENOSIS
LV : NORMAL
67 ±11
82
107
52
92
80 ±3
96 ±7
109
95 ±7
124
64 ±10
92
97
±7
100 ±7
99 ±7
120
72 ±11
82 ±12
88 ±8
97
71 ±14
102 ±8
97
53
61 ±14
C
59
56 ±15
NIF
LAO
MBF - Xe
(ml/min/100g)
LAD MBF-Xe 76 → 86
Cx MBF-Xe 82 → 86
LAO
MBF-Xe CONTROL
LAO
MBF-Xe AFTER NIF.

washout function, i.e., myocardial blood flow was displayed numerically, both in a global and regional way and expressed in ml/min/100 g myocardium; it included areas of 25 matrix points or regions perfused by large coronary branches according to the individual coronary angiogram. In addition, myocardial blood flow was displayed in an analog way as so-called flow scan, indicating flow values for each matrix point in various colors.

In order to compare regional myocardial blood flow in normal and poststenotic areas in the same patient, especially also during right atrial pacing, mainly patients with severe single vessel disease (obstructions > 75%) were included.

A typical example of the flow recording in the normal and poststenotic zone at rest before and after 20 mg nifedipine sublingually is demonstrated in Fig. 5.

Right Atrial Pacing

Both propranolol and nifedipine were studied at rest and during right atrial pacing before and after drug administration. Right atrial pacing was performed in stepwise increases of 20 beats/min up to the appearance of ST-depressions greater than 0.1 mV, associated in most of the cases with typical angina pectoris. When this level of ischemia-producing heart rate was reached, ^{133}xenon was injected into the main left coronary artery, the catheter was immediately withdrawn, and pacing was continued at the same heart rate throughout the entire flow measurement. Then, pacing was stopped and, after reaching a steady state at rest, the drug was administered. Ten minutes after administration of 20 mg nifedipine sublingually and 15 min after administration of 5 mg propranolol intravenously, pacing was again installed in the same way and up to the same heart rate as before, and the flow measurements were repeated. From heart rate and systolic blood pressure, the pressure-rate-product was calculated.

Patients

Eleven patients were studied with Nifedipine and 13 with Propranolol. In the Nifedipine-group, seven patients showed an isolated >70%-obstruction of the

---

◀ **Fig. 5.** Typical example of the assessment of myocardial blood flow with the precordial xenon clearance technique at rest before and after nifedipine (20 mg s.l.). *Top left:* left coronary angiogram in RAO-projection demonstrating a high-degree obstruction in the proximal third of the LAD. *Below:* endsystolic left ventricular angiogram in RAO projection, demonstrating normal wall motion. *Middle:* numerical printout of myocardial blood flow in areas of 25 matrix points in LAO-projection. The poststenotic area of the LAD and the normal control area perfused by the left circumflex are especially outlined. On *top of each quadrangle:* control flow values ± standard deviation; *bottom:* flow values 10 min after nifedipine. Most quadrangles show a clear increase in flow, in the poststenotic area increasing from an average of 76 ml/min/100 g myocardium to 86 ml/min/100 g myocardium, in the normal zone from 82 to 86 ml/min/100 g myocardium. *Right:* flow scan in colors in LAO-projection; *top:* at control; *below:* 10 minutes after nifedipine. Each small quadrangle represents flow of one matrix point; highest flow values are shown in red, the scale then decreases in 10% steps from red to orange to yellow, green and blue. Note that after nifedipine there is a further increase in flow in the circumflex area, but also in the normal zone, flow now being almost homogeneous throughout the heart, i.e., identical in the poststenotic and in the normal area (86 ml/min/100 g myocardium). (From: *Lichtlen PR, Engel HJ* (1979) Assessment of regional myocardial blood flow using the inert gas washout technique. Cardiovascular Radiology 2:203, Springer-Verlag, New York 1979)

LAD, two of the LCX, and two of the RCA; in the Propranolol-group, in eight patients, the LAD, and in five patients, the LCX were severely obstructed. Hence, the normal control area was perfused 15 times by the left circumflex, and nine times by the left anterior descending branch. All patients showed a normal or only mildly hypokinetic left ventricle.

Results

Average heart rate leading to significant, that is, more than 0.1 mV ST-depression and angina pectoris for the patients given Nifedipine, amounted to 139 beats/min, for those given propranolol, to 140 beats/min (Table 3). During control pacing, the nifedipine patients showed a systolic aortic pressure of 131 mmHg; during rapid atrial pacing, it decreased significantly by 13% to 114 mmHg ($P<0.05$), resulting also in a significant drop of the rate-pressure-product of 12.6% from 182 to 159 mmHg · $min^{-1} \cdot 10^{-2}$ ($P<0.05$). After propranolol, systolic aortic pressure during control pacing reached 110 mmHg and remained unchanged during rapid atrial pacing (111 mmHg, +1%, not significant); thus, the rate-pressure-product did also not change during betablockade amounting to 154 under control, and 156 mmHg · $min^{-1} \cdot 10^{-2}$ during rapid atrial pacing. Hence, nifedipine showed a clear afterload-decreasing effect in contrast to propranolol, and this was also reflected in the different behavior of the rate-pressure product.

Figure 6 demonstrates a typical example of the behavior of regional myocardial blood flow after propranolol, Fig. 7 after nifedipine. The average flow changes are summarized in Fig. 8:

At rest, before administration of the drug, regional myocardial blood flow was with approximately 54–57 ml/min/100 g myocardium significantly lower in the poststenotic, eventually ischemic area than in the normal zone, where it averaged approximately 65–70 ml/min/100 g myocardium ($P<0.005$). During control pacing, regional myocardial blood flow increased significantly in the two groups both in the normal as well as in the poststenotic zone; however, the average value reached during pacing-induced ischemia was with approximately 75 ml/min/100 g again significantly lower in the poststenotic than in the normal zone where it reached 87–90 ml/min/100 g ($P<0.005$).

It is assumed that flow in the normal zone corresponds to the myocardial oxygen consumption required for this heart rate and blood pressure. Furthermore, in a few patients, even a decrease in myocardial blood flow was observed in the poststenotic area (a typical example is shown in Fig. 7), indicating a profound drop of flow,

Table 3. Heart rate, systolic aortic pressure and rate-pressure-product (RPP) during control pacing (CP) and rapid atrial pacing after drug administration (RAP). (For details, see text)

	HR (min^{-1})	Syst. AoP (mmHg)			RPP (mmHg $sec^{-1} \cdot 10^{-2}$)		
	RAP	CP	RAP	Δ%	CP	RAP	Δ%
NTG (12)	147	122	103	– 15.6% *	180	151	– 16.2% *
NIF (11)	139	131	114	– 13.0% *	182	159	– 12.6% *
PROP (13)	140	110	111	+ 1.0%	154	156	+ 1.2%

* $P<0.05$

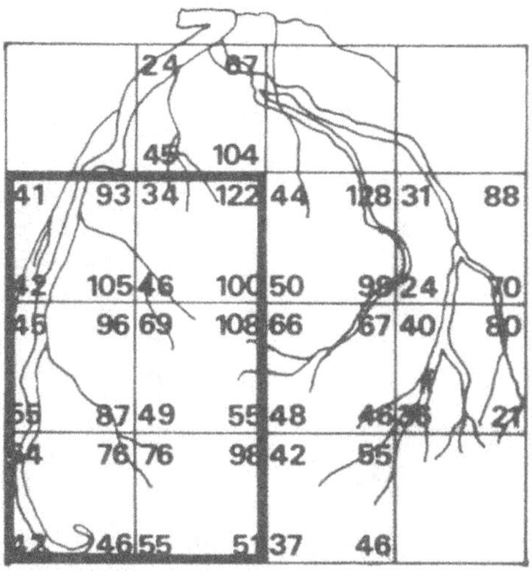

J.P. 166/80
80% STENOSIS LCx
LV NORMAL

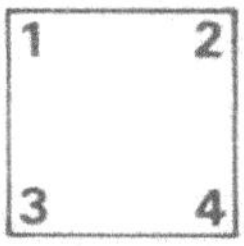

	MBF (ml/min /100g) NORMAL AREA	POST. STEN. AREA
1. CONTROL	54.8	44.8
2. RAPID PACING 150/min: A.p. ++, ST ↘ 0.1 mV	98.8	72.5
3. AFTER PROPRANOL AT REST	48.2	36.3
4. AFTER PROPRANOLOL DURING PACING 150/min, A.p. –, ST ↘ 0 mV	74.0	45.8

Fig. 6. Typical example of myocardial blood flow before and after propranolol (5 mg i.v.) in a patient with an 80% obstruction of the left circumflex branch. *Top:* schematically, the left coronary angiogram in LAO projection and superimposed coronary flow of large areas of 25 matrix points; the normal zone is outlined in black, the poststenotic zone in *stippled quadrangles.* Each quadrangle shows 4 flow values: *top left* (1): resting flow at control; *bottom left* (3): resting flow after propranolol; *top right* (2): control flow during rapid atrial pacing; *bottom right* (4): flow during the same pacing, after propranolol. Note, first, the marked decrease in resting flow after propranolol, and second, the marked reduction in flow during pacing after betablockade, the differences in flow between the normal and the poststenotic area remaining significant throughout the entire study, even after administration of the drug

especially in the endocardial layers. In this connection, it should be remembered that the xenon technique measures transmural regional flow only, each matrix point representing average flow from the endocardial to the epicardial layers. Hence, transmural flow gradients cannot be seen with this technique. After nifedipine, during rapid atrial pacing at the same heart rate as at control, a further increase in poststenotic flow was observed, whereas in the normal zone, flow decreased mildly; both the further increase and decrease in flow were, however, not

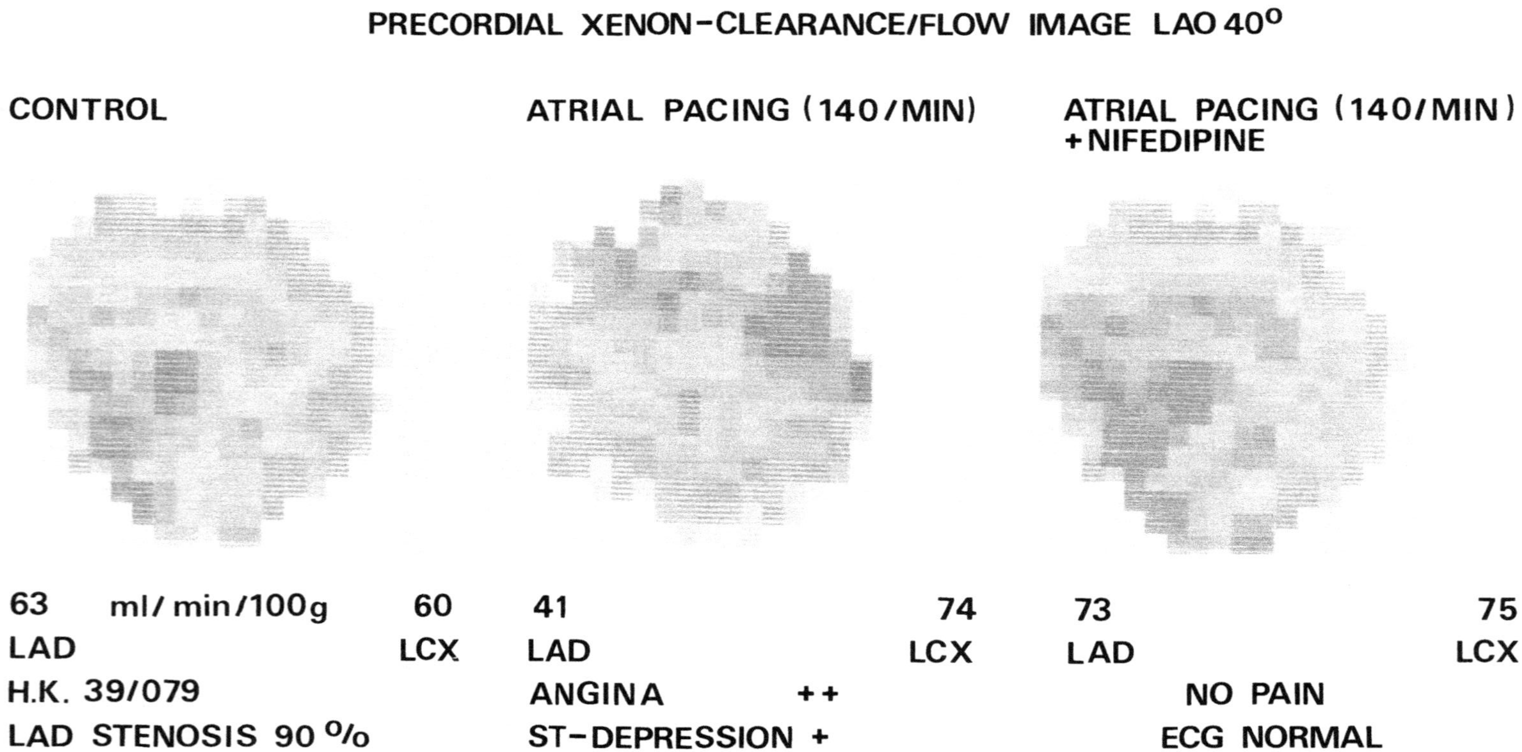

Fig. 7. Typical example of the effect of nifedipine (20 mg s.l.) on regional myocardial blood flow during pacing-induced angina in a patient with a 90% obstruction of the LAD. *Left:* flow image at rest; *middle:* flow image during atrial pacing up to 140 beats/minute; *right:* flow image during the same heart rate, yet after nifedipine. At rest, flow was almost identical in the normal and poststenotic area. During atrial pacing, flow in the ischemic zone decreased from 63 to 41 ml/min/100 g myocardium and increased in the normal area from 60 to 74 ml/min/100 g myocardium; the patient suffered from angina, the ECG showed ST-depressions >0.1 mV. Note that the flow scan shows most red matrix points in the circumflex area, whereas in the poststenotic LAD-area, yellow and green colors prevail, indicating a low flow-zone. After nifedipine (*right*), flow was homogenous throughout the heart, amounting to 73 and 75 ml/min/100 g myocardium in the area perfused by the LAD and LCX, respectively

significant. Nevertheless, this resulted in a homogeneous distribution of flow throughout the heart, the differences between normal and poststenotic flow becoming insignificant after the drug, averaging approximately 85 ml/min/100 g myocardium. Hence, in spite of a significant drop in the rate-pressure product, that is, in myocardial oxygen consumption, coronary blood flow did not decrease, in contrast, it showed a further increase, especially in the ischemic zone. Nine of the 11 patients were free from pain and showed no change in the ST-segment during the second period of pacing.

After propranolol, a completely different behavior was seen during rapid atrial pacing: both in the normal as well as in the poststenotic zone, regional myocardial blood flow, in spite of identical heart rates, showed a significant decrease ($P < 0.005$); however, this drop of flow was much more pronounced in the normal than in the poststenotic zone. In addition, there was still a significant difference of flow during pacing after Propranolol between both areas, flow in the poststenotic zone now averaging approximately 65 ml/min/100 g, and in the normal zone still more than 70 ml/min/100 g myocardium ($P < 0.005$).

Hence, regional myocardial blood flow behaved in a completely different way during rapid atrial pacing leading to ischemia following the administration of the calcium antagonist nifedipine and the betablocker propranolol. The calcium antagonist led to a further increase in flow, a "luxury" perfusion in the poststenotic area, whereas the betablocker was followed by a decrease in flow, a result of the marked reduction in myocardial oxygen demand due to the decrease in contractile force. It should be mentioned once more that the rate-pressure-product did not change after propranolol due to atrial pacing, hence the decrease in flow being entirely the result of the drop in contractility, whereas after the calcium antagonist, there was a significant decrease in the rate-pressure product due to the marked reduction in afterload. Figure 8 also shows the behavior of nitroglycerin observed in 12 patients following the same methodology. Regional flow after nitroglycerin behaves quite similarly as after the betablocker; also nitroglycerin resulted in a significant drop in normal and poststenotic flow when compared to the values obtained with control pacing, the result of a significant drop in afterload. Hence, when considering the three drugs, three different behaviors are observed: a significant decrease in normal and poststenotic flow associated with a significant decrease in the rate-pressure-product, that is, in afterload and preload after nitroglycerin, a significant decrease in normal and poststenotic flow after betablockade, however, not associated with changes in the rate-pressure-product, but only with a decrease in contractility and finally, in opposition to this behavior, a marked although not significant further increase in flow after nifedipine, especially in the poststenotic zone, in spite of a decrease in afterload, that is, in the rate-pressure-product.

Discussion

Different pharmacologic effects between drugs inhibiting the transmembrane flux of calcium and those blocking selectively the activity of beta-1- and beta-2-receptors are to be expected. Hence, the so-called calcium antagonists exhibit their strongest action on vascular smooth muscle tone and, if administered selectively into the coronary system, on cardiac muscle, i.e., contractile force, yet to a lesser

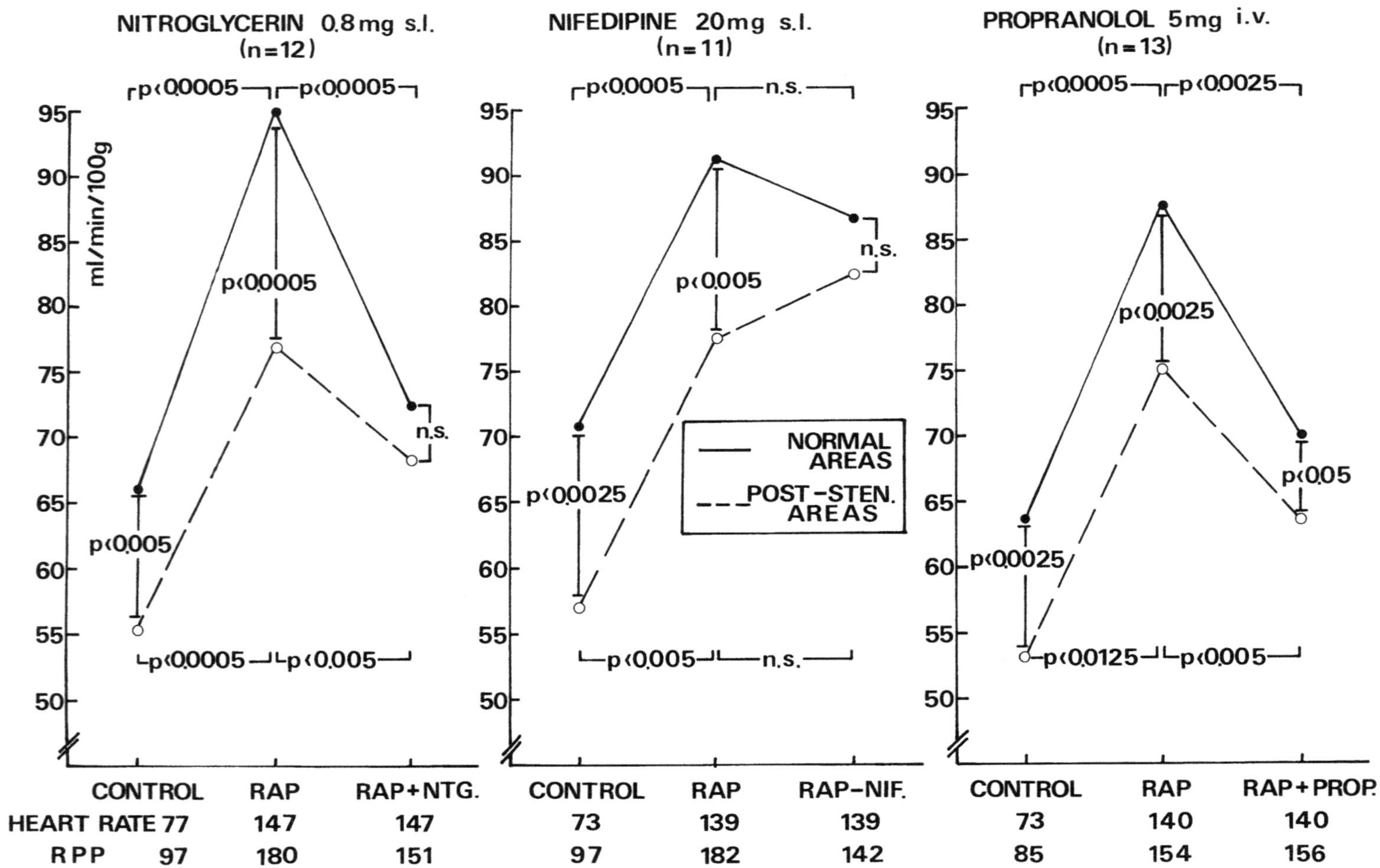

NITROGLYCERIN 0.8 mg s.l. (n=12)
NIFEDIPINE 20 mg s.l. (n=11)
PROPRANOLOL 5 mg i.v. (n=13)
ml/min/100g
p<0.0005
p<0.005
p<0.0025
p<0.0125
p<0.05
n.s.
NORMAL AREAS
POST-STEN. AREAS
CONTROL
RAP
RAP+NTG.
RAP-NIF.
RAP+PROP.
HEART RATE 77 147 147 73 139 139 73 140 140
RPP 97 180 151 97 182 142 85 154 156

degree and with varying intensity, on the conduction system. Betablockers influence vascular smooth muscle tone only minimally, especially in the coronary system. Hence, after administration of calcium antagonists, profound arteriolar relaxation both in the peripheral and coronary circulation can be observed, leading to a mild, yet significant increase in cardiac output as well as in coronary blood flow both in the normal and poststenotic, eventually ischemic zone; most importantly, they do not block the autoregulatory system, as observed with pure coronary dilators such as dipyridamole [12]. With nifedipine, when administered orally, the increase in coronary flow at rest is only mild, amounting to approximately 20%, allowing a further maximal increase following the rise of myocardial oxygen consumption provoked either by rapid atrial pacing or bicycle ergometry [12]. Furthermore, this increase in flow also occurs in the poststenotic ischemic zone without being accompanied by a steal phenomenon – an observation made with no other anti-ischemic drug, so far.

It has to be assumed that during pacing-induced ischemia, the observed impairment of the increase in transmural flow in the poststenotic area is primarily due to a low flow in the endocardial layers, this "low flow area" reaching to varying degrees into the middle layers of the myocardium. The further increase in flow after nifedipine, in spite of a decrease in O_2-demand, thus, can be explained in different ways:

1. In connection with the reduction in afterload leading to "decompression of the inner layers" and reducing the extravascular component of coronary resistance.
2. As the result of additional coronary arteriolar dilation, especially in the middle and outer layers of the myocardium.
3. Due to an additional dilation within the coronary obstruction itself, i.e., a decrease in proximal "fixed" coronary resistance (see below).

In contrast, after betablockade, an autoregulatory decrease in coronary flow both in the normal and poststenotic area takes place according to the drop in myocardial oxygen consumption brought about by the decrease in contractile force and, without atrial pacing, also in heart rate. Total coronary resistance both in the normal and poststenotic area increases mildly after betablockade, probably due to an overwhelming alpha-receptor tone [6, 12]. Furthermore, calcium antagonists exhibit a relaxing effect on high-grade coronary obstructions [11], often decreasing the degree of stenosis by more than 20%.

This corresponds to the observation claiming that a high percentage of stenoses are indeed of an eccentric nature, and considerable portions of their circumference still consist in normal wall segments able to dilate, i.e., to undergo changes in vasomotor tone [7]. It is therefore understandable that calcium antagonists are also especially effective in the treatment of Prinzmetal's angina, where a massive increase in coronary smooth muscle tone, often leading to complete spasm, is the cause of ischemia [3, 21].

◀ **Fig. 8.** Comparison of the behavior of regional myocardial blood flow during rapid atrial pacing before and after administration of nitroglycerin (0.8 mg s.l.), nifedipine (20 mg s.l.) and propranolol (5 mg i.v.). Each *panel* shows on the left control values at rest (control); *middle:* control values during rapid atrial pacing (RAP); *right:* flow values during rapid atrial pacing plus drug; *bottom:* heart rate and rate-pressure-product. *Filled circles, heavy lines* = normal areas; *open circles, broken lines* = poststenotic areas. (For details, see text)

Also with regard to the peripheral circulation, profound differences exist between calcium antagonists and betablockers. Due to the marked decrease in afterload following peripheral vasodilatation, cardiac output increases with calcium antagonists, whereas it decreases after betablockade, as a result of the depressant effect on contractility. The negative inotropic effect of calcium antagonists, which is to be expected for a drug inhibiting the calcium flux into the myocardial cell, is usually absent after systemic administration, due to the reduction in afterload resulting in an increased sympathetic drive leading to a rise of heart rate and contractility. A marked, although short-living negative inotropic effect can, however, be demonstrated after direct intracoronary injection of Nifedipine, i.e., functional separation of the heart from the peripheral circulation. It can also be seen for several minutes after oral or intravenous administration of the drug when measurements of parameters reflecting contractility are performed very early, i.e., during the first minutes after drug intake, at a moment when blood pressure has not yet started to decrease; these changes are, however, very mild, of short duration and cannot be compared to those observed after intracoronary administration [2].

Both betablockers and calcium antagonists are very effective in the treatment of angina pectoris and can even be combined, under certain conditions, to enhance their antianginal activity. Yet, due to their differing modes of action, they have preferential indications. Betablockers act best in patients with angina induced by exertion, i.e., increased myocardial oxygen consumption in the presence of a normal or only mildly abnormal left ventricle. Due to the fact that they impair wall motion, especially in the poststenotic area, they have to be handled with great caution in patients with large hypo- and akinetic areas. Calcium antagonists act best in the treatment of angina at rest, i.e., Prinzmetal's angina with and without severe underlying coronary artery disease, i.e., in the presence of a primary decrease in oxygen supply due to an increase in coronary smooth muscle tone. However, they are also effective in exercise-induced angina, mainly by lowering peripheral blood pressure, eventually also by means of a dilation of the coronary obstruction itself and may be also due to their well-known cardioprotective effect [10].

References

1. Amende I, Simon R, Hood WP, Lichtlen PR (1979) The effects of the beta-blocker atenolol and nitroglycerin on left ventricular function and geometry in man. Circulation 60:836
2. Amende I, Simon R, Lichtlen PR (1980) Early effects of nifedipine on left ventricular diastolic function in man. Circulation 62:III–259
3. Antman E, Muller J, Goldberg S et al. (1980) Nifedipine therapy for coronary-artery spasm: experience in 127 patients. N Engl J Med 302:1269
4. Engel H-J, Lichtlen PR (1980) Effect of nifedipine on regional myocardial blood flow of coronary patients at rest and during rapid atrial pacing. *In:* Puech P, Krebs R (eds) 4th International Adalat Symposium: new Therapy of Ischemic Heart Disease. Excerpta Medica, Amsterdam, p 55
5. Engel H-J, Hundeshagen H, Lichtlen PR (1980) Prevention of pacing-induced ischemia by propranolol – changes of regional myocardial blood flow and coronary sinus O_2-content. Circulation 62:III–296
6. Engel H-J, Wolf R, Hundeshagen H, Lichtlen PR (1980) Different effects of nitroglycerin and nifedipine on regional myocardial blood flow during pacing induced angina pectoris. Eur Heart J 1 [Suppl B]:53

7. Freudenberg H, Lichtlen PR (1981) Das normale Wandsegment bei Koronarstenosen. Z f Kardiol (in press)
8. Gaasch WH, Cole JS, Quinones MA, Alexander JK (1975) Dynamic determinants of left ventricular diastolic pressure-volume relations in man. Circulation 51:317
9. Hood WP, Amende I, Simon R, Lichtlen PR (1980) The effects of intracoronary nitroglycerin on left ventricular systolic and diastolic function in man. Circulation 61:1098
10. Kaltenbach M, Schulz W, Kober G (1979) Effects of nifedipine after intravenous and intracoronary administration. Am J Cardiol 44:832
11. Leutenegger F, Rafflenbeul W, Gahl K, Walpurger G, Engel H-H, Lichtlen P (1980) Quantitative Koronarangiographie: Dilatation von Koronarstenosen nach Nifedipin. Schweiz Med Wochenschr 110:1703
12. Lichtlen PR (1975) Coronary and left ventricular dynamics under nifedipine in comparison to nitrates, beta-blocking agents and dipyridamole. *In:* Lochner W, Braasch W, Kroneberg G (eds) 2nd International Adalat Symposium: New Therapy of Ischemic Heart Disease. Springer, Berlin Heidelberg New York, p212
13. Lichtlen PR, Engel H-J (1979) Assessment of regional myocardial blood flow using the inert gas washout technique. Cardiovasc Radiol 2:203
14. Lichtlen PR, Engel H-J (in press) Relation between myocardial blood flow and degree of coronary obstruction under increased oxygen demand induced by rapid atrial pacing. In: Detection of ischemic myocardium with exercise. Symposium of the European Society of Cardiology, Düsseldorf, Springer-Verlag Berlin Heidelberg New York
15. Lichtlen PR, Engel H-J, Amende I, Rafflenbeul W, Simon R (1976) Mechanisms of various antianginal drugs. Relationship between regional flow behaviour and contractility. *In:* Jatene AD, Lichtlen PR (eds) 3rd International Adalat Symposium. Excerpta Medica, Amsterdam, p14
16. Lichtlen PR, Amende I, Simon R, Hundeshagen H (1977) Left ventricular function and regional myocardial blood flow after atenolol in normals and patients with coronary artery disease. Postgrad Med J 53 [Suppl]:85
17. Lichtlen PR, Engel H-J, Hundeshagen H (1977) Regional myocardial blood flow in normal and poststenotic areas after nitroglycerin, betablockade (atenolol), coronary dilatation (dipyridamole) and calcium antagonism (nifedipine). Herz 2:81
18. Lichtlen PR, Engel H-J, Hundeshagen H (1978) Clinical application and results of the assessment of coronary blood flow by the regional precordial xenon residue detection technique. Nucl Med 17:161
19. Lichtlen PR, Engel H-J, Amende I, Hundeshagen H (1978) Effect of beta blockade on regional myocardial blood flow in coronary heart disease. *In:* Gross F (ed) Modulation of sympathetic tone in the treatment of cardiovascular diseases. Huber, Bern, p134
20. Mirsky I, Cohn PF, Levine JA, Gorlin R, Herman MV, Kreulen ThH, Sonnenblick EH (1974) Assessment of left ventricular stiffness in primary myocardial disease and coronary artery disease. Circulation 50:128
21. Schroeder JS, Russell RO, Resnekov L, Wolk M, Hutter AM, Rosati RA, Conti CR et al. (1978) Unstable angina pectoris-national randomized study of surgical vs. medical therapy: results in Prinzmetal type angina. Am J Cardiol 41:397
22. Simon R, Amende I, Lichtlen PR (1979) Das linksventrikuläre Angiogramm. *In:* Lichtlen PR (ed) Koronarangiographie. Straube, Erlangen, p249

Discussion

Koch, Berlin: I am always again surprised to hear you propose that calcium blockade by nifedipine has no effect on postcapillary vessels. Is it really conceivable that Ca antagonism is confined to the smooth muscles of the precapillary vascular section only? A vasodilatory effect on postcapillary venous human vessels has clearly been shown in in vitro experiments. We feel that the often dramatic decrease in left ventricular filling pressures that we have seen and reported in patients with ischemic heart disease during exercise after nifedipine can only be explained by a distinct effect on preload via vasodilatation of the postcapillary section of the vascular bed. I think that even the side-effect of ankle edema sometimes seen during nifedipine treatment suggests an action on postcapillary resistance and probably capacitance vessels.

Lichtlen: The question whether nifedipine acts on the venous side in humans is still a matter of debate. Personally, we have not seen any substantial effects, neither on enddiastolic pressure nor on enddiastolic volume at rest and during pacing in patients with ischemic heart disease. In addition, cardiac output increases due to peripheral arteriolar dilatation, also during exercise, which necessitates an increase in venous return. If veno-dilatation would really occur, i.e., the drug would have a direct influence on venous capacity, as for instance nitroglycerin, one would expect cardiac output to decrease. Hence, all effects on the venous side, if they occur, would have to be secondary in nature. It is indeed true that in patients with elevated enddiastolic volumes (> 20 ml/m^2), nifedipine leads to a decrease of enddiastolic volume index; however, this is a secondary phenomenon following the increase in cardiac output and improvement of left ventricular performance. Ankle edemas, which rarely occur as a side effect, have to be explained in another way, probably by increased capillary permeability.

Hjalmarson, Göteborg: I am surprised about the lack of effect often seen in my clinical practical treatment of patients with angina pectoris. That might fit with the slide shown by Dr. Ekelund of an almost unchanged amount of nitroglycerin taken over a week when on nifedipine as compared to placebo. Addition of β-blockade markedly reduced the intake of nitroglycerine. Furthermore, we have had 3 patients (of less than 20 studied invasively) with a dramatic drop in heart rate and blood pressure and subsequent need for resuscitation. Only one was on β-blockade. The dose was 10 mg sublingual in all three patients (one with acute myocardial infarction, 2 patients with angina pectoris).

Lichtlen: In our experience, patients with severe hypotension after nifedipine are extremely rare. We have never observed a case who needed resuscitation as mentioned by Dr. Hjalmarson. On the other hand, we have so far only treated very few patients with acute myocardial infarction with nifedipine. To answer the question more precisely, one really would have to know more details about these three cases which, in my opinion and in accordance with the literature, must be quite exceptional and also difficult to explain.

Beta-Blockers and Calcium Antagonists: Synergistic Effects

L.-G. Ekelund

Summary

Calcium-blockers dilatate the peripheral vessels and thereby decrease the blood pressure. The decrease in blood pressure induces a reflex increase in heart rate which could counteract the decrease in blood pressure and diminish the decrease in rate pressure product. Therefore it is logical to combine a Ca^{2+}-blocker with a beta-blocker when treating patients with coronary insufficiency or hypertension. Nifedipine could safely be combined with a beta-blocker. In a study of 21 patients with angina pectoris the addition of a beta-blocker to the calcium-blocker nifedipine further increased the working capacity with 21%. The same synergistic effect was also obtained in a new double-blind trial after six months continuous treatment with nifedipine.

In a study of 12 patients with essential hypertension nifedipine 10 mg t.i.d. alone decreased the resting blood pressure with 20 mmHg. The addition of metoprolol 100 mg b.i.d. gave a decrease in resting blood pressure with 37 mmHg. A synergistic effect was also obtained during exercise.

The few pharmacological side-effects, flushing, headache and palpitations, induced by nifedipine were completely abolished after addition of a beta-blocker.

Introduction

Calcium plays a fundamental role in the excitation and contraction of cardiac, striated and smooth muscles. Muscular contraction is the result of an interaction between parallel filaments of actin and myosin. This interaction is prevented by a protein complex, troponin-tropomyosin, which is built into the actin filament. Upon excitation, calcium ions (Ca^{2+}) are released from stores in sarcoplasmatic reticles and flow through the so called "slow channels" into the cell membrane, inhibiting the inhibitory action of the troponin-tropomyosin complex. Due to cessation of the inhibition the muscle contracts. If extracellular Ca^{2+} is prevented from penetrating the cell membrane, muscular contraction will be prevented; vascular smooth muscles will relax producing vasodilation, and cardiac muscle will contract less powerfully; a negative inotropic effect. The excitation-contraction coupling in cardiac and smooth muscle requires a flow of Ca^{2+} from external to internal stores. In striated muscles the internal stores are enough.

The transportation of Ca^{2+} through the slow channels with triggering of the internal stores could be inhibited by the so called "calcium-blockers" or "calcium-antagonists". A calcium-blocker could therefore alter cardiac rhythm, change the inotropy in the heart and act as a vasodilator [1, 7, 15, 20]. Calcium-blockers have been widely used in the treatment of patients with angina pectoris [5, 17].

This presentation focuses on the central and vascular effects of calcium-blockers. A short summary of the central and vascular effects is given in Tables 1 and 2.

Table 1. Central effects of Ca-blockade

1. Reduction of high-energy phosphate utilization by the contractile system
2. Diminution in the magnitude of mechanical tension developed without a major change in action potential
3. Lowering of the extra-uptake of oxygen related to the generated contractile force

Table 2. Vascular effects of Ca-blockade

1. Decrease in smooth muscle tension especially in the coronary arteries and resistance vessels
2. Decrease in systemic and coronary vascular resistance
3. Decrease in arterial blood pressure
4. Relief of coronary artery spasm

As calcium-blockers dilatate the peripheral vessels, they will decrease the arterial blood pressure [2, 10, 11, 21]. This decrease in blood pressure is more pronounced when the initial blood pressure is at a high level (Fig. 1). The decrease in blood pressure induces, via the baroreceptor reflex, an increase in sympathetic tone. The reflex increase in sympathetic tone results in an increased heart rate and systemic vascular resistance and an attempt to stabilize blood pressure. The reflex increase in heart rate is rather small at rest for one of the calcium-blockers, nifedipine (Table 3).

As a sign of an increased sympathetic tone, the values for renin in plasma (PRA) will almost double after administration of a calcium-blocker (Fig. 2). The increase in PRA is readily counteracted with a beta-blocker.

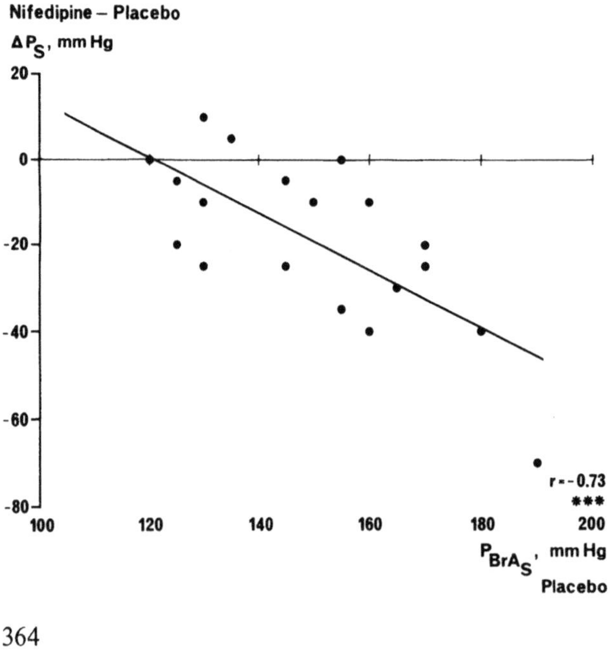

Fig. 1. Change in systolic blood pressure at rest (ΔPs) induced by nifedipine related to basal systolic blood pressure (P_{BrAs}) during the placebo period

Table 3. Heart rate and systolic blood pressure at rest and during exercise recorded after treatment with placebo, nifedipine or nifedipine + β-blocker. I, values from a study by *Ekelund, Orö* 1979; II, values from a study by *Ekelund, Ekelund, Rössner* 1981.

		Placebo	Nifedipine	Nifedipine + β-blocker
Heart rate (min^{-1}) at rest	I	72	77	67
	II	77	82	61
Heart rate (min^{-1}) during exercise	I	110	113	90
	II	123	128	92
Systolic blood pressure (mmHg) at rest	I	147	130	118
	II	178	158	141
Systolic blood pressure (mmHg) during exercise	I	170	152	133
	II	230	216	177
Exercise tolerance, total work (Watt · min)	I	770	925	1088

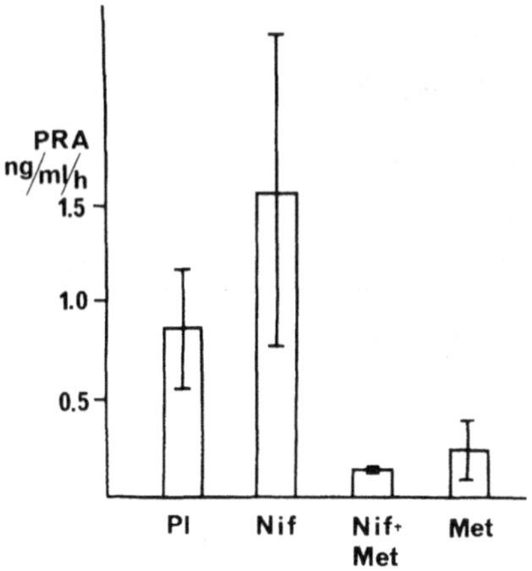

Fig. 2. Plasma renin activity (*PRA*) during treatment with placebo (*Pl*), nifedipine (*Nif*), nifedipine + metoprolol (*Nif* + *Met*) and metoprolol (*Met*)

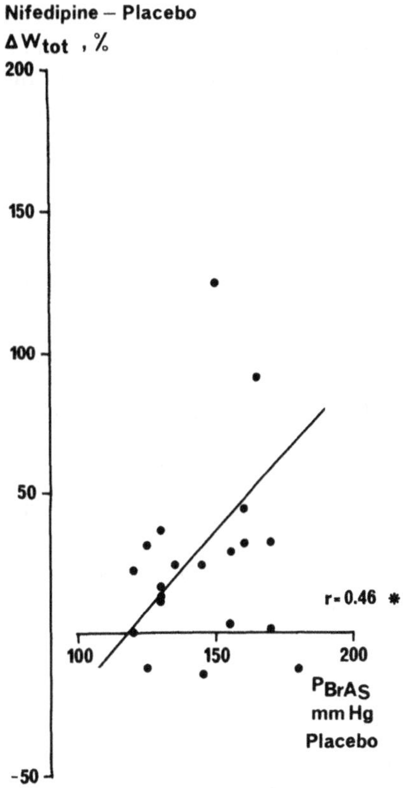

Fig. 3. Per cent change in total work (ΔW_{tot}) induced by nifedipine related to basal systolic blood pressure (P_{BrAs}) during the placebo period

Thus, the increase in sympathetic tone tends to decrease the antianginal efficiency of a calcium-blocker. Also, when treating hypertension, the increase in sympathetic tone will counteract the action of the calcium-blocker with an increase in heart rate and increase in systemic vascular resistance.

To balance the increase in sympathetic tone it seems logical to administer a beta-blocker together with a calcium-blocker. One commonly used calcium-blocker, nifedipine, could safely be combined with a betablocker without any influence on the atrioventricular conduction time according to measurements of surface ECG [5] and intracardiac measurements [18]. The other commonly used calcium-blocker, verapamil, has a much stronger influence on atrioventricular conduction and therefore it is recommended not to combine it with a beta-blocker. However, in clinical practice, this combination has been tried but no regular written report has been published about the combination.

Methods

All the values presented here were obtained either after 10 min rest in supine or during submaximal steady-state exercise in the sitting position on a cycle ergometer.

Heart rate was determined from a one minute interval of the electrocardiogram recorded on a calibrated Mingograph. The blood pressure was measured with the cuff method using calibrated manometers. To achieve the same degree of angina in tests of patients with coronary insufficiency, we used rating scales according to *Borg* to rate perceived exertion and perceived angina [4].

Results

Coronary Insufficiency. In a double-blind randomized trial 21 patients with stable coronary insufficiency were investigated with standardized exercise tests after administration of placebo or nifedipine for 3 weeks [5]. At rest there was a significant increase in heart rate (5 beats/min) and a decrease in systolic (17 mmHg) and diastolic blood pressure (11 mmHg) after nifedipine compared to placebo (Table 3).

After the two initial trials there was a period when the patients were treated with nifedipine, 10 mg t.i.d., together with either metoprolol or alprenolol.

During the period with nifedipine + beta-blocker the heart rate decreased slightly at rest but not significantly. The blood pressure was, however, markedly lowered with a decrease in systolic and diastolic blood pressure of 29 and 14 mmHg, respectively (Table 3).

Exercise. During exercise at comparable loads there was no significant difference in heart rate when comparing placebo and nifedipine periods. After nifedipine administration the systolic blood pressure decreased significantly by 18 mmHg.

The combination of nifedipine and beta-blocker induced a significantly lower heart rate, 20 beats/min and a very much lower systolic blood pressure, mean difference 37 mmHg (Table 3).

Exercise Tolerance. Total work, calculated as the product of watts and number of minutes of exercise, was greatly increased after nifedipine compared to placebo; a significant mean increase of 20%. When combining nifedipine with a beta-blocker there was an additional work output with a total mean increase of 41% (Table 3).

The increase in exercise tolerance was correlated to the initial arterial blood pressure at rest during the placebo period (Fig. 3).

Conduction Effects. The P-Q interval was carefully measured from a surface electrocardiogram. There was a relation between P-Q interval time and resting heart rate. After administration of nifedipine or the combination of nifedipine with a beta-blocker there was no change in the regression lines (Fig. 4).

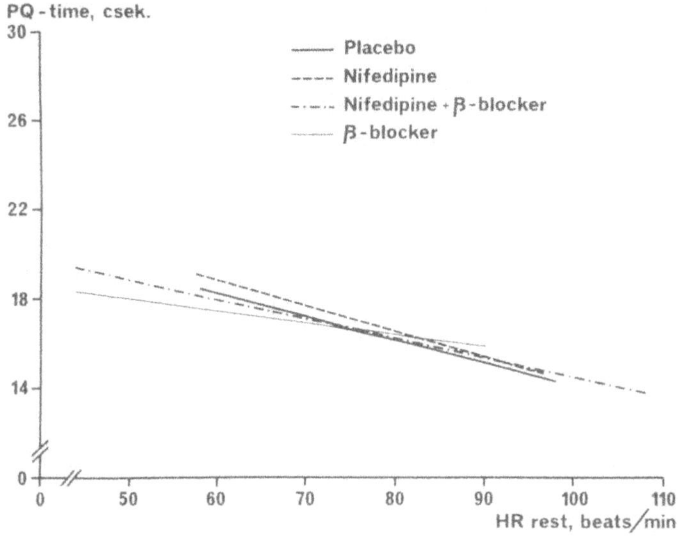

Fig. 4. Repression lines for the relation between PQ time to heart rate in the resting ECG during different treatments

Prolonged Treatment. All results were obtained after 3 weeks' treatment. To investigate if a prolonged treatment with nifedipine could induce a tolerance to the drug we performed, on 9 patients who were on continuous treatment with metoprolol and nifedipine for a mean of 6 months, a second double-blind administration of placebo or nifedipine. The beta-blocker treatment was unchanged but the nifedipine administration was stopped for two days. On the 2nd day either placebo or nifedipine (10 mg) was given and standardized exercise tests were performed. The mean increase in exercise tolerance was 23% at the first test and 31% in the second test (Fig. 5), with the same values for working capacity during placebo, mean value 78 watts. The decrease in blood pressure both at rest and during exercise induced by nifedipine was the same in the first and second study. Thus, there were no signs of adaptation to a continuous nifedipine treatment.

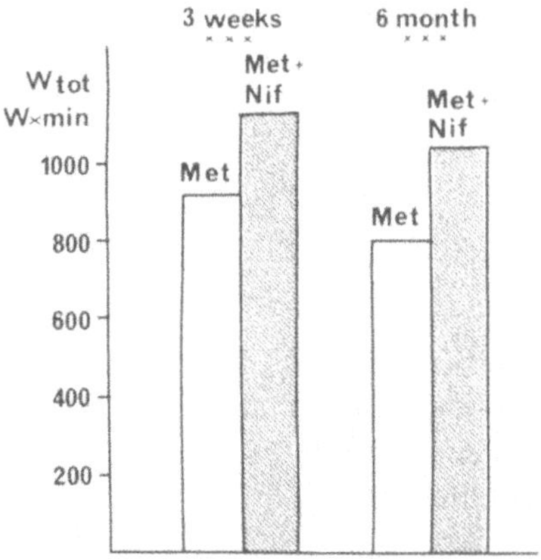

Fig. 5. Exercise tolerance, W_{tot}, in Watt×min, obtained after short and long-term treatment with nifedipine in addition to a β-blocker, metoprolol

Hypertension. In the study of hypertension we observed a group of 12 patients with slight to moderate essential hypertension [6]. Heart rate and blood pressure were measured at rest and during submaximal exercise. The patients were studied after treatment with placebo, nifedipine 10 mg t.i.d., nifedipine 10 mg t.i.d. + metoprolol 100 mg b.i.d. and lastly only metoprolol 100 mg b.i.d. Nifedipine alone decreased the resting blood pressure by 20 mmHg. In combination with a beta-blocker, metoprolol, nifedipine gave a further decrease in the resting blood pressure, mean decrease 37 mmHg systolic blood pressure (Table 3). During exercise the blood pressure decreased 14 mmHg with nifedipine alone and 53 mmHg with nifedipine + metoprolol. Thus, there was a rather marked synergistic effect when combining nifedipine with metoprolol in patients with essential hypertension.

Side-effects. Treatment with Ca^{2+}-blockers can induce several pharmacological side-effects such as flushing, head-ache and palpitations. In the studies of patients with coronary insufficiency and hypertension, side-effects of this type were found during nifedipine treatment in a small number of cases. During treatment with nifedipine + beta-blocker no patients reported such side-effects.

Discussion

In the study of patients with coronary insufficiency we found a strong synergistic effect when combining nifedipine with a beta-blocker both in respect to working capacity and decrease in blood pressure. A strong synergistic effect in combining nifedipine with a beta-blocker was also reported by *Lynch* et al. [14]. They compared in a double-blind fashion two doses of nifedipine, 10 mg t.i.d. and 20 mg t.i.d. alone and with propranolol, 240 and 480 mg per day, in a group of 16 patients with exercise-induced angina. They found a significant decrease in number of anginal episodes, consumption of nitrates and exercise-induced ST-segment depression after administration of nifedipine or propranolol alone. The combination of nifedipine together with propranolol was always significantly better than either drug alone. In their report there was no individual data for blood pressure or heart rate; only the exercise heart rate-pressure product was reported.

Several other authors have found a synergistic effect of a combination of nifedipine with a beta-blocker in the treatment of angina pectoris [3, 9, 19, 22].

In the study of patients with hypertension, the addition of a beta-blocker, metoprolol, to nifedipine, gave a more marked decrease in blood pressure at rest and during exercise than nifedipine alone. Similar results have been reported by *Aoki* et al. [2], Lederballe *Pedersen* et al. [13].

An interesting observation is the reduction of the pharmacological side-effects when a calcium-blocker is combined with a beta-blocker.

References

1. Andersson KE, Ekelund L-G, Johansson BW, Landmark K (1978) Calciumantagonists (CA-blockers). Pharmacological, physiological and clinical aspects. Proc from a meeting in Vedbaeck, Denmark, 1977. Acta Pharm Toxicol 43 [Suppl 1]
2. Aoki K, Kondo S, Mochizuki A, Yoshida T, Kato S, Kato K, Takikawa K (1978) Antihypertensive effect of cardiovascular Ca^{2+}-antagonists in hypertensive patients in the absence and presence of beta-adrenergic blockade. Am Heart J 96:218
3. Bevegård S, Flygt G, Orö L (to be published) Combination treatment in angina pectoris. Addition of nifedipine or isosorbidedinitrate to metoprolol in a double-blind study
4. Borg G (1970) Perceived exertion as an indicator of somatic stress. Scand J Rehabil Med 2:92
5. Ekelund L-G, Orö L (1979) Antianginal efficiency of Nifedipine with and without a beta-blocker, studied with exercise test. A double-blind randomized subacute study. Clin Cardiol 2:203
6. Ekelund L-G, Ekelund C, Rössner S (to be published) Antihypertensive effects of a Calcium-blocker nifedipine alone and in combination with metoprolol. A sub-acute study with exercise tests
7. Ellrodt G, Chew CYC, Singh BN (1980) Therapeutic implications of slow-channel blockade in cardiocirculatory disorder. Circulation 62:669
8. Guazzi M, Fiorentini C, Olivari MT, Bartorelli A, Necchi G, Polese A (1980) Short- and long-term efficacy of a Calcium-antagonistic agent (Nifedipine) combined with Methyldopa in the treatment of severe hypertension. Circulation 61:913
9. Kenmure ACF, Scruton JH (1979) A double-blind controlled trial of the anti-anginal efficacy of Nifedipine compared with Propranolol. Br J Clin Pract 33:49
10. Kuwajima I, Ueda K, Kamata C, Matsushita S, Kuramoto K, Murakami M, Hada Y (1978) A study on the effects of Nifedipine in hypertensive crises and severe hypertension. Jpn Heart J 19:455
11. Lederballe Pedersen O, Mikkelsen E (1978) Acute and chronic effects of Nifedipine in arterial hypertension. Eur J Clin Pharmacol 14:375
12. Lederballe Pedersen O, Mikkelsen E, Christensen NJ, Kornerup HJ, Pedersen EB (1979) Effect of Nifedipine on plasma renin, aldosterone and catecholamines in arterial hypertension. Eur J Clin Pharmacol 15:235
13. Lederballe Pedersen O, Christensen CK, Mikkelsen E, Rämsch KD (1980) Relationship between the antihypertensive effect and steady-state plasma concentration of Nifedipine given alone or in combination with a beta-adrenoceptor blocking agent. Eur J Clin Pharmacol 18:287
14. Lynch P, Dargie H, Krikler S, Krikler D (1980) Objective assessment of antianginal treatment; a double-blind comparison of propranolol, nifedipine, and their combination. Br Med J 281:184
15. Nayler WG (1980) Calcium antagonists. Eur Heart J 1:225
16. Olivari MT, Bartorelli C, Polese A, Fiorentini C, Moruzzi P, Guazzi MD (1979) Treatment of hypertension with Nifedipine, a Calcium antagonistic agent. Circulation 59:1956
17. Opie LH (1980) Drugs and the Heart. III. Calcium antagonists. Lancet 12:806
18. Rowland E, Evans T, Krikler D (1979) Effect of nifedipine on atrioventricular conduction as compared with verapamil. Intracardiac electrophysiological study. Br Heart J 42:124

19. Schmutzler H, Dorow P (1980) The combined use of acebutolol and nifedipine in patients with coronary heart disease. In: Roskamm H, Graefe K-H (eds) Advances in β-blocker therapy. Excerpta Medica, Amsterdam-Oxford-Princeton
20. Stone PH, Antman EM, Miller JE, Braunwald E (1980) Calcium channel blocking agents in the treatment of cardiovascular disorders. Part I Hemodynamic effects and clinical applications. Ann Intern Med 93:886
21. Thibonnier M, Bonnet F, Corvol P (1980) Antihypertensive effect of fractionated sublingual administration of Nifedipine in moderate essential hypertension. Eur J Clin Pharmacol 17:161
22. Vuori MI, Hämäläinen H, Pietilä TJ, Kallio VI (to be published) Assessment of exercise tolerance of cardiac patients by bicycle, tread-mill, and tread-mille plus isometric exercise with and without nifedipine

Discussion

Szekeres, Szeged: Did the combination of β-blockers and Ca-antagonists elicit conduction disturbances?

Ekelund: No, not as could be estimated from careful measurements of the surface ECG.

Opie, Cape Town: Theoretically the addition of nifedipine (10 mg b.d.) to β-blockade (atenolol 150 mg daily) should improve both angina and hypertension in patients who have both. In reality the hypertension is better controlled, as expected, but the angina is not dramatically improved (see *Opie* et al., Brit J Clin Pharm, in press). Do you have any comments?

Ekelund: We found a relation between basal blood pressure and the increase in working capacity, both with nifedipine (10 mg b.d.) alone and when combined with a β-blocker. However, also patients with a low blood pressure (100–150 mmHg systolic) showed a positive benefit, i.e. a 10%–30% increase in working capacity.

Lichtlen: Treatment of hypertension with nifedipine is presently under study by several groups. The combination with β-blockers in these patients might improve the antihypertensive effect, since the decrease in afterload should be combined with a small decrease in cardiac output and especially the tendency of heart rate to increase under nifedipine would be counteracted. In addition, the action of β-blockers on the central nervous system might be enhanced. Angina pectoris, if present in these patients, however, could under certain circumstances not be improved dramatically, as the decrease in blood pressure might enhance endocardial ischemia.

Beta-Blockers and Calcium Antagonists in Hypertrophic Obstructive Cardiomyopathy

G. Kober, R. Hopf, S. Dowinsky, M. Kaltenbach

Summary

Fifty patients with hypertrophic cardiomyopathy were treated with 480–720 mg/day of verapamil, a calcium antagonist, for a period of 1–7 years. During the preceding observation phase, some patients had received beta-blocking agents.

The annual case fatality rate was 1.3%. The clinical status of the surviving patients improved from Class 2.6 to 1.9 in the mean according to criteria of the New York Heart Association.

Objective, non-invasive parameters of myocardial hypertrophy which declined included Sokolow index, roentgenographic heart volume, left atrial diameter, and septal as well as left ventricular posterior wall thickness. Invasive studies demonstrated a reduction in left ventricular filling pressure, intraventricular pressure gradient, left ventricular muscle mass, and coronary artery diameter in improved patients. Serious side effects were not observed.

Whereas untreated or propranolol-treated patients had displayed a progression of complaints and clinical findings, a large number of patients improved in response to verapamil. This suggests that high-dose verapamil therapy represents an effective conservative approach of the treatment of hypertrophic cardiomyopathy.

Hypertrophic cardiomyopathy (HCM) with (HOCM) and without (HNCM) obstruction is characterized by increased left ventricular wall thickness of unknown origin. In some families a genetic predisposition has been identified, although underlying hemodynamic changes associated with other cardiac disease may be excluded as a causative factor. The etiologic mechanisms are currently thought to involve increased left ventricular contractility [4, 5], heightened sensitivity to catecholamines [2, 3], and, possibly, potentiated intracellular calcium activity.

Because left ventricular contractility at rest is found to be increased even in advanced stages of the disorder, the introduction of beta-blocking agents in 1964 appeared to offer a specific mode of therapy. Follow-up studies of patients receiving beta-blockers, however, failed to show the desired treatment objectives in many instances [12].

The frequent therapeutic failures during beta blockade raised considerable doubts concerning the role of increased sympathetic activity or increased responsiveness of cardiac beta receptors in the pathogenesis of hypertrophic cardiomyopathies. Indeed, the mechanism underlying the hypertrophic process may lie at the intracellular level. This would involve an increased intracellular accumulation of calcium, as has been found in other cardiomyopathies [1, 15], or a potentiation of intracellular calcium activity. These theoretical considerations, coupled with the successful application of verapamil in the hereditary cardiomyopathy of the Syrian gold hamster, provided the impetus for a trial of calcium antagonistic therapy in human HCM.

Methods

In an uncontrolled study, 50 patients with hypertrophic cardiomyopathy were treated with a daily dose of 480–720 mg of non-timed release verapamil and followed up regularly, some since as early as 1973.

Over a period of up to 24 months prior to the initiation of verapamil therapy, up to 39 patients were followed and evaluated using a range of diagnostic methods. During this phase, 15 patients were treated with various beta-blocking agents.

Patients were studied during the observation period, immediately before the onset of verapamil therapy, and at six-month intervals during verapamil treatment.

The following non-invasive parameters were evaluated every six months: subjective complaints in terms of the New York Heart Association (NYHA) classification; electrocardiographic signs of left ventricular hypertrophy in terms of the Sokolow Index, and roentgenographic heart volume. Because of technical reasons, echocardiographic determinations of left atrial and left ventricular diameter, interventricular septal thickness, and systolic anterior motion of the anterior mitral valve leaflet were performed in the first 20 patients only after verapamil treatment had been initiated.

Left and right heart catheterization was carried out in all patients prior to the start of verapamil therapy. In addition to the standard pressure measurements, the left ventricular pressure gradient during post-extrasystolic potentiation, after 0.8 mg of sublingual nitroglycerin, and after 0.5 mg of i.v. orciprenaline was registered. Left ventricular volume, left ventricular muscle mass, and coronary artery diameter at defined points [10] were also evaluated. Recatheterization was performed in 18 patients at a mean of 26 months (range 14–52 mos) after the onset of therapy; 12 were improved, 4 were unchanged, and two were worse in terms of subjective complaints and non-invasive findings.

Results

During a mean treatment period of 3.1 years, corresponding to 155 patient-years, two patients known to have ventricular premature beats died suddenly. These figures suggest an annual case fatality rate of 1.3 per cent, which is lower than the 3–5% mortality rate quoted in the literature for HCM patients receiving beta blockers [6, 13, 14].

The clinical course of the patients in terms of New York Heart Association functional class is shown in Fig. 1. Whereas 6 patients (12%) were asymptomatic before and during therapy, an improvement of up to two functional classes was noted in 37 of 50 patients (74%). Six patients (12%) remained unchanged, and only 1 patient reported a worsening of complaints by 1 functional class.

Figure 2 demonstrates the development of the Sokolow Index in terms of the duration of therapy. Thirty-nine patients could be followed for a mean of 24 months before the onset of verapamil treatment. During this period, 15 patients

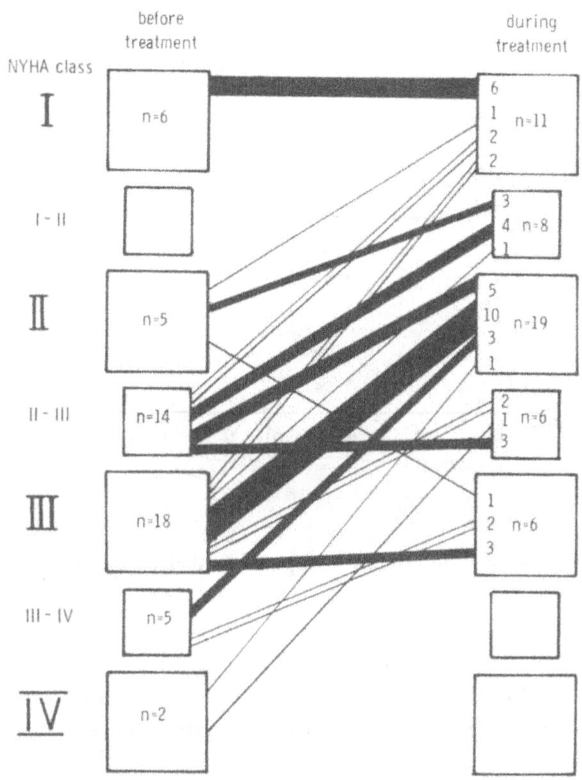

Fig. 1. Clinical course of 50 patients with hypertrophic obstructive cardiomyopathy treated for a mean of 37.3 months with 480–720 mg verapamil p.o. per day

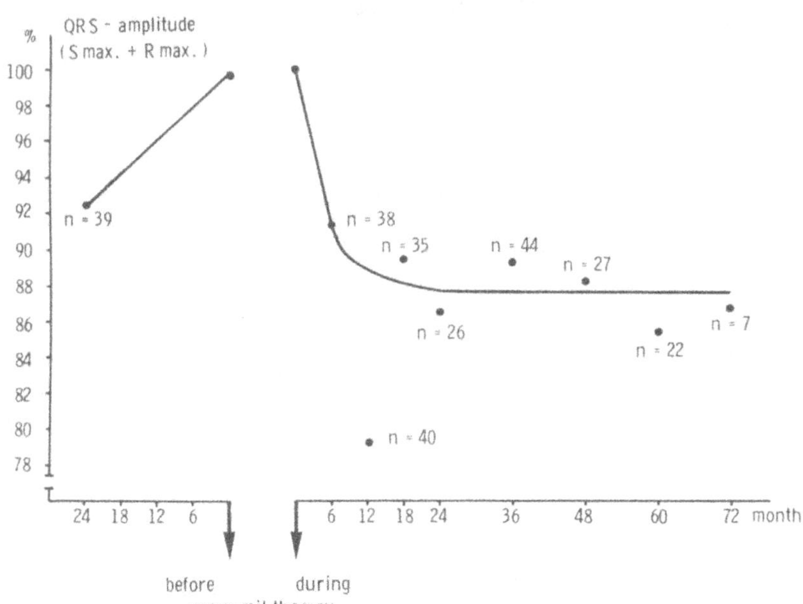

Fig. 2. Increase in the Sokolow Index before verapamil and decline after verapamil in terms of patient groups observed for different time intervals. Follow-up time: 24 months

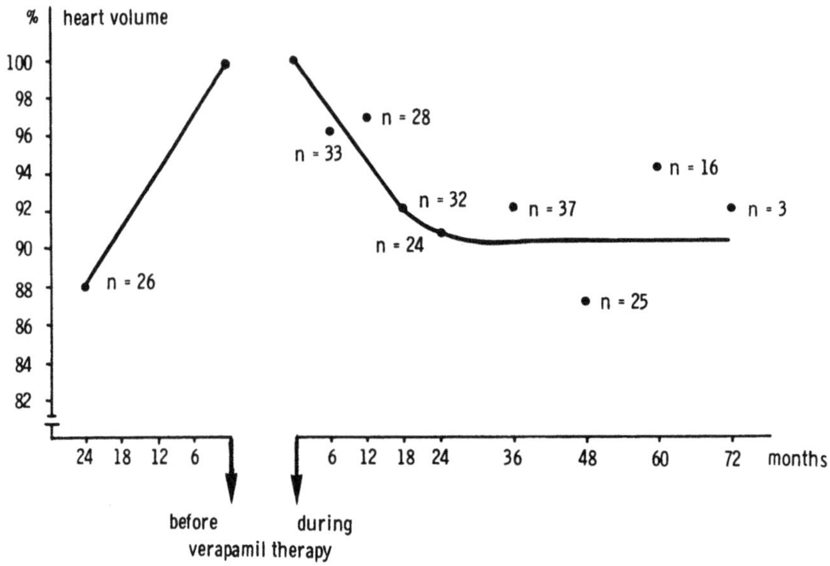

Fig. 3. Increase in roentgenographic heart volume before verapamil and decrease after the onset of therapy

received beta blockers. The Sokolow index increased from 4.6 to 5.0 mV (9%) during the pre-verapamil phase. In contrast, during verapamil treatment the Sokolow index declined in 31 of 50 patients along with most of the repolarization disturbances.

Roentgenographic heart volume (Fig. 3) in 26 patients increased from 870 to 975 ml/1.73 m², corresponding to 11%, during the pre-verapamil phase. Before the onset of verapamil therapy, 41 of the 50 patients were found to have relative heart volumes greater than 800 ml/1.73 m² (upper normal limit). After the initiation of

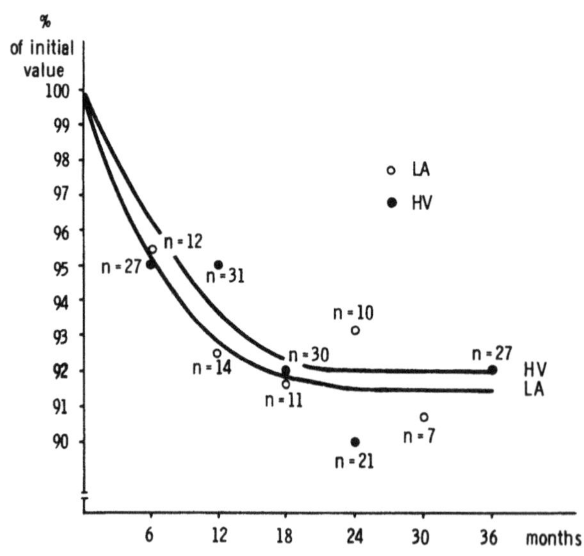

Fig. 4. Correlation between roentgenographically calculated heart volume (*HV*) and echocardiographically determined left atrial diameter (*LV*) during verapamil treatment

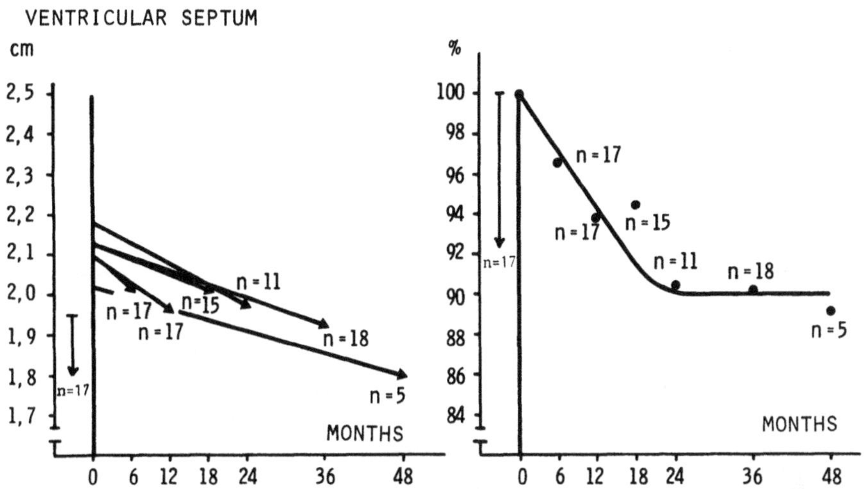

Fig. 5. Reduction in absolute (*left*) and relative (*right*) thickness of the interventricular septum in groups of patients treated for varying time intervals

calcium antagonistic treatment, heart volume declined in 35 of 50 patients up to the 24th month of therapy. These favorable results were maintained even in those patients followed up for 72 months; a mean reduction of 9% was found.

A relatively close correlation was established between roentgenographic heart volume and left atrial diameter. Both parameters showed quantitatively comparable declines during therapy (Fig. 4).

Figure 5 depicts changes in the thickness of the interventricular septum. No data prior to verapamil therapy are available. In 20 patients, the first echocardiogram was recorded during the treatment phase. Overall, septum thickness declined by a mean of 10% within 24 months.

Systolic anterior motion of the anterior mitral leaflet, which usually correlates well with the degree of outflow tract obstruction, generally did not change markedly during verapamil treatment. In individual cases, however, complete disappearance was noted.

Figure 6 presents a comparison of some non-invasive findings and different heart catheterization data for the 12 patients who improved clinically. Eight had proven HOCM (above), and 4 had HNCM (below). Mean improvement in heart volume (–14%), Sokolow index (–15%), and functional class (2.4 to 1.7) did not differ from that of the previously described total collective. The invasive data disclosed an insignificant decline in filling pressure of about 6%. The pressure gradient at rest fell by 33% and after post-extrasystolic potentiation by 28%. Muscle mass was reduced by 18%, and the diameter of both coronary arteries, correlating to the perfused left ventricular muscle mass [10], decreased by 14% and 11%, respectively.

In the two patients who had worsened clinically, no improvement in these parameters was found. Two untreated patients showed deterioration of both invasive and non-invasive findings. Despite the high doses of verapamil, serious side effects were not observed. Initial complaints such as increased diaphoresis, mild headache, dizziness, and constipation usually resolved spontaneously during the course of therapy and at no time forced discontinuation of the drug. In only one case, a first degree AV-block was observed. AV conduction time normalized when the dosage was reduced from 480 to 320 mg/day.

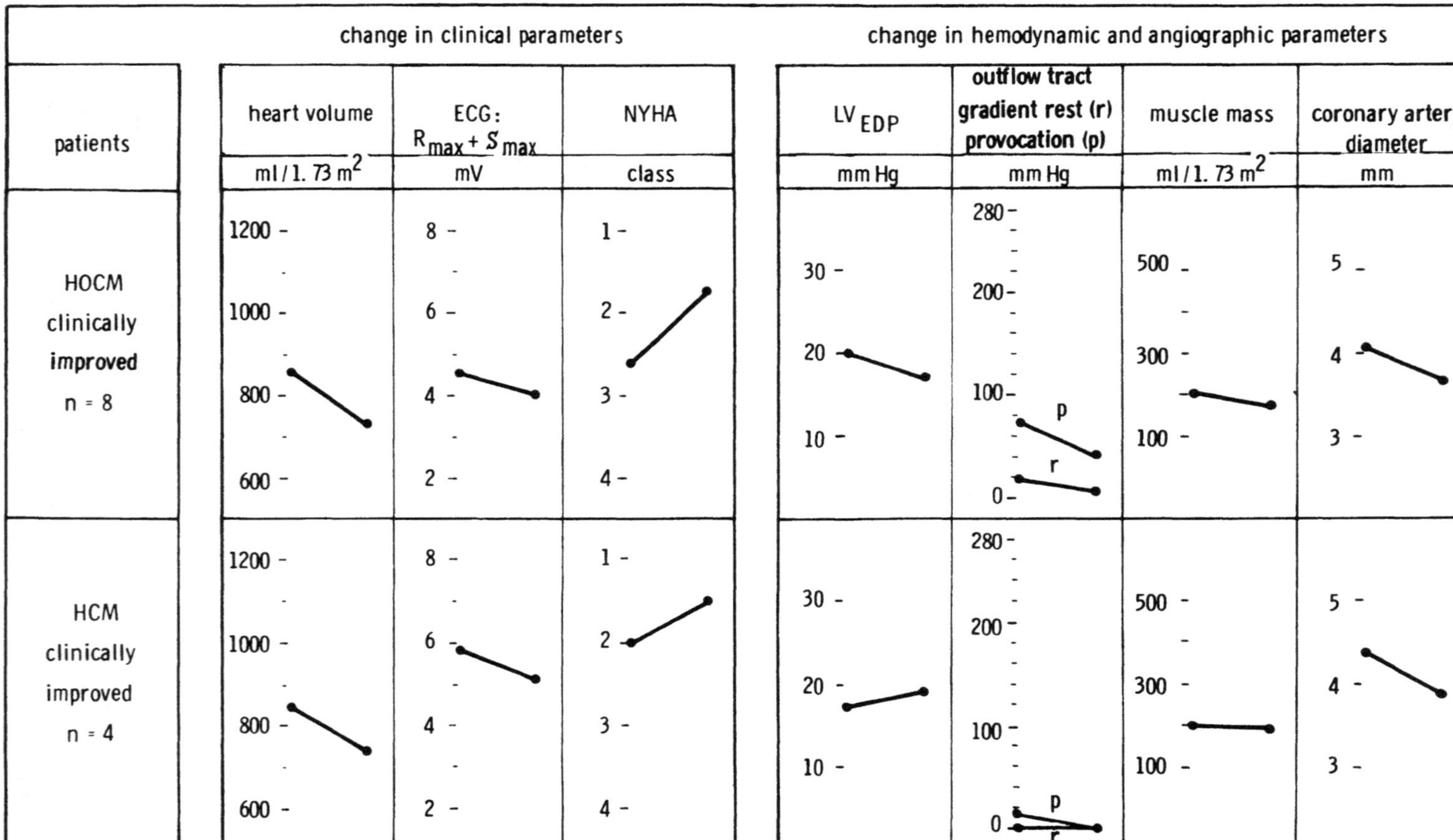

Fig. 6. Changes in invasive and non-invasive findings during verapamil therapy in 12 clinically improved patients (8 with obstructive, 4 with non-obstructive hypertrophic cardiomyopathy)

Discussion

These results obtained over a treatment period of up to seven years document the effectiveness of verapamil in the conservative treatment of patients with obstructive and non-obstructive hypertrophic cardiomyopathy. In contrast to a preceding period with beta blocking agents during which pathological findings progressed, pathological changes were found to regress during verapamil treatment.

The survival rate during verapamil treatment appears comparable to or better than that achieved with other forms of therapy. In the majority of patients, subjective complaints improved markedly. Electrocardiographic, echocardiographic, roentgenographic, and heart catheterization findings improved considerably.

In about 20% of the cases, no objective improvement during verapamil therapy was noted. It is unclear whether the etiology of the disorder in these patients is different or whether they might respond better to other agents such as beta-blockers. A third possibility is that these patients did not comply with their dosage schedules. Poor compliance in unsuccessfully treated patients is supported by serum verapamil determinations made during long-term therapy. In 14 unimproved patients markedly lower verapamil levels were obtained 12 h after the last dose than in the improved patients with comparable increases in blood levels 1 and 4 h after an oral dose of 160 mg verapamil [7].

The low incidence of side effects is confirmed by preliminary findings of an ongoing multicenter study comparing the efficacy of high dose propranolol and verapamil in hypertrophic cardiomyopathy. Termination of therapy due to significant side effects occurs less often with verapamil. The incidence of side effects not leading to discontinuation of the drug is also lower. Finally, subjective symptoms appear to respond better to verapamil [11].

References

1. Bajusz E, Lossnitzer K (1968) Ein neues Krankheitsmodell: erbliche nichtvaskuläre Myokarddegeneration mit Herzinsuffizienz. Münch Med Wochenschr 31:1756
2. Blaufuss AH, Laks MM, Garner D, Ishimoto BM, Grilley JM (1975) Production of ventricular hypertrophy simulating "Idiopathic Hypertrophic Subaortic Stenosis" by subhypertensive infusion of norepinephrine in the conscious dog. (Abstr) Clin Res 23:77
3. Braunwald E, Lambrew CT, Harrison DC, Morrow AG (1964) The hemodynamic effects of circulatory drugs in patients with idiopathic hypertrophic subaortic stenosis. In: Wolsteinholme GEW, O'Connor M (eds) Cardiomyopathies – Ciba Foundation Symposium. Churchill, London
4. Bussmann W-D (1974) Kontraktilitätsreserve des linken Ventrikels unter körperlicher und pharmakologischer Belastung bei verschiedenen Herzerkrankungen. Habilitationsschrift, Universität Frankfurt
5. Ensslen R, Schwarz F, Thormann J, Schlepper M (1977) Auswurffraktion und Reserve des linken Ventrikels bei hypertrophischer obstruktiver Kardiomyopathie. Z Kardiol 66:633
6. Frank MJ, Abdulla AM, Canedo MI, Saylors RE (1978) Long-term Medical Management of hypertrophic Obstructive Cardiomyopathy. Am J Cardiol 42:993
7. Hopf R (1980) Die hypertroph obstruktive Kardiomyopathie (HOCM) und ihre Behandlung mit Kalzium-Antagonisten. Habilitationsschrift, Universität Frankfurt
8. Kaltenbach M, Hopf R, Keller M (1976) Kalzium-antagonistische Therapie bei hypertroph obstruktiver Kardiomyopathie. Dtsch Med Wochenschr 101:1284
9. Kaltenbach M, Hopf R, Kober G, Bussmann W-D, Keller M, Petersen Y (1979) Treatment of hypertrophic obstructive cardiomyopathy with verapamil. Br Heart J 42:35–42

10. Kober G, Spahn G, Becker H-J, Kaltenbach M (1973) Weite und Querschnittsfläche der großen, epikardialen Koronararterien bei Herzmuskelhypertrophie. Z Kardiol 63:297
11. Kober G, Hopf R, Kaltenbach M et al. (in press) Long-term treatment of hypertrophic cardiomyopathy with verapamil or propranolol. Preliminary results of a multicenter study. In: Kaltenbach M, Epstein SE (eds) Verapamil treatment of hypertrophic cardiomyopathy
12. Kuhn H, Loogen F (1978) Die Anwendung von Beta-Rezeptorenblockern bei hypertrophischer obstruktiver Kardiomyopathie (HOCM). Internist 19:527
13. Loogen F, Krelhaus W, Kuhn H (1976) Verlaufsbeobachtungen der hypertrophischen obstruktiven Kardiomyopathie (HOCM). Z Kardiol 65:511
14. Loogen F, Kuhn H, Krelhaus W (1978) Natural history of hypertrophic obstructive cardiomyopathy and the effect of therapy. In: Kaltenbach M, Loogen F, Olsen EGJ (eds) Cardiomyopathy and myocardial biopsy. Springer, Berlin Heidelberg New York, p286
15. Lossnitzer K (1975) Genetic induction of cardiomyopathy. In: Schmier J, Eichler O (eds) Heart and Circulation. Springer, Berlin Heidelberg New York. (Handbuch der experimentellen Pharmakologie, vol XVI/3, p309)

Discussion

Sleight, Oxford: Have you looked at the incidence of arrhythmia in hypertrophic obstructive cardiomyopathy patients treated with verapamil, using *Holter* monitoring?

Kober: No, in this group of patients we have not done evaluations of arrhythmias before and during verapamil treatment.

Rutishauser, Geneva: How do you measure muscle mass in hypertrophic cardiomyopathy in view of

- angiocardiographically unequal local distribution of muscle around the cavity
- echocardiographically the beam not necessarily being perpendicular to any wall or a cone shaped septum?

Kober: We measure muscle mass angiographically and we know of the difficulties and problems of this procedure. Because individual patients were compared before and during treatment, there should, however, be a comparable error in both measurements.

Subject Index

Adaptability of Vascular Wall

Proceedings of the XIth International Congress of Angiology – Prague 1978
Editors: Z. Reiniš, J. Pokorný, J. Linhart, R. Hild, A. Schirger

1980. 486 figures, 205 tables. XIX, 755 pages
ISBN 3-540-09907-7

Cardiovascular Surgery

Editors: W. Bircks, J. Ostermeyer, H. D. Schulte

1981. 300 figures, 271 tables. XXII, 767 pages
ISBN 3-540-10929-3

Creatine Kinase Isoenzymes

Pathophysiology and Clinical Application
Editor: H. Lang

1981. 111 figures. XVIII, 317 pages
ISBN 3-540-10714-2

Frontiers in Hypertension Research

Editors: J. H. Laragh, F. R. Bühler, D. W. Seldin

1981. 246 figures. XXXIX, 628 pages
ISBN 3-540-90557-X
Distribution rights for Japan: Igaku Shoin, Tokyo

B. J. Harlan, A. Starr, F. M. Harwin

Manual of Cardiac Surgery

Volume 1

1980. 193 figures (183 full in color), 8 tables. XV, 204 pages
(Comprehensive Manuals of Surgical Specialties)
ISBN 3-540-90393-3

B. J. Harlan, A. Starr, F. M. Harwin

Manual of Cardiac Surgery

Volume 2

1981. 130 figures in full color. XV, 347 pages
(Comprehensive Manuals of Surgical Specialties)
ISBN 3-540-90563-4

Springer-Verlag
Berlin
Heidelberg
New York

www.ingramcontent.com/pod-product-compliance
Lightning Source LLC
Chambersburg PA
CBHW061154240926
51846CB00006B/1016
* 9 7 8 3 6 4 2 6 8 3 2 3 7 *